POWDER METALLURGY OF IRON AND STEEL

POWDER METALLURGY OF IRON AND STEEL

Randall M. German
The Pennsylvania State University

A WILEY-INTERSCIENCE PUBLICATION
JOHN WILEY & SONS, INC.
New York • Chichester • Weinheim • Brisbane • Singapore • Toronto

This book is printed on acid-free paper.

Published simultaneously in Canada.

Library of Congress Cataloging-in-Publication Data:

German, Randall M., 1946–
Powder metallurgy of iron and steel / by Randall M. German.
p. cm.
"A Wiley-Interscience publication."
Includes bibliographical references (p.) and index.
ISBN 0-471-15739-2 (alk. paper)
1. Powder metallurgy. 2. Steel—Metallurgy. 3. Iron—Metallurgy.
I. Title.
TN697.I7G47 1998
672.3′7—dc21 97-31764

Printed in the United States of America.

10 9 8 7 6 5 4 3 2 1

To Harvey Brush

CONTENTS

PREFACE

Iron and steel components made via the powder metallurgy process are an integral part of many automobiles, appliances, tools, and machines used on a day-to-day basis. Over the years, ferrous powder metallurgy has emerged as an important means to fabricate complicated components to high precision with excellent mechanical properties. Close to Penn State are some of the world's largest and most active ferrous powder metallurgy operations. Such proximity has led to several educational programs focused on the fundamental aspects of the process. This book is one outcome from that educational effort.

Many details in this treatment of ferrous powder metallurgy emerged from continued interaction with former students. Much of the university's research effort is directed toward emerging concepts. However, continued employment growth occurs in the mainstay of the industry. Iron and steel constitute 85% of metal powder production. In light of the importance of this field, a book was needed to teach the classic powder metallurgy process for the fabrication of everyday components. This work advances the field by organizing the underlying concepts useful for production personnel, engineers, researchers, managers, sales and quality personnel, students, and suppliers.

The justification of a new book is self-evident to those involved in the field. A critical need exists for a means to educate the expanding workforce. A resource is required to bridge between the introductory brochures provided by trade associations and the research papers presented at scientific conferences. Accordingly, this book was written in a logical manner to follow the typical production steps.

The introductory topics provide a foundation for understanding the current production processes. Chapter 1 starts with an introduction to iron and steels, the alloys, properties, controlling features, and main applications. Why is iron so important to humans, and what are the properties that make it a dominant engineering material? Chapter 2 shows how powder metallurgy fits with other production processes in ferrous metallurgy.

In the core of the book, various production techniques are described, showing the main options. Chapter 3 discusses the production routes for fabricating iron and steel powders and describes the measures and characteristics of these powders. Chapter 4 describes the powder shaping processes that rely on pores saturated with polymers. Much emphasis is given to injection molding in this chapter. Chapter 5 describes the compaction processes where the powder is consolidated using high pressures.

Sintering is a common concern once a compact is formed. Chapters 6 and 7 describe sintering operations, first in terms of the basic concepts and then in terms of equipment, atmospheres, production cycles, and property development. Full-density processes are taken up in Chapter 8, including hot isostatic pressing and hot powder forging. Since heat treatments are important to tailoring the mechanical properties, Chapter 9 is devoted to the heat treatment response. The finishing operations, such as plating and machining, are described in the next chapter, with particular focus on the problems encountered with residual pores.

The final three chapters look at the engineering attributes of powder metallurgy products. Chapter 11 overviews the attainable properties, with emphasis on the mechanical properties of die compacted steels. Design of components for various powder metallurgy processes is covered in Chapter 12. This treatment brings together the required performance attributes (properties), economics, microstructure, and geometric features to rationalize the selection of alloy and production route. Applications for powder metallurgy abound. Several diverse fields and components are described in Chapter 13 to help the reader understand how the principles described earlier are used in practice. Finally, appendixes are including with extensive definitions of terms and standards used in powder metallurgy.

The book focuses on current practices. Research issues and new developments are only given brief mention. Thus, die compaction of steel components receives the greatest attention. In most cases, each chapter starts with the basic principles and grows to show how the principles are applied. To enhance readability, mathematical principles have been moved out of the core text. This way the reader interested in basic concepts can elect to skip over the equations and symbols without loss of continuity. The mathematical concepts are presented in stand-alone clusters. Many figures are included to illustrate the principles, but production equipment is not emphasized. One reason for this latter decision is the rapid turnover in models and features, while the basic concepts remain enduring. Manufacturer catalogs are available with pictures of current equipment configurations and a good source of information is attainable via the internet.

This treatment will be valuable to all personnel involved in powder metallurgy. It builds from 30 years of experience in the field and conveys the special learning from many projects conducted for industry. Further, the presentation style benefits from two decades of teaching experience. Hopefully, it will be the basis for many successes.

ACKNOWLEDGMENTS

I want first to thank my benefactor, Harvey Brush. He generously endowed the Brush Chair in Materials at Penn State. Such a commitment is aligned with his interest in teaching, undergraduate education, and new concepts of learning. It has been a pleasure to have him repeatedly visit the P/M Lab to discuss our progress. During our conversations, Harvey has articulated the importance of engineering fundamentals and interdisciplinary knowledge of materials for success in engineering practice. His support for our teaching, research, and service mission has been critical to the success of the P/M Lab and to writing this book. The dedication to Harvey is a small repayment for his important contributions.

Funding for the artwork was provided by Quebec Metal Powderds and John Wiley. I appreciate the arrangements provided by Martin Thibodeau and Gregory Franklin in this regard. During the planning stage, valuable suggestions on the emphasis, organization, and content were provided by Ulf Gummeson and Deepak Madan. After the manuscript was prepared, several individuals provided review assistance, and I wish to thank Sundar Atre, Anita Cardamone, Kimberly Comstock, Anthony Griffo, Ronald Iacocca, Andrew Marzalek, Anand Lal, Yang Liu, Ken Moyer, Neal Myers, Tracy Potter, T. S. Shivashankar, Gregory Shoales, Karthik Sivaraman, Julian Thomas, Anish Upadhyaya, Tim Weaver, and Mu-Jen Yang for their help in this regard.

Micrographs, photographs, data, process descriptions, and sample components were provided by several individuals, and they are acknowledged throughout the book. Especially helpful in this regard were James Marsden, Peter Johnson, and Julian Thomas. Much assistance with the actual production mechanics was provided by Patrick Sharbaugh, Helen Edson, and Robin Smail. I want to thank them for keeping the project on track over the past 3 years.

CONSTANTS AND CONVERSION FACTORS

SYSTEM OF INTERNATIONAL UNITS

Prefixes

G (giga) = 10^9
M (mega) = 10^6
k (kilo) = 10^3
c (centi) = 10^2
m (milli) = 10^{-3}
μ (micro) = 10^{-6}
n (nano) = 10^{-9}

Fundamental Units

length = m (meter)
time = s (second)
mass = kg (kilogram)
amount = mol (mole)
current = A (ampere)
temperature = K (kelvin)
plane angle = rd (radian)
solid angle = sr (steradian)
luminous intensity = cd (candela)

Derived Units

Hz (hertz) = 1/s
h (hour) = 3600 s
min (minute) = 60 s
L (liter) = 10^{-3} m^3
t (ton) = 1000 kg
°C (celsius) = K − 273
N (newton) = kg·m/s^2
Pa (pascal) = N/m^2
bar (bar) = 0.1 MPa
J (joule) = N·m
W (watt) = J/s
C (coulomb) = A·s
V (volt) = J/C
Ω (ohm) = 1 V/A
S (siemen) = 1 A/V
T (tesla) = V·s/m^2

H (henry) = V·s/A
F (farad) = A·s/V
Wb (weber) = V·s
lm (lumen) = cd·sr

Important Constants

acceleration due to gravity = 9.8 m/s^2
speed of light = $2.998 \cdot 10^8$ m/s
Avogadro's number = $6.022 \cdot 10^{23}$ molecules
atomic mass unit = $1.661 \cdot 10^{-24}$ g
Boltzmann's constant = $1.381 \cdot 10^{-23}$ J/K
gas constant = 8.314 J/(mol·K)
elementary charge = $1.602 \cdot 10^{-19}$ C
electron mass = $9.11 \cdot 10^{-28}$ g
Planck's constant = $6.626 \cdot 10^{-34}$ J·s
Stefan–Boltzmann constant = $5.67 \cdot 10^{-8}$ $J/(m^2 \cdot s \cdot K^4)$
Faraday's constant = $9.65 \cdot 10^4$ C/mol
electric permittivity of vacuum = $8.854 \cdot 10^{-12}$ C/V
Bohr magneton = $9.27 \cdot 10^{-24}$ $A \cdot m^2$
gas volume at standard temperature and pressure = 0.0224 m^3/mol
permeability of vacuum = $7.958 \cdot 10^5$ H/m

Length Conversions

1 m = 39.4 in (inch)
1 m = 3.28 ft (foot)
1 m = 1.09 yd (yard)
1 cm = 0.394 in (inch)
1 mm = 0.0394 in (inch)
1 μm = 39.4 μin (microinch)
1 nm = 10 Å (angstrom)

Area and Volume Conversions

1 cm^2 = 0.155 in^2 (square inch)
1 m^2 = 1550 in^2 (square inch)
1 cm^3 = 0.061 in^3 (cubic inch)
1 m^3 = 35 ft^3 (cubic foot)
1 L = 1000 cm^3 (cubic centimeter)
1 L = 0.264 gal (gallons)
1 L = 1.06 qt (quart)

Amount of Substance Conversion

1 mol = $6.022 \cdot 10^{23}$ molecules

Density Conversions

1 Mg/m^3 = 1 g/cm^3
1 g/cm^3 = 0.0361 lb/in^3 (pound per cubic inch)
1 kg/m^3 = 10^{-3} g/cm^3

Temperature Conversions

To convert K to °F (fahrenheit), multiply by 1.8, then subtract 459.4°F.
To convert °C to °F (fahrenheit), multiply by 1.8, then add 32°F.

Heating and Cooling Rate Conversions

1 K/s = 1°C/s = 1.8°F/s
1 K/min = 1°C/min = 1.8°F/min

Mass Conversions

1 g = 0.035 oz (ounce)
1 kg = 2.2 lb (pound)
1 Mg = 1.1 ton (2000 pounds)

Force Conversions

1 N = 10^5 dyn
1 N = 0.225 lb force (pound force)

Pressure, Stress, and Strength Conversions

1 Pa = 0.0075 torr (millimeter of mercury)
1 Pa = 10 dyn/cm^2 (dyns per centimeter square)
1 kPa = 0.145 psi (pounds per square inch)
1 MPa = 9.87 bar (atmosphere)
1 MPa = 145 psi (pounds per square inch)
1 MPa = 0.145 kpsi (thousand pounds per square inch)
1 GPa = 145 kpsi (thousand pounds per square inch)

Energy Conversions

1 J = $9.48 \cdot 10^{-4}$ Btu (British thermal unit)
1 J = 0.737 ft·lb (foot pound)
1 J = 0.239 cal (calorie)
1 J = 10^7 erg
1 J = $2.8 \cdot 10^{-7}$ kW·h (kilowatt hour)
1 J = $6.24 \cdot 10^{18}$ eV (electron volt)

1 J = 4.83 hp·h (horsepower hour)
1 J = 1 W·s (watt second)
1 J = 1 V·C (volt coulomb)
1 kJ = 0.239 kcal (kilocalorie)

Power Conversions

1 W = 0.737 ft·lb/s (foot pound per second)
1 W = $1.34 \cdot 10^{-3}$ hp (horsepower)

Thermal Conversions

1 J/(kg·K) = $2.39 \cdot 10^{-4}$ (Btu/(lb·°F) (British thermal unit per pound per degree fahrenheit)
1 J/(kg·K) = $2.39 \cdot 10^{-4}$ cal/(g·°C) (calorie per gram per degree celsius)
1 W/(m·K) = 0.578 Btu/(ft·h·°F) (British thermal unit per foot per hour per degree fahrenheit)
1 W/(m·K) = $2.39 \cdot 10^{-3}$ cal/(cm·s·°C) (calorie per centimeter per second per degree celsius)

Viscosity Conversions

1 Pa·s = 1 kg/(m·s)
1 Pa·s = 10 P (poise)
1 Pa·s = 10^3 cP (centipoise)

Stress Intensity Conversion

1 MPa·$\sqrt{\text{m}}$ = 0.91 kpsi·$\sqrt{\text{in}}$ (kilopounds per square inch times square root inch)

Magnetic Conversions

1 T = 10^4 G (gauss)
1 A/m = $1.257 \cdot 10^{-2}$ Oe (oersted)
1 Wb = 10^8 maxwells

POWDER METALLURGY OF IRON AND STEEL

CHAPTER 1

Iron and Steel

1.1. IRON AND STEEL AS ENGINEERING MATERIALS

For most of our history, we have measured progress via the materials at our disposal—stone age, bronze age, iron age, and maybe now the silicon (microprocessor) age? Today we take for granted the array of engineered materials available for converting ideas into function. This is really a recent transition from a society where the available materials were only those found naturally, such as wood, stone, cork, or obsidian. Now we have the situation where we can calculate the required properties of strength, magnetic response, optical translucency, density, stiffness, or corrosion resistance, and turn to handbooks to find the best-suited materials. These are precisely formulated systems that are available in an array of forms with various properties. Most important, many of today's materials are not available in nature. Without dwelling on a complete list, just consider that common synthetic (human-made) materials such as aluminum, polyethylene, silicon, styrofoam, fiberglass, and stainless steel all have unique properties. They are engineered materials, with customized combinations of strength or other properties, cost, function, and availability.

Of all engineered materials, steel is a dominant actor; every year we consume more than 750 million tons of steel worldwide. On a weight basis it is the most widely used material, while on a volume basis plastics now are the top engineered material class. The dominant role associated with these two materials is evident in an automobile. Both plastics and steels are easily formed and have modest costs. In contrast, we now think of the natural materials as exotic and more costly—look at the special wood grain consoles in an expensive automobile. When various materials are analyzed in terms of usage by humans, an inverse relation emerges. Figure 1.1 plots the relative price of various metallic elements against the relative production. Scaling is logarithmic for both the horizontal and vertical axes (x and y). When scaled with respect to diamond use, iron is tens of millions times larger, but at a cost significantly lower. Note the low-cost position held by iron. It is the most widely used metallic element and has the lowest relative cost. Does the high consumption result from the low price, or does the low price reflect the high consumption? Either way, iron is situated as the baseline for cost and consumption. The cost of steel increases with the alloying level, since all other metals cost more than iron.

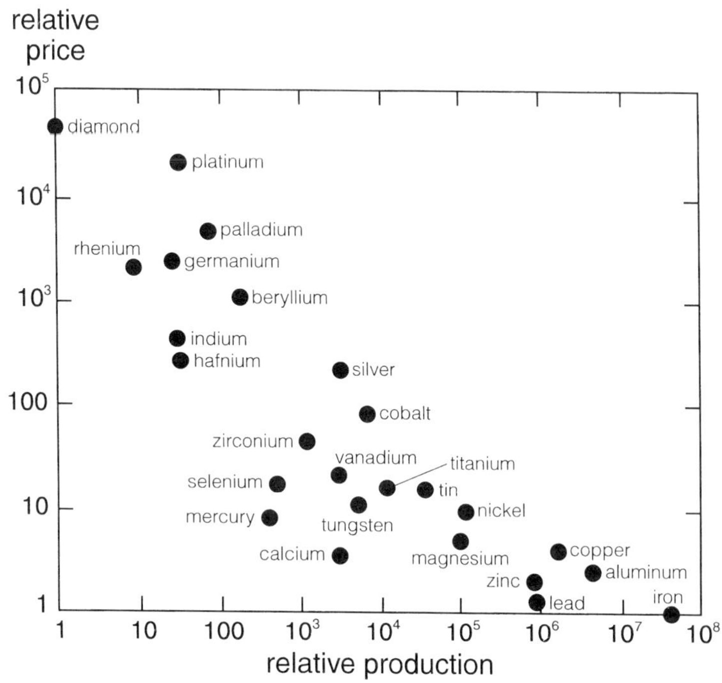

Figure 1.1. A logarithmic plot of the cost of metallic elements and their relative use by humans. Iron (Fe) occupies an unusual position, being one of the least expensive and most widely used metals (based on Jere Brophy).

1.1.1. Iron

To understand steels, there is a need to start with iron. It is the basis of all steels. Iron is one of the fundamental building blocks of matter. These building blocks are called *elements*; most people have heard of elements such as hydrogen, oxygen, chlorine, neon, nickel, iron, copper, silver, gold, zinc, silicon, lead, carbon, and uranium. All alloys, polymers, ceramics, intermetallics, and other compounds start with the elements. There are more than 114 presently known elements, but many are not useful because of instability or radioactivity. Others are very rare. Thus, we have maybe 60 elements in common use. Iron is element 26, indicating that it has 26 protons and 26 electrons, with about 30 neutrons. Protons are positive charges located at the atomic nucleus with the neutral neutrons, while electrons are negatively charged and spin around the nucleus, like planets around the sun.

As indicated in Figure 1.2, which shows the periodic table of the elements, iron is neighbored by manganese (element 25) and cobalt (element 27). This periodic table starts with element 1, hydrogen (H) with one proton, and builds to the elements with more protons than uranium (U, element 92), the last of the stable elements. All the elements beyond uranium are fabricated in high-energy physics experiments. The chemical symbol for iron is Fe (capital F and lowercase e). This symbol originates

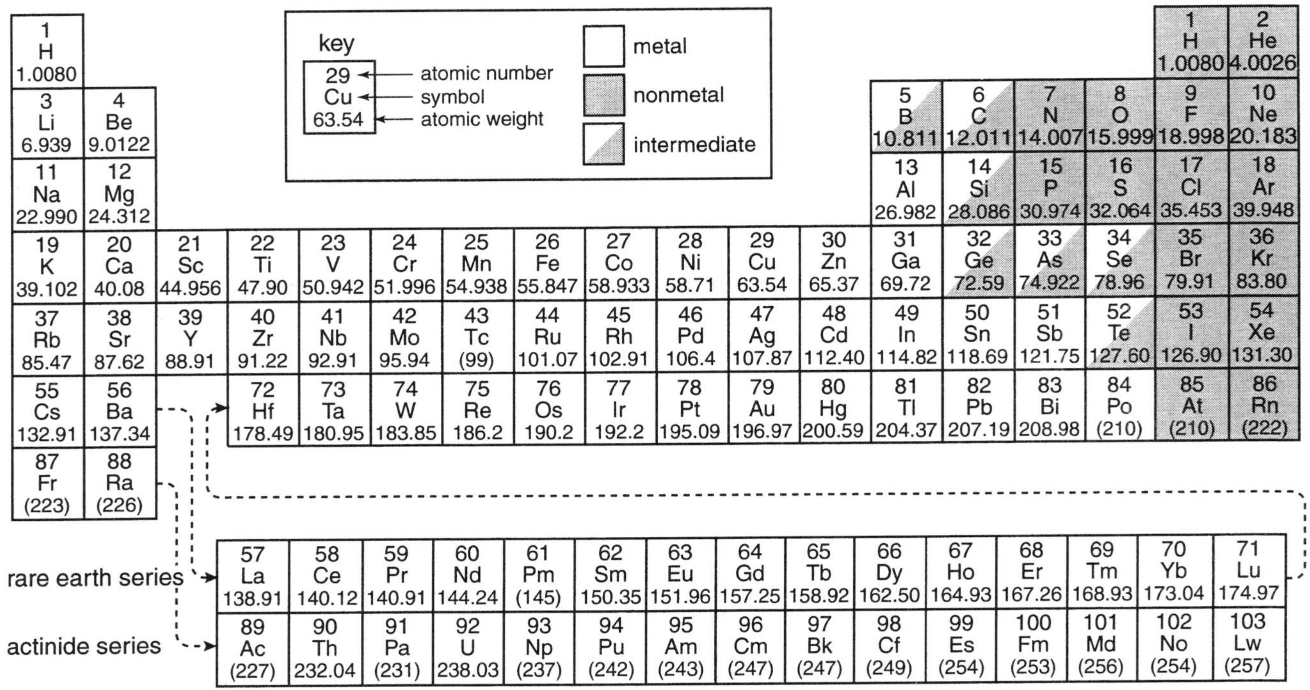

Figure 1.2. The periodic table of elements. Iron (chemical symbol Fe) is element 26 and is neighbored by manganese (Mn) and cobalt (Co) to the left and right. The chart builds from the lowest number of protons and electrons at hydrogen (shown twice), as indicated by the atomic number. Included in this table is the atomic weight, giving the grams needed to obtain one mole of atoms (by definition one mole consists of $6.02 \cdot 10^{23}$ atoms).

TABLE 1.1. Properties of Pure Iron

Chemical symbol = Fe
Atomic number = 26
Number of protons = 26
Number of electrons = 26
Typical number of neutrons = 30
Atomic size = 0.28664 nm
Average atomic weight = 55.85 g/mol
Melting temperature = 1537°C (2800°F)
Boiling temperature = 2735°C (4955°F)
Room-temperature density = 7.87 g/cm^3 (0.284 lb/in^3)
Heat capacity = 460 J/kg·°C [0.101 Btu/(lb·°F)]
Thermal conductivity = 80.2 W/m/°C [46.4 Btu/(ft·h·°F)]
Thermal expansion coefficient = $11.76 \cdot 10^{-6}$/°C ($6.5 \cdot 10^{-6}$/°F)
Yield strength = 133 MPa (19 ksi)
Ultimate strength = 266 MPa (38 ksi)
Poisson ratio = 0.3
Elastic modulus = 200 GPa (29,000 ksi)
Elongation to failure = 45%
Hardness = 21 HRB
Electrical resistivity = 9.7 micro ohm·cm

from its early Latin name *ferrum*. Often the alloys of iron are called *ferrous alloys*. In turn, it would seem that everything else metallic that is not based on iron is nonferrous.

Iron is the fourth most abundant element on Earth and composes much of the core. Table 1.1 summarizes the key attributes for iron. It is a solid at room temperature and does not melt below 1537°C (2800°F). If heated to a higher temperature, it will not boil below 2735°C (4955°F). The strength tells how much force per unit area the material can withstand before first deformation (yield strength) and maximum before failure (tensile strength). Ductility is measured by the elongation to failure. It tells how much stretch occurs before failure and is measured by the change in length to failure from the initial length. Hardness is the resistance to deformation and is measured on various standardized scales, here the Rockwell B scale. These, and several other terms, will be defined as we progress through the first two chapters.

In nature we rarely find pure iron; rather, it is found in various compounds with oxygen or sulfur, known as *oxides* and *sulfides*, or more complicated ores. Thus, it is converted into the pure element using metallurgical reduction steps; the most famous reduction product is pig iron, which contains carbon as a major contaminant. Fortunately, iron ores are somewhat abundant and the energy required to extract iron is relatively low, making it possible to form iron with reasonable ease and low cost. In its pure form, iron is lustrous, silvery, soft, and ductile. Actually, this combination of features is unremarkable, and many plastics are stronger and more easily formed at a lower cost per unit volume. Thus, alloying has evolved as an important concept in modifying iron for improved properties such as higher strength, hardness, or corrosion resistance.

1.1.2. Steels

Steels are alloys based on iron. Indeed, most ferrous products are steels. In contrast, iron is weak and used only for selected magnetic applications. Various elements are added to iron to change its properties, with carbon being very important. There are thousands of steels with a wide range of properties. Further, beyond alloying, heat treatments can adjust many properties. To understand the powder metallurgy (P/M) processing of iron and steel, we will first spend some time on understanding the basic structure and properties of ferrous systems. Then, attention will flow to the powder metallurgy process and its application in forming iron and steel structures.

1.2. TRADITIONAL IRON AND STEEL PRODUCTION

The production of ferrous materials is a large industry. Traditionally, all steels derive from the blast furnace, which reduced iron oxides and other minerals to produce pig iron. Steel production begins by reacting iron ore with carbon from charcoal, coke, or coal. A general production scheme is outlined in Figure 1.3. Today, an

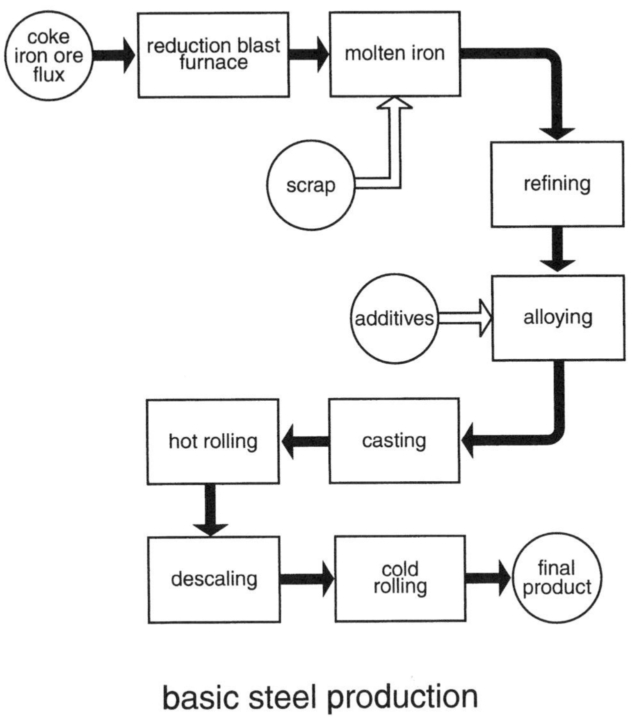

Figure 1.3. An overview of basic steel production. Iron ore is reduced by carbon sources to produce molten pig iron that is subsequently purified, alloyed, and formed into continuous products such as sheets.

important aspect of steel production is the recycling of scrap steel, thereby avoiding primary reduction.

Steel production technology became widespread by the 1600s, and made significant steps in the second half of the 1800s through the invention of Henry Bessemer. He produced steels with higher melting temperatures and lower carbon contents as compared with previous efforts. These advances in steel production were followed by the development of new open-hearth processes, killed steels, and other metallurgical advances. The important point is that basic steel technology has developed over hundreds of years and is very sophisticated. During this time, several instances are known where steel production was used to make powder. Most famous were the efforts involving direct reduction of oxides to form metals. Today, a major trend in steel production is to skip the blast furnace using the minimill concept. Here steel scrap is melted in an electric arc furnace directly to form high-quality steel that is continuously cast and worked into slab, billet, bar, rod, wire, tube, beam, plate, pipe, or other useful shapes. These operations are highly automated and have recently made dramatic changes in the steel industry.

Most steels are produced as long shapes. Rods, slabs, and beams are drawn, worked, rolled, and in other ways brought down to dimensions required for common applications. Some of these shapes are used directly in construction, stamping, machining, or forging operations to generate the final microstructure, properties, and components. Unfortunately, there is a loss of productivity as the final shape becomes more complex. While automobile body panels can be stamped from sheet steel, there is much more difficulty in fabricating gears or pulleys. More complicated shapes require discrete manufacturing steps, and this transition from continuous manufacturing, evident in continuous casting, to forming single objects is where component costs climb. The usual protocol is to fix the raw steel (possibly an ingot or casting) into a machining center and to remove unwanted mass. It is a slow process that uses expensive equipment, with a high material loss and unattractive cost for many large production operations.

It is at this point that powder metallurgy becomes an attractive alternative, since there is little mass loss and shaping is very rapid. In contrast, most attempts to use P/M to compete with steel mill products fail, since the economics of making powder and shaping that powder into plate, foil, rod, or tube shapes proves costly, often without technical advantage. Alternatively, P/M excels at the fabrication of discrete components with complex shapes intended for uses at high stresses with tailored properties.

1.3. CHARACTERISTICS OF IRON

1.3.1. Crystallography

Iron is an unalloyed element. As a solid, iron forms a crystal where the atoms align in periodic structures as diagrammed in Figure 1.4. Periodic structures occur when we neatly pack cannon balls in front of the armory or stack brick to build a wall.

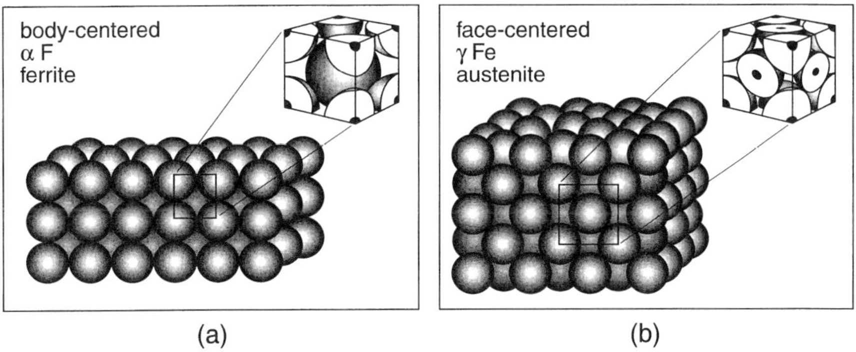

Figure 1.4. Atoms are approximated by spheres. As a solid, iron forms one of two crystal structures where the atoms sit in specific relations to each other. The small unit cells repeat over and over to make up the bulk crystal, called a *grain.* The two common forms of iron are (a) body-centered cubic alpha iron (ferrite) and (b) face-centered cubic gamma iron (austenite).

Atoms can align in any of 14 periodic structures, but iron atoms exist in only two common arrangements. Normally, the crystal structure of iron changes between these two arrangements at specific transformation temperatures, but these temperatures can be manipulated by alloying and rapid cooling. This proves very important in understanding heat treatment, since it is the manipulation of the crystal structure via heating and cooling that allows dramatic shifts in the properties of steels.

The two important crystal structures of iron are body-centered cubic and face-centered cubic. These are both sketched in Figure 1.4. Simple cubes with atoms on the corners of the cube represent the basic crystal building blocks, just as bricks make up the building blocks of a wall. The body-centered cubic structure has an atom on each corner and one in the center of the cube. For iron this structure is also known as *ferrite* or *alpha iron,* α-Fe. The alternative structure is *austenite* or *gamma iron,* γ-Fe. It consists of the same corner atoms, but now one centered on each face. For pure iron, the body-centered cubic structure is stable up to 910°C (1670°F) and again from 1392°C (2538°F) to the melting temperature of 1537°C (2800°F). Although the same crystal structure as α-iron, the high-temperature body-centered cubic form is designated as *delta iron,* δ-Fe. Between 910° and 1392°C, the stable crystal structure is face-centered cubic. This change in crystal structure with temperature is called *polymorphism.* Iron is a polymorphic element. Other elements exhibit similar polymorphic shifts: cobalt, uranium, titanium, manganese, tin, and even carbon (graphite and diamond). Changes in the crystal structure influence the properties. Alloying is important because it not only provides strength and corrosion resistance but also allows shifts in the stable crystal structures and properties. This becomes particularly important when discussing the heat treatment of steels.

One other change occurs for iron that is not due to a crystal structure change. This is the magnetic response. Pure iron is magnetic up to about 770°C (1420°F), but the crystal structure change to face-centered cubic does not occur until 910°C (1670°F). The nonmagnetic form of body-centered cubic iron was formerly known as *beta iron,* β-Fe, but this nomenclature is rarely used anymore. The transition temperature from magnetic to nonmagnetic is the Curie temperature, and is a determinant of the maximum operating temperature for a permanent magnet.

1.3.2. Crystal Defects

Crystals are not ideal. They contain missing atoms, extra atoms squeezed into crevices, pores, surfaces, and even lines and planes of defective arrangement. Each of these defects plays an important role in the powder metallurgy of iron and steel. The missing atoms leave vacant crystal lattice sites, while the extra atoms crowded into the lattice are called *interstitials.* Vacancies are important to sintering, while interstitials are important to strengthening. Line defects, called *dislocations,* control compaction and strength of iron powders. Figure 1.5 shows a simple crystal structure and two ways in which dislocations exist. Edge dislocations are extra planes of atoms that pass part way through the crystal, while screw dislocations are similar to the central axis of a spiral parking garage.

Both dislocations move when sufficient stress is applied to break neighboring atomic bonds. It is the onset of dislocation motion through the crystal that dictates strength. A curious aspect of dislocations is that only a few atomic bonds are broken

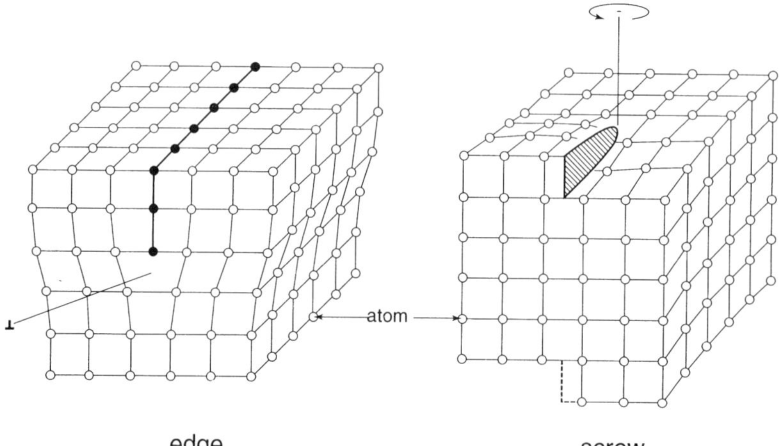

Figure 1.5. *Dislocations* are line defects in crystals that move under stress. The atom position is indicated by the small circle. There are two types of dislocations: edge and screw. The *edge* dislocation appears as an extra partial plane of atoms squeezed in the structure; the *screw* dislocation is the pivot point for a spiral of atomic planes. The yield strength generally corresponds to the point where dislocations move during stressing.

to cause a shape change in the crystal—those bonds near the dislocation. Motion of the dislocation through the crystal starts at the yield strength, giving a sliding of atomic planes called *slip*. The stress for dislocation motion is usually less than that required to fracture the steel. All the atomic bonds are broken across the test sample at the fracture stress. Thus, yield strengths are lower than tensile strengths, unless the material lacks ductility.

When two crystals come together, there is a surface of disrupted atomic bonding that is a grain boundary. Grain boundaries are favorite sites for impurities. Since impurities usually lower the grain boundary strength, segregation to grain boundaries usually embrittles a steel. A small grain size implies a large amount of grain boundary area, and improves resistance to segregated impurity effects. Accordingly, grain size has an impact on mechanical properties, especially ductility and toughness. Finally, pores are common in powder metallurgy steels. These are bulk defects remaining from the powder consolidation process. Usually, as the porosity increases the strength and other properties decrease.

1.3.3. Key Points

To summarize, iron naturally oxidizes, and forms rust in the presence of oxygen and water. Consequently, iron is found in nature as oxides or other compounds because of its exposure to water and air. In the first stage of steel production, these compounds are broken down to yield pig iron, a common first step in steelmaking. Pure iron has few properties of interest as an engineered material, but when alloyed, it is possible to create a wide array of properties that can be tailored to specific applications. By itself, pure iron exhibits a polymorphic transformation; it changes crystal structure. It is a body-centered cubic crystal structure up to 910°C (1670°F). At higher temperatures it converts to a face-centered cubic structure, before reverting to the body-centered cubic structure at 1392°C (2538°F) before melting. The difference in these crystal structures impacts on alloying, properties, and response to heat treatment. Iron is the base for all ferrous alloys, and steels constitute one of the largest classes of engineered materials. In powder metallurgy we usually start with iron powder, but other powders are mixed with the iron to form a steel during sintering. Thus, ferrous powder metallurgy starts with iron, but ends with steel.

1.4. PROPERTIES OF IRON AND STEELS

Steels account for about 90% of all metals used on Earth, due to the combined low cost and useful properties. Likewise, steels dominate powder metallurgy. The simplest steel consists of iron and carbon. Carbon is inexpensive and very potent in terms of strengthening. These simple alloys are called *plain carbon steels,* and usually range from 0.2 to 0.8% carbon. In this book the compositions will be given in weight percent, unless specifically stated otherwise, so 0.8C implies 0.8 wt% carbon.

TABLE 1.2. The Main Alloying Classes of Steels

Alloy Class	Alloying Level	Key Uses	Use in P/M
Iron	None	Magnets	Somewhat
Plain carbon steel	<0.8% carbon	Moderate strength, general purpose	Somewhat
Low-alloy steel	≤5%, some carbon	High strength	Very large
Stainless steel	≥12% chromium	Corrosion resistance	Moderate, growing
Tool steel	30–50%, >0.4% carbon, includes carbide forming metals	Wear resistance, tooling, cutting, drilling, machining	Limited
Cast iron	High carbon levels	Large structures, castings, automobiles engines	Not used
High-heat alloy	≤25%, high chromium and aluminium contents	High heat, jet engines, furnaces, heating elements	Very limited
Special alloy	≤50%	Special applications, electronics, glass seals, filters	Limited
Silicon alloy	≤8% silicon	Electrical transformers	Limited
Magnetic alloy	≤50% nickel, many special compositions	Magnetic components	Moderate, growing

Table 1.2 provides a first introduction to the classes of ferrous alloys and comments on the relative usage in powder metallurgy.

1.4.1. Alloy Types

Plain carbon steels might contain other ingredients besides carbon, largely for ease of processing, such as silicon or manganese. When more than 0.8% carbon is present, the steels are brittle. However, higher carbon levels are used in a few applications, such as for knife blades, cutting tools, and other high-hardness structures. In the nomenclature of steels, these are designated by the carbon level in 100s of percent. Thus, a plain carbon steel with 0.4% carbon would be designated as 1040, where the first two digits designate a simple steel with no alloying beyond carbon and the last two indicate the carbon level at 0.4% (40 out of 100 parts of one percent), indicated as Fe–0.4C in this book.

Alloys with about 0.8% carbon are very special and are termed *eutectoid steels*. Less carbon gives a hypoeutectoid steel (below the eutectoid), and more carbon gives a hypereutectoid steel (above the eutectoid). As detailed in Chapter 9, there is a dramatic shift in hardening reactions with carbon levels above and below the eutectoid composition.

Low alloy steels represent the next hierarchial step. These contain less than 5% total alloying elements, but even that quantity of addition increases the cost. Besides

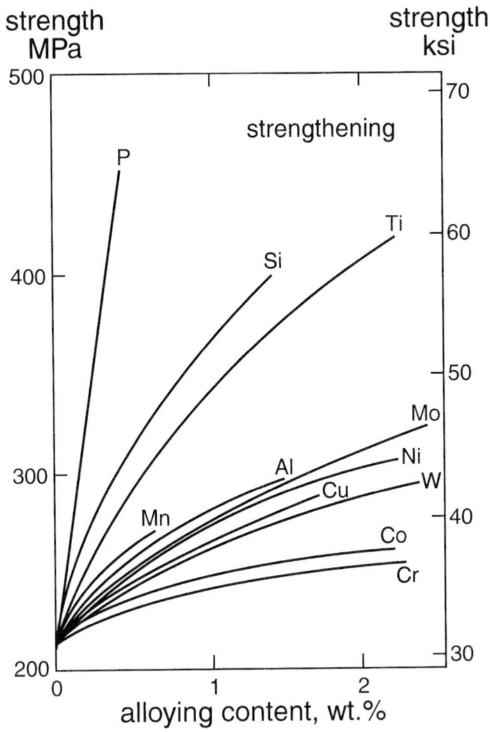

Figure 1.6. Strength versus amount of alloying addition for the body-centered cubic form of iron. Some elements are more potent in strengthening iron.

carbon, common alloying elements are chromium, manganese, molybdenum, nickel, silicon, vanadium, and tungsten. Figure 1.6 plots the strength of iron as various alloying additions are made. Note the potent effect from phosphorus and the weak effect from cobalt and chromium. In powder metallurgy, the common low-alloy steels use copper and nickel more than any other ingredients. Although copper is widely used in P/M steels, it is avoided in normal ferrous metallurgy. This is because copper segregates from molten iron, but this is not a problem in powder metallurgy since sintering is performed much below the melting point of iron. In powder metallurgy, steel alloys with low concentrations of copper or nickel are the most popular low-alloy compositions. Sometimes these also contain molybdenum or other alloying ingredients.

The stainless steels consist of several different subclasses, but a common feature is the presence of chromium. It is the chromium that provides corrosion and oxidation resistance associated with the stainless characteristic. As illustrated by the saltwater corrosion data in Figure 1.7, at least 12% chromium is required for protection against corrosion. Depending on the specific class of stainless steel, other alloying ingredients might include nickel, manganese, carbon, molybdenum, or copper. Usually if nickel is present in large quantities, then carbon is absent. These are the austenitic steels, meaning that the stable crystal structure at room temperature is face-centered cubic. Other grades are martensitic or ferritic, which might contain carbon, but have lower total alloying. Applications using powder metallurgy stainless steels are increasing, with many customized compositions becoming available for improved corrosion resistance.

Tool steels are known for their high alloying levels and high hardness. Typically carbon levels are very high, possibly up to 1.5%. The high carbon content induces the formation of carbides that provide hardness, strength, and wear resistance. The carbides are usually formed with a refractory element, typically molybdenum (M-grade tool steels) or tungsten (T-grade tool steel). Other designations say something about the heat treatment needed for high hardness; O grades are oil-quenched, D grades are cold-worked, H grades are hot worked during production to build up strength, S grades are shock resistant, and A grades are air-quenched. In P/M there is most interest in the M and T grades. To obtain the desired high hardness requires considerable alloying, so tool steels prove expensive as compared with other steels. Premium tool steels are fabricated from powders to reduce waste and to allow high vanadium alloying levels.

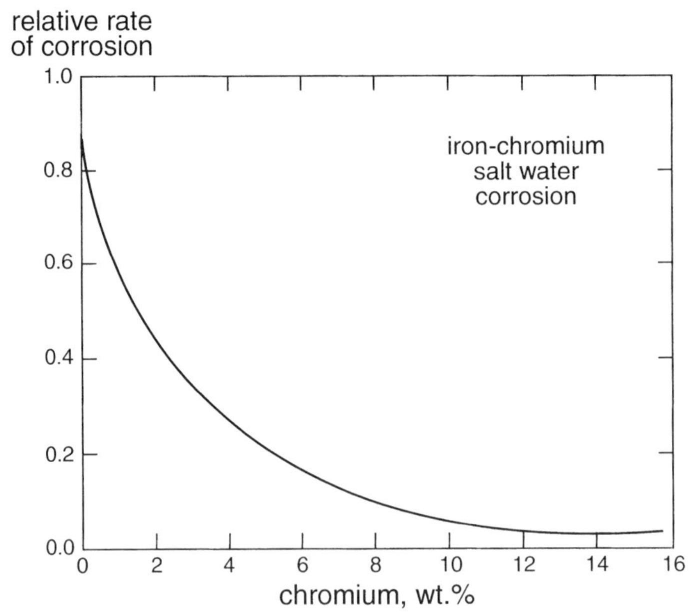

Figure 1.7. Corrosion resistance in iron is improved by chromium additions. This plot shows the relative rate of corrosion (in seawater) of iron with increasing levels of chromium. Usually 12% or more chromium is used to form a stainless steel. The actual rate of corrosion depends on the environment and many service details, including the temperature, agitation, surface area, and concentration of chlorine and other agents that accelerate corrosion.

TABLE 1.3. Composition, Name, and Mechanical Properties of Common Steels

Designation	Type	Composition wt% (Balance Iron)	Density (g/cm^3)	Elastic Modulus (GPa)	Yield Strength (MPa)	Tensile Strength (MPa)	Elongation (%)	Hardness
Iron	Pure	0	7.9	200	133	266	45	21 HRB
Malleable	Cast-iron	2.5C–1Si–0.5Mn	7.3	172	231	364	12	75 HRB
4340	Low-alloy	1.8Ni–0.7Mn–0.7Cr–0.2Mo–0.4C	7.9	200	1675	1875	10	50 HRC
M2	Tool	6W–5Mo–4Cr–2V–0.3Mn–0.3Si–0.9C	8.0	210	1650	1860	10	65 HRC
250 maraging	Special	18Ni–8Co–5Mo–0.4Ti	7.8	210	1750	1820	10	49 HRC
316L	Stainless	17Cr–13Ni–2Mo	8.0	193	280	550	60	85 HRB
420	Stainless	13Cr–0.2C	7.7	200	1365	1610	8	51 HRC
A286	High-heat	26Ni–15Cr–2Ti–1.3Mo–0.2Al	7.9	200	735	1022	25	28 HRC
Invar	Special	36 Ni	8.1	145	280	490	43	75 HRB

Other ferrous alloys exist, but these are used less frequently in powder metallurgy. These include various grades of cast iron, welding grades of high-strength steels, ultra-high-strength steels, and special steels with custom magnetic, electrical, or thermal properties. This latter area is small, but a growing activity for powder metallurgy. There are even alloys designed for reinforcing concrete, winding transformers cores, forming sheetmetal automotive panels, sealing glass, and for applications dependent on specific thermal expansion, electrical, oxidation, or other properties. Most of these are special and not emphasized in this book. Clearly, the list of applications is long and the number of steels in use is well more than 10,000. Table 1.3 summarizes some steels by giving examples from different alloy types, including the common name, composition, theoretical density, and mechanical properties. At this point the attention will be put to a few important alloying effects, starting with plain carbon steels.

1.4.2. Iron–Carbon Steels

Iron and carbon are the basis for many steels. Carbon is not very soluble in ferrite (α-Fe) at room temperature, yet it is soluble in austenite (γ-Fe or austenite), the high-temperature form of steel. This change in solubility makes for some useful strengthening via heat treatment. Figure 1.8 shows the phases, carbon solubility, and melting behavior of iron versus carbon solubility and temperature. The maximum solubility of carbon in ferrite is 0.025% at 727°C (1340°F), while the maximum solubility in austenite is 2.1% at 1148°C (2098°F). This nearly 100-fold difference in carbon solubility makes it possible to strengthen ferrite by quick cooling iron–carbon alloys from austenite. Carbon dissolves in austenite, and on quick cooling becomes frozen in the steel as austenite tries to form ferrite. The net result is a distortion of the crystal structure and formation of a hard phase known as *martensite,* as distorted

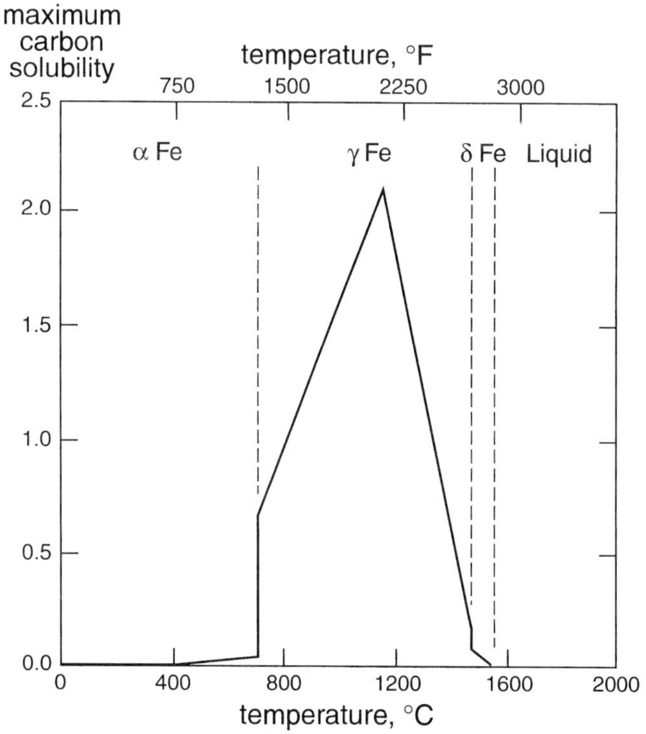

Figure 1.8. The solubility of carbon in iron changes with temperature. Iron has a low solubility for carbon in the body-centered cubic α-Fe phase. However, carbon is very soluble in γ-Fe, the face-centered cubic form that is stable at higher temperatures. The very high-temperature form of iron, δ-Fe, is the same crystal structure as α-Fe.

ferrite with dissolved carbon. If a high-carbon austenite is slow-cooled, then the carbon precipitates out of solution, just as snow forms in cooled clouds, to form a carbide Fe_3C called *cementite*. It is a new crystal structure, consisting of 6.67 wt% carbon (25 atomic percent). Thus, for a given carbon level there are different states possible for the carbon and this leads to a range of mechanical properties.

The various phases of iron and carbon have been the subject of much study. Many phases were isolated and studied long ago and bear the names of the first investigators. This is the source of the names such as bainite, austenite, and martensite. Later, when heat treatments are discussed, we will see other names to describe the phases in the iron–carbon system.

1.4.3. Alloying Effects

Most alloying elements are used to improve specific properties of iron. Associated with the two phases, ferrite and austenite, there is a natural tendency for an alloying element to prefer one phase over the other. This leads to an idea of phase stabilization. For example, the addition of carbon expands the austenite region. Thus, carbon is an austenite stabilizer. Further, some alloying elements associate with carbon more than iron; thus they form compounds called *carbides*. Metals such as niobium, titanium, vanadium, tungsten, molybdenum, and chromium fall into this category. These carbide formers are common ingredients in tool steels, where carbides provide strength and wear resistance.

Of all possible alloying elements, only a few are widely used in P/M steels. These are as follows with rough composition ranges:

- Carbon, typically up to 0.8%, except tool steels where up to 1.5% is possible.
- Chromium, used sparingly at 0.5% except stainless or tool steels with up to 20%
- Copper, if used, then usually between 2 and 10%.
- Molybdenum, used sparingly up to 1%, except in tool steels.
- Nickel, up to 2% for structural steels; for stainless steels up to 12% is possible, special cases up to 50% for magnetic properties or thermal expansion control.
- Phosphorus, used for magnetic alloys up to 0.8%.
- Silicon, used mostly in special electrical or magnetic steels at 3–6%.

The actual alloying level depends on the application and dominant properties, such as strength, corrosion, wear resistance, magnetic strength, thermal expansion, electrical conductivity, or high-temperature oxidation resistance.

1.4.4. Mechanical Properties

As a rule of thumb, alloying additions that increase strength and hardness in a steel usually decrease ductility. Likewise, porosity lowers all of the mechanical properties. In powder metallurgy it is important to understand these properties since the dominant applications are for mechanical components.

1.4.4.1. Hardness. Hardness is measured by impressing an indentor test tip into the surface of the sample. As shown in Figure 1.9, the test tips consist of standard balls, cones, or pyramids. Under a constant load the tip deforms the steel such that a high hardness corresponds to a small impression. Most hardness testers automatically apply the load and measure the impression length or depth to calculate the hardness. Several hardness scales exist; the Rockwell scales are used most in ferrous P/M. Values on the Rockwell B range (HRB, measured using a 1.6-mm ball with 100-kg load) are characteristic of most sintered steels, while values on the higher Rockwell C range (HRC, measured using a diamond cone and 150-kg load) are encountered for tool steels and heat-treated steels. A high hardness gives greater resistance to scratching and wear. On the other hand, a low hardness may suggest easy scratching.

Each hardness test is named after the inventor who first described the test geometry, loading, and measurement technique. Note that the Rockwell A, B, and C scales

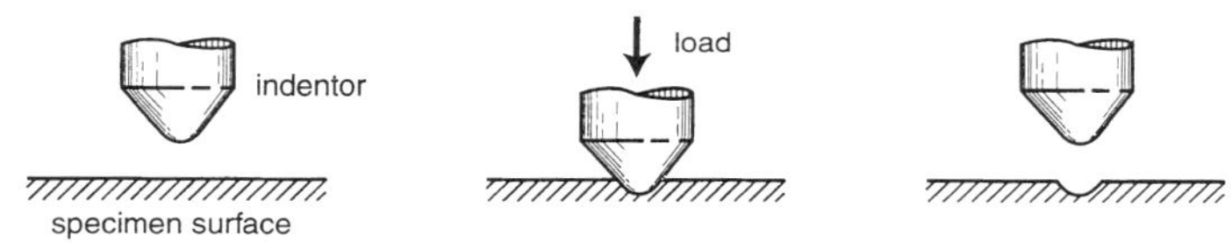

test	indenter	shape of indentation: side view	shape of indentation: top view	load	formula for hardness number
Brinell	10 mm sphere	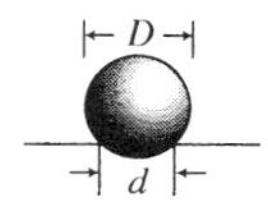	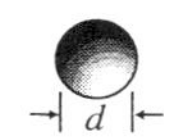	P (variable)	BHN = $\dfrac{2P}{\pi D[D-\sqrt{D^2-d^2}]}$
Vickers	diamond pyramid	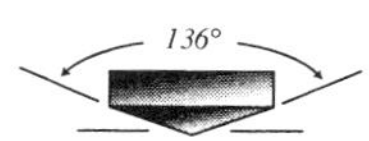	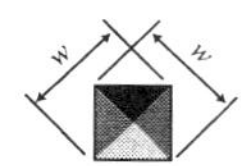	P (variable)	VHN = $1.72P/w^2$
Knoop microhardness	diamond pyramid	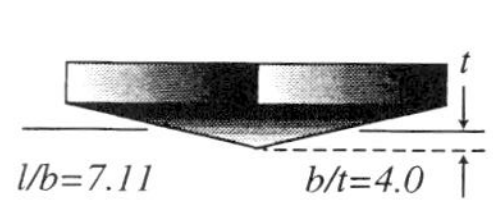	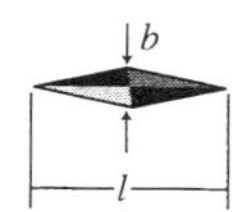	P (variable)	KHN = $14.2P/l^2$
Rockwell A, C	diamond cone	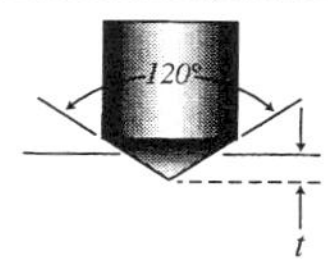	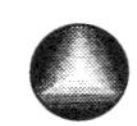	60 kg, 150 kg	HRA = 100-500 t; HRC = 100-500 t
Rockwell B	1.6 mm diameter sphere	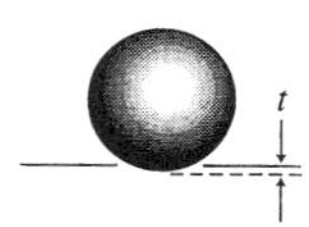		100 kg	HRB = 130-500 t

EQUATION BOX 1.1

In ferrous powder metallurgy, bulk hardness is measured on the Rockwell B and C scales (HRB and HRC). The B scale is most applicable to sintered steels, while the C scale is most applicable to heat-treated steels. These indents are clearly visible on the surface of the steel. On the other hand, microhardness measurements are small and are made using a low load and microscopic measurement of the indent. In P/M, most microhardness measurements are on the Vickers scale (VHN). It is common to use microhardness to follow surface carburization reactions or to measure the hardness of individual phases. Two equations give an ability to convert microhardness measurements into equivalent bulk hardness measurements:

$$\mathrm{HRC} = \frac{0.44\ \mathrm{VHN} - 63.2}{1 + 0.0044\ \mathrm{VHN}}$$

and

$$\mathrm{HRB} = \frac{4.25\ \mathrm{VHN} - 201}{1 + 0.0306\ \mathrm{VHN}}$$

Both equations are more accurate than the typical hardness test.

all rely on the depth of penetration. These tests cannot be considered nondestructive since there is an impression remaining in the steel after testing. On the other hand, the Vickers test is a widely used microhardness test. The measurement is performed using a small indentation measured in a microscope. It is useful in isolating reactions or surface hardening zones in P/M compacts.

1.4.4.2. Tensile Test. Tensile tests use a sample of sintered material and pull it to failure. This destructive event measures the force and stretch during testing. Stress is simply the load divided by the cross-sectional area at the center of the test bar. Strain is the change in a dimension divided by the original dimension. The strain is only measured over the central sample length where the cross section is constant.

Figure 1.9. *Hardness* is a resistance to indentation. Various tests and measurement scales are encountered in metallurgy. As indicated in this figure, each test has a specific indentor geometry, load, and means to quantify the hardness.

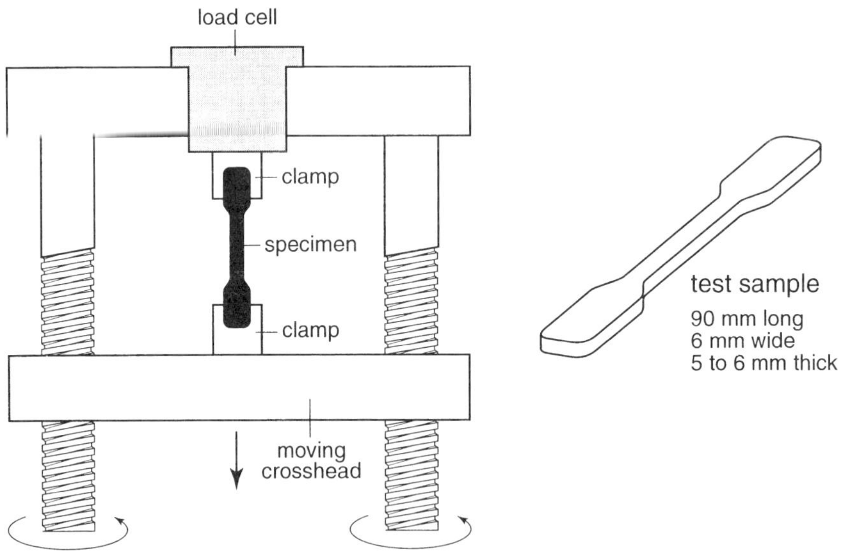

Figure 1.10. The tensile test sample and a typical test where the enlarged ends of the tensile sample are gripped and pulled to cause the sample to fail. Recordings of the load and displacement during the test are used to calculate the stress and strain.

Stress and strain are important measures of tensile behavior. Figure 1.10 draws a standard powder metallurgy tensile sample. For testing it is gripped at the two enlarged ends and fails in the center where the cross-sectional area is the least and the stress is the largest. Strain is calculated as the change in length divided by the original length (usually 25 mm or 1 in) over the center of the tensile sample.

The tensile test starts with an unloaded sample. As the sample is stretched, the load required to induce each level of strain is measured. Initially, the tensile sample stretches in proportion to the applied load. If the load is removed during this portion of the test, the sample will return to its original length. We term this elastic deformation and the proportionality between the stress and strain the *elastic modulus* or *Young's modulus*. The behavior is exactly like a spring that stretches on loading, but returns to its original length when the load is removed. For most full-density steels, the elastic modulus is about 200 GPa (30,000 ksi). This means that a strain of 0.1% (change in length from 1.000 to 1.001) requires about 200 MPa (30 ksi). This is a very stiff spring. Because the sample volume is essentially conserved during tensile testing, its thickness shrinks as it stretches. The proportionality between the thickness shrinkage and length elongation is Poisson's ratio, and for most full-density steels it is slightly less than 0.3.

With continued loading and stretching, the tensile sample permanently deforms, leading to plastic deformation. A plot of the stress versus strain results in curves that might match one of those shown in Figure 1.11. It is from such curves that many important mechanical properties are extracted. The yield strength is the stress

EQUATION BOX 1.2

Several terms associated with the mechanical response of a steel are related; stress, strength, pressure, and elasticity all have the same units. Stress depends on the loading situation, but strength is a material property. Formally, stress and strength are defined by force over area. In metric units this is N/m^2 (newtons per square meter), which is a pascal (Pa). For English units it is pounds per square inch (lb/in^2 or psi). Since steels have high strengths, usually multiples of thousands of pounds per square inch are used: ksi (meaning kpsi) or tsi (meaning tons per square inch) or MPa (meaning million pascals). Stress σ is given by the formula

$$\sigma = \frac{F}{A}$$

where F is the force or load and A is the cross-sectional area. Tensile strength is the maximum force per unit area in tensile testing, but more useful is the yield strength, which is the maximum load safely applied without deforming the material.

Strain is given by the change in length Δl over the original length l_0

$$\epsilon = \frac{\Delta l}{l_0}$$

Although the test sample elongates, perpendicular to the loading axis, it contracts. The ratio of contraction to elongation is termed *Poisson's ratio* ν. It is formally defined as follows:

$$\nu = -\frac{\epsilon_P}{\epsilon}$$

where ϵ is the strain defined above, and ϵ_P is the perpendicular strain (actually a contraction and thus the negative sign) in the perpendicular direction. Usually ν is near 0.3 for full-density P/M steels. This means that an axial elongation of 10% will result in a shrinkage of the perpendicular dimension of 3%.

Elasticity is the proportionality between stress σ and strain ϵ (for stresses below the yield strength) and is measured by the elastic modulus E, which is also know as *Young's modulus*

$$E = \frac{\sigma}{\epsilon}$$

This last equation, known as *Hooke's law,* is most important to design since it relates the load (stress) to dimensions of a component. A typical value for E is 200 MPa (30 Mpsi) for a full-density steel.

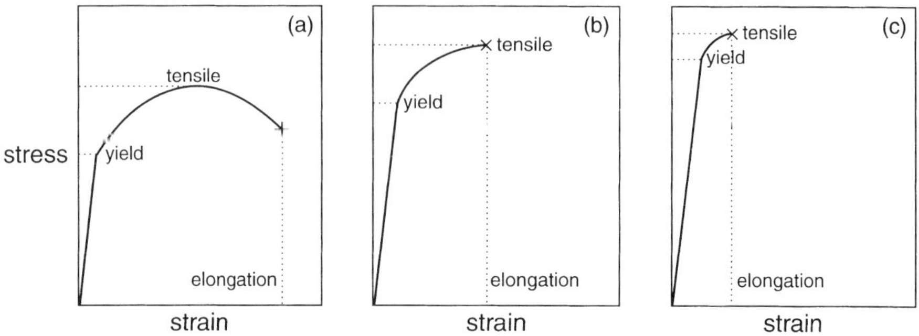

Figure 1.11. Example stress–strain curves for tensile tests. Many mechanical properties are evident from these plots, including the elastic modulus (slope of the straight-line initial portion), yield strength, ultimate tensile strength, and elongation to failure. The plot in (a) is typical for a steel with very low strength, but high ductility, while (b) corresponds to intermediate levels of ductility and strength, and (c) is characteristic of high strength and low ductility. In the latter cases, the ultimate tensile strength is the point of fracture, which is more typical in powder metallurgy.

where the sample shifts from elastic (recoverable) stretching to plastic (permanent) deformation. It typically occurs for steels at strains of 0.4–0.6%, being higher for the stronger steels. By standard convention, the reported yield strength corresponds to a plastic strain of 0.2%, where observable deformation has taken place. Steel yield strengths range from 200 to 1000 MPa (from 30 to 140 ksi), but in a few special alloys can be even 3 times higher.

Often another strength is reported, called the *tensile strength* or *ultimate tensile strength.* This is the highest stress encountered in the tensile test, and for many P/M steels corresponds to the stress at failure. In the very ductile steels, the stress at fracture is lower than the tensile strength. The tensile strength always equals or exceeds the yield strength, but as porosity increases these two values converge. Thus, for some brittle steels, no yield strength is reported, since the yield strength is the same as the tensile strength and fracture strength.

A common measure of ductility is the elongation given as total stretch to failure divided by the initial center test length. It has no units, so it is given as a percent. Another measure of ductility is the change in cross-sectional area at the point of failure, measured as the change in area divided by the original area. This is the reduction in area, and it, too, is reported as a percent. Some sintered steels have almost no ductility, while others are capable of 20–60% elongation.

1.4.4.3. Transverse Rupture. Another strength test designed for low-ductility materials, such as ceramics, is widely used in powder metallurgy. It is a failure strength test, but involves bending instead of pulling. This transverse rupture test is sketched in Figure 1.12. Three contacts are made by cylinders (two on the bottom) with the rectangular sample, which looks like a domino. The maximum load is recorded, and the sample dimensions and testing span are used to calculate the stress

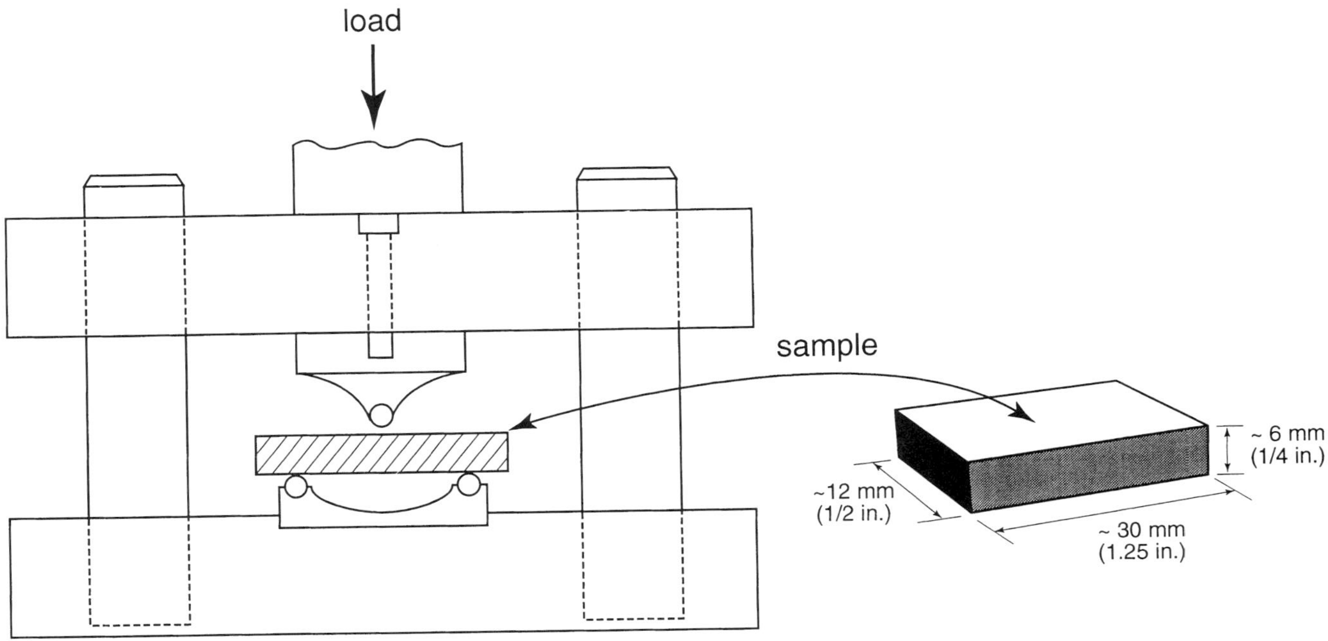

Figure 1.12. The transverse rupture test is a means for measuring the strength in bending for brittle materials. It consists of a rectangular sample and three cylindrical contacts. Failure loads are recorded for the sample, and the maximum stress is reported as the transverse rupture strength, which is about 1.5–2.0 times the tensile strength. The nominal sample dimensions are shown; actual sizes depend on the test standard.

EQUATION BOX 1.3

The *transverse rupture strength* is a common measure of the strength of low ductility P/M materials, especially the strength before sintering. It is measured on a rectangular specimen in three-point bending, with fracture on the surface opposite the upper ram. The transverse rupture strength σ_T is calculated from the specimen geometry and failure load as

$$\sigma_T = \frac{3PL}{2WT^2}$$

where P is the breaking load, T is the thickness, W is the width, and L is the distance between the lower load supports. This formula assumes no ductility in the compact. Should the deflection before failure exceed 0.25 mm (about 4%), then the calculated values are in error. In such cases, it is best to resort to tensile testing.

The elastic modulus E can also be estimated from the deflection before bending or fracture as follows:

$$E = \frac{PL^3}{48\delta I}$$

where δ is the deflection and I is the moment of inertia for the sample and equals

$$I = \frac{WT^3}{12}$$

This approach gives an approximate elastic modulus. More accurate values are obtained by tensile tests or ultrasonic velocity measurements.

needed to cause failure. Typically the transverse rupture strength is about 1.5–2.0 times the tensile strength, since the loading is in bending instead of tension.

1.4.4.4. Impact. Impact toughness is important when a material must withstand sudden loading, as might be encountered in accidents. As shown in Figure 1.13, the test is performed using a swinging pendulum with a test bar placed at the bottom of the pendulum swing. If the sample is tough, meaning that it requires considerable energy for failure, then it consumes much of the pendulum's kinetic energy. Consequently, the pendulum will not swing very high after breaking the sample. Occasionally a sample is so tough that the pendulum is halted without breaking the sample. On the other hand, if the sample has a low toughness, then the pendulum will swing to a high height after fracture.

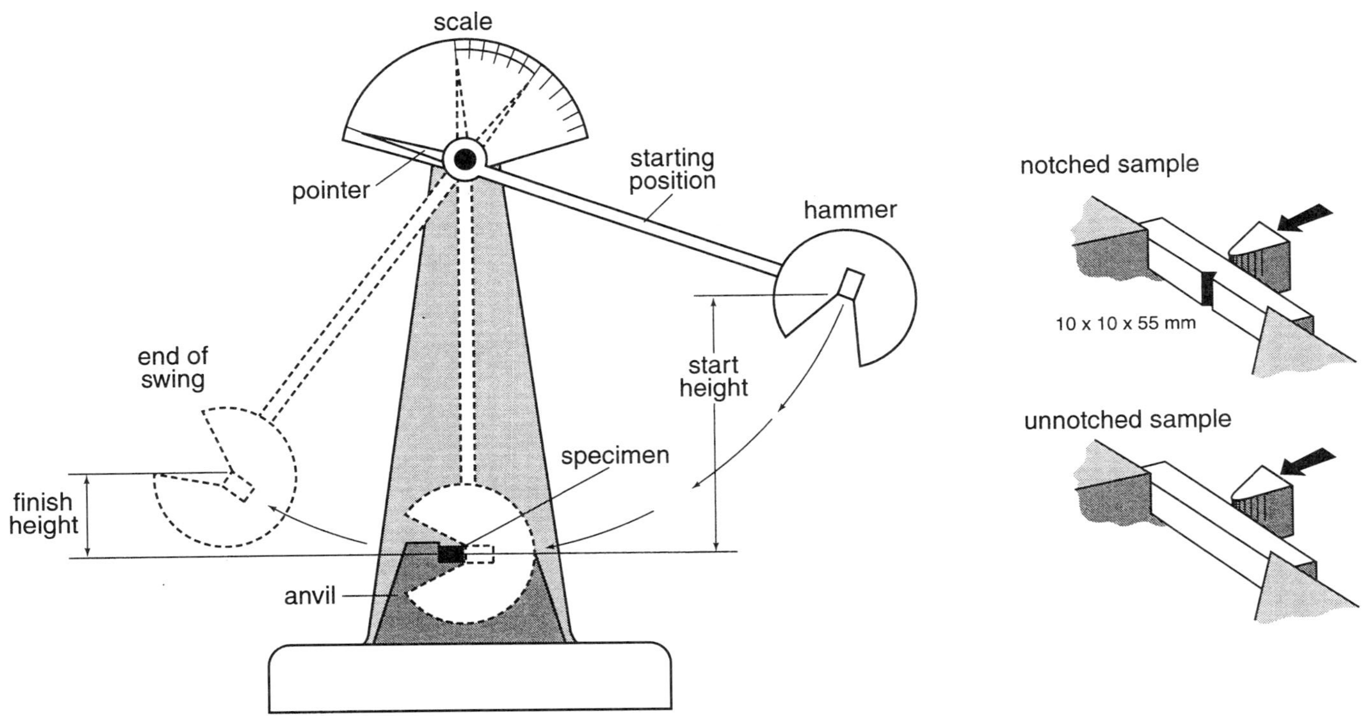

Figure 1.13. Impact toughness is measured by the energy loss for a pendulum that swings through and fractures a square cross section of the test sample. For P/M steels, the sample is usually smooth, while most testing outside P/M uses a notched bar. The test is commonly called the Charpy impact test.

There are several test geometries, but the most typical for powder metallurgy is the unnotched Charpy bar. It is a simple rectangle 55 mm long with a width and thickness of 10 mm (2.1 × 0.4 in). Many porous P/M products suffer a low toughness [1 J (less than 1 ft · lb)] and prove unacceptable for sudden loading situations. Some P/M steels can be tailored for high toughness, giving values in the 30–60 J range (22–44 ft · lb), and P/M stainless steels have exceeded 300 J (220 ft · lb). In traditional ferrous metallurgy, it is common to notch the Charpy bar to ensure failure at the center and to lower the toughness, reflecting a more conservative design parameter. Because many P/M compacts have pores, the toughness is low and notching is usually avoided to obtain higher readings.

1.4.4.5. Fatigue. Finally, an important mechanical property for many ferrous steels is fatigue resistance. This is a measure of the stress that a material can see repeatedly without failure. Tests usually cycle loading and unloading to a known peak stress and measure the number of cycles to failure. Although there are several variants (tensile, bending, and rotating), the most widely used in ferrous P/M is the rotating bending fatigue test. Figure 1.14 sketches the idea, where the test sample

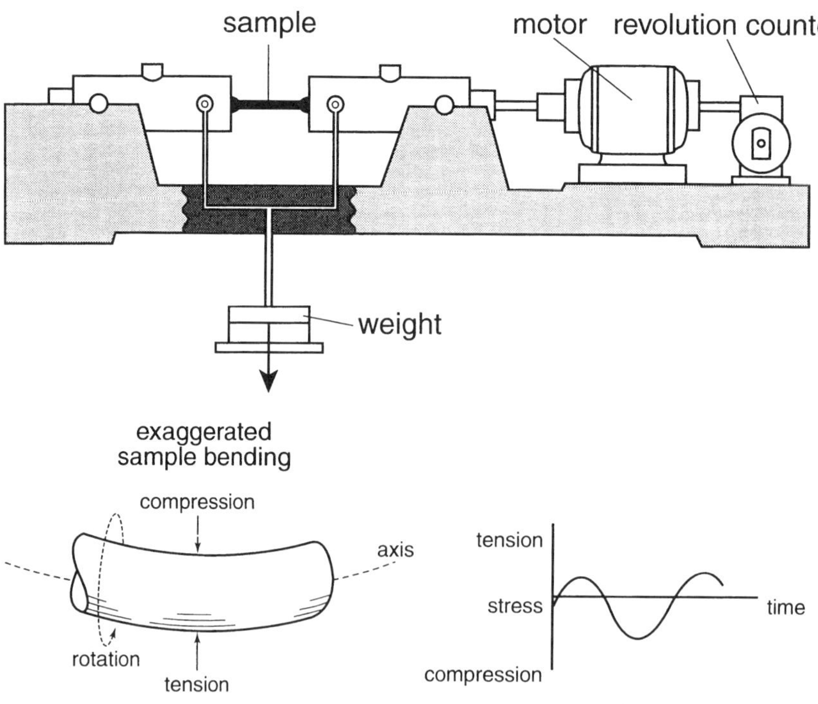

Figure 1.14. The rotating-beam fatigue test relies on a cylindrical sample loaded to cause tension on one surface and compression on the opposite surface. Since the beam rotates, each point on the surface sees a cycle of tension and compression. The fatigue strength reports the stress that the sample can withstand without failing in 10 million cycles.

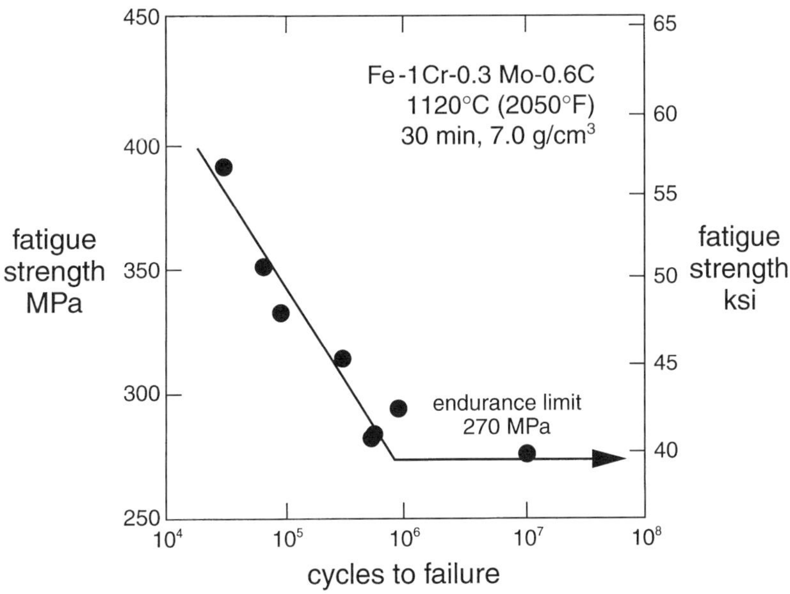

Figure 1.15. Example fatigue data for a die-compacted alloy consisting of iron, chromium, molybdenum, and carbon. The rotating-beam test was conducted at various stresses, and this stress is plotted as a function of the number of cycles before fracture. The endurance strength of 270 MPa (39 ksi) corresponds to the survival of at least 10 million cycles. (Data courtesy of Satoshi Uenosono.)

is loaded to cause the lower surface to be in tension and the upper surface to be in compression. On each rotation any point on the surface then sees a stress cycle that involves repeated tension and compression. The average stress at which the steel can sustain 10 million loading cycles without failure is reported as the fatigue strength. It is also called the *endurance limit.* Figure 1.15 plots data for the stress and number of cycles to failure for a sintered steel. At the higher stresses, the number of cycles to failure is low. However, below 270 MPa (39 ksi) the material can withstand essentially an infinite number of loading and unloading cycles without failure.

Failure in fatigue is a combination of crack initiation and propagation, and sophisticated design equations are available to predict component life in cyclic loading. The propagation phase of fatigue is very evident on the fracture surface as small steps. Figure 1.16 shows these microsteps in a scanning electron micrograph of a fatigued P/M material. Fatigue strength is important for components that will be in service for long times, such as in automobile engines or office machines. The fatigue strength is always below the yield strength and is often in the range from 30 to 50% of the tensile strength. Fatigue strength is lower, with angular pores and high porosity levels. Although most reports are for the stress giving 50% survival, the more appropriate design parameter is a lower stress corresponding to 99 or 99.9% survival.

EQUATION BOX 1.4

Fatigue is a failure process resulting form millions of small damage events associated with each cycle of loading. A well-established body of mathematics provides a base for predicting the life of a component under any combination of stress conditions. As evident in the micrograph of Figure 1.16, each stress cycle contributes to a small advance of a crack. If the size of the crack is a and the number of stress cycles is N, then the step size is determined by the stress intensity on each cycle. As a simplified version of the Paris law model is given as

$$\frac{da}{dN} = C\left(\frac{\Delta\sigma}{E}\right)^m$$

where $\Delta\sigma$ is the change in stress on each loading cycle and E is the elastic modulus. The coefficient C depends on many parameters, including the material strength, loading conditions, and crack geometry, and m is the power-law exponent. One consequence of porosity is that the crack takes a bigger step on each stress cycle, giving a shorter component life.

1.4.4.6. Selection Criteria. For most applications, only a few factors need be specified to ensure successful function. For an application where wear is a concern, a high hardness is most important. Alternatively, if buckling is a concern, then a high yield strength is important. If abuse might occur in service, then impact toughness is the most important property. When the material is subjected to cyclic loading, such as in consumer and industrial products (for example, automobile transmission gears, and hand held power tools), then the fatigue life is dominant and a high tensile strength is important. Thus, depending on the application, the required properties will vary.

As a first introduction to P/M steels, Table 1.4 gives several mechanical properties for die-compacted and sintered iron. The samples have a low carbon content, as evident by the low strength at full density. This listing shows a clear picture of how strength, ductility, elastic modulus, and the Poisson ratio change with density. Thus, several important properties have been introduced and already the negative effect of porosity is evident.

Besides the mechanical properties, ferrous P/M alloys are used on the basis of magnetic, electrical, corrosion, oxidation, and thermal characteristics. Just as there are many tests for mechanical properties, there are also various measures of these properties. Other texts on engineered materials and their selection help reconcile the properties, tests, and design rules. Most of the P/M steels are used for mechanical components; thus our emphasis here will be on mechanical properties. Similar rules concerning processing and porosity apply to other properties.

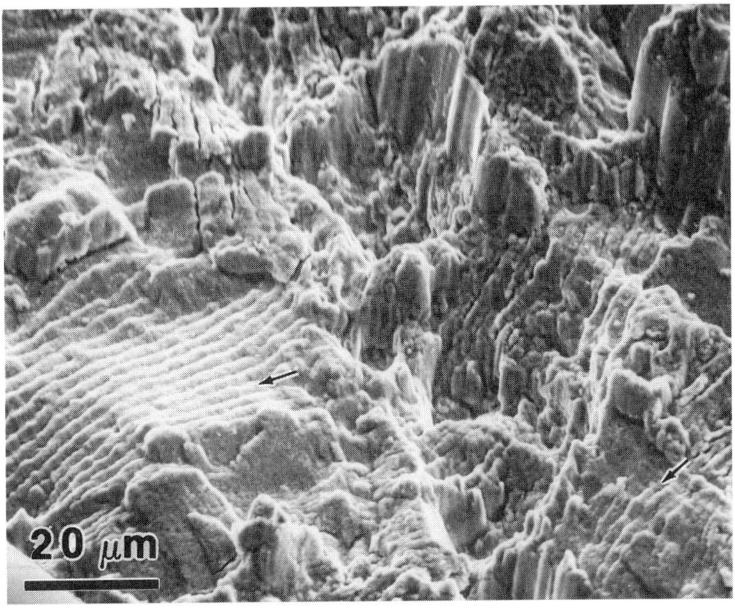

Figure 1.16. Fatigue striations on the fracture surface of a failed hot isostatically pressed component. These steps indicate the crack advance on each cycle of loading and unloading. Although the steps are only 1.5 μm in size, after thousands of cycles, the crack has moved all the way across the sample. (Micrograph courtesy of Suresh Golwalker.)

1.5. TYPICAL APPLICATIONS FOR FERROUS ALLOYS

In analyzing the world's use of engineered materials, there is a rationalization as to the best opportunity to use steels and the best opportunity to use powder metallurgy. It is the overlap between those two opportunities that becomes the focus for this

TABLE 1.4. Properties of Iron Powder Compacted and Sintered to Various Densities

Density (g/cm^3)	Elastic Modulus (GPa)	Tensile Strength (MPa)	Elongation (%)	Poisson's Ratio	Impact Toughness[a] (J)
5.8	72	84	2	0.18	0
6.2	90	126	3	0.20	2
6.6	110	166	5	0.21	4
7.0	130	225	9	0.23	8
7.4	160	280	28	0.26	40
7.87	205	335	34	0.28	60

[a] Unnotched.

book. Where are steels the material of choice? As is evident, the answer is in structures where stiffness, strength, and toughness are major concerns. There are other materials that are stronger, stiffer, but few that offer a high stiffness (elastic modulus) with high strength in a tough material that can withstand impact without failure at the same cost as steels. Ceramics often exceed steels in stiffness and strength, but fail to have toughness. However, steels are heavy materials and cannot match the low weights of composites, aluminum, and titanium. So the better applications are stationary or where cost is more important than weight. Steels are not very attractive where thermal or electrical conductivities are important, and they have only moderately good corrosion or high-temperature properties. Thus, the most successful applications for P/M steels rely on the combination of low sensitivity to weight and where there is a need for strength, stiffness, toughness, and low cost, especially in the fabricated condition.

The categories already introduced in Table 1.2, showing the major types of steels, also provide a glimpse at the many uses for steels. These include automotive components, such as bumpers, engine blocks, body panels, and transmission gears. Industrial tools and machines are fabricated from steels, mainly because of the favorable strength:cost ratio. Railroad rails, water pipes, oil and gas lines, and electrical transmission towers have always been large users of steels. Today, most larger buildings use steel beams and supports as main structural members, and the common nail is directly fabricated from steel wire. We construct very large structures by joining steel components via welding; these include submarines, bridges, pipelines, ships, and storage tanks. Special steel alloys are used in transformers, but by far the major uses of steel are related to structural applications, situations where the strength is a major concern. For most of these applications, there is no interest in powder metallurgy. However, in the home, steel components are typically smaller and used in many devices ranging from computer disk drives to kitchen utensils. These smaller, complex-shaped objects become expensive to form from wrought steels. Accordingly, shaped objects that require precision, good strength, and high production quantities are often produced from metal powders. Thus, we see emphasis on tools, gears, cams, levers, latches, and other mechanical components that require the special attributes of P/M.

SUGGESTIONS FOR FURTHER READING

Ashby, M. F., *Materials Selection in Mechanical Design*, Pergamon Press, Oxford, UK, 1992.

Bain, E. C., and H. W. Paxton, *Alloying Elements in Steel*, 2nd ed., ASM (American Society for Metals), Materials Park, OH, 1966.

Boyer, H. E., and T. L. Gall (eds.), *Metals Handbook Desk Edition*, ASM, Materials Park, OH, 1985.

Edwards, L., and M. Endeau, *Manufacturing with Materials*, Butterworth-Heinemann, Oxford, UK, 1995.

Lankford, W. T., N. L. Samways, R. F. Craven, and H. E. McGannon (eds.), *The Making, Shaping and Treating of Steel*, 10th ed., United States Steel, Pittsburgh, PA, 1983.

Shackelford, J. F., *Introduction to Materials Science for Engineers*, 2nd ed., Macmillan, New York, NY, 1988.

CHAPTER 2

The Powder Metallurgy Process

2.1. INTRODUCTION

2.1.1. Product Features

Powder metallurgy (P/M) is a processing technique that involves the production of metal powders and conversion of these powders into useful engineered structures. In the classic process, the steps involve powder compaction in rigid dies and the subsequent heating of the compacted powder to bond the particles together. At the beginning, the powder starts as a fluidlike substance, where it is neither a liquid nor a strict solid. This fluidlike character initially makes P/M similar to casting. A dramatic difference is that in P/M the powder flows into a die cavity at room temperature. Under pressure the powder is densified by deformation, with the formation of weak friction bonds between the particles. After compaction the powder takes on the properties of a solid. Although weak after compaction, the solid structure can be heated to a temperature where the particles sinter to one another to form strong bonds. This sintered product has many benefits—it holds the shape imparted by the compaction die, but builds up strength in sintering. Many variants exist to this basic process.

Where is powder metallurgy the most attractive processing option? When the component has moderate thickness and size, with a weight probably in the 1–1000-g (0.002–2.2-lb) range. The process capability matches well to shapes with moderate levels of complexity and moderately smooth surfaces, but not too complex or too smooth. Ferrous powder metallurgy can then be understood by the intersection of various technological characteristics. It is best applied to strong, stiff, low-cost, and tough structures that have a low weight sensitivity, where the shape is somewhat complex, with a weight compatible with holding the object in your hand. The applications are less concerned with corrosion, oxidation, and wear, and more concerned with surface finish, shape complexity, and good mechanical properties at a low manufacturing cost.

The production of metallic components by compaction and sintering in P/M is a large industry that produces many shapes and materials. Of these materials, steels are the dominant metallic species. Few powder metallurgy components are very large. The shapes range from simple cylindrical bearings to very complex structures such as hollow golf-club heads and helical seat gears. Because of a high tooling cost, most P/M production is targeted at structures fabricated at high rates (measured typically in parts per year), such as for automobiles, lawnmowers, home appliances, business machines, computers, and other electromechanical structures. Some P/M components, such as tantalum capacitors, bronze bearings, and molybdenum heat sinks, are produced at rates as high as 40 million per day. More typical to ferrous P/M are production rates of a few million per year, often for automotive applications.

2.1.2. Historical Perspective

The P/M process had several successes in ancient times (probably as early as 3000 BC) with the fabrication of coins and implements in ancient Egypt, gold–platinum powder compacts by the Incas, an iron pillar in India, copper coins in Germany, and platinum crucibles in the 1800s in England and Russia. The modern era is traced to Coolidge's experiments in the early 1900s (working for Edison) and the production of tungsten lamp filaments for the light bulb, a process that is still used today. By the 1930s several products emerged, including porous bronze bearings, tungsten radiation containers, copper–graphite electrical contacts, cemented carbides for metal cutting, and the first steel mechanical components. By the 1950s the industry became organized and the field emerged as a viable production technology for selected components. Early P/M was applied mostly to low-stress, moderately complex components as an alternative to machining.

By the 1960s, interest in the net-shaping attribute of powder metallurgy emerged. This idea relies on the attainment of final dimensions in the powder forming step, by that avoiding machining. Aerospace uses grew with the development of full-density fabrication techniques, such as hot isostatic pressing. In the 1970s there emerged recognition of the rapid solidification rates possible in powder atomization, allowing the production of new compositions with novel properties. In the more recent years, many new alloys, property combinations, and production technologies have emerged.

2.1.3. Current Situation

Today, the P/M industry is made up of many companies, each with specialization in alloys, applications, and production techniques. The largest activity is associated with ferrous alloys, some iron, and many steels. Thus, the focus of this book is on the largest segment of powder metallurgy. Most P/M products are for applications in automobiles. In spite of the many successes, P/M is always under attack from alternative forming technologies and alternative materials—mainly fine blanked steels, cast steels, die-cast zinc, injection-molded plastics, injection-molded ceramics, extruded aluminum matrix composites, and machined steels.

Thus, ferrous alloys constitute the dominant powders, and structural automotive components constitute the dominant applications. Attainment of the desired mechanical properties requires alloying to form high-strength steels during sintering. The combination of low production costs and high sintered strength dominates the P/M selection criteria. Since high density is a precursor to high strength, the trend is toward higher densities to improve the mechanical properties. Production technologies that allow high final densities include high-temperature sintering, powder forging, powder rolling, injection molding, infiltration, hot isostatic compaction, pneumatic forging, and warm compaction. For this book, the emphasis will be on the basic process of die compaction followed by sintering, and several similar concepts. They prove most competitive in the production of moderately complex components at high production rates, with dimensional tolerances near those associated with machining.

2.2. OVERVIEW OF PROCESSES AND STEPS

The P/M processes can be divided between two extremes. One extreme is encountered in injection molding and other processes that start with loose powders. These are sintered to nearly full density at high temperatures. Here the powder is shaped in the forming step, but not compacted. Most important, the particles are densified in sintering, not in shaping. The other extreme is encountered in traditional die compaction, where densification occurs in pressing. Here, sintering has little impact on densification and is used largely to bond the particles. Between these two extremes are various technologies where sintering and compaction are combined for densification. Some examples occur in hot pressing, hot isostatic pressing, and hot forging. This categorization is illustrated in Figure 2.1, where key steps in the conversion of powder into a component are outlined. Pressure-based densification relies on

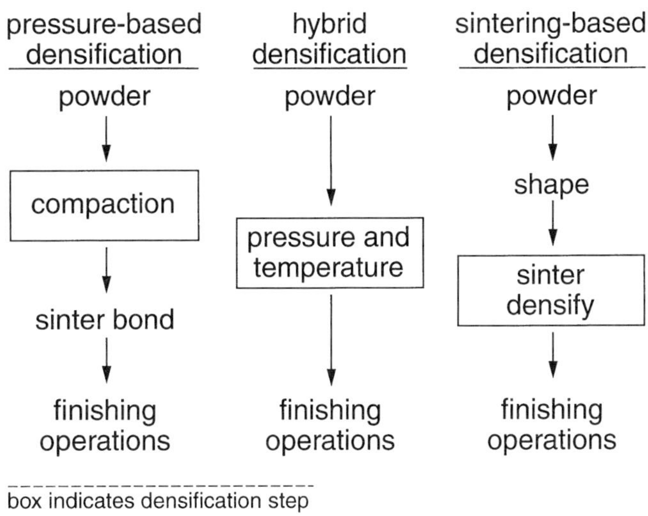

Figure 2.1. A schematic categorization of P/M production processes, differentiated on the basis of where densification occurs in the cycle, ranging from the forming step to the sintering step, with hybrid technologies that simultaneously compact and sinter.

lower sintering temperatures, larger particle sizes, and higher forming pressures to densify the powder. Alternatively, sintering-based densification relies on low forming pressures, high sintering temperatures, and small powders. The hybrid densification techniques apply temperature and pressure simultaneously to obtain a pore-free product.

Table 2.1 provides a first basis for categorization of P/M processes as to shaping versus compaction, and sinter bonding versus sinter densification, with several variants where both are performed simultaneously. The first column of this table gives the common name for the process. In each of these, densification occurs in either the forming step, the sintering step, or both. For example, die compaction compresses the powder to produce a density 2–3 times the initial powder density. Alternatively, injection molding applies a relatively low pressure that shapes the powder, so the particles are not deformed nor densified. These forming steps are characterized by typical stresses or pressures and forming times. Die compaction occurs in roughly one second, while slip casting requires a half hour or so to form a compact.

With respect to the sintering step, some technologies rely on low temperatures to bond the particles, while others rely on high temperatures to densify the structure. Sintering densification results in a volume change, while bonding results only in a strength gain. Again, each process is characterized by a typical temperature and time. Various P/M fabrication techniques rely on a combination of pressure and temperature; for example, hot isostatic compaction might involve 200 MPa (30 ksi) at 800°C (1470°F) for one hour to densify a steel powder. In each case the cited conditions are only examples; there is much variation between production facilities and materials.

A variety of P/M processing techniques are applicable to iron powders. However, when analyzed as to actual production, two are dominant and will constitute a major focus for this book: die compaction and injection molding. On a relative use basis, the press and sintering technology is very large on a tonnage basis, consuming about 85–90% of all ferrous powders. Next is injection molding, which constitutes about 5% of all ferrous powder sales. The next largest is hot forging, followed by other technologies listed in Table 2.1. Each of these production routes has certain advantages and disadvantages, usually leading to selected component shapes, sizes, and applications. As an example, for steels that will see high fatigue stresses, full density is necessary; thus, hot isostatic compaction or hot forging are recommended. Forging is easier to automate for high production volumes, has a short cycle time, and provides a high final strength. Accordingly, forgings are used for connecting rods where millions are produced every year, while hot isostatic compaction is the choice for lower-volume production of high-performance aircraft landing gears. In any selection of a P/M process, consideration of the production economics is necessary.

2.3. TECHNICAL ADVANTAGES

A common problem is in determining where and when to use P/M. Clearly the high material use and minimized machining make component production attractive. As

TABLE 2.1. Categorization of P/M Processes Based on Densification Characteristics

Process	Forming Step	Typical Condition (MPa)	Time Frame	Sintering Step	Typical Condition (°C)	Time Frame	Sequence
Binder-assisted extrusion	Shaping	20	10 s	Bonding	1250	1 h	Extrude, then sinter
Cold isostatic processing	Densification	420	10 min	Bonding	1100	30 min	Press, then sinter
Die compaction	Densification	700	1 s	Bonding	1120	30 min	Press, then sinter
Double press, double sinter	Densification	700	1 s	Bonding	1120	30 min	Repeat both steps
High-temperature sintering	Densification	700	1 s	Densification	1250	60 min	Press, then high-temperature-sinter
Hot extrusion	Densification	800	1 s	Densification	850	1 s	Simultaneous
Hot forging	Densification	100	1 s	Densification	1000	1 s	Simultaneous
Hot isostatic compaction	Densification	200	1 h	Densification	800	1 h	Simultaneous
Hot pressing	Densification	10	1 h	Densification	850	1 h	Simultaneous
Injection molding	Shaping	14	15 s	Densification	1250	2 h	Shape, then sinter
Loose powder sinter	Shaping	0	10 s	Densification	1300	2 h	Pour powder into cavity and sinter in that cavity
Roll compaction	Densification	700	1 s	Bonding	1000	30 min	Roll, then sinter
Slip casting	Shaping	0.1	30 min	Densification	1300	1 h	Shape, then sinter
Spray forming	Densification	10	0.1 s	Bonding	800	1 s	Splat particles onto substrate
Tape casting	Shaping	0.1	10 s	Densification	1250	30 min	Shape, then sinter

TABLE 2.2. Comparison of P/M and Competitive Metalworking Techniques

Technique	Advantages Versus P/M	Disadvantages Versus P/M
Casting	Widely used technique; small to large part size range; low setup cost; low tooling cost	Not useful for many materials; flashing, parting line, pores; chemical segregation; single-use molds; recycle of sprues and runners
Cold forging	Faster production; higher strength; good surface finish	Lower precision; shorter tool life; limited materials
Extrusion	Long parts; smooth surface; rapid production	Constant cross section; lower precision; high energy consumption; simple shapes; short tool life
Hot forging	High mechanical properties; large, complex shapes; fast production; tailored properties	Flash and material waste; poor dimensional control; inclusions and blemishes; rapid tool wear
Machining	Available materials and shapes; used widely in industry; high precision; short lead time, no tooling; small production volumes	Waste; low productivity in some cases; nonuniform properties; high cost; labor-intensive
Stamping	Flat, high precision; high production rates; large area parts; precise feature location	Single-level, thin shapes; limited materials; waste, edge roughness, small features prove difficult

an offsetting factor, steel powders cost more on a weight basis than wrought steel. Generally, P/M techniques have an advantage with respect to microstructure control, material use, product homogeneity, and mass production. These advantages are best seen in alloys customized to the P/M process, for example, iron–copper–carbon alloys.

Table 2.2 compares powder metallurgy with other production technologies. Listed here are the classic means for forming steel components via deformation, machining, or casting routes. Only selected attributes justify the expense of the P/M route. When these attributes are missing, other routes should be used. There is some competition between metalworking technologies. Each production route is improving and pushing capabilities to secure better quality, performance, or cost. Accordingly, the relative strengths and weaknesses change continuously.

The components fabricated by P/M range over a wide scale in terms of size, mass, or other attributes. Table 2.3 describes the current range where most P/M occurs. This table suggests the following: P/M is competitive as to materials, properties, and general engineering attributes, but it needs to extend the size and mass ranges, part complexity, and surface finish to sustain a competitive position.

Another important visualization of P/M is through the attribute plots as presented by Ashby. These show the technology ranges for various production techniques. An overlay of powder metallurgy and competitive technologies leads to the observation that P/M operates in a busy arena. Figure 2.2 is a plot of the typical component complexity (measured by the information content) versus size (measured by the mass). Most P/M products range from 1 g to 1 kg (0.002 to 2.2 lb), although examples are known where a single component weighs 1000 kg (about 1 ton) or 0.02 g (about $7 \cdot 10^{-4}$ oz). Likewise, P/M products are often complex, but not too complex. Here the complexity is measured by how much information is needed to specify the part on the engineering drawing. The more dimensions, angles, surface finishes, and other details required to describe the component, the larger its information content. As a generalization, P/M products have 15–1000 callouts on the drawings.

Note how many viable production technologies operate in this arena. Powder metallurgy competes with various casting, deformation, machining, plastic molding, and die casting techniques. It sometimes succeeds against these competitive technologies based on improved surface finish, longer tool life, lower expense, and higher final properties. However, when the production volumes are low (less than a few hundred), it proves difficult to justify P/M and machining is typically the lowest-cost process. At intermediate production quantities (a few hundred to a few thousand), the usual choice is casting followed by machining of critical dimensions. Powder metallurgy is most useful for production quantities over a few thousand from moderately complex components.

In much the same manner, the various engineering materials can be compared with respect to properties. Here steels succeed because of strength, cost, stiffness, fracture resistance, and ability to operate in moderately corrosive or warm environments. Thus, the following technological conceptualization summarizes the advantages of P/M:

- Moderately complex shapes (at least 15 but less than 1000 callouts on the engineering drawing).

TABLE 2.3. Typical Ranges of P/M Production

Characteristic	Range	Status	Opportunity
Complexity, tolerances, or dimensional specifications	15–1000	Competitive with other technologies, far behind electronic systems	Increase shape complexity and precision with new shaping techniques
Mass, g	1–1000	Competitive with many other technologies	Smaller devices, more complex shapes, biomedical
Tolerances, mm	0.1–1	Weakness compared with polymers and machining	Improve precision using more production controls
Roughness, μm	0.2–2	Area of weakness; particle size and shaping are limits	Need improved surface smoothness, smaller powders, better finishing

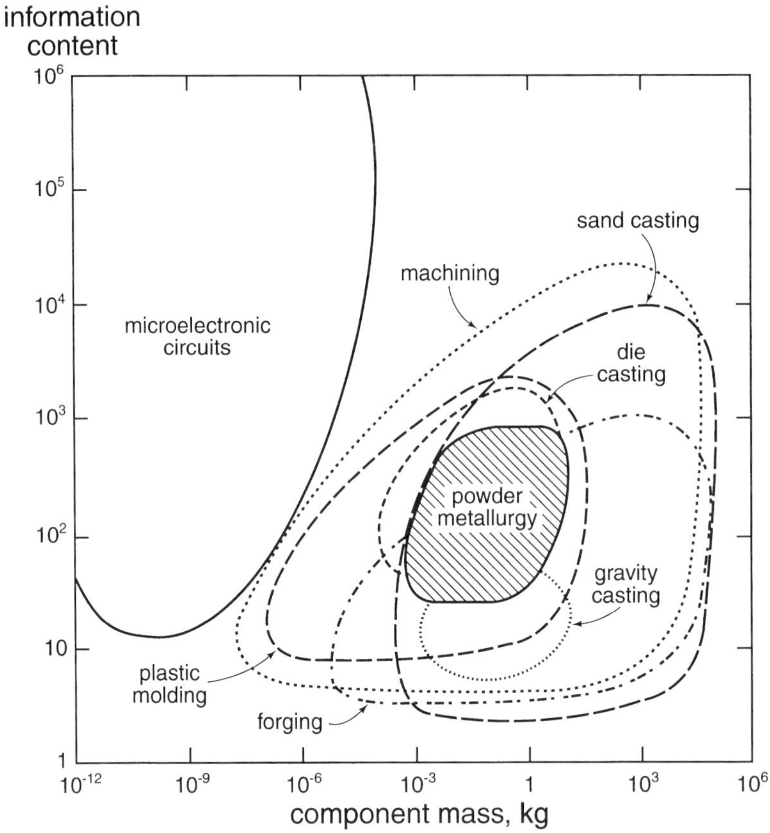

Figure 2.2. This diagram plots various component fabrication technologies with respect to the component mass and the information content of the component (number of callouts on the engineering drawing). Powder metallurgy exists in a field with many competitive technologies. (Based on data from Michael Ashby.)

- Created at higher production quantities (a few thousand per year to a few million per day).
- In a mass range from 1 to 1000 g.
- Where strength and other mechanical or physical properties require metallic materials.
- Surface finish, tolerances, and other requirements do not greatly increase production expenses.

In comparison with other technologies, powder metallurgy is at a disadvantage with respect to the raw-material cost, tool cost, and processing complexity. The greatest difficulty derives from residual pores. Properties suffer from residual porosity, making P/M less attractive as compared to other metalworking technologies.

Thus, the continual push is for higher-density structures with higher levels of shape complexity, but at modest production costs.

2.4. EXAMPLE APPLICATIONS

Technical libraries, trade magazines, internet web pages, and company brochures are rich with examples of powder metallurgy applications. The range is so large that it proves difficult to provide a brief listing. Recent examples include wristwatch cases, laminated knife blades, automobile engine timing components, computer disk-drive magnets, wood cutting tools, metal drawing dies, hair clippers, automotive gears, brake sensors, and shotgun trigger guards. Of the many possible applications, those associated with the automobile are dominant. Figure 2.3 traces the use of metal powders in automobiles produced in North America since 1976. These applications include components ranging from shock absorber compression valves to connecting rods. Many successes are associated with the engine, where P/M components are used for the water and oil pumps, valve seats, lifters, sensors, rockers, power steering pump, bearings, sprockets, flanges, injectors, and even fuel filters. Likewise, the complex electromechanical systems associated with steering, braking, power transmission, and emission controls create many P/M component opportunities. Most of these components are formed from steels. In North America, this accounts for 72% of the consumption of ferrous powders; in Europe, 80%; and in Japan, the fraction is higher at 88%. Thus, automotive applications are a dominant aspect of ferrous

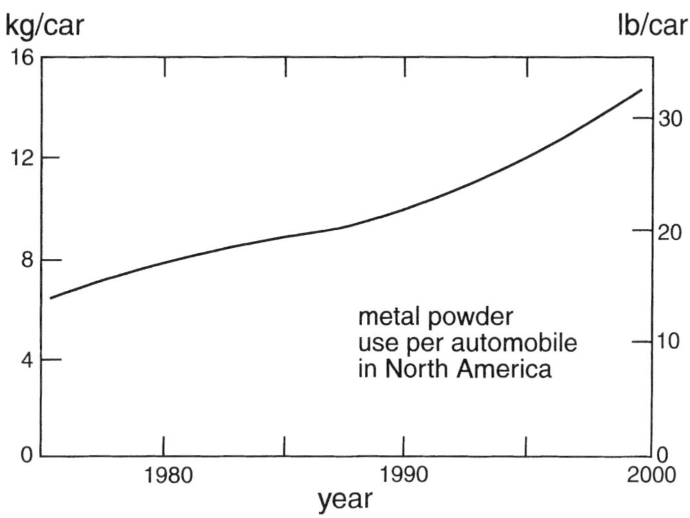

Figure 2.3. The use of metal powders in automobiles produced in North America shown against the model year. The increased use of metal powders and the increased production of automobiles are two primary contributors to the industry growth.

P/M. Figure 2.4 shows a few example components to help identify the typical shapes and sizes associated with powder metallurgy. Continued growth in automotive applications depends on improved performance, typically via higher strengths, higher densities, and greater precision.

One application area being developed is for use in the automobile engine exhaust manifold. The target is a long-life, maintenance-free system that exhibits oxidation resistance against attack by hot engine gases; resistance to corrosion by salt, water, and other fluids; and strength retention to high temperatures. Various stainless steels, iron–aluminum, and alloy steels have been examined for the application. The severe service requirements tax the existing P/M materials and fabrication technologies.

Although automotive applications are dominant, often great excitement exists because of more common and easily recognizable applications. The wristwatch case is one that few people recognize as a P/M product. Figure 2.5 shows an example of an injection-molded stainless-steel watch case. Likewise, common items such as eyeglass hinges and the orthodontic braces children wear for straightening teeth are fabricated from stainless-steel powders. Indeed, even amalgam dental fillings are formed from powders; here, silver alloy powders are used. Thus, human uses for metal powders are extensive, more so than generally recognized.

One recent description of a new knife blade showed a microstructure similar to the damascene steel of 500 BC. In the traditional process, compositions of high and low carbon steel (hard and tough) were laminated to form a steel that produced exceptional properties. The steel was highly regarded for its exceptional behavior as a sword. The modern version of these steels involves filling a capsule with two grades of steel powder as individual layers, then hot-consolidating these powders

Figure 2.4. A collection of automotive parts produced by powder metallurgy techniques. (Photograph courtesy of Julian Thomas.)

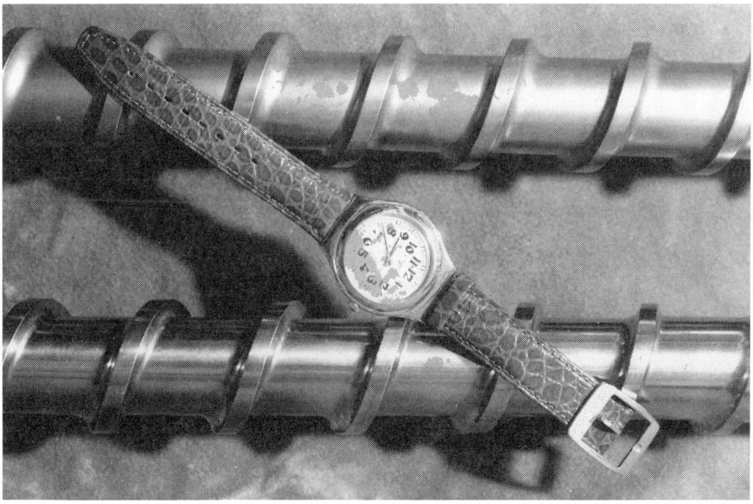

Figure 2.5. This stainless steel watch case was produced by powder metallurgy using injection molding. (Photograph courtesy of Julian Thomas.)

into a billet, followed by twisting, layering, or otherwise laminating the billet into structures with 100 or more layers. A mixture of two stainless steels, one with high strength and the other with high toughness, is a favorite for swords, knives, and other cutting tools.

Not all iron powder is consolidated. Applications exist for the powder in the loose form. This includes food supplements (the reduced iron in vitamin pills and flour), wrist weights and other exercise equipment, abrasive and shot blasting grit, welding electrodes, mirror coatings, paint, and even radar evading surface coatings.

2.5. INDUSTRY STRUCTURE

As sketched in Figure 2.6, the P/M field is divided into three primary levels with several adjunct industries. The first group of companies is the suppliers. They produce the raw materials, largely powders, sintering gases, lubricants, and alloying additives. A few large companies dominate this portion of the industry. Powder producers use iron, steel, or scrap metal to fabricate powders with specific purities, sizes, shapes, and compressibilities. Depending on the application, the attributes of the powders might be very different. The powder producers are generally large and probably the most profitable components of the industry. Closely associated with the powder vendors are the suppliers of the consumables, including process gases, tools, lubricants, heat treating fluids, alloying additives, and impregnation fluids. In each of these industry segments there are relatively few companies, so they benefit from large-scale production.

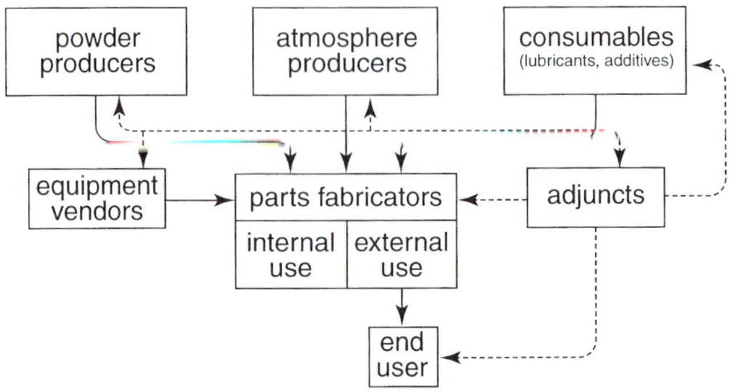

Figure 2.6. An outline of the industry structure, showing the primary components of suppliers of powders, gases and consumables, the parts production industry, and end users, with many service adjuncts.

More than 1000 facilities are involved in parts production, and they represent the core focus of the technology. Many are smaller companies, with less than 100 employees, that fabricate to specifications created by the users. A few component fabricators have more than 1000 employees, usually scattered over several sites. A large segment of component production is for captive uses. This means that the parts are produced for internal use at a P/M operation placed inside a larger plant. P/M production is part of an overall fabrication scheme where the P/M parts are assembled or combined with other components. Many companies that fabricate business machines, automobiles, consumer products, filtration systems, medical equipment, electronic components, firearms, electrical connectors, industrial components, farm equipment, hand tools, or magnetic systems have internal P/M operations to satisfy their own needs. Thus, the fabrication of P/M components has two categories of operations—those focused on production of products for internal uses, and those focused on production of products for external customers.

This then leads to the third major industry segment—the users. Many users own the production operation, for reasons of proprietary technology, cost control, production scheduling, or special needs. Nevertheless, the larger segment is those companies that purchase components on the open market. A specification is generated and circulated to possible vendors. Through a process of elimination, final production is assigned to one to three vendors. Many of these components have limited production lifetimes. The product might be discontinued, or a lower-cost vendor might take over production. Obviously, component production is one of the more competitive aspects of P/M.

Associated with this core industry is an important complement of companies that produce the production tools. These are companies that specialize in building compaction presses, tooling sets, injection-molding machines, forging presses, sizing presses, heat treatment furnaces, sintering furnaces, impregnation systems, cleaning equipment, densification equipment, and robots for plant automation. Each vendor operation has a profile on equipment size, shape, or application where they have great expertise. Accordingly, most of the equipment vendors are moderate-sized with a core of specialists that thrive on construction of custom equipment designed for the parts production operations.

Finally, a large array of adjunct operations support the industry. These include consultants, patent lawyers, technology transfer specialists, testing centers, universities, training operations, government agencies, professional and trade associations, equipment dealers, publishers, and research institutions. These operations vary from one or two individuals up to government supported research operations with 500 employees. This industry partitioning helps to understand the diverse participation at P/M trade exhibits and conferences.

2.6. GROWTH AND ECONOMIC TRENDS

Most powder metallurgy relies on uniaxial die compaction. The powders used in die compaction are produced by a few dozen suppliers around the world. For 1997, 62% of the iron powders were produced in North America, while Europe accounted for 21% of the production and Japan accounted for 17%.

In traditional die compaction, there are many fabricator operations, exceeding 400 in North America alone, but only a handful are even moderately large. These 400 companies have sales near $1.5 billion per year in North America. A smaller group of operations participates in advanced powder metallurgy technologies, and these constitute about $500 million in annual sales in North America. Keep in mind that ferrous powder metallurgy represents nearly 80% of the tonnage of traditional powder metallurgy activity and automotive applications constitute about 80% of the traditional ferrous powder metallurgy sales. It is estimated that by the year 2000 there will be an average of 15 kg (35 lb) of powder metallurgy product in each automobile fabricated in North America, but about half that for those automobiles fabricated in Japan and Europe.

The situation is similar in Japan, where three industrial giants hold 60% of the total parts production market. A few other companies have the next 30% market share, and the balance (10%) is divided between almost 100 parts manufacturers. A similar profile exists in Europe, with a few large and hundreds of smaller parts producers. In the balance of the world, powder metallurgy is not very prevalent. Considering the relation between automobile production and P/M, most activity is found in regions that produce automobiles.

The sales per employee in traditional powder metallurgy averages $104,000 (standard deviation of $20 thousand), which is low compared with competitive technologies. For example, traditional metal producing industries have more than $250,000 in annual sales per employee. One industry problem traces to the deep reliance on automotive sales and the associated constant price and profit pressure. Thus, P/M companies are constricted, leading to an industry-average profit margin of 4% of sales.

Powder metallurgy has sustained an average growth rate of 8–10% in recent years. In years when automobile production falls, there is a similar dip in P/M activity. Figure 2.7 plots the iron powder shipments for 50 years in North America, showing an average growth rate of 7% per year. However, the traditional iron-based press and sinter industry has grown at a sustained 4–6%/year level. Recently, more rapid growth has come from the advanced P/M systems targeted at jet engines, biomedical implants, sporting equipment, electronic systems, wear structures, instrumentation, medical diagnostics, electrical contacts, and batteries. For the advanced materials and applications, there is emphasis on full density processing, including hot isostatic pressing, injection molding, powder rolling, hot forging, and hot pressing. Consequently, little communication exists between traditional ferrous P/M devoted to automotive applications and advanced P/M devoted to high-performance applications. Indeed, these two communities have separate technical conferences and trade groups.

With respect to metal powder sales in North America, Figure 2.8 compares the relative production of iron and steel powders with other common metal powders. In both Japan and Europe the annual consumption of iron powder is about one-half

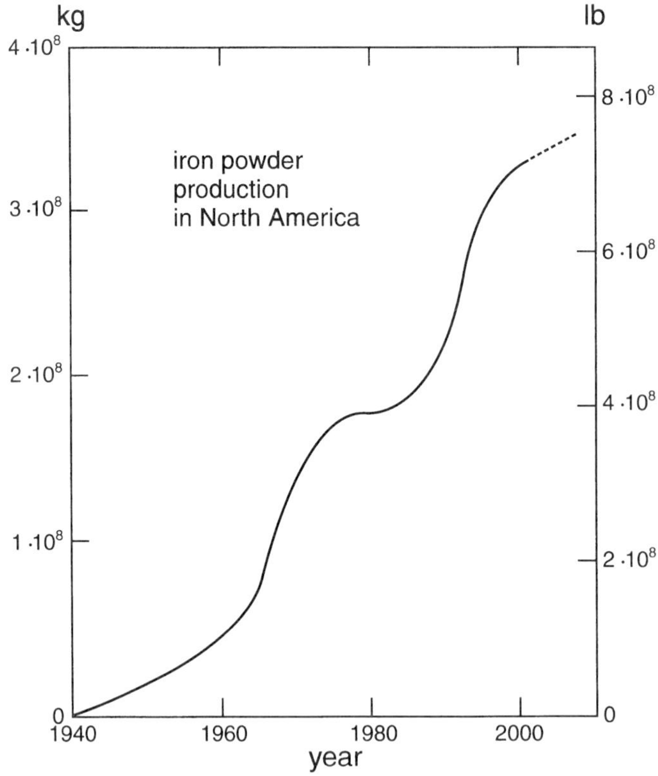

Figure 2.7. North American shipments of iron powder for P/M versus year. On the basis of tonnage, the average compound growth rate is 7% per year. For 1997, the worldwide consumption was just under $6 \cdot 10^8$ kg.

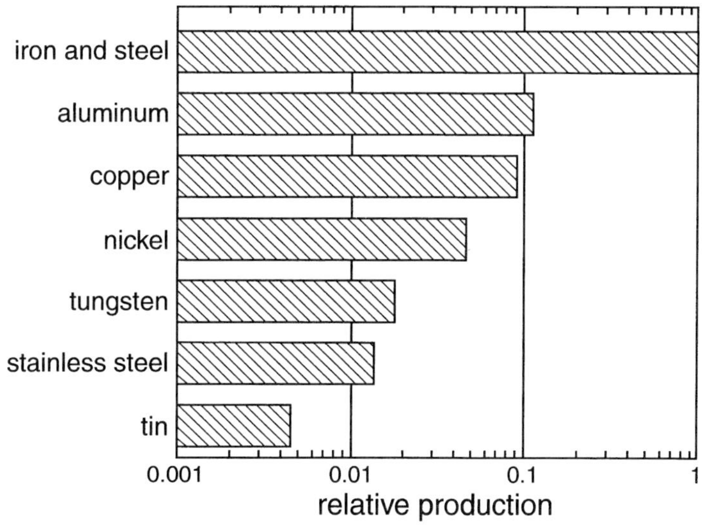

Figure 2.8. Comparison of iron and steel powder consumption (relative use of unity) and other metal powder consumption on a logarithmic mass scale.

and one-third that of North America, respectively. Large tonnage applications for the iron and steel powders will come largely from increased use in automobiles. Several projects are in progress that could substantially increase iron powder consumption. These include new single-press connecting rods, various joints and transmission components, main bearing caps, cam lobes, bearing races, transmission gears, and cylinder liners. These opportunities provide an optimistic future for P/M. On the other hand, electric automobiles use relatively small quantities of P/M components. Should these become widely accepted, P/M production might suffer.

2.7. SUMMARY

In one form of ferrous powder metallurgy, the powder is compressed in a die and subsequently sintered. The final dimensions are set largely by the die. The higher the compaction pressure, the higher the final density and properties. Minimal densification occurs in sintering, largely because the powders are large and the sintering temperature is low.

In contrast, if the powder is small and the sintering temperature is high, then densification occurs in sintering. Higher temperatures lead to pore elimination in sintering. Here no external stress is required as the powder compact densifies and strengthens during sintering.

When both temperature and pressure are applied simultaneously, rapid densification occurs. High temperatures weaken steels, resulting in easier compaction. The combined action of temperature and pressure produces full density for most steel powders. However, the greater complexity to the processing cycle couples to the high equipment and tooling expenses to create barriers to widespread use of hot deformation in processing steel powders.

SUGGESTIONS FOR FURTHER READING

Ashby, M. F., *Materials Selection in Mechanical Design*, Pergamon Press, Oxford, UK, 1992.

Edwards, L., and M. Endean, *Manufacturing with Materials*, Butterworth-Heinemann, London, UK, 1990.

German, R. M., *Powder Metallurgy Science*, 2nd ed., Metal Powder Industries Federation, Princeton, NJ, 1994.

Jenkins, I., and J. V. Wood (eds.), *Powder Metallurgy and Overview*, The Institute of Metals, London, UK, 1991.

Klar, E., (ed.), *Powder Metallurgy*, Vol. 7 of *Metals Handbook*, ASM International, Materials Park, OH, 1984.

CHAPTER 3

Iron and Steel Powders

3.1. OVERVIEW

Iron powders are generated in many facilities around the world, and at each facility there is a different production concept. Accordingly, a wide array of powder sizes, shapes, chemistries, and other attributes are possible. Such a diversity in powders is justified only by the array of applications and their differing demands on the powders. For example, spherical powders used in photocopiers are not useful for die compaction. Likewise, gas-atomized powders used in hot isostatic compaction are not useful for vitamin pills, and the carbonyl iron powders used in radar absorption cannot be plasma-sprayed.

To address such diversity, this chapter on powder production is organized into four divisions. First, we need to have a language for discussing powders and their characteristics—size, shape, packing, and flow. The second section describes powder production techniques with information on the powder characteristics associated with each. Third, the powder characteristics will be contrasted and compared. Finally, we turn attention to the selection of powders for various applications. This latter step will show how certain powder characteristics are desirable for specific applications.

For introduction purposes, consider the powder production information given in Table 3.1. As we progress in the chapter, these techniques will be described. Nevertheless, as a starting concept the diversity of powders and characteristics should be evident. Such diversity has its merits. Each application has identified powder characteristics that are most suitable. Some are satisfied by small powders, while others require large powders, irregular powders, or sponge powders. By the end of this chapter you should be able to understand how powders are made and their attributes, and to rationalize the powder characteristics, production processes, and applications.

3.2. TESTS OF POWDER CHARACTERISTICS

Usually ferrous P/M deals with particles that are larger than smoke (0.01–1 μm), but smaller than sand (0.1–3 mm), and most of the common particles have diameters similar to that seen with human hair (25–200 μm). Scanning electron microscopy (SEM) is a useful tool for observing particles, largely because it does not have optical microscopy's limitation in focusing on three-dimensional objects. Figure 3.1

TABLE 3.1. Comparison of Ferrous Powder Production Techniques

Process	Particle Size Range (μm)	Particle Shape	Packing Density	Alloying	Cost
Carbonyl decomposition	2–20	Spherical	Moderate	No	Moderate
Centrifugal atomization	50–1000	Spherical	High	Yes	Variable
Electrolytic	5–300	Irregular, dendritic	Low	No	Moderate
Gas atomization	10–300	Spherical	High	Yes	High
Mechanical comminution	2–500	Angular	Moderate	Some	High
Melt spinning	200–1000	Flake	Moderate	Yes	High
Millscale reduction	20–800	Sponge	Low	No	Low
Oxide reduction	30–500	Sponge	Low	No	Moderate
Plasma atomization	5–80	Spherical	High	Yes	Very high
Spark erosion	0.1–20	Spherical	Moderate	Some	Very high
Water atomization	5–400	Irregular, ligamental	Low to moderate	Yes	Moderate

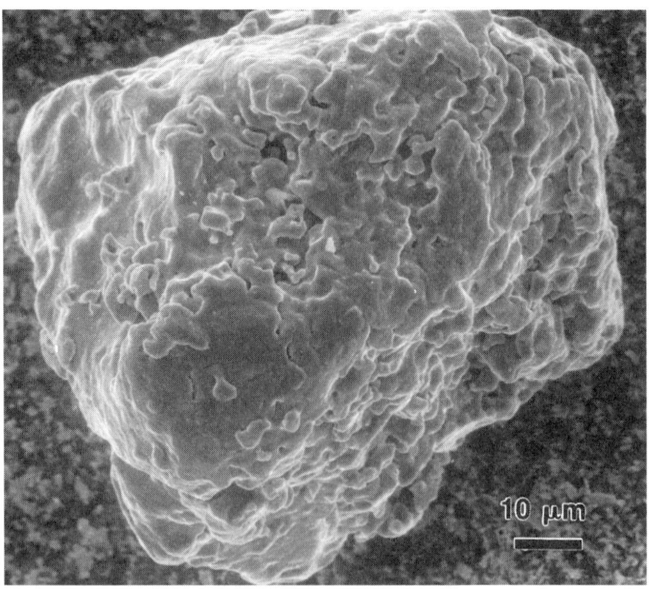

Figure 3.1. Scanning electron micrograph of a single water-atomized iron particle, showing the size, shape, and surface texture attributes associated with a typical powder designed for compaction. (Photograph courtesy of James Dunlap.)

is a SEM micrograph of a single iron particle. It shows the particle size, shape, surface texture, and hints of internal porosity. Several tests are available to fully specify the attributes of a powder, including measures of the particle size, surface area, packing density, interparticle friction, flow, internal structure, and impurities.

Iron and steel powder production deals with large quantities of materials; often production rates are measured in tons per minute. Testing the powder for quality and conformance to manufacturing specifications requires collecting a sample that might constitute only a handful of powder. Assurance this sample represents the production lot requires random selection of many small samples and then blending these to make up the test sample. Formal procedures are available, largely focused on ensuring the small sample represents the production lot.

3.2.1. Particle Size

Usually particle size is based on a characteristic feature. Just as humans can be sized by many measures—weight, height, waistline, or arm span—particles also can be sized by maximum dimension, mass, volume, surface area, minimum diameter, or a randomly selected dimension. This becomes important because some techniques rely on one measure while other techniques rely on a different parameter. A tall basketball player might have a small waistline, and a fibrous particle might have a small mass. Thus, a single size measure might be misleading. A characteristic size for each powder is usually linked to its mass or its cross-sectional size.

Most ferrous powders are measured for size by one of two sizing techniques. The classic approach, applicable to larger particles, is termed *screening* or *sieve analysis*. A square grid of evenly spaced wires creates a mesh, just like the screen on a window. The mesh size is determined by the number of wires per inch. A large mesh means a small opening size and vice versa. For example, ''200 mesh'' implies 200 wires per inch, which gives an opening size of 75 μm between the wires. Mesh sizes cannot go to very small opening sizes. However, sieve analysis is usually applied only to particles larger than about 325-mesh or 45 μm. Historically, smaller sizes are termed subsieve, although there are now screens with smaller openings. The opening sizes for standard screens are given in Table 3.2. Usually, the manufacturing tolerances allow for a ±4% variation on the opening sizes for the small mesh numbers (e.g., 35 mesh at 500 μm can range from 480 to 520 μm), but the allowed variation grows to nearly 7% at the high mesh numbers (at 325 mesh the opening can range from 42 to 48 μm). Consequently, screen or sieve analysis is not very accurate.

Screen analysis begins with a stack of screens with decreasing mesh openings as illustrated in Figure 3.2. The smallest opening size is at the bottom. The powder is loaded onto the top screen and the screen stack is agitated. For particle size analysis, a sample size of 100 g is usually sufficient when using 20-cm-diameter screens. After vibration for 30 min, the screens are unstacked and the powder in each size interval is weighed and the interval percent calculated for each size fraction. The powder passing through a mesh is designated by a minus (−) sign, and that retained on a mesh is designated by a plus (+) sign. For example, the designation

TABLE 3.2. Standard Sieve Sizes

Mesh Size	Opening (μm)
18	1000
20	850
25	710
30	600
35	500
40	425
45	355
50	300
60	250
70	212
80	180
100	150
120	125
140	106
170	90
200	75
230	63
270	53
325	45
400	38
450	32
500	25

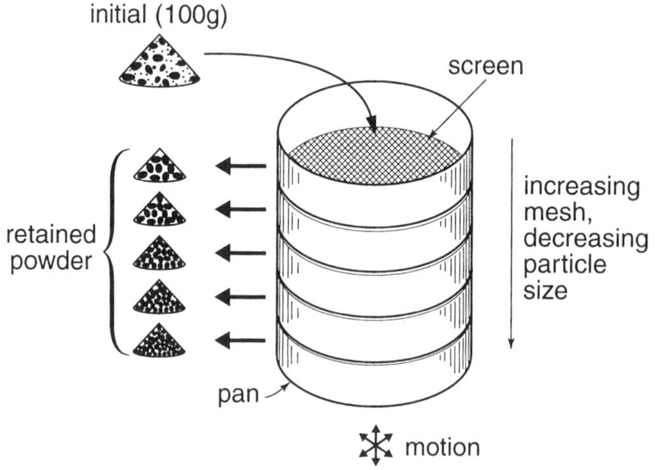

Figure 3.2. Sieve analysis for particle size relies on a stack of screens with decreasing openings that capture particles based on size.

EQUATION BOX 3.1

Screens are designed to have a geometric progression in sizes. The standard step is an opening that changes by almost a constant factor described as

$$2^{1/4} = 1.19 = \left(\frac{\text{mesh opening } i}{\text{mesh opening i + 1}}\right)$$

indicating a step size of $2^{1/4}$ between each mesh. This corresponds to an increase in the opening by a factor of 1.19. For example, the ratio of the 60-mesh opening size to the 70-mesh opening size (250–212 μm) is 1.18, the ratio of 170-mesh opening size to the 200-mesh opening size (90–75 μm) is 1.20, and the ratio of 325-mesh opening size to 400-mesh opening size (45–38 μm) is 1.18. Nominally, each screen opening in a sieve analysis should be 19% larger than the next screen. This amounts to a mass ratio between classes of nearly 1.7.

of −100/ + 200 mesh for a powder indicates that it passed through a 100-mesh opening but not a 200-mesh opening. This powder would be between 150 and 75 μm in size.

Screening is the most widely recognized particle size technique, but it is being displaced by automated size analyzers. One reason for new size analysis tools is the large degree of error associated with the technique. Industry study groups find up to 20% variation in reported particle size due to differences in sieves, techniques, screen damage, and even operators. A controlled test of common iron powders by screening using the same sample and screens found errors from 0.6 to 4.7% in the −325-mesh fraction at 12 different factories. The errors were generally smaller for larger mesh sizes, ranging from 0.6 to 2.3% differences for the −80/+100-mesh fractions. Under best conditions, the industry average is probably no better than 1% accuracy on the stated amount of powder at a given size. So, if the analysis says 45% of the powder is −200 mesh, then we anticipate the true value is between 44 and 46%.

Screening is also useful in obtaining sized powders. This procedure, termed *classification*, is used to remove certain particle sizes or to tailor particle size distributions. Figure 3.3 is an illustration of a −325-mesh stainless-steel powder used to fabricate controlled pore size filters. Note that some particles are much larger than the 45 μm associated with a 325-mesh powder. This is because they passed through the mesh along their length, so all we really know is that in one orientation the particles are below 45 μm. In several applications, ranging from flame spray to injection molding, the ability to process a powder is dependent on controlling the particle size. Consequently, screening and other classification techniques provide a means for removing specific size fractions.

Over the past several years there has been a progressive shift to laser light scatter-

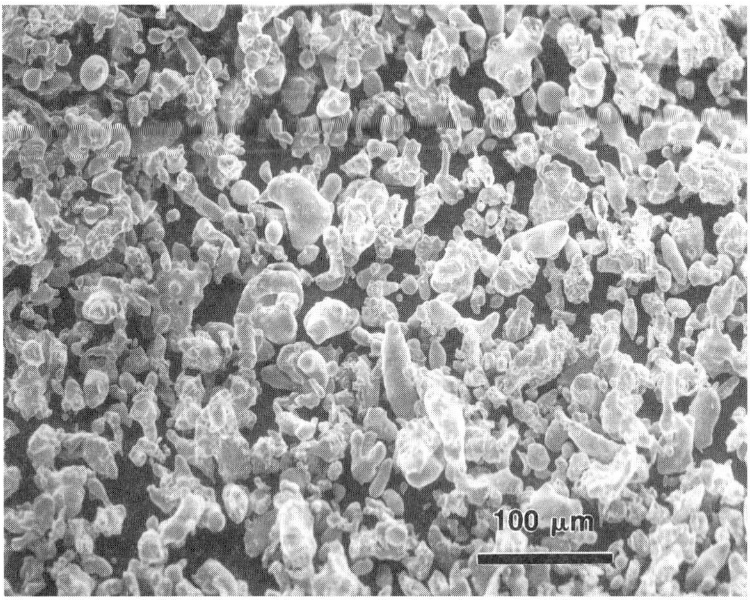

Figure 3.3. An example of a water-atomized stainless-steel powder screened to −325 mesh (below 45 μm). Several particles in this SEM picture have lengths exceeding 45 μm, showing the inaccuracy of screening.

ing techniques for automating the measure of particle size. Sometimes, this is done as part of the powder fabrication control scheme, using the instantly measured particle size as a basis for adjusting the operation. In laser light scattering analysis, the particles are dispersed in air or water. Size is determined by the light scattering from each particle as it passes through a detection zone such as sketched in Figure 3.4. Most devices are highly automated and can size particles smaller than possible with sieving, using less powder, providing a faster and more precise analysis. Because of the many advantages in speed, accuracy, size range, and automation, laser light scattering is essentially the benchmark for particle size analysis.

3.2.2. Particle Size Data

The particle size distribution shows what range of particles existed in the sample. It can be presented as a histogram, or frequency plot, showing the amount of powder in each size increment. An example distribution is plotted for a steel powder sieve analysis in Figure 3.5. The histogram is generated by plotting the incremental weight percentages of the total powder mass versus the screen opening size on which the powder was collected. On such plots, the mode is the peak or most frequent particle size; −140/+200 mesh (106 to 75 μm) in this case. Most of the time the particle size is on a logarithmic scale to give better visualization of the smaller powders. Standard screens have a nearly constant size change on a logarithmic scale.

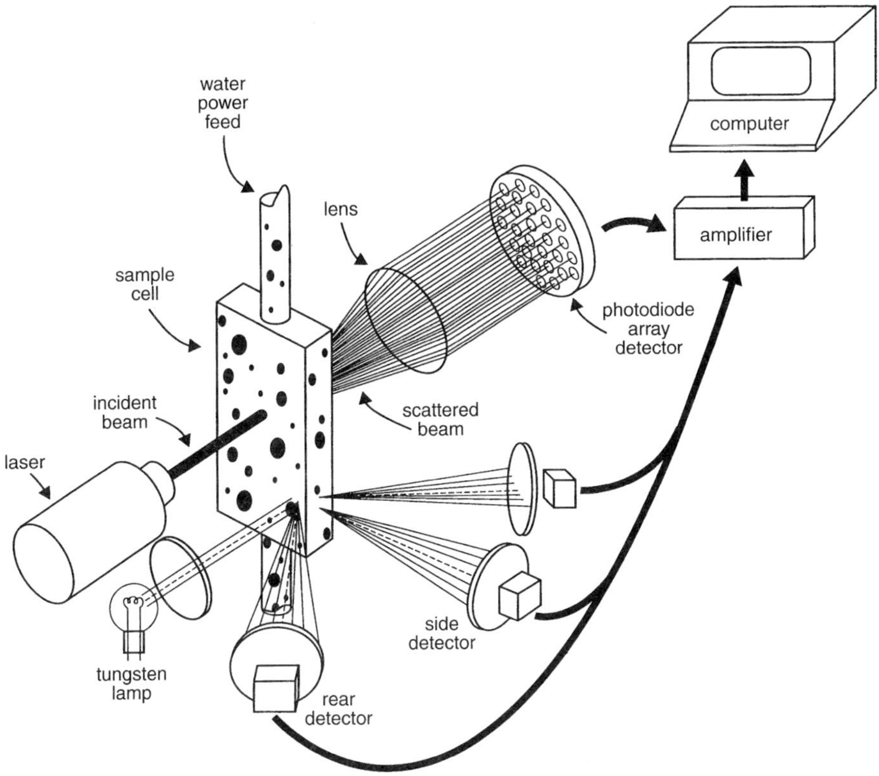

Figure 3.4. Particle size analysis based on laser and light scattering relies on dispersion of the powder in a feed fluid (air or water) and optical scattering from the particle. The detector array automatically determines the particle size from the angle and intensity of scattering.

Another form of presentation is based on the cumulative particle size. This is generated by adding the percentages accumulated in each interval. The cumulative sum (as a percentage of the total) is plotted against the upper particle size for the interval. It is commonly provided by the automated size analyzers. This results in a smooth cumulative particle size distribution, showing the amount of powder smaller than any given size. Most important is the median size that corresponds to the particle size at the 50% value. For the powder previously plotted in Figure 3.5, the cumulative particle size distribution is given in Figure 3.6. The median particle size is 77 μm. Note that this is the size where half the powder weight is smaller and half the powder weight is larger. It is different from the mean or average particle size, which proves difficult to calculate from sieve data, but is easily calculated by the automated size analyzers.

Figure 3.7 compares the particle size distributions for some available iron powders. Note that there are significant differences in particle size and the breadth of

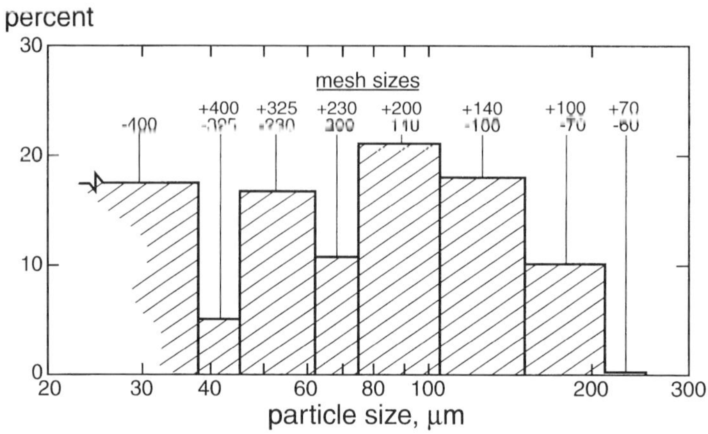

Figure 3.5. Histogram of the particle size distribution for an iron powder. The size scale is on a logarithmic basis, and the percent of total weight retained on each screen is plotted against the size range. Details about the lower size fraction and the powder below 400 mesh or 38 μm are sketchy, and these are usually termed the *pan fraction* to indicate the powder that passed through all screens.

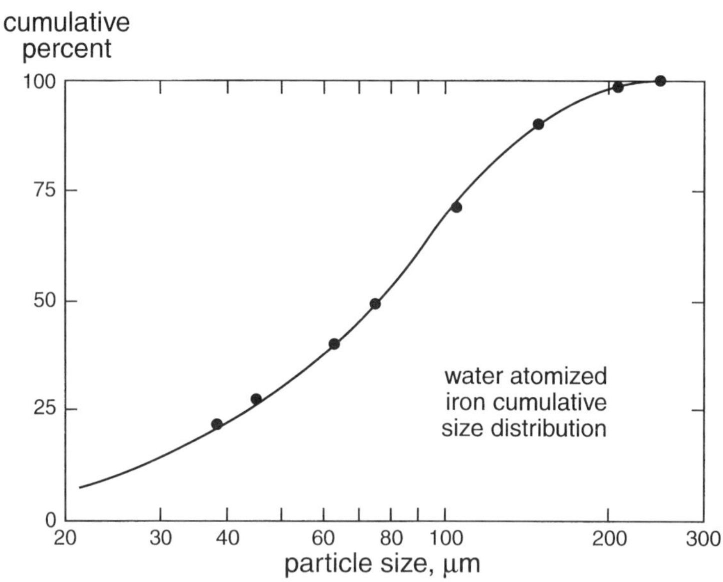

Figure 3.6. A smooth curve cumulative particle size distribution obtained from the histogram data in Figure 3.5. Each point is determined by adding the weight collected below the mesh of that size, and the points are connected by a smooth curve. Typical statistics would include the particle size at 10% cumulative mass, D_{10}, which is 24 μm, median particle size or D_{50} at 77 μm, and D_{90} corresponding to the size for 90% of the powder mass, which is 148 μm.

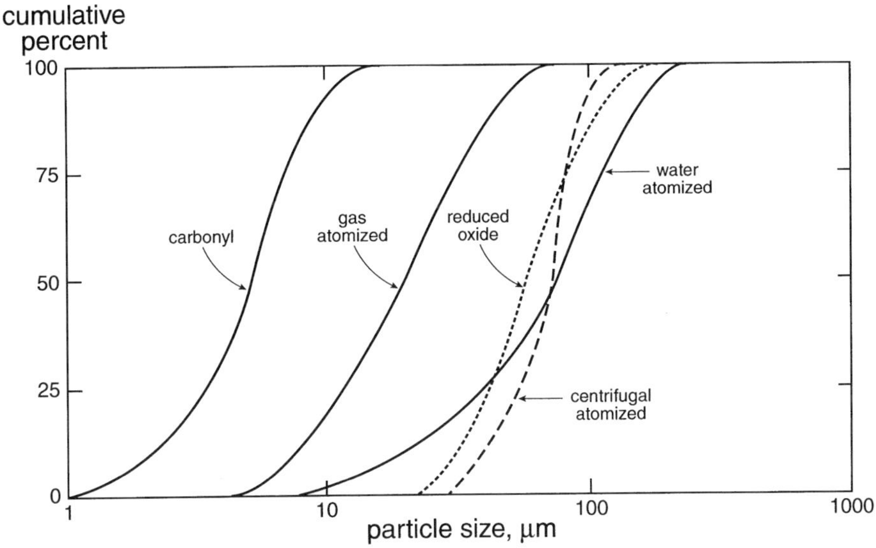

Figure 3.7. Cumulative particle size distributions of some common iron powders.

the size distribution. This plot shows the size differences between common powder production techniques. Depending on the application, one size will be preferred over another.

Today, there are also exploding wires, vapor condensation, or plasma microatomization techniques that produce iron and steel powders below 1 μm, but these are still in the development stage. These powders are off to the left of the size chart shown in Figure 3.7.

As noted earlier, most particle size analysis data are imprecise. There are problems with selecting an accurate sample, and most particles are not spheres with single dimensions. Problems increase with smaller powders, because these agglomerate and become difficult to analyze using standard techniques. It takes only one large particle to equal the mass of thousands of small particles. Thus, the use of size distributions based on mass are very sensitive to agglomeration. Under the best of conditions, particle size distributions can be reproduced with a 1–3% error. As an example, Figure 3.8 illustrates monthly tests and calibrations of a laser particle size analyzer using carbonyl iron and gas-atomized stainless-steel powders. The figure plots the measured particle sizes at the 10, 50, and 90% points on the cumulative particle size distributions (D_{10}, D_{50}, and D_{90}). These are for the same powders, same operator, and size analyzer, showing that typical errors are probably a few percent of the quoted values.

3.2.3. Particle Shape

Particle shape varies significantly between iron powders. Further, shape influences packing, flow, and powder compaction. Because measuring particle shape is difficult,

EQUATION BOX 3.2

Observations on the width of the particle size distribution have found instances where the powders seem better behaved. These "magic" or "special" distributions are usually quantified by some measure of the distribution width. One of the most famous comes from the cumulative particle size distribution expressed by Andreasen as follows:

$$W = C\, D_M^q$$

where W is the cumulative weight of powder below the size D_M (size based on a cumulative weight distribution) and C is an empirical constant. The exponent q provides a gauge to the packing and flow; values between 0.5 and 0.67 are special.

Another view is to use the spread of the weight distribution S_W as a measure of the powder behavior. It is calculated from the D_{90} and D_{10} particle sizes, corresponding to the particle sizes at 90 and 10% on the cumulative distribution, which are separated by 2.56 standard deviations:

$$S_W = \frac{2.56}{\log_{10}(D_{90}/D_{10})}$$

Normally, S_W values near 2 and 7 indicate powders with better handling characteristics.

qualitative descriptors are used to convey the particle shape. Figure 3.9 is a collection of electron micrographs that illustrate the variations possible in particle shape. The best particle shape descriptors are simple words, especially words that invoke widespread images. A full set of shape descriptors includes the following: spherical, angular, irregular, dendritic, sponge, cubic, polygonal, flake, ligamental, acicular, and rounded. Particularly useful are words that invoke images of food—carrots, oranges, potatoes, pears, grapes, and so on.

Particle shape varies with particle size and manufacturing technique. However, most particle size analyses assume a spherical shape. This is often invalid, as evident in Figure 3.1. Not only is the particle shape irregular, but evident in the cross-section can be large differences in pore structure. Cross-sectioned images are prepared by mixing the powders with fresh epoxy and allowing the mixture to harden. Subsequently the epoxy mount is polished and examined in a microscope to obtain images of the inside of the particles. Internal contaminants and pores in powders, as seen via cross-sectional analysis, change their response to compaction and sintering.

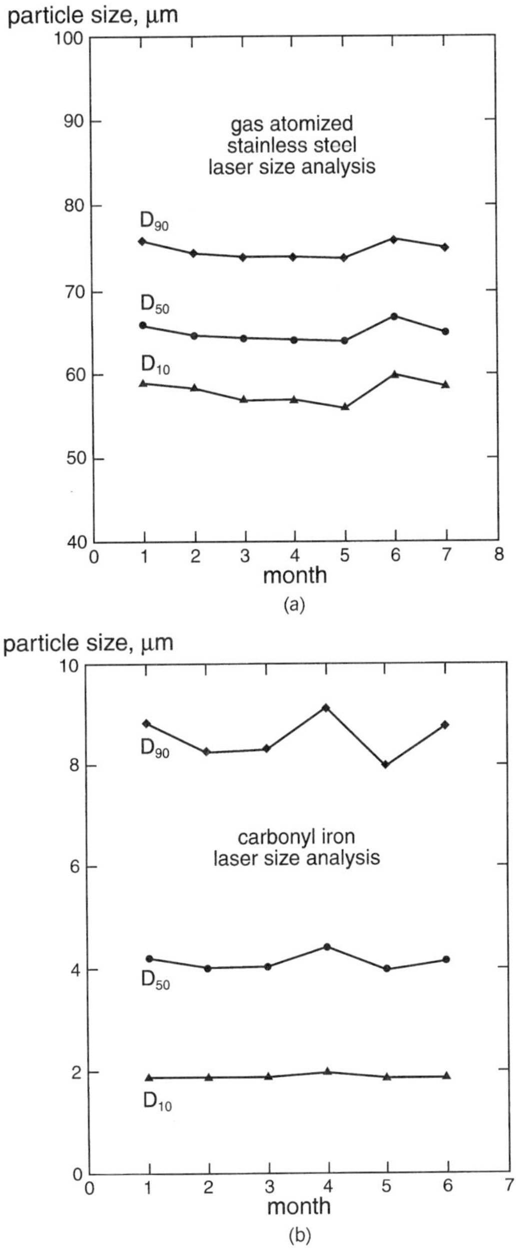

Figure 3.8. Illustrations of the repeatability of laser-based particle size analyzers, showing the particle sizes corresponding to 10, 50, and 90% on the cumulative particle size distributions for (a) gas-atomized stainless-steel and (b) carbonyl iron powders. (Data courtesy of Pamela Downs and Beth Carson.)

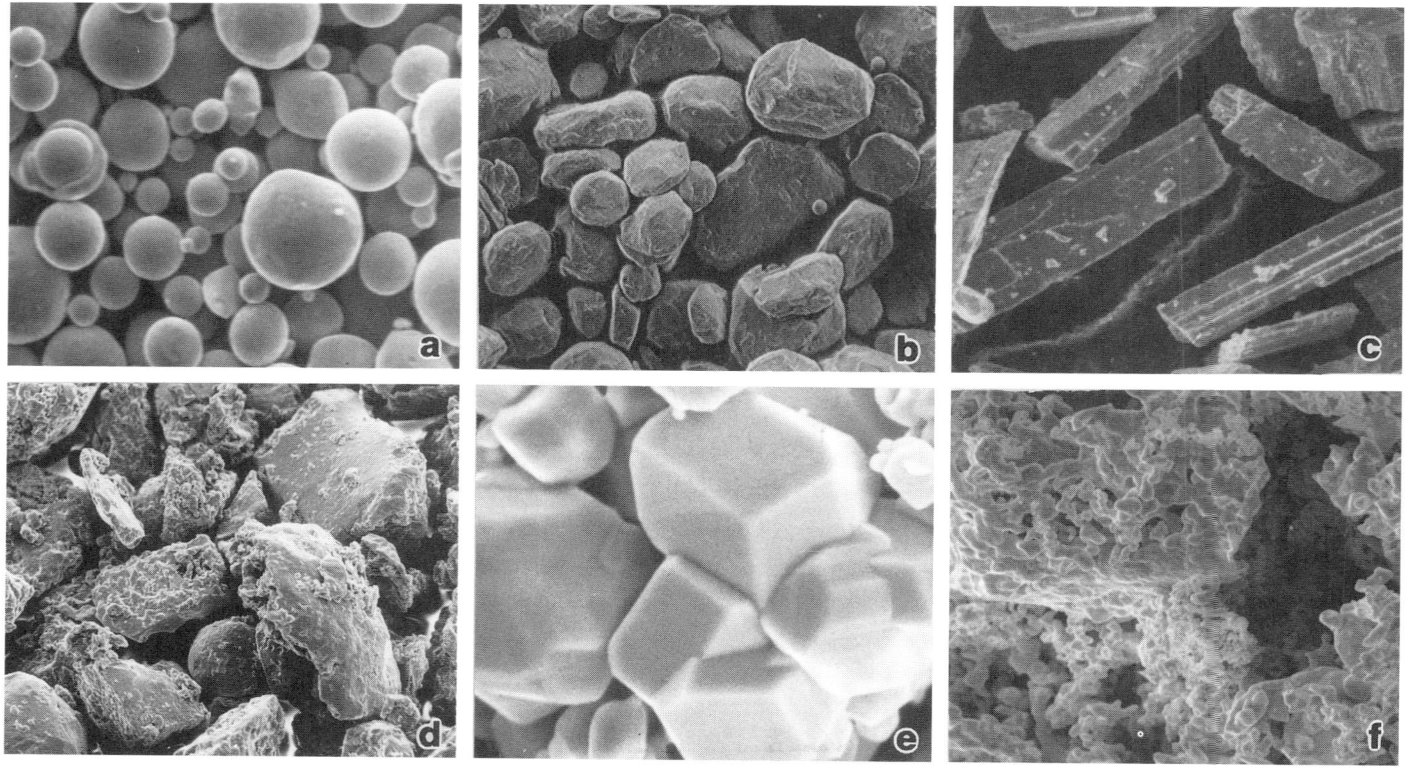

Figure 3.9. SEMs showing the shape diversity in powders. These six pictures only begin to illustrate the diversity, but show how simple terms can be used to describe particle shape: (a) spherical; (b) rounded; (c) angular; (d) irregular; (e) polygonal or cubic; (f) sponge.

EQUATION BOX 3.3

The concept of an equivalent spherical diameter is used frequently to convert particle size data into a diameter. The actual measurement might be the projected particle shadow or area A, mass or volume V, or surface area S. Equivalent spherical diameters are calculated from these measures, assuming that each parameter corresponds to a sphere, giving the following for the projected area, mass, and surface area measurements:

$$D_A = \left(\frac{4A}{\pi}\right)^{1/2}$$

$$D_V = \left(\frac{6V}{\pi}\right)^{1/3}$$

$$D_S = \left(\frac{S}{\pi}\right)^{1/2}$$

Another parameter is the *aspect ratio*, defined as the largest dimension of the particle divided by the smallest dimension. A sphere has an aspect ratio of unity, while a fiber might have an aspect ratio of 200.

Ratios of the particle sizes have been used to characterize shape. One such parameter is termed the *sphericity* ψ. It is defined as the ratio of surface area of a sphere having the same volume as the particle divided by the actual surface area:

$$\psi = \frac{\pi D_V^2}{S}$$

This requires independent measures of the equivalent diameter based on particle volume and particle surface area.

3.2.4. Surface Area

Surface area is sometimes used to indirectly measure the particle size. However, surface area tells nothing about the distribution in particle size or other differences between powders. The two main analysis techniques for measuring surface area are gas adsorption and gas permeability.

Gas adsorption starts with a clean powder surface achieved through a vacuum bakeout. This clean surface is exposed to an adsorbing gas, most typically by chilling the powder to liquid nitrogen temperatures (77 K, or -196°C, or -384°F) where nitrogen willingly condenses on the powder surface. The larger the surface area, the

EQUATION BOX 3.4

The specific surface area S and the average particle size on an equivalent sphere basis D_S are related to the density as follows:

$$D_S = \frac{6}{\rho_T S}$$

where ρ_T is the theoretical density. Convenient units for this expression are density in grams per cubic meter (g/cm^3), particle size in micrometers (μm), and surface area in square meters per gram (m^2/g).

greater the amount of absorbed nitrogen. The measurement gives the BET specific surface area, named after Brunauer, Emmett, and Teller, who developed the concept. These instruments are highly automated, but the measurement is usually applied only to the very small, subsieve powders.

An alternative is to flow a gas through a tube filled with powder. The more pressure needed to force gas through the tube, the higher the surface area. The technique is now largely out of use in favor of gas adsorption, but occasionally we still encounter the Fisher subsieve sizer designation of the powder size. It is calculated from the surface area measured via the gas flow rate assuming the particles are all spheres.

Surface areas for iron powders are expressed in square meters per gram and range from 0.01 m^2/g for the large powders used in compaction, to 0.15 m^2/g for carbonyl iron used in injection molding. New iron powders (termed *nanoscale*) are available with surface areas of 7–10 m^2/g, meaning that one gram of powder has more surface area than most kitchen floors.

3.2.5. Packing and Flow

Small powders or irregular-shaped particles do not pack or flow very easily. The small particles have a high surface area, and the irregular particles have many surface asperities. Sharp corners are a source of friction between particles that inhibit flow and packing. Both attributes are important to powder shaping, since the powder needs to fill out a die cavity rapidly and uniformly. The packing density decreases because of poor flow past neighboring particles. *Apparent density* of a powder is the density (mass divided by volume) when the powder is in the loose state without agitation. It is usually measured using the Hall or Arnold meter. These are pictured in Figure 3.10; the Hall flowmeter is used most often, since it measures both the flow time and apparent density. On the other hand, the Arnold meter does not require powder flow since it uses a ring containing powder to directly fill a 20-cm^3 cylindrical

Figure 3.10. Two apparent density meters. The Hall flowmeter consists of a funnel and measuring cup, where the time for 50 g to pass through the funnel is used for the flow time and the density of the powder in the cup is used to measure the apparent density. The Arnold meter simulates powder loading into a die (20-cm^3 hole that is shown full here) and provides an apparent density relevant to die compaction. (Photograph courtesy of Julian Thomas.)

volume. That technique best correlates with the filling of die cavities in powder compaction operations.

Tap density is the highest density that can be achieved by vibration of a powder without the application of external pressure. For the tap density, a graduated cylinder of powder is vibrated for 3000 taps. Afterward the volume is measured, giving the tap density as the weight divided by that final volume. It is higher than the apparent density.

Theoretical density corresponds to the handbook density for a powder, the density when there is no porosity present. Because of contamination, most powders have a measured true density below the handbook value. The true density of a powder is measured by a device called a *pycnometer*, which pushes gas into the pores between the particles. For example, the theoretical density of iron is 7.86 g/cm^3, yet many iron powders exhibit pycnometer densities of 7.70–7.83 g/cm^3.

The *angle of repose* is another friction index. It is the angle from the horizontal observed by pouring a powder into a pile as shown in Figure 3.11. A high angle of repose indicates considerable difficulty for particles to flow past one another. A related test is the *flow time*, measured by the time for 50 g of powder to fall through the Hall flowmeter funnel. Small, subsieve powders do not flow and are designated as non-free-flowing. Short flow times indicate free-flowing powders while long times suggest high interparticle friction. The apparent density and flow times are easily obtained with the Hall flowmeter by combining a precision volume cup with

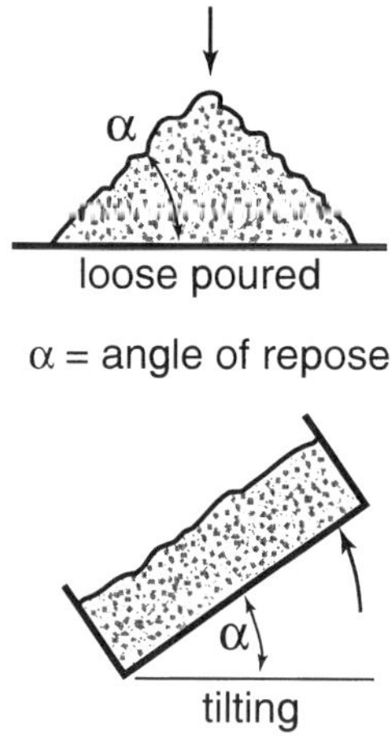

Figure 3.11. The angle of repose is determined by the naturally formed angle from the horizontal axis for a powder and relates to the friction between particles; easy flowing powder has a low angle of repose.

the funnel. The apparent density is the weight of powder divided by the cup volume. For most ferrous powders, the typical apparent density is 30–60% of the theoretical density, or usually 2.3–4.7 g/cm^3. A high packing density is generally desirable.

Finally, there is a concern with small powder separation during routine handling. The dusting index is a measure of the problem. Its determination involves letting a powder pass through a funnel and fall into an enclosure about 50 cm (about 19 in) below the funnel. Dust is collected from air samples continuously taken from the container at the bottom of the free-fall. A high dust content can be a precursor to many problems, ranging from health hazards, explosive hazards, poor dimensional control, and even variable mechanical properties. Polymers are added to many iron powders to better agglomerate alloying additions (such as graphite) to the particles and to reduce dusting. Very low concentrations prove useful, as demonstrated in Figure 3.12 using a polyvinyl glycol coating. This polymer is favored because it is safe and water-soluble.

3.2.6. Compressibility

A main activity in P/M is the fabrication of complex shapes by compaction. Accordingly, the compaction behavior is an important characteristic. Compressibility measures the density of a powder under an applied load. As shown in Figure 3.13, the density increases continuously with increasing pressure, but each iron powder has a different response to pressure. For compressibility, the typical test geometry is a simple cylinder or rectangle. The die is loaded with powder and the density is measured after compaction at a predetermined pressure of 414 MPa [corresponding to 30 tons/in^2 (tsi) or 60,000 psi]. That is the green density and is the basis for expressing the compressibility. Iron powders designed for good compaction give green densities in the 6.6–7.1-g/cm^3 range. A related parameter used in tool design is the *compression ratio*, defined as the compaction density divided by the apparent

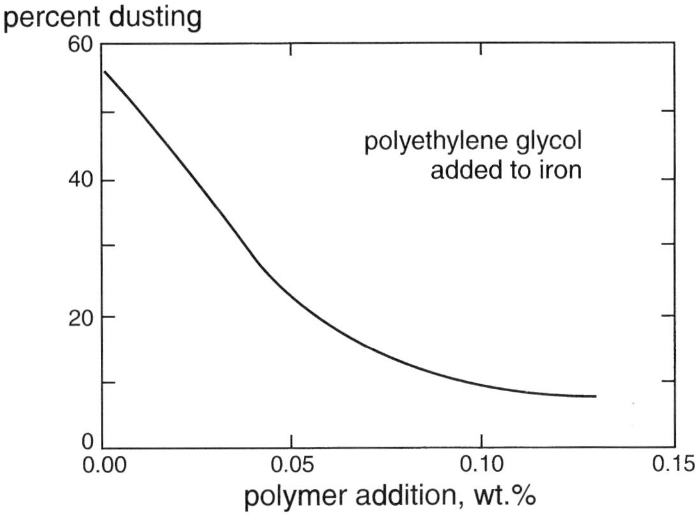

Figure 3.12. Reduction of powder dusting is possible by coating a small quantity of a polymer on the surface, as evident here for an iron powder treated with polyethylene glycol. (Data courtesy of Yves Trudel.)

density. This tells the punch travel needed to attain a desired density from a powder once a die cavity is filled.

Related to compressibility, there might be a need for information on green strength. In P/M the term *green* refers to the compressed or shaped condition prior to sintering. The green strength, meaning the strength after forming, but before sintering, is calculated from the stress needed to break a compacted test bar in bending. The transverse rupture test is most commonly applied to measure green strength. This uses a rectangular sample. Figure 3.14 plots the green strength versus compaction pressure for water-atomized and reduced-oxide iron powders. Note the larger difference in green strength as compared with green density. Generally, a green strength of 10 MPa (1.4 ksi) is acceptable for most forming processes.

Additionally, a springback measure might be recorded. The sample width is compared to the die width, and the excess sample width divided by the die width is termed the *springback*. This is the sample expansion on removal from the die and gives a guide to pressing difficulties and possible cracking of green parts on ejection. Values for iron powders range up to 0.5%, but more typical is a springback of 0.2% on the width. A few companies even go as far as to sinter the test bar and report a final dimensional change as referenced to the die size. Sometimes this might be growth or swelling, but in others it will be shrinkage; either will be expressed as a percentage dimensional change.

3.2.7. Composition

Three types of powders are in use: elemental, premixed, and prealloyed. *Elemental* iron powders are mostly iron and are characterized by some impurities. The common

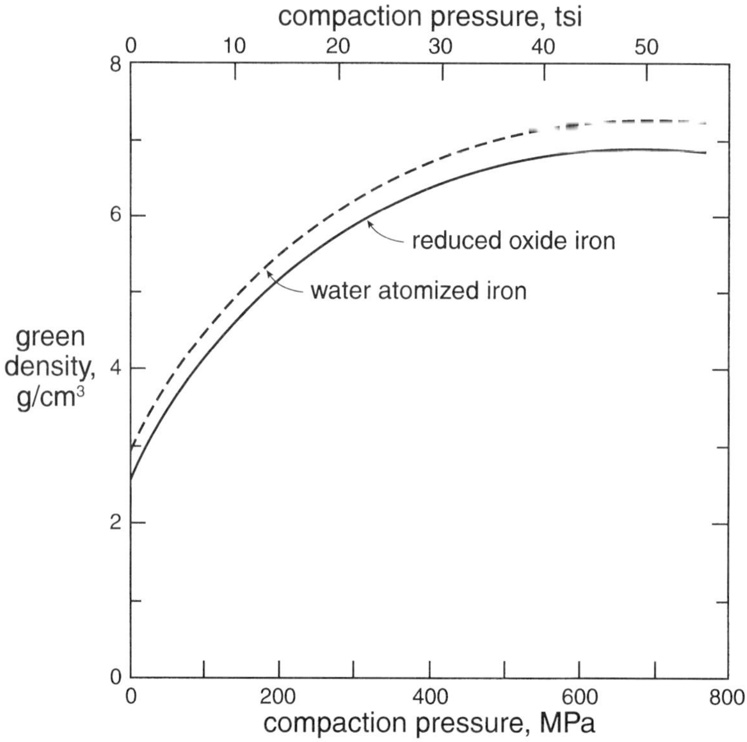

Figure 3.13. Green density versus compaction pressure for two common iron powders.

EQUATION BOX 3.5

The compression ratio (final density divided by apparent density) is used to calculate the fill setting in uniaxial die compaction. Let C_R be the compression ratio, H_F be the fill height, and H be the specified component height; then

$$H_F = HC_R$$

provides the calculation of the fill height if the final density and final dimensions are known. For example, if the final height is to be 12 mm at a density of 6.9 g/cm^3 using a powder with an apparent density of 2.9 g/cm^3, then the fill height will be $12 \cdot 6.9/2.9$ or 28.6 mm.

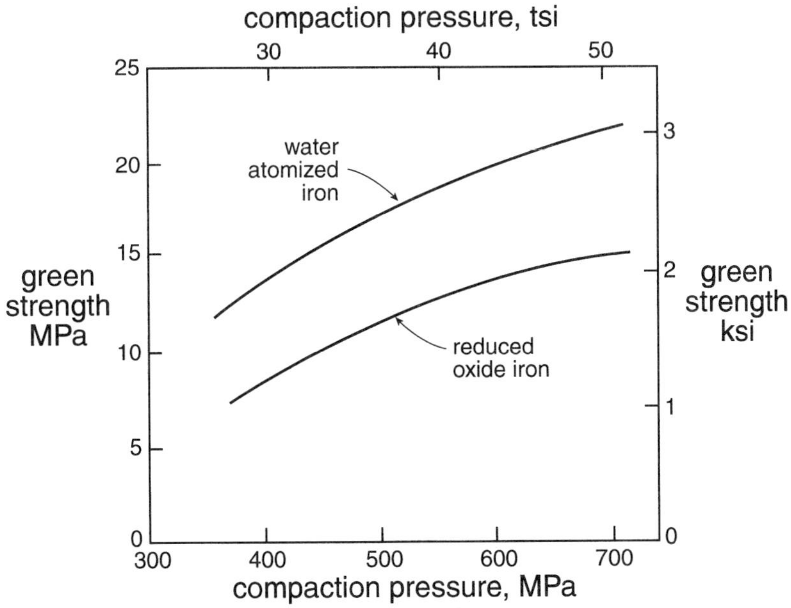

Figure 3.14. Green strength, the strength after compaction, for the same two iron powders used for Figure 3.13. The large difference in green strength relates to the better mechanical interlocking of the water-atomized particles.

impurities are carbon, oxygen, nitrogen, manganese, silicon, and sulfur. *Premixed* powders are combinations of two or more distinct powders (e.g., iron, copper, graphite) that form an alloy during sintering. Impurity levels and proper compounding of the mixture are primary concerns with the premixed powders. *Prealloyed* powders are microcastings with multiple elements in a predetermined ratio. For the prealloyed powders, attention is given to measuring the alloy composition and impurities.

Beyond composition, surface impurities are measured by heating the powder in hydrogen to 1050°C (1920°F) for 60 min, and determining the weight loss from evaporation of contaminants. This is termed a *loss on reduction*. Inert contaminants are a concern in some powders. These are usually silica, alumina, or titania carried over from the original steelmaking process. Sometimes they are assessed by metallographic crosssections, where an inclusion size count is done for many particles. An alternative is to assess the weight of inclusions by measuring the acid dissolution behavior, meaning that the particles are dissolved in acid and the weight of undissolved ceramic contaminants is reported as the weight percent of acid insolubles. This gives no information on the inclusion size and shape, which is an advantage of the metallographic analysis.

Depending on the concern over bulk chemistry, composition can be measured using any of several standard chemical techniques; combustion, wet analysis, emission or flame spectroscopy, atomic absorption, X-ray diffraction, X-ray fluorescence, or neutron activation analysis. Emission spectroscopy and X-ray fluorescence are used for routine analysis, with lower sensitivities of approximately 10 ppm (*ppm* is the standard abbreviation for parts per million, so 10 ppm means ten out of a million). The decision on appropriate test and tolerance ranges depend on the application; usually composition is measured by the powder fabricator.

3.2.8. Data Sheets

Ferrous powders are shipped with data sheets from the manufacturer that disclose the composition, particle size distribution, and other characteristics of the powder. Figure 3.15 shows an example of a water-atomized iron powder. Often values on the data sheet are not verified at the parts production facility. However, knowledge is required to decide whether the powder meets the desired attributes. Typically, the goal is to control the process to provide a standardized product. Characterization of the starting powder ensures consistent feed into a process.

All powder should be accompanied by a Material Safety Data Sheet (MSDS). These provide information on safety, health, and hazards. Some very small particles can ignite and burn in air. The MSDS reports any possible problems associated with storage and handling, including cautions on breathing, explosion, fire, toxicity, and disposal. Usually these are ignored until an urgent situation arises. Thus, good manufacturing practices dictate the MSDS sheets and vendor certifications be filed in locations easy to reach, just in case a problem arises.

In the characterization of a powder, the test should be meaningful, accurate, and easy to perform and replicate. Priority is given to the powder chemistry. Beyond bulk chemistry, contamination is determined using the hydrogen weight loss test. Particle size is the next most important parameter, especially for precise component fabrication. Additionally, packing and flow tests are usually reported on the data sheet.

3.3. PRODUCTION OF FERROUS POWDERS

Iron powder production is a worldwide activity. Major powder production operations exist in the United States, Canada, Japan, China, Germany, Italy, India, Sweden, Belgium, and Russia. The capacity to make iron powder generally exceeds the worldwide demand, so powder prices have remained stable. The highest-volume production techniques—water atomization and oxide reduction—also produce the lowest-cost powders.

3.3.1. Mechanical Techniques

A conceptually simple means for making a powder is to grind, file, or hammer a piece of steel. Eventually the steel will disintegrate into powder. Such a production route was employed to make the first powders. Today, many processes rely on mechanical comminution as part of the production sequence. However, it is a poor direct route to the production of iron or steel powders. Steel is strong and tough, so it naturally resists grinding and milling. Thus, to start making powder from raw

QMP

QUEBEC METAL POWDERS LIMITED
1655 MARIE-VICTORIN
TRACY, QUEBEC, CANADA
J3R 4R4
TEL.: (514) 746-5050
FAX: (514) 743-0223

ISO 9001 Since Depuis 1993

CERTIFICATE OF ANALYSIS

CUSTOMER
G-GXD
PENN STATE UNIVERSITY
ATT: DR. A. GRIFFO 814-865-1308
118 RESEARCH BUILDING WEST
UNIVERSITY PARK
PA 16802-6809

YOUR ORDER NUMBER ▶ FCH FCH

OUR ORDER NUMBER ▶ S-33879

GRADE	LOT NO.	TOTAL WEIGHT SHIPPED
ATOMET 1001	63174	50 LB

NOMINAL COMPOSITION

	PARTS BY WEIGHT	PARTS BY WEIGHT
IRON	100.00	
COPPER		
GRAPHITE		
LUBRICANT		

CUSTOMER SPECIFICATION

FOR TECHNICAL INFORMATION CONTACT
FRANCOIS CHAGNON

PRODUCT ANALYSIS

SCREEN ANALYSIS OF STEEL POWDER

SIEVE SIZES U.S.		OPENING MICRONS	% BY WEIGHT
ON	+60	+250	TRACE
-60	+70	212	0.4
-70	+100	150	10.3
-100	+140	106	18.1
-140	+200	75	21.1
-200	+230	63	10.9
-230	+325	45	16.6
-325		-45	22.6

PHYSICAL PROPERTIES

APPARENT DENSITY		2.92	g/cm³
FLOW	HALL	28	s/50g

METALLOGRAPHY

INCLUSION LENGTH	NUMBER (/100mm²)

CHEMICAL ANALYSIS OF STEEL POWDER

	% BY WEIGHT
CARBON	0.003
OXYGEN	0.09
SULPHUR	0.007
MANGANESE	0.19

SINTERED PROPERTIES
ATMOSPHERE Nitrogen/Hydrogen :90/10
TIME 25 MINS AT 1121 °C

TEST MIX COMPOSITION (PARTS BY WEIGHT)	COMPACTING PRESSURE t.s.i.	GREEN DENSITY g/cm³	GREEN STRENGTH p.s.i.	TRANSVERSE RUPTURE STRENGTH p.s.i.	DIMENSIONAL CHANGE %	HARDNESS RB
0.9C-2.0Cu-0.5	29.5	6.8		153,500	+0.27	83
STANDARD LOT 1461/7				149,000	+0.27	83

Typical combined carbon: 0.75%

DATE

Product Development Supervisor

Figure 3.15. An example powder data sheet, providing key information on composition, particle size, packing, and flow, as well as final performance.

steel would be energy-intensive and costly. However, when scrap is available, such as oxide millscale or machining chips, then mechanical grinding is useful. These materials are ground to form powders that are subsequently treated to remove contaminants. The chips from machining provide an abundant source of cheap powder, giving the powder photographed in Figure 3.16. More typically, milling or grinding are part of another powder production process. For example, after annealing, the cake of lightly agglomerated powder is broken into particles by milling.

The mills for grinding powders consist of balls, hammers, impellers, and meshing jaws to mechanically impact on the material, causing it to fracture into powder. Milling via mechanical impaction of falling hard balls in a rotating chamber is a classic approach to grinding brittle materials. These are known as *ball mills* or *jar mills*. A most common technique uses a cylindrical jar filled with balls and the material to be milled. The jar is rotated at a speed where the balls are lifted to fall on the material. As the particle size decreases, further grinding becomes less productive. Generally, the louder the mill, the more efficient the grinding operation. For optimal milling, the diameter of the grinding balls should be approximately 30 times the diameter of the charge material. The volume of balls should be about half the jar volume, and the fill of the input material should be about 25% of the jar volume. Fluids or protective atmospheres are used to reduce contamination.

Variants on ball milling include jet milling, attritor milling, vibrator milling, and hammer milling. Each of these techniques uses a mechanical means to reduce the particle size. For example, specialized powders are created with the coldstream technique where coarse particles captured in high-velocity gas jets are allowed to

Figure 3.16. This angular powder was fabricated by milling machining chips to form small particles that were subsequently hydrogen-reduced to soften and purify the powder. (Photograph courtesy of Kevin Christian.)

Figure 3.17. SEM of a small iron powder produced by jet milling, a self-impact attritioning process. (Photograph courtesy of Mu-Jen Yang.)

collide with a cold target. The coarse input powder is chilled and accelerated in the feed gas using pressures of approximately 7 MPa (about 1000 psi). The product is generally above 10 μm in size with a rounded but irregular shape. A low temperature (liquid nitrogen temperature) reduces the metal's ductility and thereby assists attritioning. The technique is used in fabricating a few speciality steel powders.

One variant relies on two streams of the same powder, which are directed at one another. Momentum carries the particles together into a high-velocity impact. The intent is to avoid contamination from a target material. A micrograph of an iron powder produced this way is shown in Figure 3.17. The particles are rounded from the impact attritioning.

3.3.2. Electrolytic Techniques

Another early technique for fabricating iron powder was in an electrolytic cell. These have essentially disappeared from P/M, because of cost and environmental contamination. However, 50 years ago electrolytic production was dominant. It is an electroplating process (similar to chromium plating) where the deposit is a spongy mass that can easily be dried and ground into powder.

Electrolytic powder production begins with dissolution of iron at the anode. Under the applied voltage, iron goes into solution as an ion and passes to the cathode. An iron sulfate solution is circulated between the anode and cathode that carries the electrically charged ions. Under an applied voltage, powder is precipitated at the cathode. To optimize powder production, the cell is operated at conditions that form a porous deposit on the cathode. Periodically, the deposit is removed, washed, dried, ground into fine powder, sieved, and annealed to remove any hardening from the milling step. These powders are often dendritic or spongelike in shape and have sulfur as a main contaminant. An early attraction to electrolytic powder was its purity. Although some specialty iron powders are still fabricated this way, the process has largely disappeared and is no longer significant for iron, although it remains a main process for producing other metal powders.

3.3.3. Oxide Reduction and Reaction

A classic form of iron powder is formed by oxide reduction. The conceptual flowchart is given in Figure 3.18. There are several variants, ranging from continuous to batch operations, with the reducing agent being hydrogen or coke. Powder fabrication can even involve multiple reduction passes, with milling between each pass. A new direct reduction technique relies on fluidized beds to convert iron ore into iron powder.

One of the highest productivity operations relies on inexpensive millscale, a waste product from steel production. It is reduced using heat and hydrogen to produce iron powder. A sketch of the process is given in Figure 3.19. Reduction starts with

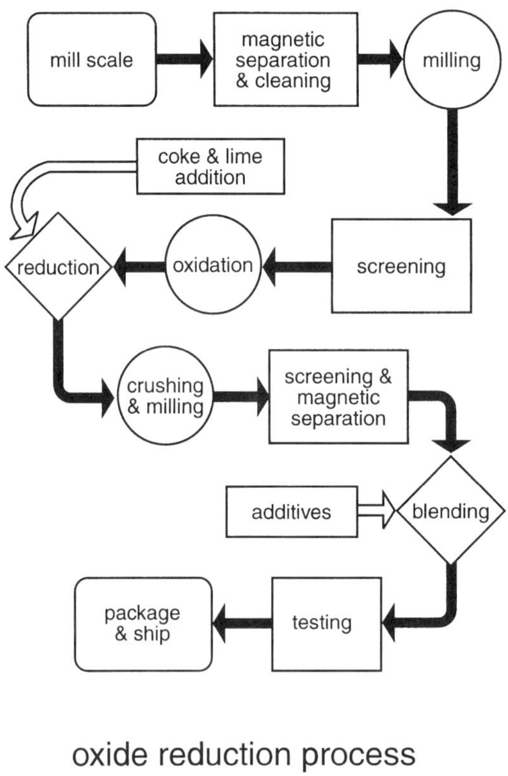

Figure 3.18. A generic flowchart for the oxide reduction process for forming iron powder. Here the starting material is scale from a steel mill, which is converted into powder by a sequence of milling and reduction steps.

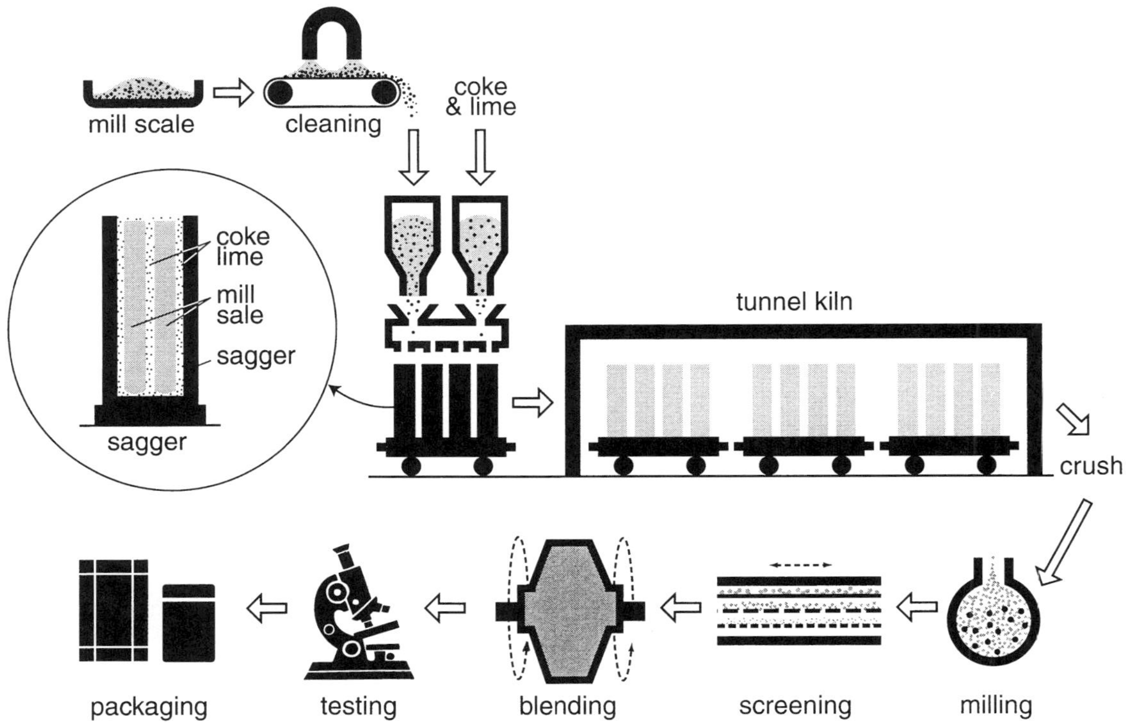

Figure 3.19. A visual illustration of the classic oxide reduction process for fabrication of sponge iron powder. The time in the tunnel kiln might be days to fully convert the oxide into iron. This product is milled, screened, and possibly reduced again before shipment.

a purified iron oxide that is sorted by magnetic means. These oxides are milled into small powders and reduced (the oxygen is removed by reaction with hydrogen) to form pure iron. Oxide reduction is achieved using hydrogen, but carbon monoxide also works. In one variant, coke and lime are mixed with iron ore to give the desired reduction. In another process, high-carbon steel scrap is melted and formed into shot, which is mixed and milled with millscale (iron oxide). Carbon is removed via heating in air, and the powders are milled, reduced, and annealed.

The reduction of an oxide, such as FeO, in hydrogen occurs by the transfer of oxygen from the solid oxide to form steam,

$$\mathrm{FeO\ (s)} + \mathrm{H_2\ (g)} \rightarrow \mathrm{Fe\ (s)} + \mathrm{H_2O\ (g)}$$

For the above reaction, solids are designated by (s) and gases are designated by (g). To keep the reaction going, the steam [H_2O (g)] is continually removed from the reaction zone. Time and temperature are the primary controls of oxide reduction. The time to complete the reaction is dramatically shorter at high temperatures, but the resulting particles sinter to form a cake. Consequently, reduction times of days at low temperatures are employed to minimize sintering. The reduction reaction can use pulverized slag from steel mills or other scrap iron products. Typically, reduction is done in the 700–1000°C (1290–1830°F) temperature range, with subsequent mechanical attritioning to control particle size. A low reduction temperature produces a small particle size. A large volume change occurs, so the final product is a sponge such as shown in Figure 3.20. In many powders the oxide is not fully reduced,

Figure 3.20. A scanning electron micrograph of sponge iron powder fabricated from millscale via hydrogen reduction. (Photograph courtesy of Anthony Griffo.)

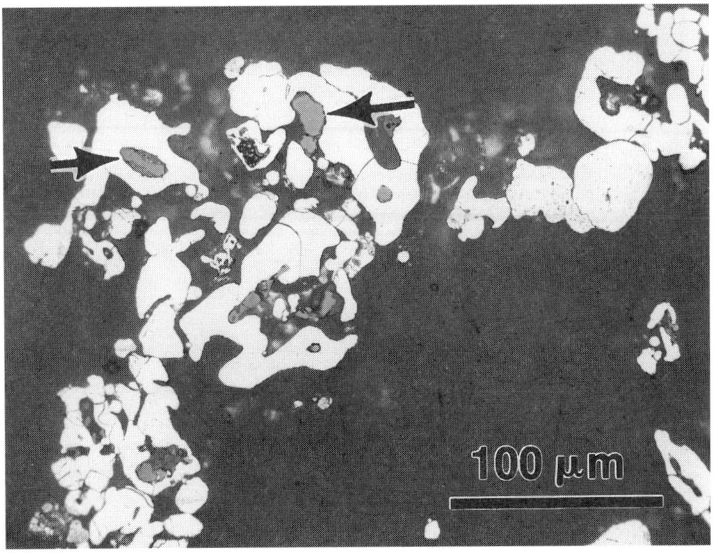

Figure 3.21. Cross-sectional optical micrograph through a sponge powder. Pores are apparent and the residual inclusions are evident as highlighted by the arrows. (Photograph courtesy of James Marsden.)

leaving internal inclusions as evident in Figure 3.21. Note the two arrows pointing to the inclusions. The reduced powder sponge is milled to form particles of the desired size.

3.3.4. Carbonyl Decomposition

Small iron particles are fabricated by the combination of vapor decomposition and condensation using a carbonyl intermediate. Large quantities of these powders are used for electronics, radar, and injection molding. They are formed by reacting porous scrap iron with carbon monoxide at 200°C (390°F) and 200 atmospheres (atm) of pressure to form iron pentacarbonyl $Fe(CO)_5$. The simultaneous pressurization and heating add to the process expense. The carbonyl molecule is cooled to a liquid at temperatures below room temperature (−20°C or −4°F) and fractional distillation is applied to purify it.

Reheating the liquid in the presence of an ammonia catalyst leads to vapor decomposition, giving a spherical iron powder. The particles grow by a sequence of surface reactions, producing growth rings like those seen in an onion or tree. Figure 3.22 is a high-magnification cross section through carbonyl iron particles where these growth rings are evident. The particles are small, with a purity near 98% and a nearly spherical shape as shown by the scanning electron micrograph in Figure 3.23. It is possible to control the reaction conditions to vary the size from about 0.2 to 20 μm, with 4–8-μm spheres being most common. Higher purity carbonyl iron powders are reduced in hydrogen after synthesis.

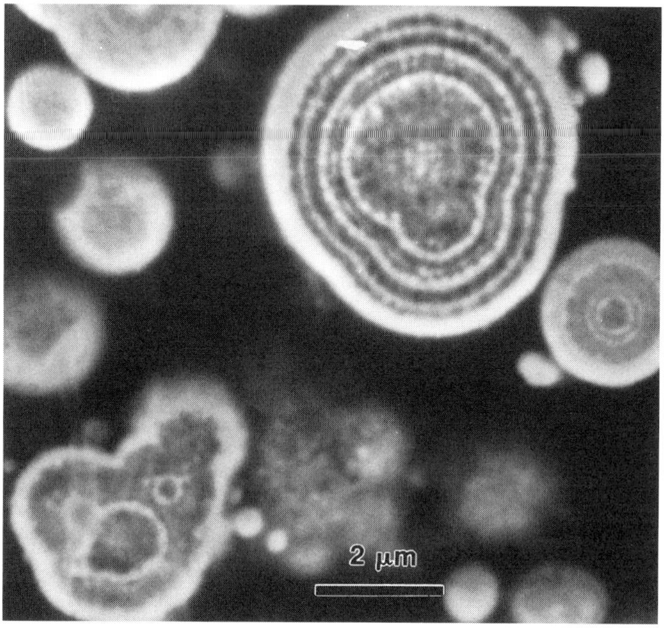

Figure 3.22. Growth rings in a cross-sectioned iron particle formed by the carbonyl decomposition process. This picture shows the layers of growth that take place during synthesis of the powder. (Photograph courtesy of Robert Gong.)

Figure 3.23. SEM of carbonyl iron powder, showing a mild level of agglomeration and a nearly spherical shape. The particle size tends to average near 5 μm.

3.3.5. Atomization

Atomization techniques are very productive and continue to displace other iron powder production routes. There are several variants to the basic concept, all sharing the idea of delivering energy to a stream of molten iron or steel. The impacting fluid causes the melt stream to disintegrate into droplets that solidify to become particles. The average production rate for commercial atomization units is 400 kg/min (880 lb/min) of iron powder. Both elemental and prealloyed powders are formed by such processes. It is attractive because of the applicability to several alloys and easy process control.

3.3.5.1. Gas Atomization. For some compositions, especially stainless steels, tool steels, electronic alloys, or aerospace alloys, the desired particle shape is spherical. Pressurized gas flows out of a nozzle to impact a stream of molten metal to form a droplet spray that solidifies into powder. The greater the energy in the gas, the smaller the powder. Most typically the gas is nitrogen or argon.

The gas atomization concept is sketched in Figure 3.24. In this instance, melting is performed under a vacuum in an induction melter. This is not necessary for many steels that can be melted directly in air, but vacuum melting is required to meet the purity demands of many biomedical, aerospace, and electronic alloys.

The expanding gas around the molten metal stream causes a dramatic disruption of the melt stream. The resulting droplets are accelerated to high velocities and solidify before reaching the walls of the chamber—requiring a large chamber. To ensure that the melt is not frozen in the nozzle, it is heated much over its melting temperature. This also allows the particles to spheroidize before solidification. Final particle size is determined by several factors, with higher gas pressures and gas flow rates being necessary for producing small particles. Most powders are in the 15–300-μm range, but smaller powders are possible. Figure 3.25 is a scanning electron

EQUATION BOX 3.6

In gas atomization the key determinant of the average particle size D is the ratio of the metal flow rate to gas flow rate [these are normalized by recording them as kilograms per minute (kg/min)]. The resulting relation is simple for most atomizers

$$D = \alpha \frac{dM_M/dt}{dM_G/dt}$$

where M_M is the metal mass and dM_M/dt is the metal mass flow rate, and M_G is the gas mass and dM_G/dt is the gas mass flow rate. Secondary factors captured in the α factor include the nozzle design, gas type, metal temperature, and pressures, but the mass flow rate ratio is usually the most important factor. Note smaller particles require more gas consumption.

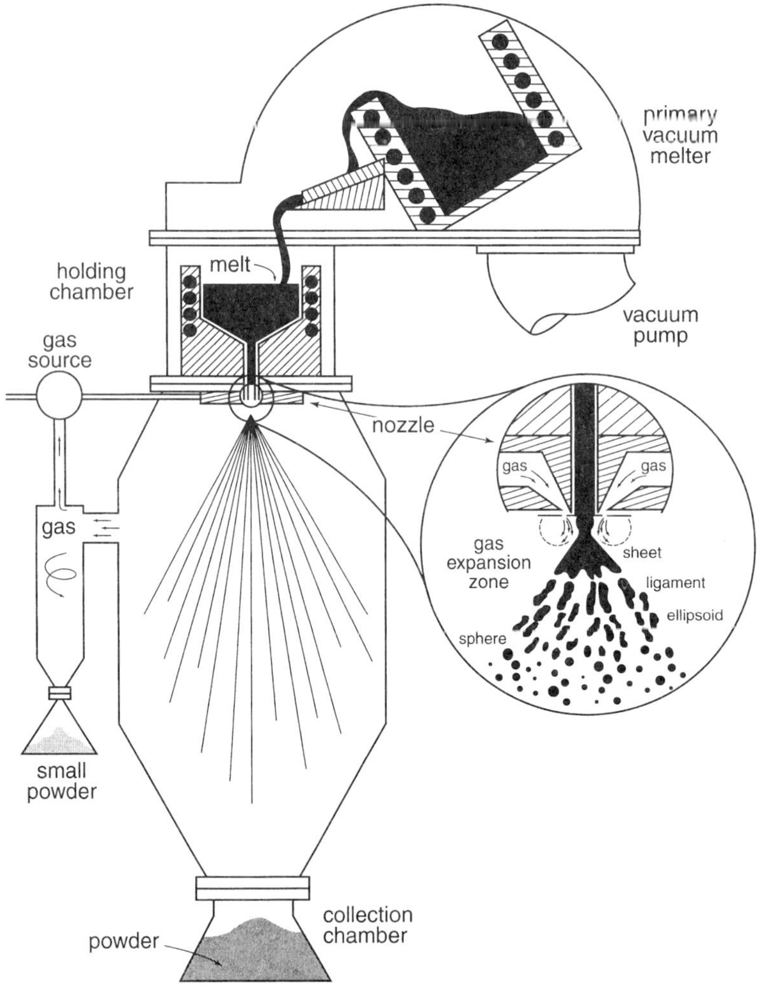

Figure 3.24. Gas atomization is a production process that gives spherical steel powders. A melt is poured through a gas expansion nozzle to form droplets that solidify during flight. The breakup of the melt stream is shown in the enlarged view of the nozzle region. Melting can be performed in air for alloys less sensitive to oxidation.

micrograph of a gas-atomized stainless-steel powder. Such powders have excellent packing and flow properties, and exhibit apparent and tap densities within 60–65% of theoretical. Sometimes the powders can be packed to densities of up to 75% of theoretical, if there is no agglomeration and a wide size distribution.

3.3.5.2. Water Atomization. Water atomization is the most common technique for producing elemental and alloy ferrous powders. Two examples of water atomizer

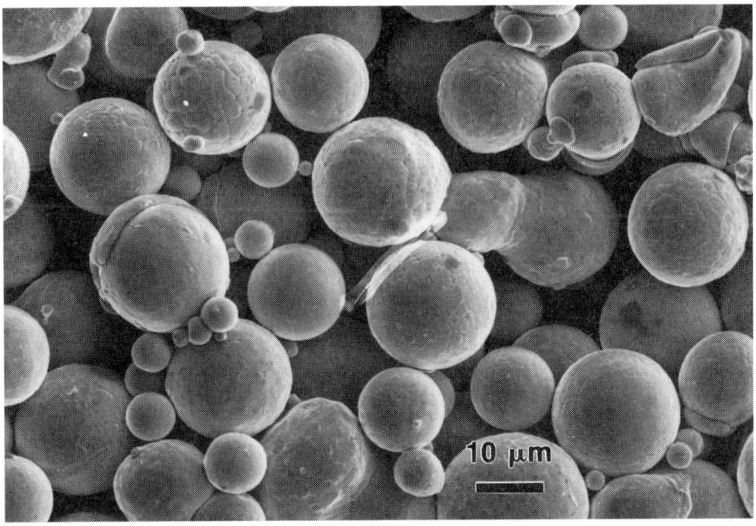

Figure 3.25. SEM of a gas-atomized stainless-steel powder. The powder is nearly spherical in shape, but on close examination some agglomerates are evident and even some cases where one droplet has impacted onto a solidified particle. (Photograph courtesy of Christine Kipphut-Furstoss.)

powder are shown in Figure 3.26. The upper photograph is the typical compaction grade powder with many bumps and nodules on the surface. In the lower picture, a very small, ligamental shaped stainless-steel powder is evident, one useful for injection molding. As contrasted with gas atomization, powder produced by water atomization is not spherical.

A simple production scheme is sketched in Figure 3.27. At the core of water atomization, high-pressure water jets are directed against the melt stream, forcing disintegration of the melt into droplets that solidify into irregular shapes. The water can be directed by a single jet, multiple jets, or an annular ring. Although similar to gas atomization in concept, water atomization is geared to much larger production quantities, resulting in lower powder costs. A melt might consist of 60 tons of iron and would be atomized at a rate of 5 kg/s (11 lb/s).

Several variants exist for water atomization. One very popular approach uses a molten iron–carbon byproduct from processing titanium–iron–oxide ore. As diagramed in Figure 3.28, the molten metal is transported to a water atomization facility where the melt is granulated by water jets, dried, and subsequently decarburized (carbon is extracted by heat and atmosphere reactions) to produce iron powder. Oxidation is useful in removing the carbon, but a final stage reduction is necessary to remove oxygen. Since the reduction process causes the particles to sinter-bond, subsequent milling is required to obtain discrete particles. Finally the product is mixed with alloying additions and lubricants and packaged.

A flowchart for an alternative route using a low-carbon steel starting material is

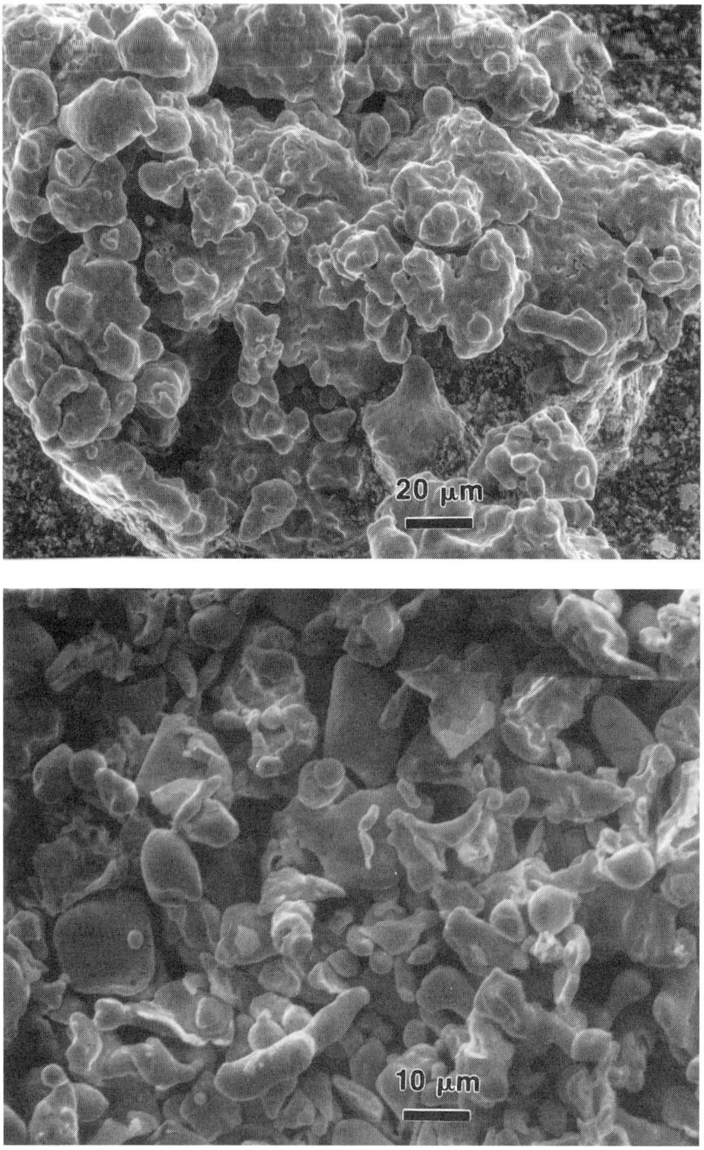

Figure 3.26. Two examples of water-atomized powders. The upper powder has considerable nodular character and is pure iron, while the lower powder shows a ligamental character and is stainless-steel.

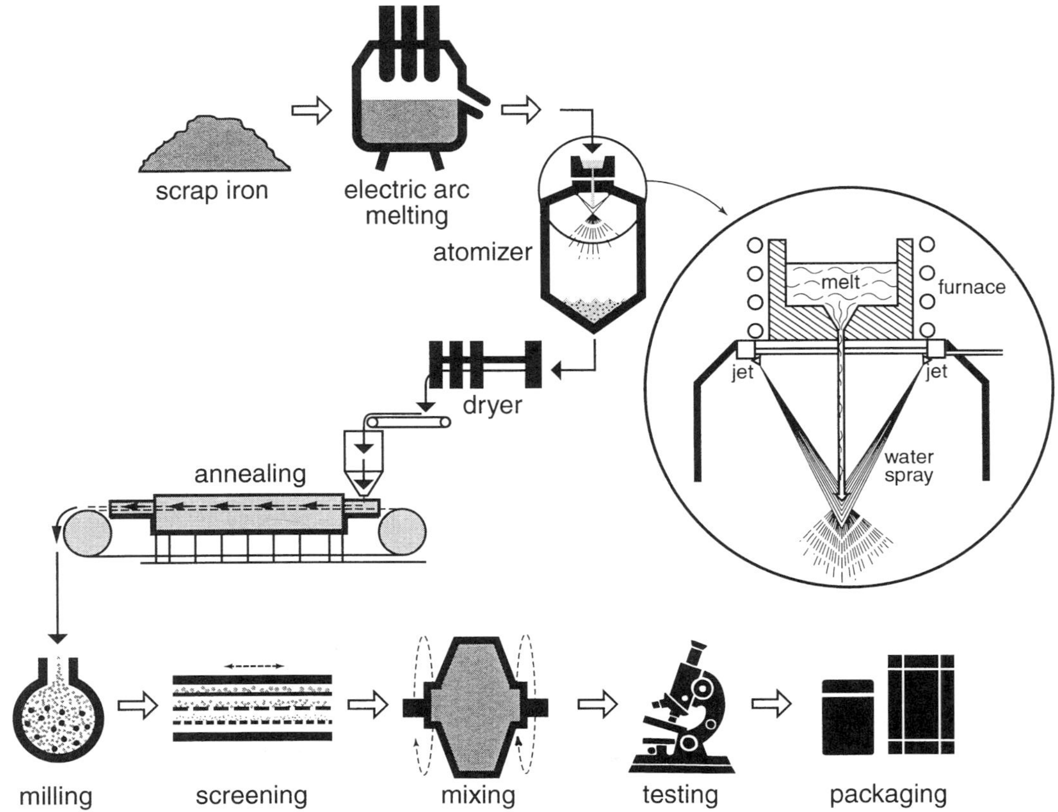

Figure 3.27. An overview of water atomization for the production of iron powders from scrap metal. Melting occurs in an electric arc furnace, with water jets providing the disintegration of the melt stream needed to form the powder. Because of oxidation in atomization, the product is annealed in a reducing atmosphere before final milling, sieving, and mixing.

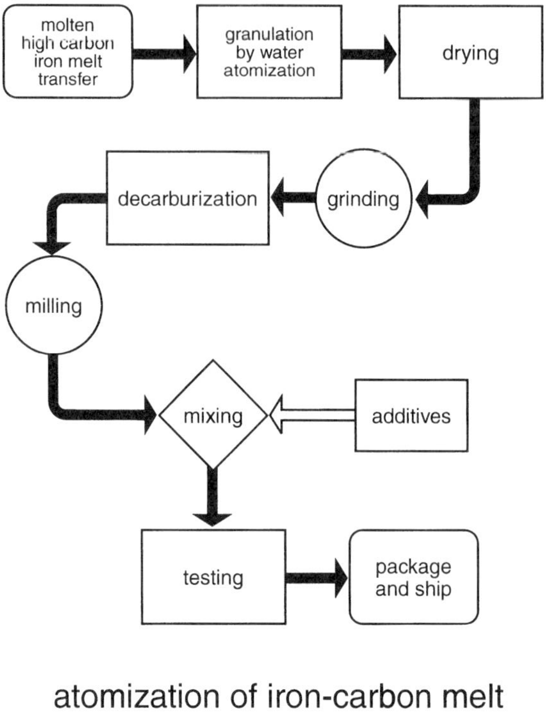

Figure 3.28. A process flowchart for the atomization of a liquid high-carbon iron using water, with subsequent decarburization to make an iron powder.

shown in Figure 3.29. Here the melt is water-atomized, dried, annealed, milled, classified, mixed, and packaged. No decarburization step is required in this case. Each production facility has a variation on these themes. At some steel mills, iron powder is produced using molten metal obtained directly from the melt shop, but this is mostly employed for internal products.

During atomization, the typical flow-rate ratios are about 5 : 1 (5 kg of water per kilogram of steel powder). Because of rapid cooling, the powder shape is irregular and rough, with some oxidation. On occasion, oils or other liquids have been used instead of water to gain better control of particle shape and oxidation. Like gas atomization, pressure is the main process control variable in water atomization. Higher water pressures result in higher water velocities and smaller particle sizes. Pressures up to 150 MPa (20 ksi) have been used in water atomization, resulting in particles in the 5-μm size range. This proves most useful for forming low-cost powders used to make filters and small injection-molded objects. However, the bulk of the iron powder in production is targeted to an average particle size near 100 μm, with sizes up to several hundred micrometers.

3.3.5.3. Centrifugal Atomization. Several specialized atomization processes use a spinning disk or other means to impart centrifugal force to a melt. Droplets are thrown off as a fine spray that solidifies into spheres. As one example, the rotating-electrode process is used to make large iron and steel powders. A sketch of the apparatus is given in Figure 3.30. The electrode passing along the horizontal axis is consumed by an arc from the tungsten cathode as it spins on its axis.

An example of the rotating electrode process spherical powder is shown in Figure 3.31. The consumable electrode must be made from the desired alloy, since there is no alloying in the thin melt layer on the end of the electrode. Because the electrode is spinning at rates of up to 50,000 revolutions per minute, molten material is thrown off into droplets, the size depending on the electrode size and spinning rate. Usually, the process is performed under inert gas to protect the powder from oxidation. Other variants include various forms of spinning disks, cups, or crucibles.

Benefits of centrifugal atomized powders are the purity, spherical shape, and

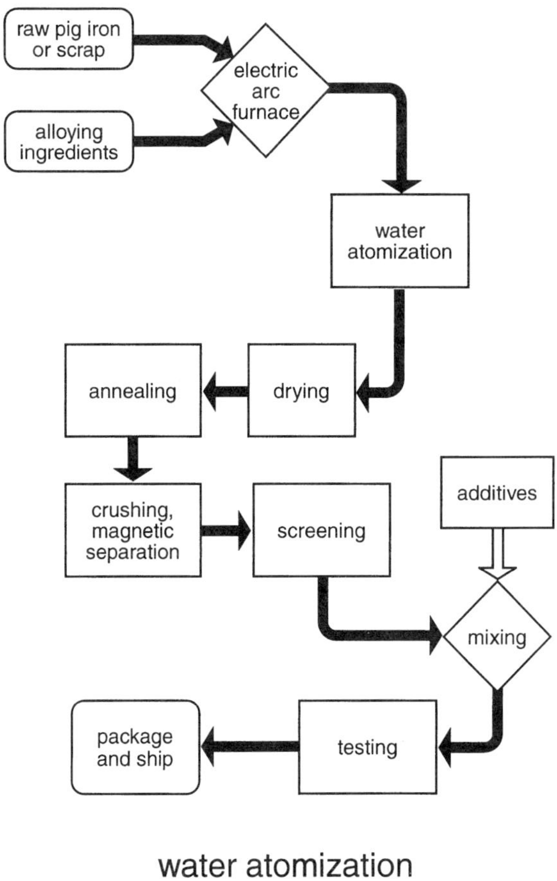

Figure 3.29. Flowchart for the water atomization of an iron input material with electric arc refining, atomization, and final preparation of a soft iron or steel powder.

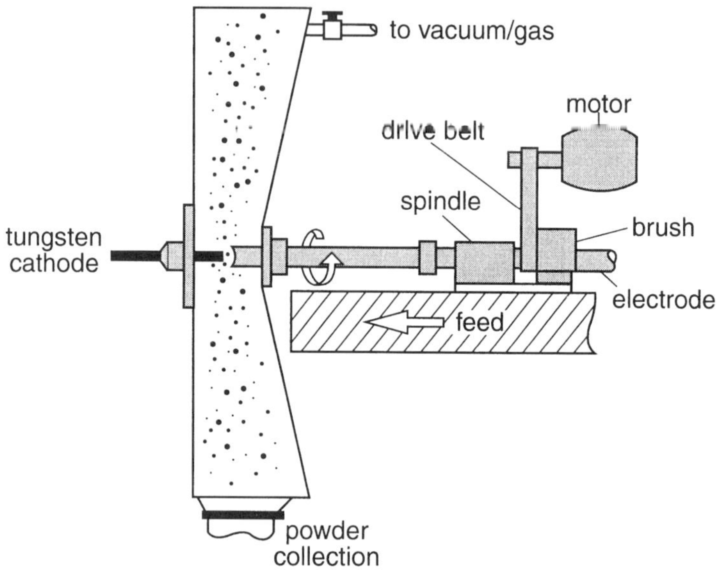

Figure 3.30. Sketch of the rotating electrode process for forming spherical powders from electrodes of iron or steel. As the tungsten cathode creates a melt on the end of the rotating electrode, the melt is spun off as droplets that solidify into powder.

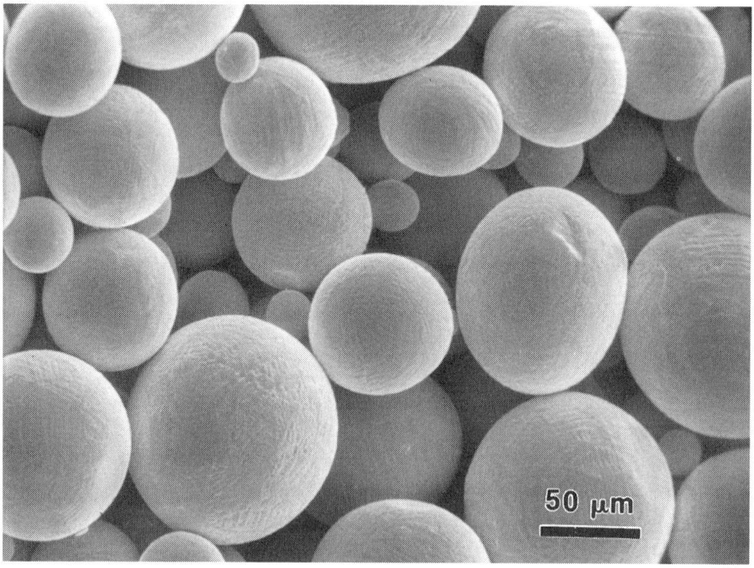

Figure 3.31. A SEM picture of a steel alloy powder formed by centrifugal atomization, showing the nearly perfect spheres of powder. (Photograph courtesy of Clem Karfs.)

EQUATION BOX 3.7

The energy delivered to the melt during centrifugal atomization is the key factor determining the final particle size. Shear increases in centrifugal atomization as the spinning rate increases. For most centrifugal atomizers, the average particle size D is inversely dependent on the rate of electrode rotation ω

$$D = \frac{A}{\omega}$$

where A is a factor that includes the metal density, melt surface tension, melt viscosity, and other operating parameters such as the diameter of the electrode.

uniform apparent density. The disadvantages are a low production rate, high equipment and processing costs, and coarse particle sizes.

3.3.6. Other Production Processes

Progress has been made in forming novel iron powders directly from the vapor phase. Such approaches are in the early production stages. One technique vaporizes

Figure 3.32. Nanoscale iron powder produced by electrical arc erosion of wires under liquid argon. Note that the particle size is near 100 nm while the particle shape is nearly spherical, but the powder is highly agglomerated. (Photograph courtesy of Ronald Iacocca.)

Figure 3.33. Splat quenched ribbon was subsequently milled to form platelets of the amorphous (no crystallization in solidification) iron–neodymium–boron powder. (Photograph courtesy of Ronald Iacocca.)

the iron in a residual pressure of argon (0.015 atm pressure). Since the temperature falls off sharply with distance from the vapor source, eventually the vapor becomes cooled to the point where very small iron particles condense in the argon. Such vapor-nucleated particles range from 50 to 1000 nm in size (1000 nm = 1 μm). The shape is usually faceted or cubic, and agglomeration is a common problem.

A related method relies on arcing a wire under liquid argon. Small droplets result from arc erosion of the wires, and rapid quenching in the liquid argon freezes the droplets as small particles. Figure 3.32 is a high-magnification scanning electron micrograph of approximately 100 nm of iron powder formed by this process.

It is possible to splat and quench melt onto a spinning disk or wheel to form rapidly solidified and amorphous iron alloys. The most common approach relies on squirting molten alloy onto a rapidly spinning [20,000-rpm (revolutions per minute)] copper disk. The melt freezes on the disk and is then thrown off by centrifugal forces, producing a ribbon of 25–100 μm thickness. This rapidly solidified ribbon is mechanically attritioned to generate an angular powder such as shown in Figure 3.33. This powder is an amorphous iron–neodymium–boron composition that is very useful in magnets.

One process for forming intermetallic powders that is gaining favor involves the reaction synthesis of mixed iron and aluminum powders to form iron aluminides. If the reaction is controlled, then the heat produced by the reaction is extracted to ensure the formation of compound powders. This reaction synthesis process typically involves mixing the constituent powders into a loosely packed bed. The bed is rotated and heated to initiate a reaction. If heat is properly extracted from the reaction, then the product is an intermetallic powder such as FeAl or Fe_3Al.

3.4. CHARACTERISTICS OF COMMON FERROUS POWDERS

An examination of the scanning electron micrographs in this chapter reveals that iron powders are not the same. Besides the differences in surface texture, particle size, and particle shape, internal differences exist as well. Further, Figure 3.34 is an illustration of two water-atomized iron powders that were cross-sectioned. These pictures were taken through an optical microscope to show the difference in internal porosity and inclusion contents. One powder has internal pores that would impact on reactivity, compressibility, and dimensional changes in sintering. Depending on the processing goals, the porous powder might be desirable or undesirable. Lack of knowledge on powder characteristics and their impact on processing is a common problem in P/M.

Tests on ferrous powders show clear differences in their size and other attributes.

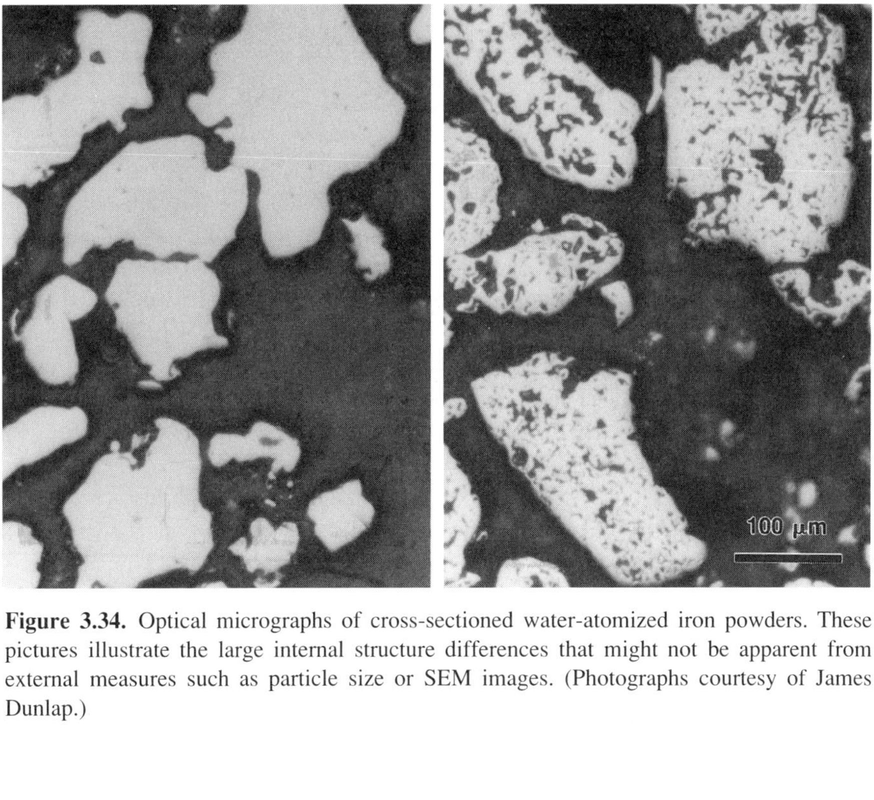

Figure 3.34. Optical micrographs of cross-sectioned water-atomized iron powders. These pictures illustrate the large internal structure differences that might not be apparent from external measures such as particle size or SEM images. (Photographs courtesy of James Dunlap.)

TABLE 3.3. Example Metal Powders and Characteristics by Various Processes

Powder	Process	Shape	Median Size (μm)	Oxygen Level (%)	Apparent Density (g/cm³)	Tap Density (g/cm³)	Flow Time (s)
Iron	Water atomization	Irregular	75	0.11	2.9	3.4	26
Iron	Water atomization	Irregular	104	0.15	2.9	3.5	26
Iron	Water atomization	Irregular	106	0.08	2.9	3.1	26
Iron	Gas atomization	Spherical	66	0.15	4.5	5.0	9
Iron	Carbonyl	Spherical	5	0.38	2.7	4.3	NF[a]
Iron	Carbonyl	Spherical	4	1.0	3.6	4.5	NF[a]
Iron	Oxide reduction	Porous	77	0.39	2.6	—	29
Iron	Oxide reduction	Porous	50	1.0	1.9	3.0	35
Iron	Centrifugal atomization	Spherical	75	0.10	4.7	5.0	14
4600 steel	Water atomization	Irregular	70	0.10	2.9	3.0	26
Stainless steel	Gas atomization	Spherical	12	0.10	3.8	4.7	38
Stainless steel	Water atomization	Irregular	60	0.20	2.6	3.7	30
Stainless steel	Water atomized	Rounded	10	0.46	3.8	4.5	NF[a]
Tool steel	Gas atomization	Spherical	10	0.01	5.0	5.5	NF[a]
Tool steel	Water atomization	Nodular	70	0.10	1.8	2.4	50

[a] Indicates no flow = NF.

Table 3.3 summarizes the characteristics of several commercial iron and steel powders. Chemistry is an important variable. Low-alloy steel, stainless-steel, and tool steel powders are all available. In each alloy class, the oxygen content is a good measure of the powder production skills. Also, particle size and particle shape are widely different. For some applications, a low-packing-density, sponge powder is most desirable, such as for brakes. Further, high porosity powders are useful for situations in which swelling occurs during sintering, especially in iron–copper alloys. The water-atomized iron powders can vary considerably, but industry standardization keeps most of the offerings close in terms of powder characteristics.

Usually, the higher-purity powders exhibit the highest compressibility. The classic water-atomized powders do not pack to a high density, but do undergo extensive deformation during compaction that results in a high green strength. The gas-atomized and centrifugally atomized powders exhibit high packing densities due to the spherical shapes. However, they prove difficult to compact and have low compressibilities and negligible green strengths. Generally, spherical powders are used where packing density is the primary concern, since compaction is difficult and green strengths prove low. Thus, carbonyl and small atomized powders are most useful

EQUATION BOX 3.8

In some powder systems, a powder mixture improves the packing density. The concept of a high-packing-density bimodal powder mixture starts by considering a container filled with the larger particles. The packing density increases as small particles are added to fill the voids between the large particles. Eventually, enough small particles are added to fill all the spaces between the large particles, producing the maximum packing density. The maximum packing density composition corresponds to a weight fraction of large particles X^* that depends on the amount of void space between large particles, which equals $1 - f_L$, where f_L is the fractional packing density of the large particles

$$X^* = \frac{f_L}{f^*}$$

with the packing density at the optimal composition f^* given as

$$f^* = f_L + f_S(1 - f_L)$$

and the fractional packing density for the small particles is f_S. For many powders, the maximum works out to two-thirds large powder and one-third small powder. This idea can be extended to multiple-mode mixtures of three, four, and even five powders. Continuous particle size distributions can also be tailored to give high packing densities. Easy-flowing powder mixtures with just 5% porosity are possible using five powder blends.

in injection molding, tape casting, and slip casting. Often powders are mixed to customize the packing, flow, compaction, shrinkage, or other engineering attributes.

Some grades of ferrous powders are prealloyed, including low-alloy steels, tool steels, and stainless steels. Alloying additions are required for powders that will be used in applications where certain properties are a concern—magnetic, corrosion, strength, machinability, or electrical. Usually the alloying additions increase the price and reduce the compressibility of the powder.

3.5. POWDER SELECTION

Consider the diverse sizes, shapes, and other characteristics of the powders listed in Table 3.3. Such diversity at first will be a barrier to selection of the right powder for a specific application. This section addresses the issues on powder selection and reconciliation with the application. Table 3.4 categorizes the powder requirements

TABLE 3.4. Iron Powder Requirements for Various Applications

Application	Particle Size Range (μm)	Particle Shape	Major Considerations
Bonded magnets	5–100	Rounded	High purity, magnetic response
Chemical reactions	20–100	Rounded	High purity
Cold isostatic pressing	45–300	Rounded, irregular	Compressibility, green strength
Die compaction	20–200	Irregular	Compressibility, packing
Electrophotography	100–200	Spherical	Magnetic properties, shape
Flame cutting	40–200	Rounded, ligamental	Good flow, low impurity level
Forging	20–150	Irregular	Purity, compressibility, strength
Friction materials	50–200	Irregular	Irregular shape, sponge
Hot isostatic pressing	75–300	Spherical	High purity, high packing density
Injection molding	1–10	Spherical	High packing density, sintering
Magnetic sorting and testing	20–100	Rounded, flattened, irregular	High purity, magnetic response
Pharmaceuticals and foods	2–50	Rounded, spherical	Dispersion
Radar absorption	1–10	Spherical	Small grain size
Rolling	1–40	Irregular, rounded	High green strength, high purity
Shot peening	100–3000	Rounded	High hardness
Stone cutting	500–1400	Angular	High-hardness shot for abrasive cutting
Thermal spray	45–200	Rounded, spherical	Uniform flow, high purity
Welding electrode coating	60–400	Irregular	Low oxygen

for several applications. Since there are many applications, there are many viable powder production techniques and powder characteristics.

Recognize that many powders are used directly and are not consolidated. For example, the iron powders used in pharmaceuticals and foods are often a reduced or carbonyl product. (Ever looked at the cookies, cakes, or flour label and thought about that reduced-iron ingredient?) Interestingly, these grades are not as pure as the iron powder used in magnetic components. Other direct uses for iron powders include wrist weights, magnets for sorting seeds, and shot for impact hardening or abrasion of a component. Iron powders used for electrode coatings in welding are usually water-atomized, with low sulfur and phosphorus contents. A rounded or irregular particle shape is most desirable. Flame, plasma, and other thermal spray techniques require specific powder sizes. Too large a particle size, and the powder will not melt during passage through the torch; but too small a particle size, and the particles evaporate. Figure 3.35 provides a view of the desired particle size ranges for several powder applications. Many of these processes are detailed later in this book.

More than for any other applications, iron powders are used to form structural P/M components by die compaction or injection molding, followed by sintering. Even in these two applications, the powder requirements are significantly different. Injection-molding grades are nearly spherical and very small. A high packing density is needed, so this drives the shape toward a sphere, and sintering shrinkage is desirable, which dictates a small particle size. In contrast, die compaction and powder forging applications require powders with irregular particle shapes, good compressi-

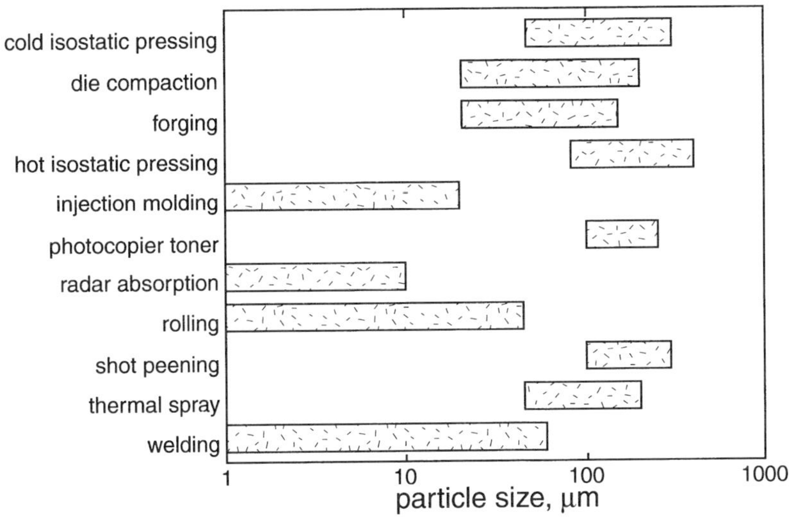

Figure 3.35. A plot of the particle size ranges used for various applications. This size difference is part of the reason for the existence of so many iron powder production processes.

bility, and larger particle sizes. Accordingly, powders useful for one application cannot be substituted for another application.

SUGGESTIONS FOR FURTHER READING

Allen, T., *Particle Size Measurement*, 5th ed., 2 vols., Chapman & Hall, London, UK, 1997.

Beddow, J. K., *Particulate Science and Technology*, Chemical Publishing, New York, NY, 1980.

German, R. M., *Powder Metallurgy Science*, 2nd ed., Metal Powder Industries Federation, Princeton, NJ, 1994.

German, R. M., ''The Importance of Powder Characteristics in Powder Injection Molding,'' *Rev. Particulate Mater.* **1,** 109–160 (1993).

Ghadiri, M., F. A. Farhadpour, R. Clift, and J. P. K. Seville, ''Particle Characterization: Size and Morphology,'' in *Powder Metallurgy—an Overview*, I. Jenkins and J. V. Wood (eds.), The Institute of Materials, London, UK, 1991, pp. 56–75.

Lawley, A., *Atomization: The Production of Metal Powders*, Metal Powder Industries Federation, Princeton, NJ, 1992.

Lowell, S., and J. E. Shields, *Introduction to Powder Surface Area and Porosity*, 2nd ed., Wiley, New York, NY, 1992.

Rhodes, M. J., (ed.), *Principles of Powder Technology*, Wiley, New York, NY, 1990.

Svarovsky, L., *Powder Testing Guide*, Elsevier Applied Science, London, UK, 1987.

Watts, J. F., and T. J. Carney, ''Surface Chemical Characterization of Powders,'' in *Powder Metallurgy—an Overview*, I. Jenkins and J. V. Wood (eds.), The Institute of Materials, London, UK, 1991, pp. 76–91.

CHAPTER 4

Powder Shaping

4.1. SHAPING VERSUS COMPACTION

Several approaches exist to the overall process known as powder metallurgy. Die compaction at high pressures is the dominant approach, where densification and shaping occur simultaneously. After compaction, the structure will be near 85% dense. In the subsequent sintering step, the particles are bonded, but little sintering densification occurs. Strength increases in sintering by a factor often near 100.

An alternative is to use low pressures and high binder contents to shape a powder. Generally, the binder acts as a glue to hold the particles into the desired shape. Most binders are polymers, which might be mineral oil or polyethylene. In shaping, the porosity ranges from 40 to 60% (ignoring the polymer that is sacrificial), so sintering is performed at temperatures where densification occurs. Small powders assist in densification during sintering. Thus, key differences between compaction and shaping are in the forming pressures, initial porosity levels, particle sizes, and sintering temperatures. Compaction relies on high pressures and the porosity after compaction is low, while shaping is the opposite. The third alternative, full-density processing, relies on the simultaneous application of pressure and temperature and is usually applied to larger structures aimed at high performance levels.

The differences between compaction and shaping technologies impact on production costs. Key considerations are the component size, desired final density, and shape complexity. A general categorization of shaping technologies is as follows:

- *Binder-assisted extrusion*—long structures, small powders, constant cross section, relatively simple shapes
- *Injection molding*—complex, small components, high-performance materials
- *Slip casting*—very large structures, constant wall thickness, low precision
- *Tape casting*—flat sheets, very simple shapes, small powders

This chapter focuses on powder shaping techniques that rely on small powders and polymeric binders that are extruded, cast, or molded in various forming devices. The polymer is required to lubricate particle flow, since the forming stresses are insufficient to deform the powder. Many variations exist, but for this chapter, three will be detailed: extrusion, casting, and molding. These forming approaches use similar powder–binder mixtures. An emerging area of rapid prototyping and rapid tool production is discussed as a new option. All of these have similar concerns with powder characteristics, binder formulations, dispersants, mixing, viscosity, and polymer extraction prior to sintering.

4.2. BINDERS

4.2.1. Basic Attributes

The *binder* is a temporary vehicle for packing a powder into the desired shape and holding the particles in that shape until the beginning of sintering. Accordingly, although the binder should not dictate the final composition of the molded steel, it has an influence on the process. The binder is mixed with the powder to form feedstock. Unfortunately, no binder is perfect, so selection of an appropriate binder varies between situations. Most binders are multiple-component mixtures that contain a major ingredient that dictates the basic properties along with additives used to tailor the properties. After shaping the binder holds the particles in place until the particles are heated in sintering. Finally, the binder is eliminated, hopefully without contaminating the powder.

Table 4.1 lists the attributes of an ideal binder for shaping ferrous powders. These attributes include concerns with toxicity, cost, flow, and ease of manufacturing. As a primary requirement for easy shaping, the binder must wet the powder surface, so various chemicals are employed to bridge between the powder and the binder. For ferrous alloys, the most common coupling chemicals are stearates—including stearic acid, zinc stearate, and ethylenebisstearamid. Stearic acid is a low cost polar

TABLE 4.1. Desired Binder Characteristics

Flow characteristics
Low inherent viscosity at the shaping temperature
Strong and rigid after shaping
Low molecular weight
Powder interaction
Wets and adheres to powder
Chemically passive
Stable during mixing and shaping
Binder burnout
Noncorrosive, nontoxic decomposition product
No residue on burnout
Decomposition temperature above mixing temperatures
Decomposition below sintering temperature
Manufacturing
Inexpensive, available, safe, and low-cost
Long shelf life, low water absorption, no volatile components
High strength and stiffness

EQUATION BOX 4.1

Powder–binder mixtures flow by shearing. Shear stress τ is the force per unit area that causes the mixture to flow in a die. The shear strain γ is the relative motion of the feedstock over the surface. The shear strain rate $d\gamma/dt$, better known as the *shear rate*, is the change in shear strain divided by time. Effectively, it is how fast the mixture moves. Fluid resistance to shearing is termed *viscosity* η, defined as

$$\tau = \eta \left(\frac{d\gamma}{dt}\right)^m$$

where m is the exponent used to characterize the flow. For low-molecular-weight liquids like water, the viscosity depends on temperature and pressure, but not on the shear rate. Such liquids are termed *Newtonian*. Powder–binder mixtures exhibit more complex behavior, usually with a viscosity that depends on the shear rate and temperature, generally decreasing at higher shear rate due to shear thinning. In many powder–binder forming operations there is a yield strength to the mixture that must be exceeded before flow occurs. In this case, the shear stress is that over the yield strength.

molecule with a molecular structure of $CH_3(CH_2)_{16}COOH$, a density of 0.85 g/cm^3, and a melting temperature of 65–75°C (149–167°F). Also, ethylenebisstearamid consists of two stearate-type molecules joined by carbon–nitrogen bonds:

$$CH_3{-}(CH_2)_{16}{-}CO{-}NH{-}CH{=}C{=}CH{-}NH{-}CO{-}(CH_2)_{16}{-}CH_3$$

This molecule proves very effective in lubrication and bonding, and has the further attribute of burning out of a steel compact with minimal contamination. Such additives reduce the mixture viscosity, allowing more solid in the feedstock. *Viscosity* is the resistance to flow. A high-viscosity mixture is difficult to shape, so low-viscosity systems are desirable. For slip casting, the mixture has a viscosity similar to paint, while for extrusion or injection molding the mixture viscosity is higher, and resembles toothpaste.

Once a component has been shaped, polymer extraction usually involves heating without disturbing the powder shape. Thus, an important binder requirement is easy removal during sintering with minimal contamination. To prevent property changes over time, the binder cannot interact with the powder or ambient environment. If it were not for rusting, water would be widely used as a binder ingredient for many steels.

TABLE 4.2. Example Binders Used in Powder Shaping

Extrusion
56% water, 25% methyl cellulose, 13% glycerine, 6% boric acid
72% water, 12% hydroxypropyl methyl cellulose, 8% glycerin, 4% ammonium polyacrylate, 4% ammonium stearate
65% polyethylene glycol, 30% polyvinyl butyryl, 5% stearic acid
Injection molding
69% paraffin wax, 20% polypropylene, 10% carnauba wax, 1% stearic acid
75% peanut oil, 25% polyethylene
50% carnauba wax, 50% polyethylene
55% paraffin wax, 35% polyethylene, 10% stearic acid
Slip casting
96% water, 4% sodium lignosulfonate, trace calcium nitrate
93% water, 4% agar, 3% glycerine
99% water, 1% ammonium alginate
97% water, 3% polyvinyl alcohol
Tape casting
77% water, 9% polyacrylate emulsion, 9% glycerin, 3% ammonium polyacrylate, 2% ammonium hydroxide
80% toluene, 13% polyethylene glycol, 7% polyvinyl butyral
47% mineral spirits, 24% isopropanol, 8% polyvinyl butyral, 8% dibutyl phthalate, 5% polyethylene glycol, 5% stearic acid, 3% menhaden fish oil

4.2.2. Binder Examples

Table 4.2 provides example binder formulations for extrusion, injection molding, slip casting, and tape casting. There are many possible variants, but these summarize common formulations as applied to powder shaping.

Generally the binders are either thermoplastics or water-based gelation systems. The term *thermoplastic* implies that the mixture flows easily on heating, but becomes rigid on cooling. Candle wax is a common thermoplastic. Wax–polymer mixtures are most widely employed in powder shaping technologies, because of low cost, easy processing, and lack of hazards. A common example would be a mixture of two-thirds paraffin wax and one-third polypropylene, which would be molded between 130 and 150°C (265 and 300°F). After molding, the wax can be extracted by slow heating, immersion in solvents, or evaporation by heating in a vacuum. Variants on this binder are widely employed in injection molding and some extrusion operations. Water as a binder is used in freeze–firing cycles, where a slurry of powder and water is frozen in the die.

A related group of binders relies on high water contents and polymers that form gels on cooling or drying. The gels are derived from soaps, sugars, or starches. After evaporation of the water, the remaining gel holds the particles in place until sintering. These are preferred for some extrusion, tape casting, and slip casting operations. Gelation forms a single, large molecule that extends throughout the component.

Once formed, this massive three-dimensional network molecule resists flow, much like dessert gelatins. Cellulose-type gelation binders with water or glycerine are widely used in powder extrusion. Similar concepts are being developed for injection-molding binders.

The binder system is a necessary, yet sacrificial, constituent in powder shaping. Clearly, several formulations are in use, and several candidate binder systems can be used with any powder or shaping technique. Because of the transient role, low-cost binders dominate the field. They are responsible for placing the particles into close packing, but should be essentially invisible otherwise. In spite of its transient role, the binder is an important aspect of particle shaping.

4.3. RHEOLOGY

Powder shaping processes require specific viscous flow characteristics of the powder–binder mixture. The overall study of polymer flow falls into a field of study known as *rheology*, with a primary emphasis on viscosity, which measures the resistance to flow. For powder shaping technologies, the molten binder provides viscous flow attributes to the mixture.

Viscosity can be measured by a variety of techniques. For most powder–binder mixtures, the easiest measurement is via the *melt flow index*, which is the mass of feedstock passing through a cylindrical capillary under a prescribed pressure during 10 minutes. Other measures based on mixing torque or capillary extrusion tests are possible. Viscosity increases with the addition of powder. Figure 4.1 shows an example of this behavior for carbonyl iron powder mixed with molten paraffin wax in varying ratios. As the steel content approaches a critical loading of 60 vol%, the viscosity becomes essentially infinite and the mixture is too stiff to shape. A high viscosity makes shaping difficult, but a low viscosity usually results in a low strength that contributes to component distortion and a loss of precision.

A second property is the elasticity. An elastic material subjected to low stress has a shape memory. Once the stress is removed, it returns to the original shape. Elasticity is the basis for a bouncing ball—it deforms and returns to its original shape. Powder–binder mixtures exhibit both viscous and elastic characteristics and are termed *viscoelastic*.

Viscous materials flow by shearing, just like painting. Thick, viscous mixtures are difficult to mold (viscosities over ~200 Pa · s), but well dispersed systems have lower viscosities. Typically, high viscosity mixes are very sensitive to the forming rate and are avoided when possible. An example is the children's toy Silly Putty—it stretches easily at low deformation rates and can be stretched without failure, but when stretched rapidly, it becomes brittle and breaks instead of deforming. Thus, a binder is required with low sensitivities to forming rates and the powder–binder mixture must be homogeneous to allow forming of uniform products with the least difficulty.

For low-molecular-weight liquids like water, the viscosity depends only on temperature and pressure, not on the rate of stirring. Powder–binder mixtures exhibit

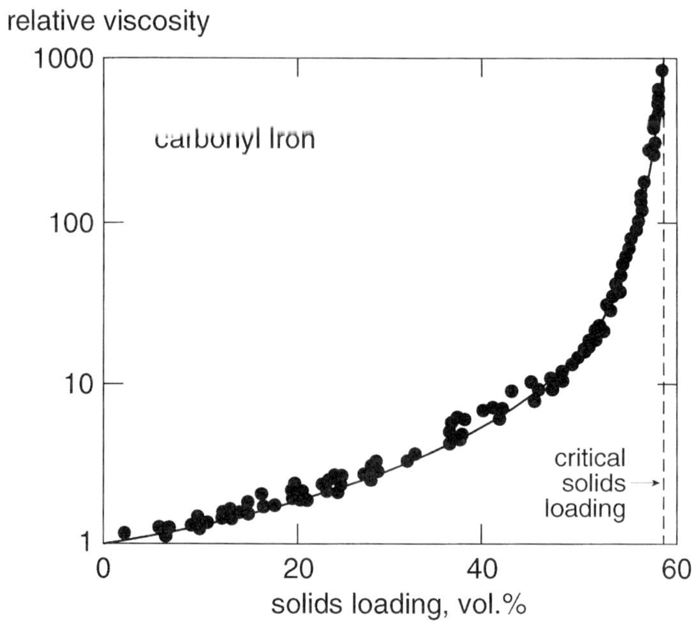

Figure 4.1. Viscosity is a measure of the resistance to flow. For powder–binder mixtures the viscosity is dependent on the amount of powder, measured by the relative solids loading in the feedstock. As illustrated in this plot for iron–wax, viscosity increases dramatically as the solids loading approaches a critical value.

more complex behavior. Most thermoplastic binders (wax–polymer systems) have a viscosity that decreases with temperature, but increases with the powder content. Accordingly, design of a feedstock for shaping is an involved task that requires extensive rheological testing to find a formulation suitable for each of the shaping technologies.

4.4. MIXING

An optimal viscosity range exists for each shaping technique. Accordingly, the binder is customized for the powder and shaping technology. Once the formulation is known, a feedstock mixture is prepared for shaping. Best mixing occurs with a high shear, but not to the point where the work of mixing damages the particles or overheats the binder. A properly mixed material will consist of a homogeneous powder dispersion in the binder, with no internal porosity or agglomerates. Inhomogeneities result in nonuniform viscosities, uneven shaping, and difficulties in sintering.

Most powder–binder mixtures are compounded in batch mixers similar in construction to that sketched in Figure 4.2. For the common binders, mixing is done at

EQUATION BOX 4.2

The change in powder–binder mixture viscosity η_M with solids loading Φ depends on a property termed the *critical solids loading* Φ_C and usually follows a behavior where

$$\eta_M = \frac{\eta_B}{[1 - (\Phi/\Phi_C)]^2}$$

η_B is the viscosity of the pure binder at the same temperature. The binder viscosity changes with temperature once it is molten, and because the thermal expansion of the binder is much larger than of the steel powder, this, too, increases the effective binder:powder ratio. The consequence is the mixture viscosity depends on temperature in a very sensitive manner

$$\eta_M = \eta_O \exp\left(\frac{Q}{RT}\right)$$

as a generalize sense of the temperature dependence, where Q is an apparent activation energy, R is the gas constant, and T is the absolute temperature. The factor η_O reflects several other factors. The critical solids loading corresponds to the powder content (on a volume basis) where the mixture will not flow. Note that the mixture viscosity divided by the binder viscosity at the same temperature is termed the *relative viscosity*, and it is unity when the solids loading approaches zero. Additionally, it becomes infinite as the solids loading approaches the critical value.

a temperature where the polymer is molten. Heat is added to lower the viscosity. Usually the feedstock is mixed in a sequence that might start with cold metal and polymer powders that are mixed as powders, followed by addition of a liquid (water) or heat to obtain a molten binder. Mixing continues for a selected time; then the mixer is discharged and the cycle repeated.

An alternative is to employ continuous mixing, where powder and binder are added to one end of a twin-screw mixer that continuously extrudes the mixed product. Figure 4.3 is a schematic of the twin-screw mixing concept, where the feedstock is constantly sheared and conveyed by two corotating screws. Twin-screw mixers produce a higher mixture homogeneity and less polymer degradation as compared to batch mixers.

To reduce contamination, the mixer construction materials need to be hard. High-vanadium-content tool steels and boride cermets have proved to be most wear-resistant. Heating is performed using internal heaters or double-walled mixing vessels with externally heated oil or steam circulating around the chamber. It is estimated

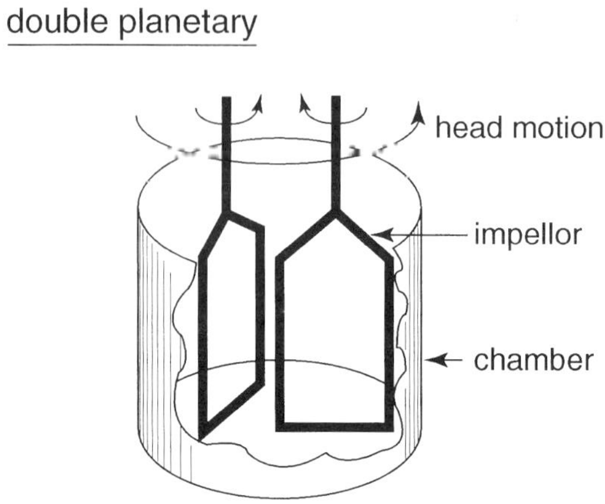

Figure 4.2. Double planetary mixers are commonly used in forming feedstock for various powder shaping technologies. The two blades rotate around individual axes while the whole upper head rotates, giving a swirling motion. Most operate under vacuum to avoid air entrapment and use a temperature-controlled jacket.

that 2 MJ of energy (about 0.5 kW · h) is needed to mix 10^4 cm^3 of feedstock (about 100 lb). On exit, the mixed feedstock is formed into pellets, usually via extrusion through a noodle plate. Figure 4.4 shows typical pellets ready for loading into a molding machine. These both were formed using continuous mixers.

In some operations, mixing is combined with pelletization. Alternatively, the mixture can be mechanically crushed or sheared into small fragments suitable for introduction into the feed hopper of a molding machine. Recycled powder–binder feedstock is remixed with virgin material to ensure full material use.

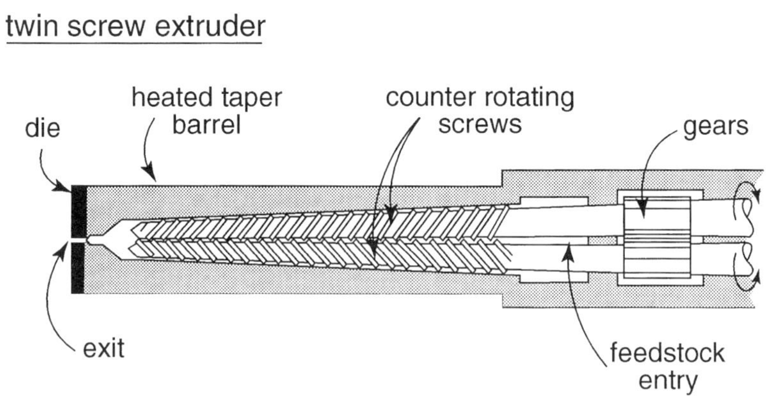

Figure 4.3. Twin-screw mixers give very effective compounding and extrusion of feedstock with short mixing times. They employ two intermeshing screws that counterrotate to create a zone of high shear while extruding the product.

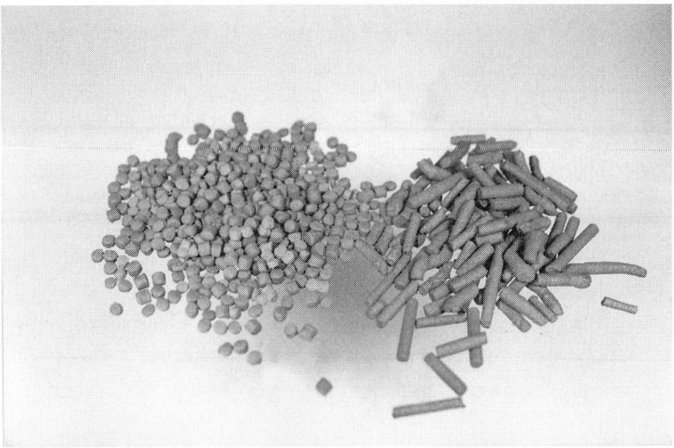

Figure 4.4. Photograph of pellets and extrusion products of steel powder and polymer binders. This form of feedstock is ready for loading into any of the forming machines. (Photograph courtesy of Julian Thomas.)

Feedstock formulations for powder shaping operations prove highly variable. Table 4.3 summarizes a few common systems for injection molding, giving the powder, binder, solids loading, and data on the viscosity, molding temperature, and strength after molding. The green strengths are usually sufficient for handling the components, but most fracture if dropped. The feedstocks in this table represent a balance of several considerations. Sufficient binder is needed to fill all voids between particles and to lubricate particle sliding during molding. But an excess of binder is not useful. Best success is attained with mixture viscosities similar to that of toothpaste. That viscosity depends on several factors. At too high a solid: binder ratio there is insufficient binder to fill all void space between the particles and the feedstock is impossible to mold. Improper mixing leads to inhomogeneous feedstock and molding defects. High shear is required in mixing to force binder among all particles intimately. Viscosity tests are performed using rheometers, where precise measures of temperature, pressure, and flow are recorded to extract viscosity data. Control of viscosity and solids content is necessary for the production of uniform and precise components.

4.5. POWDER EXTRUSION

Several shaping approaches are applied to steel powders for the fabrication of porous tubes, rods, or filters. Extrusion is most appropriate for generating long, thin structures. As illustrated in Figure 4.5, the powder–binder mixture is placed in a plunger or screw device equipped with a sized nozzle. Usually the mixture is evacuated to avoid bubbles in the extrudate. The product shape depends on the shape of the nozzle (sometimes called an *extrusion die*). In simple cases, the shape is a drilled hole, but

EQUATION BOX 4.3

One of the most useful concepts in understanding and comparing feedstocks is the solids loading Φ. It is the volumetric ratio of solid powder to the total volume of powder and binder, assuming no voids

$$\Phi = \frac{W_P/\rho_P}{W_P/\rho_P + W_B/\rho_B}$$

where W_P and W_B are the weight fractions of powder and binder, respectively, and ρ_P and ρ_B are the densities of the powder and binder, respectively. Usually the solids loading is expressed on a volume percentage basis, and a value near 60% is typical for many powder shaping operations. Volumetric comparisons are useful when examining powders of differing densities, but for manufacturing purposes feedstock formulation is by weight. For example, to obtain a 60 vol% powder loading where the binder and solid have densities of 1.0 and 7.5 g/cm^3, 91.8 wt% powder and 8.2 wt% binder are required.

As an alternative, given the solids loading, we can calculate the weight fraction of powder. For a given volumetric solids loading Φ, the weight fraction of powder W_P can be calculated as

$$W_P = \frac{\rho_P\,\Phi}{\rho_P\,\Phi + \rho_B(1 - \Phi)}$$

where ρ_P is the powder theoretical density and ρ_B is the binder theoretical density. Because the weight fractions of powder and binder sum to unity, then $W_B = 1 - W_P$ as a simple calculation. For these calculations the solids loading and weights are on a fractional basis.

Corresponding to each solids loading is a powder–binder mixture density ρ_M that depends on the theoretical powder density ρ_P and binder density ρ_B,

$$\rho_M = \Phi\rho_P + (1 - \Phi)\rho_B$$

For powder shaping, the solids loading must be below the critical condition, with a slight excess of binder.

in other cases the tooling includes center holes, splines, or other features. Spiders are used to bridge cores to the die center. This means that the support is upstream from the actual die and is in the form of a set of vanes that minimize flow disruption. The feedstock mixture is forced under pressure through the nozzle, giving a product with a constant cross section that replicates the die profile. The angle of entry into the die is usually about 25° and the die length is usually 2–4 times its diameter.

TABLE 4.3. Example Powder–Binder Mixtures for Injection Molding[a]

Powder	Binder (wt%)	Solids Loading (vol%)	Density (g/cm^3)	Shaping Temperature (°C)	Strength (MPa)
4 μm Fe	94 PE–6 OA	59	4.95	110	6
4 μm Fe	60 PW–40 PE	58	4.90	120	5
4 μm Fe	55 PW–45 PP-5 SA	61	5.12	150	22
4 μm Fe-2Ni	90 PA–10 PE	58	4.52	180	20
10 μm stainless steel	55 PW–45 PP-5 SA	67	5.60	130	15
15 μm stainless steel	90 PA–10 PE	62	5.33	190	20
12 μm tool steel	69 PW–20 PP-10 CW-1 SA	52	4.46	102	10

[a] Abbreviations: CW = carnauba wax; EVA = ethylene vinyl acetate; OA = oelic acid; PA = polyacetal; PE = polyethylene; PP = polypropylene; PW = paraffin wax; SA = stearic acid.

Pressures needed to form the shape are low, often in the range of 1 MPa (140 psi). The binder holds the particles in place during handling and is subsequently removed prior to sintering.

Related concepts use pressure to force the feedstock into closed dies. Compression molding is the most common approach, where the extruded feedstock is pressed into a die to form the desired shape. This die pressing occurs at a high pressure, so the powder–binder mixture takes on the detailed shape of the cavity. In contrast, transfer molding forces the powder–binder mixture into a die cavity using gas pressure. For these shaping processes, the binders are often based on cellulose, wax, or water, but might include polyvinyl alcohol, paraffin or mineral oils, and lubricants such as stearic acid. The forming machines are very similar to the die-compaction presses, which will be introduced in the next chapter. However, the difference in the pore space between

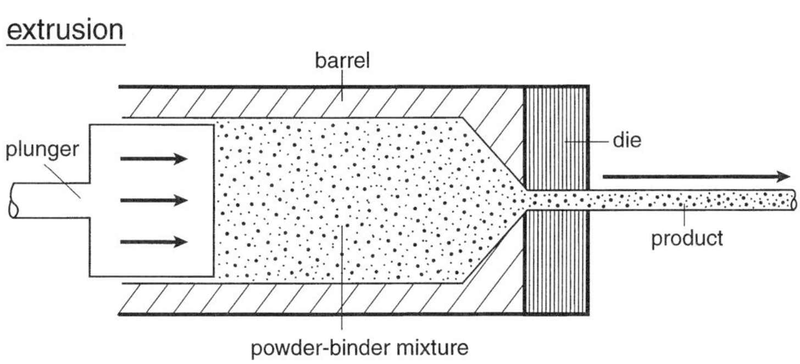

Figure 4.5. The extrusion process forces a slurry of powder and binder through a constriction to shape a continuous product such as a rod or tube.

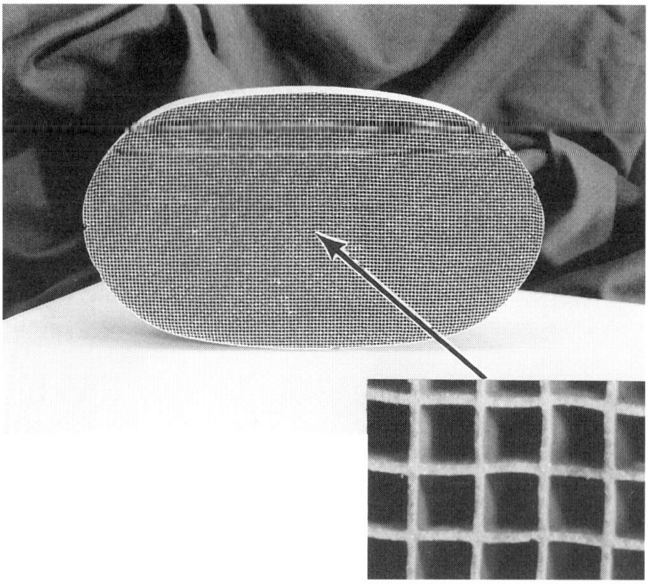

Figure 4.6. A honeycomb formed by extruding a ferrous powder through a complicated die with a methyl cellulose and water binder. (Component courtesy of David Weiss and photograph by Julian Thomas.)

particles is saturated with the polymer to aid powder flow in the cavity. Accordingly, the pressed density is low in comparison with die compaction.

Some products are used in very large volumes based on the extrusion, compression, or transfer molding concepts. These include capacitors, microelectronic substrates, porous tubing, welding rods, and catalytic converters for automobiles. Figure 4.6 shows a segment of a high-temperature alloy extrusion that is in the shape of a honeycomb. This was formed using a carbonyl iron powder mixed with alloying additions. Since the 1940s, porous stainless-steel filters have been fabricated using powder extrusion processes based on gelation of water-soluble cellulose binders.

4.6. POWDER CASTING

4.6.1. Slip Casting

The viscosity of the feedstock for extrusion is high compared to the powder–binder mixtures used in slip casting and tape casting. Here, particles are dispersed in a low-viscosity binder that might contain a solvent. The slurry has a viscosity similar to thick paint, typically between 10 and 50 Pa · s. This mixture is poured into a porous mold, usually fabricated from plaster of Paris. Figure 4.7 illustrates the process. A low viscosity is ensured by using an excess of water as the solvent. Once poured into the mold, the water is adsorbed by capillary action into the porous mold, resulting in a progressive increase in the viscosity of the remaining slurry. A typical plaster of Paris mold has pores of approximately 0.1 μm size. In this slip casting process, the slurry remains in the mold for several minutes and even several hours for thick components. However, this time can be decreased by applying a vacuum to the outside of the mold, pressurizing the slurry, or by subjecting the mold to centrifugal or rotational forces. During this time the water wicks into the mold, leaving the particles behind as a hard surface deposit on the walls of the mold. After drying, the shaped powder is removed from the mold and sintered. The green strength after drying might be 7 MPa (1000 psi).

A typical binder for slip casting is a mixture of water, alginate, and surfactants. Other binder systems rely on water solutions of cellulose, ammonium polymethyl methacrylate, polyvinyl alcohol, or polymers dissolved in common solvents, including water or alcohol. In the water-based systems, the acidity is controlled to minimize the slip viscosity.

The most popular application for slip casting is the fabrication of ceramic bathroom fixtures. For steels, the largest use is in rapid fabrication of prototype shapes and pilot tool sets. For large shapes, slip casting is a very cost-effective shaping approach, since there is almost no equipment cost and the tooling is inexpensive. The powder–water–binder mixture is simply poured into a porous die with no external pressure. Densification occurs in sintering.

Dimensions can be controlled to a tolerance of approximately 1%, but since gravity plays a significant role in determining the green structure, the top-to-bottom gradients cause distortion. With experience, this can be controlled by increasing the uniformity of particle deposition in casting.

There are three variations of the slip casting approach: *pressure filtration*, which relies on the same feedstock and mold but applies pressure to accelerate the wicking of water into the porous mold (proper selection of the filter pore size allows the liquid to pass while the particles are deposited on the surface as a cake; the higher

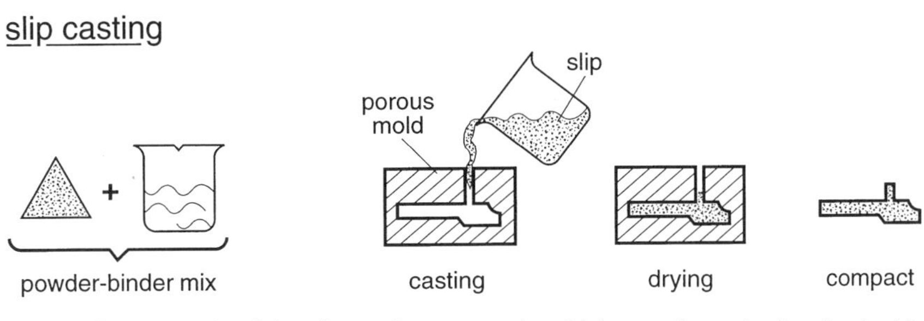

Figure 4.7. Schematic of the slip casting process in which a steel powder is mixed with a low-viscosity binder and cast into a porous mold. The pores in the mold remove the excess water in the binder by a wicking action.

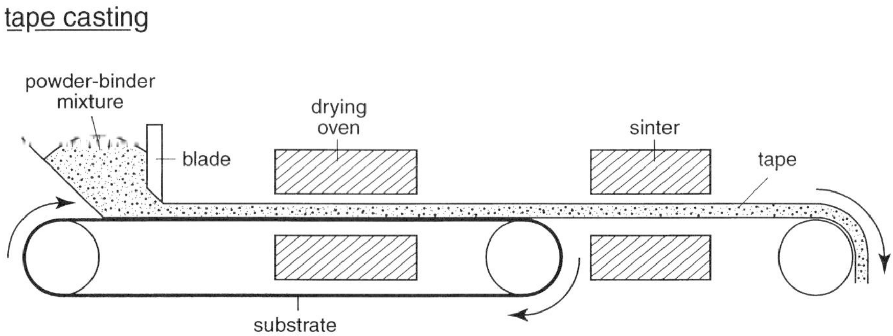

Figure 4.8. In tape casting, a thin layer of powder–binder mixture is formed on a moving Mylar substrate, which carries the tape into a drying oven prior to sintering. The technique is used to form thin sheets.

the pressure, the faster the component forms from the slip); *freeze casting*, which chills the slurry in the mold until the water is frozen; and *centrifugal casting*, a variation in which spinning generates the pressure.

4.6.2. Tape Casting

Tape casting uses a similar low-viscosity mixture of powder–water–polymer, but the forming step is via deposit of the mixture on a moving Mylar film. As the water evaporates, the particles are glued together into a thin sheet. It is used to generate homogeneous long, thin, and flat structures such as sheets. Many of the characteristics of the binder and powder are the same as encountered in slip casting. After forming, the solvent or water is evaporated, leaving residual binder to hold the particles in place.

Tape casting is widely used to form battery electrodes, brazing layers, microelectronic substrates, and small production levels of steel sheets. As shown schematically in Figure 4.8, in tape casting the powder–binder slurry is fed onto a moving paper or plastic sheet that passes through a controlled opening. A doctor blade levels the slurry to form a continuous thin sheet. Subsequently, one constituent of the binder is evaporated, leaving bonds of binder behind to provide strength during separation of the tape from the substrate before final sintering. Binders for tape casting are similar to those discussed earlier. Acrylics, waxes, polyvinyl alcohol, and polyvinyl butyryl are common components.

4.6.3. Freeze Casting

A novel shaping technique relies on water and freezing to form a shape. The water contains polyvinyl alcohol or latex. After casting into hard tooling, the powder–binder mixture is frozen in the tool set. Subsequently, the ice is extracted by vacuum evaporation (freeze-drying or sublimation) from the compact, leaving behind a polymer component to hold the particles in place for subsequent handling. Sublimation of the water is slow and requires a few hours or more. Finally, the powder is sintered to a high density. The process is known by several names, including adiabatic forming, freeze–firing, and quick set. This freeze–sublime–sinter approach is mostly used for large structures; stainless steel is a preferred material. Current production targets components more than 1 kg (2.2 lb) in mass with modest production volumes—typically a few per day. This is dictated by the time to freeze the component in the tooling, which limits the cycle time to about one hour.

4.7. INJECTION MOLDING

4.7.1. Process Overview

Injection molding is a productive and widely used technique for shaping plastics. An evolution in powder shaping has been to form steel feedstock customized for injection molding. Small steel powders are used to obtain good sintering densification. The polymer provides the lubricity for flow into the mold cavity, where heat is extracted to cause the polymer to freeze the particles into the desired shape. Subsequently, the polymer is extracted and the powder is densified by sintering at a high temperature. Thus, injection molding inherently relies on sintering densification and small powders.

Although the process is more than 70 years old, it has been applied to steels only since the 1970s. Examples of ferrous products formed by injection molding include stainless-steel orthodontic brackets for straightening teeth, high-magnetic-permeability iron for controlling computer disk drives, small stainless-steel gears for electric hand tools and toothbrushes, stainless-steel surgical tools such as scalpels, steel electrical connectors, steel handgun components, stainless-steel automotive air-bag actuator mechanisms, and iron–nickel alloy microwave filters for high-frequency computer chips.

There are several variants to the injection-molding technology, but all essentially consist of four basic operations: formation of a powder–binder mixture as feedstock, shaping the feedstock in an injection-molding machine, extraction of the binder, and sintering. The process is best applied when the components are complex in shape, with concern over production costs and properties in the context of tight tolerances and high production volumes.

The heart of the molding step is depicted in Figure 4.9. This sketch shows a cross section through a molding machine. Cold feedstock is loaded into a hopper where it is transported and compressed by the molding screw. Outside the screw is a heated barrel that melts the binder. At the end of the barrel is a nozzle that fits against the tool set. Tooling is constructed to hold the molten feedstock into the desired shape, with flow passages from the nozzle via a sprue–runner–gate system. Cooling in the tooling is assisted by water passages buried in the die. To counteract the pressures associated with mold filling, the tool cavity is clamped in the closed position until ejection.

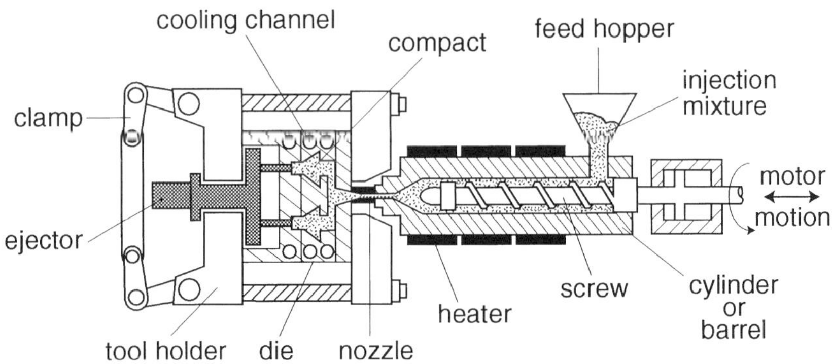

Figure 4.9. Cross section of an injection-molding machine, showing the feedstock flow path from the feed hopper to the compact.

The feedstock is mixed to form a homogeneous material consisting of typically 60 vol% powder and 40 vol% polymer phase. The best powders are the carbonyl iron and gas-atomized steels, but in some instances water atomized steels can be used if the particle size is small.

In its simplest form, molding consists of heating the feedstock to a sufficiently high temperature such that the binder melts, then forcing this melt into a cavity where it cools and assumes the shape of the tool cavity. A schematic cycle is illustrated in Figure 4.10. The cycle starts by clamping the mold closed, injecting preheated feedstock into the cavity, packing out the cavity until the gate solidifies, and preparing the next charge while ejecting the cooled component. The melt must have sufficiently low viscosity to flow freely into the mold. At the same time, the component must be produced without defects at minimum cost in the shortest possible cycle time.

4.7.2. Key Parameters

In the molding cycle, temperature and pressure are programmed to deliver the feedstock to the die without the introduction of defects. This begins with melting of the feedstock in the heated barrel of the molding machine. Usually a reciprocating screw is used to homogenize, pressurize, and pack the mixture. For limited applications, a plunger or air supply is used to pressurize premeasured quantities of feedstock, so there is no screw for mixing during heating. The actual molding stroke occurs with a forward thrust of the screw or plunger in the barrel to inject a shot of molten feedstock into the tooling. The flow from the nozzle passes through the sprue, runner, and gate prior to filling the die cavity. Figure 4.11 shows a test component with the sprue, runner, and gates still attached to the component.

Since the tooling is cooler than the feedstock, the feedstock loses heat during mold filling with a concomitant increase in viscosity. If molding is performed slowly in cool tooling, the flow resistance increases along the flow path, requiring a continual pressure increase until the cavity is filled. Many defects arise with such an approach, so it is not viable. There are two options to compensate for the increasing viscosity with cooling—mold very quickly or heat the tooling. The latter option greatly slows production. Consequently, fill times are short to minimize feedstock cooling.

Figure 4.12 plots three pressure–time curves for molding. One corresponds to the pressure in the hydraulic system, the second to the pressure in the feedstock in front of the screw, and the third to the pressure in the mold cavity. This figure shows only the initial pressurization that occurs over a split second. The total molding cycle is typically less than a minute, with the bulk of the time involved with cooling after this mold filling stage. If heated tooling is used, then cooling periods are long and production rates suffer. On the other hand, cold tooling causes flow resistance and increases the frictional pressure loss in the system from the hydraulic system to the mold cavity. Compensation for this problem is accomplished by rapid pressurization.

After the mold is filled, heat is extracted from the feedstock through the cool

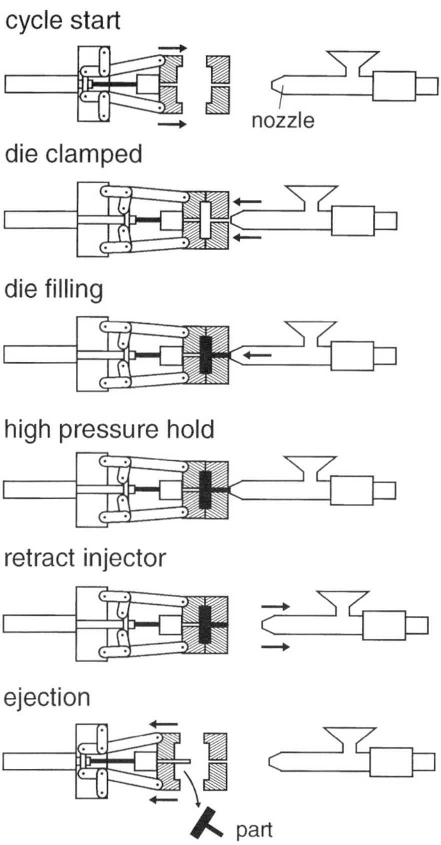

Figure 4.10. A schematic of the molding cycle for a toggle-clamped mold. The cycle starts by closing the mold. Preheated feedstock is injected into the cavity, where it is pressurized to compensate for thermal contraction. After the gate freezes, the injection unit is retracted while preparing another charge. Finally, the cooled component is ejected and the cycle repeated.

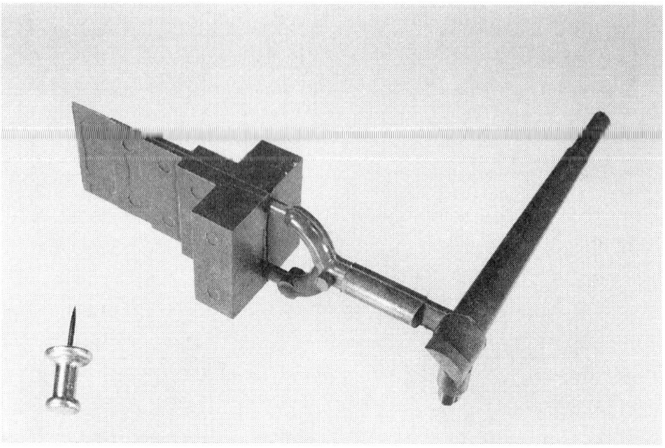

Figure 4.11. A molded steel test component with the gates, runner, and sprue attached. After molding, these pieces are cut from the component and recycled to minimize waste. (Photograph courtesy of Julian Thomas.)

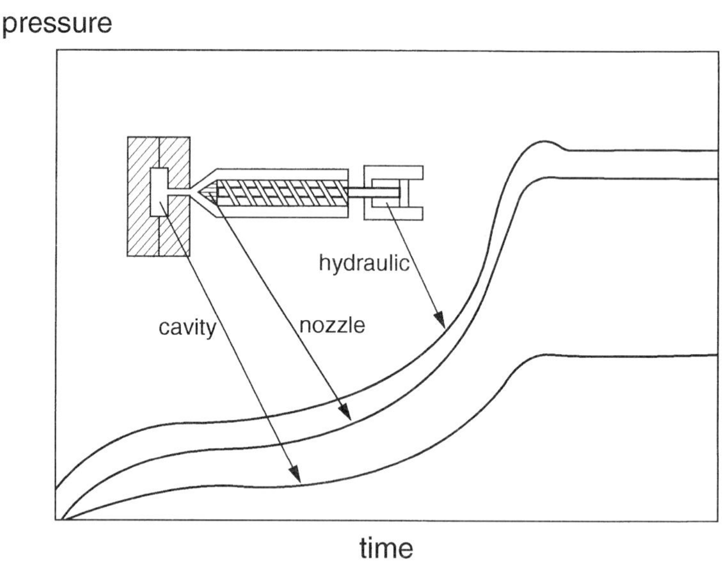

Figure 4.12. Pressure curves showing what happens in the hydraulic system, in the feedstock at the nozzle, and in the tool cavity. These pressure–time curves illustrate the loss of pressure along the flow path during mold filling.

die. Finally, when sufficiently chilled, the cavity is opened to eject the hardened component. At this point the component is rigid, which means that the temperature is below the flow temperature for the binder. Although the process may appear simple, in practice there are many potential errors. For example, a high solids content raises the viscosity, requiring high pressures during molding. This pressure alters the viscosity and contributes to flashing and separation of binder from the powder. As noted earlier, viscosity is sensitive to temperature and shear rate. Thus, a continual change occurs in the flow rate as the feedstock fills out the mold. If the molding conditions are improperly set, the feedstock will be placed in the die with density gradients that lead to dimensional warpage in sintering. Uniform particle packing in the mold is necessary, since this gives uniform dimensions after sintering. Molding cannot repair errors introduced by inadequate mixing. In turn, sintering is unable to compensate for inadequacies in mixing or molding. Hence, care is required at each step, to ensure defect-free components.

The molding parameters are highly sensitive to the feedstock characteristics and mold design. Consequently, no single set of conditions is ideal. For example, the cycle time will vary from approximately 5 to 60 s, with a typical value of 20 s. Often, the mold fill rate is near 1.5 cm^3/s (0.1 in^3/s), with peak pressures as high as 60 MPa (9000 psi) and a feedstock temperature below 200°C (400°F).

After molding, the binder is removed from the compact by debinding. There are many different options for binder extraction. Thermal debinding is the easiest to envision, where the compact is slowly heated to decompose and evaporate the binder. A popular alternative is to immerse the compact in a solvent that dissolves some binder, leaving sufficient polymer behind to hold the particles in place for subsequent handling. The remaining binder is extracted by heating during the early part of the sintering cycle. An option of increasing popularity involves catalytic phase erosion of the binder. Most of the binder is extracted by a catalytic vapor, with final binder removal being part of sintering. This approach uses polyacetal as the polymer and a nitrogen atmosphere doped with nitric acid vapor to catalyze depolymerization. Typical debinding rates are 2 mm/h.

In sintering, the void space remaining after removal of the binder is eliminated by the particles shrinking together. Isotropic powder packing in molding allows for predictable and uniform shrinkage, so tooling is oversized to give the desired final dimensions. Sintering can be incorporated directly onto a thermal, catalytic, or solvent debinding cycle. It is typically performed in a protective atmosphere or vacuum at a peak temperature that causes rapid elimination of the pores previously filled with binder. For steels and stainless steels, these temperatures are in the 1120–1350°C (roughly 2050–2460°F) range for times of 30–120 min, with linear shrinkages of 12–18%.

A typical component fabricated by metal injection molding is a trigger guard for a sporting shotgun as pictured in Figure 4.13. This guard is fabricated from a low-alloy iron–nickel steel, with a weight of 70 g. To make the alloy, a mixture of 5-μm iron and 8-μm nickel powders is used with a final carbon level of 0.5%. These are combined with wax and polyethylene to form feedstock that can be molded at 58 vol% solids. During molding, the nozzle temperature in the molding machine is

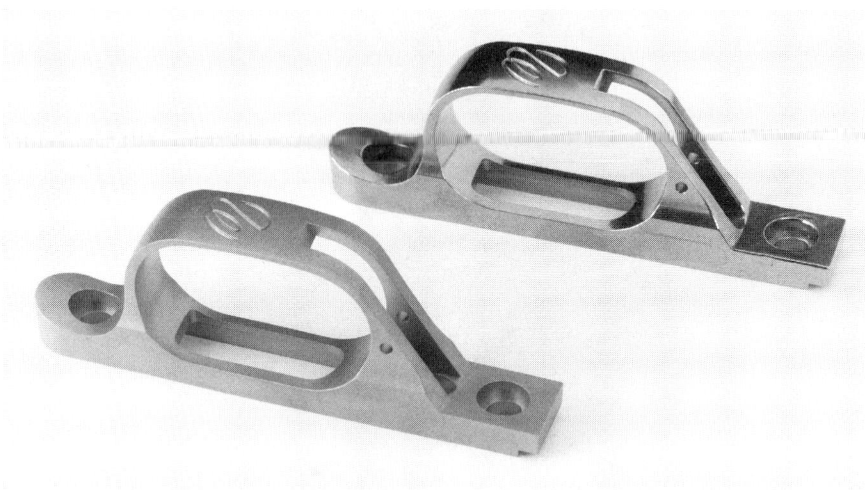

Figure 4.13. A trigger guard formed by injection molding a mixture of iron and nickel powders with a wax–polymer binder. (Photograph courtesy of Metal Powder Industries Federation.)

175°C (about 350°F) with a die temperature of 40°C (about 100°F). The maximum pressure applied during mold filling is 20 MPa (200 atm or 3000 psi), and a pressure of 8 MPa (80 atm or 1200 psi) is held on the feedstock during cooling. In fabricating this component, the mold filling time is rapid (about 0.5 s), but the mold cooling time is 18 s, with a total time of 37 s between parts. The sintered and heat-treated density was 7.55 g/cm^3, giving a tensile strength of 1655 MPa (240 ksi) and yield strength of 1480 MPa (215 ksi). Injection-molded steel components are more complex in shape than encountered in any other form of P/M.

4.7.3. Tooling

4.7.3.1. Tool Construction. The tool materials used in injection molding are similar to those encountered in many other metalworking, plastic injection molding, and powder metallurgy operations. Table 4.4 identifies some of the common tooling materials. Along with composition, this table identifies the typical applications. Table 4.5 lists the engineering properties of these alloys. The material for tool construction depends on the anticipated number of molding cycles and the required wear resistance. For molding tools, P20 is a common material, because of the combination of strength and cost. Yet where wear is a concern, higher-hardness tool steels are the material of choice. Rapid prototype tool materials, including epoxy, aluminum, zinc, or bismuth, have been used in pilot production because of easy machining. Cemented carbides are useful where wear is a primary concern, but machining is expensive and these tools are easily damaged. Tool material cost varies by a factor of 10 between these alloys.

TABLE 4.4. Construction Materials for Injection-Molding Tools

Material	Composition	Suggested Applications
410 stainless steel	Fe–14Cr–1Si–1Mn–0.3C	Corrosion-resistant cavities, cores, inserts
440C stainless steel	Fe–18Cr–1Si–1Mn–1C	Wear-resistant, small inserts, cavities, cores
H13 tool steel	Fe–5Cr–1.5Mo–1Si–1V–0.4Mn–0.4C	Larger or intricate cavities, high toughness, low wear
M2 tool steel	Fe–6W–5Mo–4Cr–2V–0.3Mn–0.8C	Core and ejector pins
P20 steel	Fe–1.7Cr–0.8Mn–0.5Mo–0.4V–0.35C	General purpose, hot runner, large cavities
Cemented carbide	WC–10Co	High wear, compressively loaded small inserts

Tool fabrication occurs in a machining center via progressive removal of material from an oversized block. Most machining is computerized, but there is still the necessity to hand finish critical components or dimensions in the tool set. A final surface roughness of 0.2 μm [8 microinches (μin]) is typical, but smoother finishes are used in selected applications. The desired tool hardness is typically more than 30 HRC, which is satisfied by many heat-treated stainless steels or tool steels. Under normal conditions, an injection-molding tool set can mold up to one million parts. With soft tool materials like aluminum, the life is less, at 1000–10000 cycles.

4.7.3.2. Tool Features. The tool set has the desired component cavity, and further consists of the pathway for filling the cavity with ejectors for extracting the component from the cavity. In most instances, the tool set consists of a single cavity. This cavity is oversized to allow for component shrinkage during sintering. Around the cavity are several tool parts needed for opening and closing the cavity, ejecting the component, aligning the die sections, moving inserts, cooling the component,

TABLE 4.5. Properties of Some Common Tooling Materials

Material	Density (g/cm^3)	Hardness (HRC)	Elastic Modulus (GPa)	Tensile Strength (MPa)	Elongation (%)	Impact Energy (J/cm^2)	Thermal Conductivity [W/(m·°C)]	Thermal Expansion (ppm/°C)
420 stainless steel	7.8	50	200	1760	10	16	25	11
440C stainless steel	7.8	57	200	1965	2	7	24	10
H13 tool steel	8.0	45	210	1440	15	18	26	13
M2 tool steel	8.2	61	210	1600	10	10	21	10
P20 steel	7.8	30	200	965	20	32	29	13
Cemented carbide	14.6	75	630	2800	0	2	112	4

and locating the sprue, runner, and gate. Figure 4.14 sketches a molding tool set in profile to help locate the ejector pins, ejector plate, and features such as slides that ensure proper closure of the die components.

A primary concern in designing a tool set is component shrinkage. The shrinkage factor is the change in a dimension divided by the original dimension. Since the target is the final component size, each dimension of the tool cavity is oversized to accommodate this shrinkage.

In the typical tool set, the feedstock flow path is from the molding machine nozzle into the sprue, along the runner, through the gate, and into the mold cavity, as illustrated in Figure 4.15. A vent is located at the end of the flow path to allow air to be pushed out of the cavity by the feedstock. This flow path is surrounded by various clamping plates, alignment and locating pins, and ejector components. Figure 4.16 is an expanded view of a tool set.

Alignment of the components and their proper sequencing and smooth motions are important to successful molding. Most of the tool components are available as premachined packages, so only the cavity needs to be custom-machined. Also, within this geometry are the cooling or heating networks designed to control mold tempera-

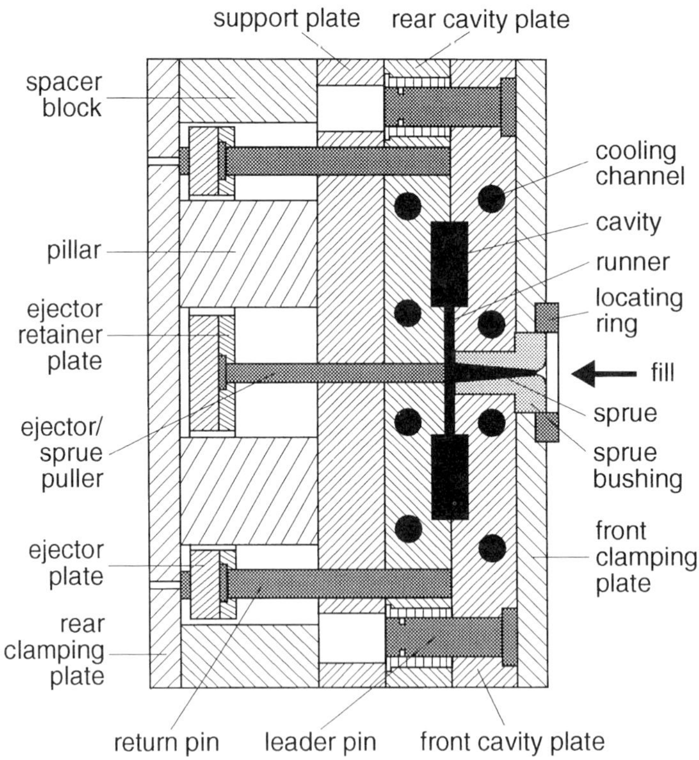

Figure 4.14. A cross section of a tool set showing the relative location of the die cavity, clamping plates, and ejector mechanism.

EQUATION BOX 4.4.

A primary concern in designing injection-molding tools is component shrinkage. On a volume basis, the typical feedstock contains 60% solid and 40% binder. To attain the desired final component size, the linear shrinkage during sintering may be 15%. The shrinkage in dimensions, known as the *shrinkage factor Y*, is calculated from the solids loading Φ and sintered fractional density ρ/ρ_T

$$Y = 1 - \left(\frac{\Phi}{\rho/\rho_T}\right)^{1/3}$$

where ρ is the final density and ρ_T is the theoretical density for the steel. This assumes isotropic shrinkage in sintering. For example, if a 13.8 mm dimension shrinks to 12 mm, then the shrinkage factor is 0.13 or 13%, calculated as the change in a dimension divided by the original dimension.

Since the target is the final component size, each dimension of the tool cavity is oversized to accommodate shrinkage. If the desired final dimension is L_f at a fractional sintered density ρ/ρ_T from a feedstock with a fractional solids content of Φ, then the initial dimension of the tooling is given in terms of the tool cavity expansion factor Z

$$Z = \frac{1}{1 - Y}$$

where the shrinkage is obtained from above. As an example, if the shrinkage factor Y is 0.15 (15% shrinkage), then the tool expansion factor Z is 1.1764. Thus, to obtain a 12 mm final dimension, a tool dimension of 14.11 mm (12 times 1.1764) is required. Note that the shrinkage from 14.11 mm to 12 mm is 2.11 mm, so the measured shrinkage factor is 2.1 mm divided by 14.11 mm, or 0.15 (15%).

ture. One key advantage of injection molding is the ability to fabricate complex shapes that cannot be produced by alternative techniques. Accordingly, complex tool designs are a necessary aspect of injection molding. An example of a slide design to insert a hole is given in Figure 4.17. When the mating halves of the tool cavity close, the slide is activated to insert the post. After molding around the post, the slide moves as the cavity opens, leaving the hole parallel to the parting line.

The number of cavities in the tool set depends on the number of components to be fabricated, the shot capacity of the molding machine, tool fabrication costs, and the available clamping force. Tool sets with up to 40 cavities have been used for high-volume production. Most injection-molded steel components are generated in tool sets with 1–16 cavities. A single cavity is satisfactory for low production quan-

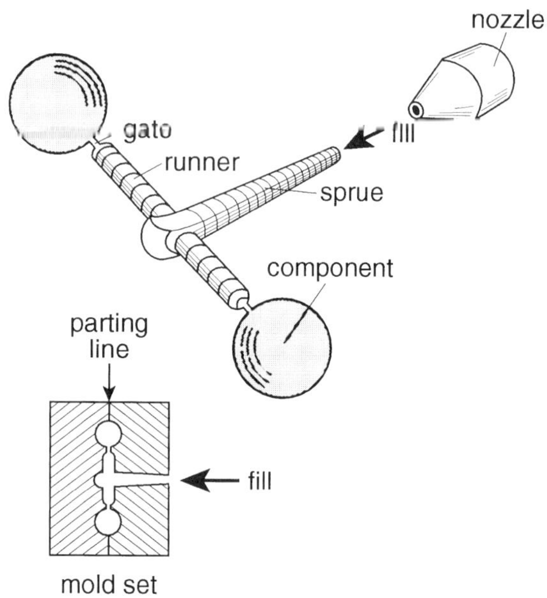

Figure 4.15. Filling from the molding machine nozzle involves flow along the sprue, then into the runner system to fill the cavity through the gate.

tities, below about 200,000 parts. As higher productivity is required, more cavities may be justified, since production increases are gained without the purchase of new molding machines. As the number of cavities increases, the cost of manufacturing the tool set increases, but the net production cost per component decreases. Since tool cost is distributed over the production quantity, there is a minimum total cost that depends on the total production quantity.

4.7.3.3. Ejection. After molding, the component is cooled in the die cavity. Once the part has strength and pressure has relaxed, it is ejected from the cavity. Cooling causes the binder to contract, and this results in a progressively lower pressure, eventually allowing ejection. To accomplish ejection, pins in the die body move forward and push the component from the cavity. If inserts, internal cores, or threads are put into the shape, these must be retracted to allow free ejection. Sometimes core pins, inserts, or other features are placed in the cavity before filling and these items become encapsulated in the part, ejecting out on each cycle.

Ejector pins blemish the component, since they concentrate ejection force on a soft material. Figure 4.18 shows a photograph of blemishes associated with injection molding, including an ejection pin mark **e**, parting line **p**, and gate blemish **g**, typical defects associated with injection molded steels.

4.7.4. Molding Machines

4.7.4.1. Types. The three common molding machines are reciprocating screw, hydraulic plunger, and pneumatic. Table 4.6 summarizes the attributes of a few such

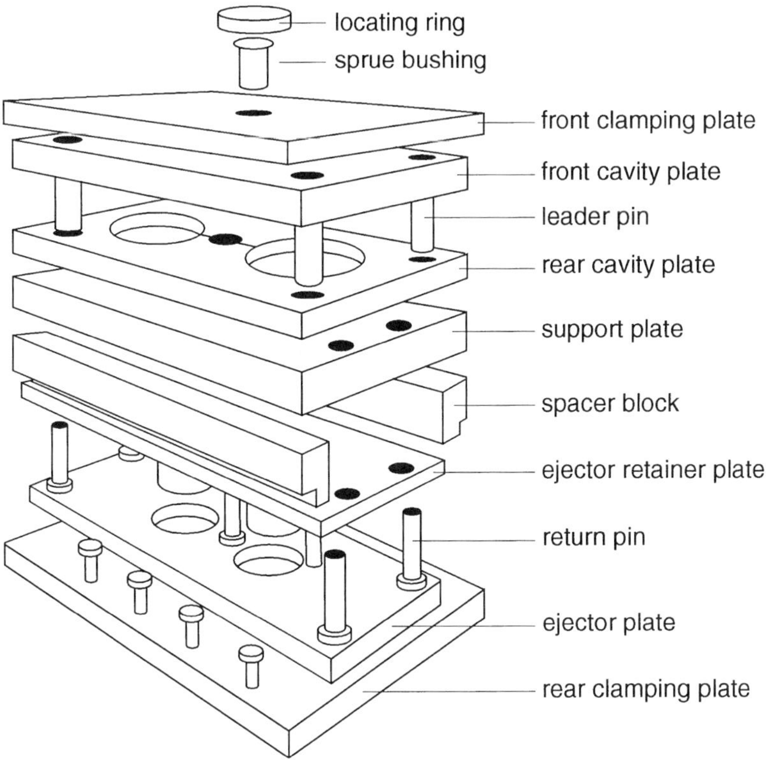

Figure 4.16. An expanded view of a typical injection-molding tool set.

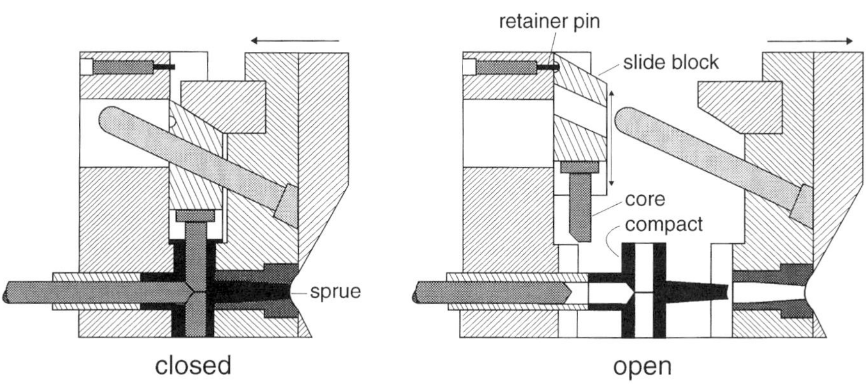

Figure 4.17. Schematic of how tooling can be constructed to allow a sliding core to form features parallel to the fill direction. An angled post is used to actuate the core during opening and closing of the mold.

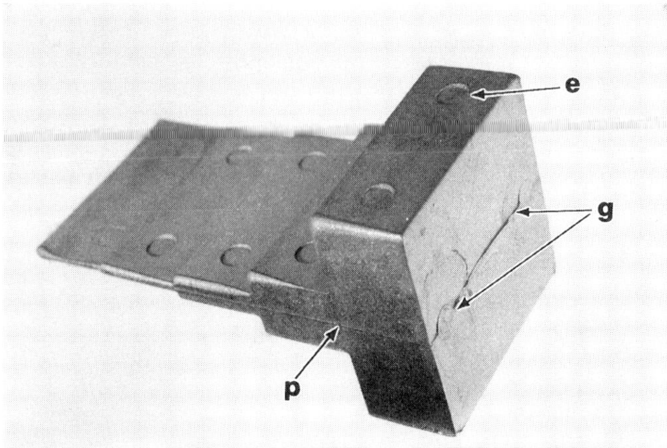

Figure 4.18. Pictures of typical blemishes on an injection-molded steel, showing the indentation from an ejector pin, gate trace, and parting line. (Photograph courtesy of Julian Thomas.)

TABLE 4.6. Example Attributes of Steel Injection-Molding Machines

Attribute	Low Pressure	Moderate Tonnage	Intermediate Tonnage	High Tonnage
Type	Pneumatic	Reciprocating-screw	Reciprocating-screw	Reciprocating-screw
Claming force, ton	0.7	25	75	200
Clamp type	Pneumatic	Hydraulic	Toggle	Hydraulic
Platen size, mm	480	260	530	900
Drive motor, kW	2	5.5	15	30
Screw diameter, mm	None	18	28	50
Injection volume, cm^3	3000	35	70	350
Fill rate, cm^3/s	0.5	46	66	100
Injection stroke, cm	None	9	11	14
Plastication capacity, kg/h	None	12	20	40
Screw speed, mm/s	Not applicable	110	300	110
Molding pressure, MPa	0.5	190	230	200
Shortest cycle time, s	4.8	2	1.2	3.6
Maximum temperature, °C	150	200	200	200
Control	Open-loop	Adaptive	Closed-loop	Adaptive

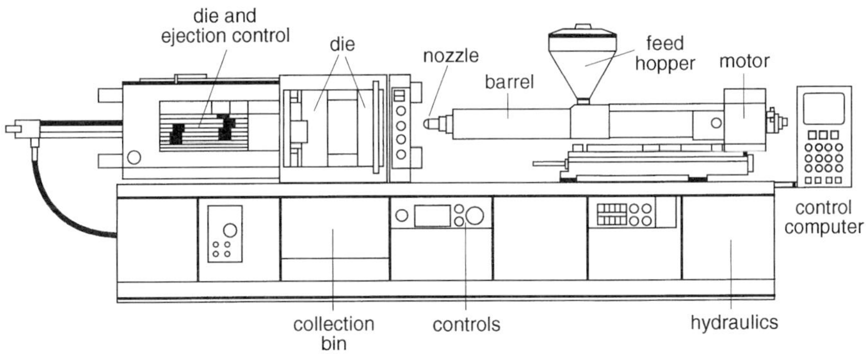

Figure 4.19. Typical horizontal injection-molding machine that functions with a reciprocating screw for forcing the heated feedstock into the die cavity. Main portions of the molding machine are shown.

molding machines. Pneumatic machines apply gas pressure to move heated feedstock into the mold. They are inexpensive and effective for small components where internal flaws are not objectionable. However, voids form because the feedstock is under low pressure, which fails to compensate for feedstock shrinkage on cooling. In a hydraulic molding machine, a plunger rams heated feedstock into the mold. Excess pressure is generated to compensate for the volume contraction normally encountered by feedstock cooling. This pressurization is important to forming defect-free compacts, but the control systems usually are not suitable for forming complicated shapes.

High-volume injection molding uses a horizontal reciprocating screw located inside a heated barrel. The screw is designed to compress and transport the feedstock to the die, and it becomes a plunger during mold filling. Figure 4.19 is a typical layout of a horizontal machine. The tooling is clamped in the center of the machine. The granulated or pelletized powder–binder feedstock is placed in the hopper for metering into the injection barrel. This is the beginning of the molding operation.

4.7.4.2. Motions. The heart of molding lies with the motions of the reciprocating screw. Figure 4.20 is a picture of a wear-resistant screw used for injection molding. It has a helical pitch, the design of which is adjusted for the viscosity of the feedstock, but generally it consists of gradual section changes along its length. Screw rotation is controlled via a hydraulic motor, and heat is supplied by external heaters on the barrel.

An important role of the screw rotation is to deair the feedstock and prepare the next charge for injection. This is termed *metering*, where the screw rotates to pressurize feedstock to the nozzle. During metering the screw becomes a mixer to ensure uniform powder–binder distribution and uniform heating. The screw has a check ring behind the tip that acts as a valve to allow feedstock flow into the front of the barrel, as shown in Figure 4.21. This valve closes during mold filling and forces

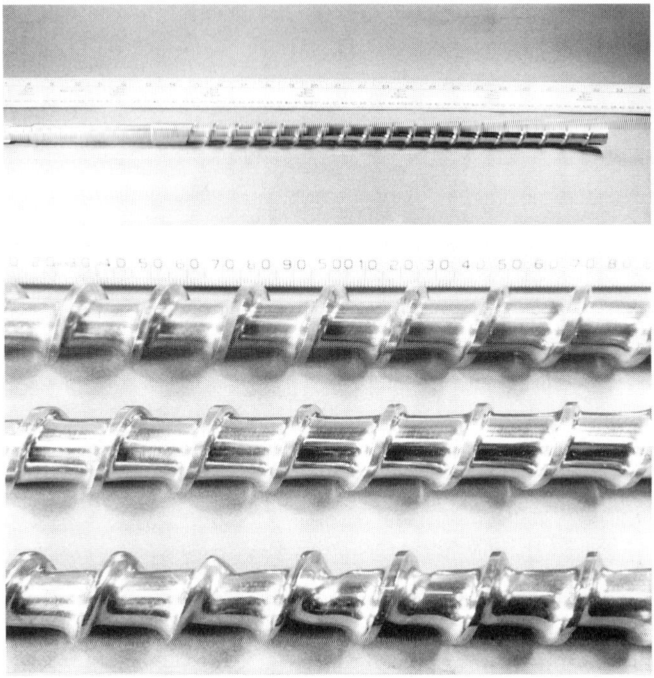

Figure 4.20. This photograph shows a typical screw used in injection molding; note the deep wear grooves on the lower screw, characteristic of wear using a low-hardness screw. (Photograph courtesy of Julian Thomas.)

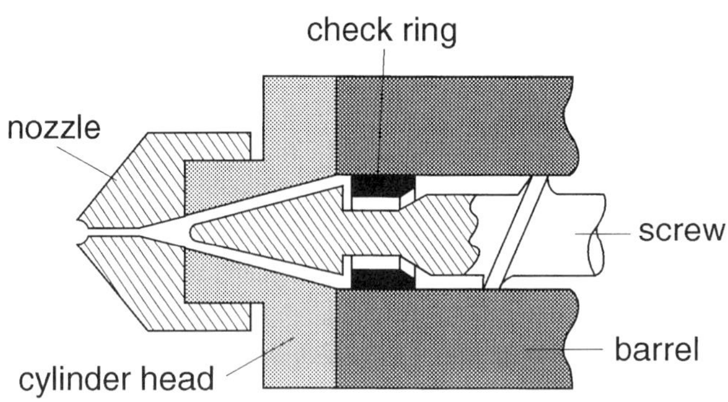

Figure 4.21. Cross-sectional view of the nozzle region, indicating the placement of the check ring on the tip of the screw. The check ring seals to prevent leakage of feedstock back down the barrel during injection.

EQUATION BOX 4.5

Feedstock flow in molding depends on the applied pressure and viscosity. For a cylindrical runner, the volumetric feedstock flow rate Q varies with the runner diameter d to the fourth power according to Poiseuille's equation

$$Q = \frac{P\pi d^4}{128\eta L}$$

where P is the applied pressure on the feedstock, L is the runner length, and η is the feedstock viscosity. The rate of mold filling is very sensitive to the injection pressure and runner diameter. Usually, a high feedstock flow rate is needed to fill the die before the feedstock cools.

Most desirable is cavity filling without turbulence. This requires a low Reynolds number, probably less than 100 for most cases. The Reynolds number Re is calculated

$$\text{Re} = \frac{Vd\rho_M}{\eta}$$

where ρ_M is the feedstock density and V is the velocity at the gate. If the gate is cylindrical, the feedstock flow rate Q and gate diameter d can be used to calculate the velocity, since

$$Q = V\frac{\pi d^2}{4}$$

Such concepts help in design tooling and molding cycles that avoid defects.

molten feedstock to flow into the die cavity through the nozzle at the end of the barrel. Effectively, the screw becomes a plunger during mold filling.

Control of the screw rotation, displacement, and pressure is important to the fabrication of precise components by injection molding. A closed-loop control system with a quick-response servohydraulic valve is required for screw position and pressure control. During the fill stroke, the volume of feedstock injected into the mold depends on the cross-sectional area of the screw and the stroke length. A typical screw diameter is 22 mm, but might range from 15 to 40 mm (0.6 to 1.6 in) depending on the machine capacity.

The barrel holds the rotating screw and is surrounded by heaters that control the mixture temperature. It has multiple heater zones to ensure temperature control during mold filling. Because cold feedstock is abrasive, the first heater zone is geared to rapidly heat the feedstock, and subsequent zones might be at lower temperatures.

The materials used in constructing the screw and barrel are critical to long service without contamination. Hard materials and close tolerances are required to reduce abrasive wear. The tool steels containing vanadium carbide or borides prove most durable. Similarly, other machine components in the flow path can exhibit considerable wear, resulting in a loss of machine control.

All the molding steps are controlled by a computer that can correct errors during molding. Besides the molding machine, coordination is required with the peripheral operations needed in automation schemes. For example, robots are used to stage the compacts for debinding. Other automation features include conveyor systems, parts and tooling storage with automated retrieval, and continuous feedstock preparation and component debinding steps.

4.7.5. Molding Cycles

A typical sequence for the injection pressure and screw position are plotted in Figure 4.22. Prior to filling, the screw rotates in the barrel, while external heaters bring the feedstock to a temperature where the binder is molten. The dwell time for the feedstock in the barrel must be sufficient to ensure thorough and uniform heating. Then, during molding, the sequence starts with the screw plunging forward in a split second. This is traced by the screw displacement curve in Figure 4.22. Three pressure curves are included to show pressure loss from the source (hydraulic pressure) into

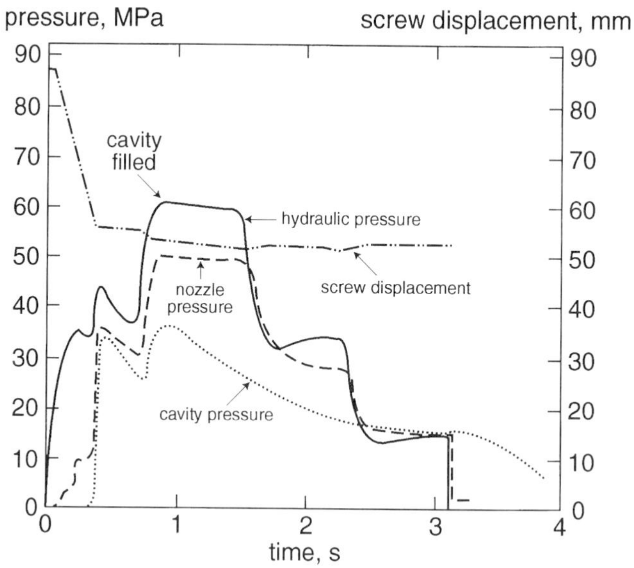

Figure 4.22. An overview of the coordinated actions involved in injection molding, showing the screw position, hydraulic pressure, nozzle pressure, and cavity pressures during a single cycle. (Data from Karl Hens.)

TABLE 4.7. Typical Molding Parameters

Parameter	Typical Range	Trigger Guard
Barrel temperature, °C	100–200	160
Nozzle temperature, °C	80–200	175
Mold temperature, °C	20–100	40
Screw rotation speed, rpm	35–70	35
Peak injection pressure, MPa	0.1–130	20
Packing pressure, MPa	0–10	8
Fill time, s	0.2–3	0.5
Packing time, s	2–60	3
Cooling time, s	18–45	18
Cycle time, s	8–360	37

the cavity. A rapid rise in hydraulic pressure induces feedstock flow into the mold. Once the cavity is filled, the gate freezes and there is little effect of hydraulic pressure on the cavity pressure. During filling, the nozzle pressure is intermediate between these two.

To compensate for the flow resistance and pressure gradients in the molding system, the screw position and hydraulic system pressure vary during molding in a coordinated manner. After mold filling, pressure is held during the packing stage. The objective is to compensate for the thermal contraction of the feedstock during cooling. Finally, when the binder has sufficiently cooled to hold the component shape and withstand the ejection forces, the die is opened and the component is ejected.

The time for molding relates to the cavity size, filling time, and cooling time. It can be as short as 5 s or as long as 1 min. Initial feedstock flow requires the molding temperature be higher than the softening point of the binder. This is usually between 50 and 200°C (122 and 392°F). Too low a temperature results in short shots (incomplete mold filling), while an excessively high temperature degrades the binder or causes flashing or powder–binder separation, and requires prolonged cooling. Molding pressure affects the mold filling rate and is usually limited by the machine design. Short shots occur if either the molding temperature or pressure is too low. At the higher pressures, the die is forced open along the parting line, giving flash. Voids are captured in a part if the feedstock shoots across the cavity without pushing out all the air. This is termed *jetting* and is very undesirable. Jetting occurs with excessively high fill rates, so most molding machines are programmed for a fill rate that avoids jetting, while keeping the fill time short.

Table 4.7 gives the range of molding conditions and one example of a specific component and its molding conditions. That component is a steel trigger guard for a rifle as shown earlier, in Figure 4.13. It is formed from carbonyl iron powder and a binder based on paraffin wax, polypropylene, and stearic acid. Once molded, the binder phase is extracted and the compact is sintered to near full density. These steps are covered in subsequent chapters on sintering.

4.7.6. Debinding

There is a wide array of options for binder extraction. Thermal debinding was the first technique applied to steels, but is not capable of high precision. This is because the binder softens and allows the particles to move from the desired shape during heating. To balance distortion against burnout, one must slowly heat the component to progressively decompose the binder. Other options include wicking, vacuum evaporation, solvent extraction, and catalytic depolymerization. Table 4.8 lists these variants and gives a sense of the debinding conditions and times. Besides the speed of binder extraction, there are important differences in component precision with each technique. For the options listed in Table 4.8, the measured rates are compared on an equivalent basis of 5 μm carbonyl iron powder and 10 mm section thickness.

A popular debinding approach for good dimensional control is to immerse the component in a solvent that dissolves some binder, leaving some polymer behind to hold the particles in place for subsequent handling. This requires a two-phase binder, with one phase insoluble in the solvent. The remaining binder is thermally extracted as part of the sintering cycle. Newer binders are water-soluble, so the debinding solvent is water. If a gelation binder can be used, debinding then becomes a simple task of drying the water out of the component. This concept has been demonstrated in the laboratory and might move to production in the near future.

TABLE 4.8. Comparison of Debinding Techniques and Times[a]

Binder System	Debinding Technique	Conditions	Time
Wax–polypropylene	Oxidation	Slow heat to 150°C, hold, heat to 600°C in air	60
Wax–polyethylene	Wicking	Slow heat to 250°C, hold, heat to 750°C in hydrogen	4
Wax–polymer	Supercritical	Heat in Freon vapor at 10°C/min to 600°C under 10 MPa pressure	6
Wax–polyethylene	Vacuum extraction	Slow heat while passing low-pressure gas over compacts, heat to sintering temperature	18
Water–gel	Vacuum sublime or freeze-drying	Hold in vacuum to extract water vapor from ice	8
Oil–polymer	Solvent immersion	Hold in ethylene dichloride at 50°C	6
Water–gel	Air-drying	Hold at 60°C	10
Polyacetal–polyethylene	Catalytic debinding	Heat in nitrogen doped with nitric acid vapor 150°C	6

[a] Section thickness = 10 mm, particle size = 5 μm, solids loading = 60 vol%.

EQUATION BOX 4.6

From various modeling efforts, the net debinding time depends on the component thickness

$$t = \frac{A\, H^2}{(1 - \Phi)^3\, D}$$

where t is the debinding time and H is the compact thickness. The particle size is given by D, while the solids loading is given by Φ. Such models predict that the debinding time varies with the square of the section thickness. If the component thickness doubles, then debinding time increases by a factor of 4. With most debinding there is a benefit from higher temperatures, because the polymer moves faster, but the component can distort. Also, note that a larger particle size reduces the debinding time.

Major growth is occurring in the use of catalytic phase erosion for debinding. As illustrated in Figure 4.23, a batch reactor is used to depolymerize the binder. A favorite binder is based on polyacetal, which is depolymerized in a nitrogen atmosphere doped with nitric acid. Most of the binder is attacked by a catalytic vapor, with the residual binder removed during heating to the sintering temperature.

Debinding is highly variable and depends on the binder system, techniques, temperature, and section thickness. Solvent and catalytic debinding ensure the best dimensional control, since the binder is kept rigid during extraction, and for these techniques penetration rates of 1–2 mm/h are common. An important development has been the design of continuous debinding and sintering configurations. Hence, high-productivity injection-molding operations now include a debinding chamber on the front of a continuous sintering furnace. This gives high efficiency and excellent dimensional control.

4.7.7. Computer-Aided Design

Most features associated with injection mold tool design and machine operation are supported by computer packages. These software packages have embedded design rules for the standard features. The use of computers is an important aspect of current efforts to optimize production. Such tools are most beneficial at the quotation stage, where different materials, construction techniques, and even component designs can be examined before commitment to manufacturing. For example, cost models include the feedstock price, tool size, and machining costs. Included in this calculation is consideration of the tool complexity, tolerances, surface finish, number of cavities, number of ejector pins and vents, and complexity of the parting plane.

Computer-aided design only slightly lowers production and design costs, but

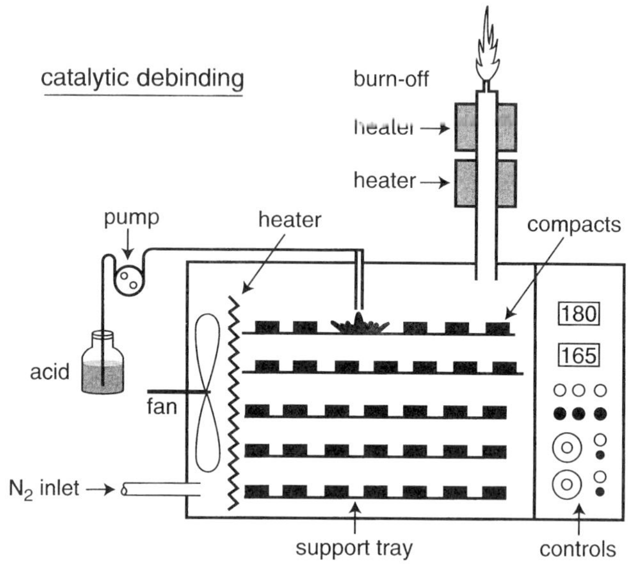

Figure 4.23. A schematic layout of a catalytic phase debinding oven. The components are placed on trays and heated in a nitrogen atmosphere, and the polyacetal is depolymerized by the addition of nitric acid vapor to the nitrogen.

greatly decreases mistakes and the time required to design tooling. Many computer models exist, but each needs to be tuned to the special features of injection molding. Figure 4.24 provides an example of a prototype stainless-steel injection-molding simulation where the fill front is shown at various stages of molding, here for a surgical tool. Vents would be located at the last portions of this cavity filled by the feedstock. The weld line forms away from the gate where the divided flow fronts converge. For these simulations to operate successfully, a database with the feedstock

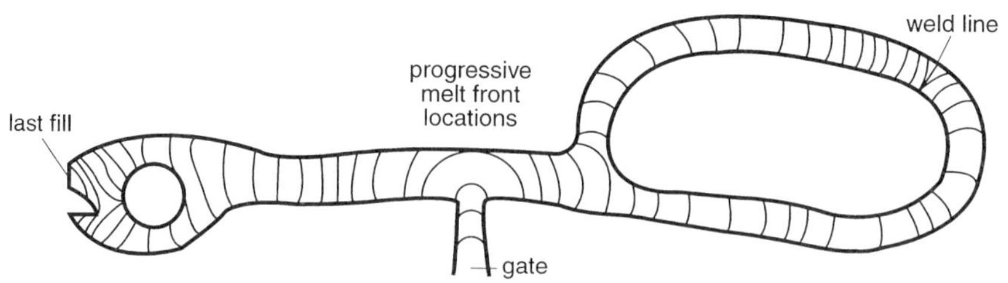

Figure 4.24. Computer-simulated flow lines associated with the filling out of a surgical hand tool, showing the position where a weld line defect might occur and the desired location of a vent at the end of the flow path for a single-gated tool. (Data from Rajiv Tandon.)

properties is needed. Unfortunately, many of these parameters are known only approximately.

4.8. RAPID PROTOTYPES AND TOOLS

Over the past 10 years a wide variety of powder–binder processes have emerged for quick production of prototype parts or tool components. Rapid prototyping is well developed for polymer components, but is poorly developed for steel powders. Much new effort is being directed to correction of that deficiency. Current concepts include direct laser sintering, slip casting into plastic cavities, and computer-controlled forming. The latter concepts are leading the way, largely because they avoid distortion, are rapid, and are very low in cost.

One of the leading concepts is to use a computer-controlled laser to cure a polymer–powder mixture. Areas to be bonded are scribed by the laser. Layer after layer builds with laser scribing, leading to the production of a component part without tooling. It is the same type of powder–polymer structure encountered in other shaping techniques. The leading process is termed *selective laser sintering*. As illustrated in Figure 4.25, a bed of polymer-coated powder is laid down as a thin layer. A laser beam scribes this layer according to instructions from a coordinated computer-generated pattern. After each scan a fresh layer of powder and polymer is introduced. This powder is then scanned and scribed to bond with the underlying layer. Progressive layers are coordinated to create a three-dimensional object consisting of thousands of bonded powder layers. By proper coordination to a computer image, the final sintered object has the shape complexity of the image. Of course, the shape needs to be oversized to allow for sintering shrinkage. An alternative is to debind the polymer and infiltrate the pores with a liquid metal such as bronze or copper.

Figure 4.26 shows a closeup of a tool set and a side view of a laser-scanned powder structure showing the striations due to each laser scribing process. This technique, termed *stereolithography*, has most appeal for rapid prototyping and rapid tool generation. Early demonstrations relied on polymer powders, but more recent versions have succeeded in creating steel tools, one layer at a time. When coupled with laser image scanning, the technology will provide the three-dimensional analog of a photocopy machine. Insert the shape you need copied, and a replica shape will be created via a combination of laser image scanning and selective laser sintering.

Several competing technologies are working on the same problem. These use heated nozzles that extrude a heated polymer–powder feedstock. The nozzle path is controlled by a computer to generate a solid object like the stored image. The feedstock is exactly the same as that used in injection molding. An alternative is to cut layers from a tape-cast powder sheet and to stack these layers into a three-dimensional object. The cutting can be by a computer-controlled robot or laser beam. Like stereolithography, the stacked layers lead to a step feature in the final object. Another route uses ultraviolet lasers to cure a photopolymer resin. The laser is guided by a computer image of the final part. Steel powder is slip-cast around this plastic part. Figure 4.27 is a picture of one success by such a process. This tool set for

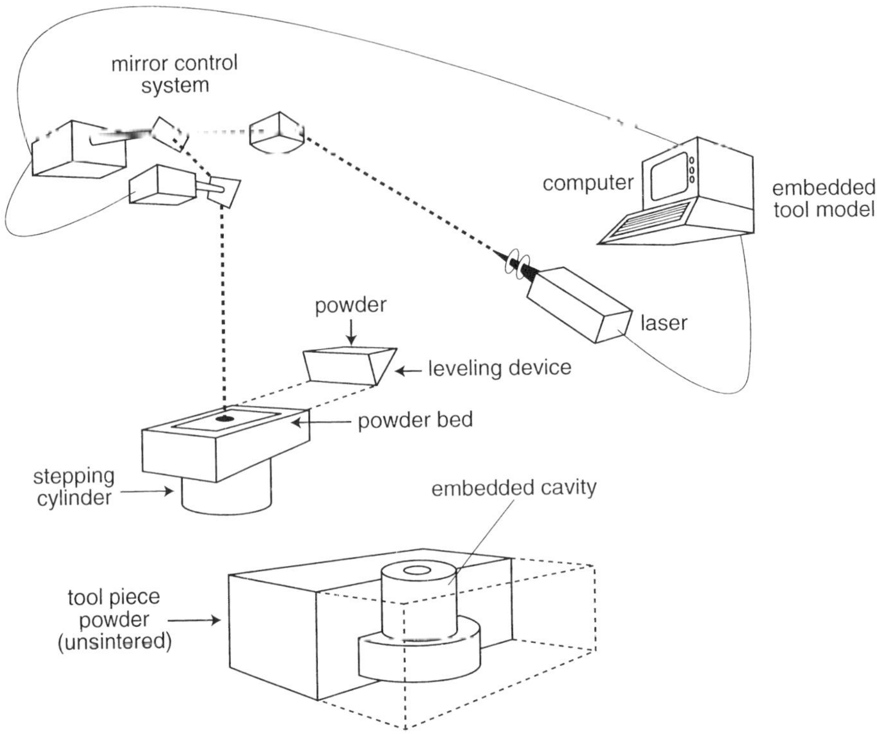

Figure 4.25. Rapid prototype and tool construction relies on laser scribing of a polymer-coated powder. The computer-controlled sweep of the laser beam causes localized bonding of the particles in each layer and between layers. Emerging from this is a three-dimensional structure that is the basis for rapid production of prototype structures.

injection molding a camera housing was fabricated in 16 h. Although the field is just emerging into practice, it is clear that such successes will have dramatic impact on times needed to construct tools.

4.9. SUMMARY COMMENTS

Powder shaping techniques have many variants, reflecting different combinations of powders, binders, forming, binder burnout routes, and sintering. In most cases, four steps are required: formation of a powder–binder feedstock, shaping of that feedstock using customized tooling, extraction of the binder, and sintering densification. Feedstock formulation involves small powders and selected binders that are mixed in a ratio dictated by the desired rheology. Shaping may be by any of several techniques, but powder lubrication by the polymer is key to fabrication of the desired

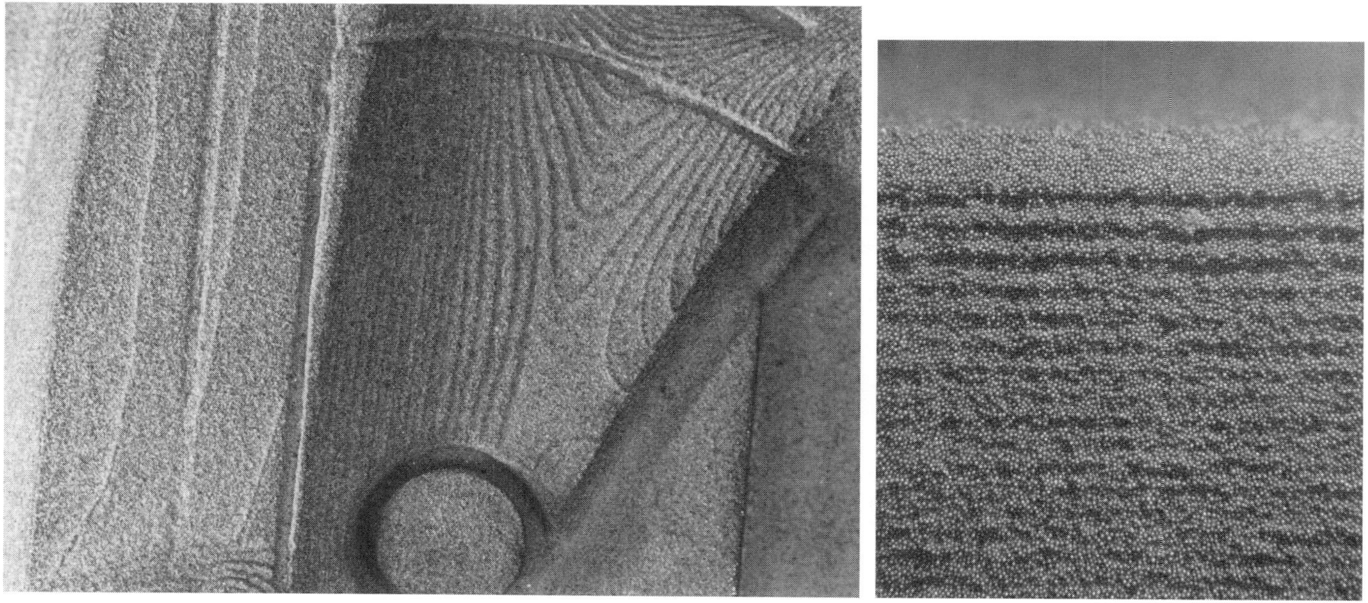

Figure 4.26. A tool component formed from steel powder using stereolithography and a closeup of the powder surface perpendicular to the laser beam. This view shows the layered particle structure from sequential bonding by the surface laser motion.

Figure 4.27. A tool insert formed in 16 h using slip casting techniques as a demonstration of rapid tool construction based on powder shaping concepts. (Component courtesy of Simeon Bojilov.)

shape. Low-pressure molding concepts, such as slip casting, require an excess of binder. On the other hand, high-pressure shaping is possible with high-viscosity mixtures that contain more powder. If necessary, the polymer viscosity can be controlled by either heat or solvent. A heated polymer allows viscous flow of the mixture to aid forming. When cooled, the binder holds the particles in place. Alternatively, an excess of binder or solvent (water for slip casting) allows a lower pressure, but is slower because of the time needed after shaping to remove the excess liquid. After shaping, the binder is removed (debinding) and the remaining powder structure sintered; these last two steps can be combined into a single thermal cycle. Since significant shrinkage is associated with sintering densification, the final dimensions rely on uniform shaping to hold tight final tolerances.

Injection molding is now the most widely employed shaping technique for steel powders. The molding process competes with many other shaping technologies. It overcomes the property limitations inherent in plastics, the shape limitations of traditional powder compaction, the costs of machining, the productivity limits of isostatic pressing and slip casting, and the defect and tolerance limitations of casting.

SUGGESTIONS FOR FURTHER READING

Angermann, H. H., and O. Van Der Biest, ''Binder Removal in Powder Injection Molding,'' *Rev. Particulate Mater.* **3,** 35–69 (1995).

German, R. M., *Powder Injection Molding*, Metal Powder Industries Federation, Princeton, NJ, 1990.

German, R. M., and A. Bose, *Injection Molding of Metals and Ceramics*, Metal Powder Industries Federation, Princeton, NJ, 1997.

Iacocca, R. G., "A Critical Assessment of the Characterization Tests Needed to Support Powder Injection Molding Component Fabrication," *Rev. Particulate Mater.* **2,** 269–313 (1994).

Harnby, N., M. F. Edwards, and A. W. Nienow (eds.), *Mixing in the Process Industries*, Butterworths, London, UK, 1985.

Hwang, K. S., "Fundamentals of Debinding Processes in Powder Injection Molding," *Rev. Particulate Mater.* **4,** 71–103 (1996).

Paul, B., and S. Baskaran, "A Review of Particulate Materials Processing in Additive Free-form Fabrication," *Rev. Particulate Mater.* **4,** 105–120 (1996).

Steinlage, G. A., R. K. Roeder, K. P. Trumble, and K. J. Bowman, "Centrifugal Slip Casting of Components," *Ceramic Bull.* 92–94 (May 1996).

Wang, F. F. Y., (ed.), *Ceramic Fabrication Processes*, Academic Press, New York, NY, 1976.

CHAPTER 5

Compaction

5.1. OVERVIEW

Iron powder is compressed to high densities by the application of pressure. However, in the as-pressed state the compact has only modest strength and requires sintering to attain strength. Unlike the shaping techniques covered in the last chapter, compaction deforms the particles. Associated with this particle deformation is a decrease in porosity and increase in particle bond strength. Many compaction options exist, including hard and soft tooling, heating the die and powder, using rolling mills, and even shock waves. However, the mainstay of powder metallurgy uses reasonably high pressures and hard tooling to deform the particles at room temperature into a desired shape under uniaxial stress. This means that the pressure is applied along one axis, from the top and bottom.

Table 5.1 provides a categorization and comparison of compaction options in powder metallurgy. This table includes the main processing variants that use pressure to densify powder in compaction. Listed are the nominal operating parameters, including pressing directions, shape complexity, and relative use. Differences in the compaction pressures and temperatures help categorize the options. In full-density techniques, both pressure and temperature are applied simultaneously, as treated in Chapter 8.

Tooling for die compaction is rigid, but in cold isostatic compaction rubber bags are used to hold the powder. The pressurization rate is usually moderate, but in explosive compaction the deformation rates are very high. Most processes are *batch*, meaning that components are fabricated one at a time, versus the production of sheet, rod, or tube by continuous processes such as rolling or extrusion. Uniaxial compaction is most prevalent. But part complexity is limited by the need to remove the compact from the die. Most of the shape complexity is in the profile seen from the pressing direction, while the complexity perpendicular to pressing is less. Polymers are used as lubricants in compaction, while high polymer concentrations are required for shaping processes such as extrusion and tape casting. In die compaction, the polymer concentration is 0.5 wt% in most cases. The component precision largely reflects the stiffness of the forming tools. More precise components are fabricated in hard tooling. Finally, as a guide to the relative use of each technology, the final entry in Table 5.1 notes the dominant use of die compaction.

Attention in this chapter is on die compaction, with comments on other ways of

TABLE 5.1. Compaction Options

Process	Cold Isostatic Compaction	Cold Forging	Die Compaction	Explosive Compaction	Roll Compaction	Warm Compaction
Pressure	Moderate, 400 MPa	Very high, >800 MPa	High, 700 MPa	Very high, >1 GPa	Low	High, 700 MPa
Temperature	Ambient	Ambient	Ambient	Very high	Ambient	Warm
Tooling	Soft	Hard	Hard	Soft	Hard	Hard, heated
Deformation rate	Low	High	High	Very high	Low	High
Continuous	No	No	No	No	Yes	No
Pressing direction	3	1	1	1	1	1
Shape complexity	Moderate to high	Moderate	High	Low	Low	Moderate
Polymer level	None	None	Low, 0.5%	None	Low, 0.1%	Low, 0.6%
Precision	Low	Moderate	High	Low	High	High
Use	Moderate	Low	Extensive	Very low	Moderate	Low

compressing powder. Die compaction involves the pressurization of the powder in a rigid tool set. A conceptualization of a die compaction cycle is given in Figure 5.1. Powder is loaded into the cavity by gravity from a feed shoe that sweeps over the die opening. A few important features of the tool set become evident: the die body, lower punch, upper punch, and feed shoe. The lower punch is in the fill position during powder entry, but then drops to a lower position to create room for the upper punch to enter the die cavity. In the most common pressing cycle, the two punches work toward the center of the compact to densify the powder: double-action pressing. After application of the peak pressure, the upper punch is pulled from the die while the lower punch pushes the compact out of the die. After this point the cycle repeats.

This description provides a first basis for examining the various factors involved in die pressing, including the lubricants, powders, tools, and presses. The rules for densification and compact strengthening are applicable to many processes. However, differences exist in the shapes possible by each route. This will be emphasized later in the book when component design for P/M is addressed.

5.2. LUBRICANTS

5.2.1. Objectives

Friction between the die wall and the powder during pressing is a fundamental problem. As the compaction pressure increases, ejection of the compact from the tool becomes difficult. Consequently, lubricants are used to ease ejection and minimize die wear.

Polymer lubricant particles, such as that seen in Figure 5.2, are mixed with the iron powder prior to pressing. The alternative is to apply a lubricant to the die wall between compaction strokes. Spray units have been designed to charge the lubricant particles with static electricity for electrostatic adherence to the die wall. In this

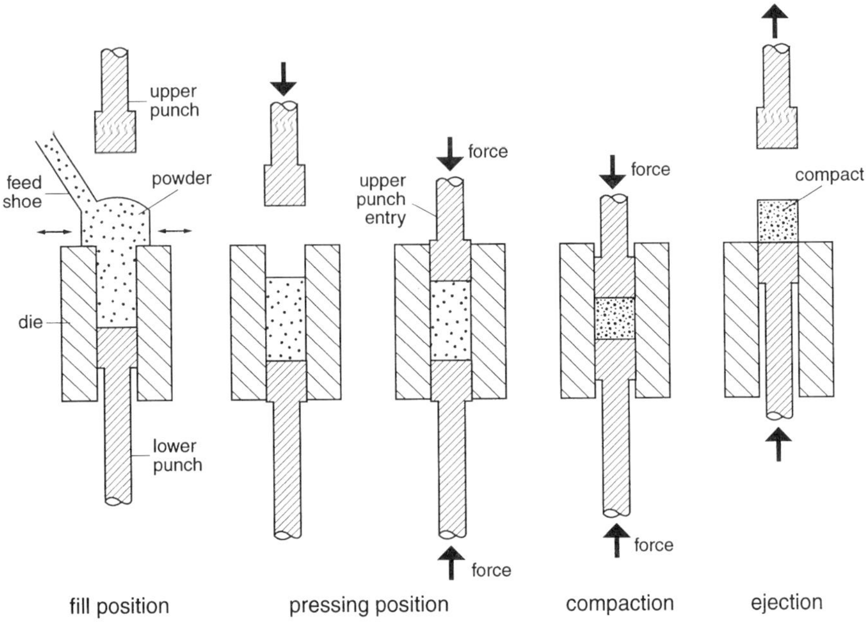

Figure 5.1. Die compaction involves a cycle of powder flow into the die cavity, with the lower punch at the fill position. Subsequently, the metered portion of powder is dropped to the pressing position, and the upper and lower punches press toward the center. After compaction, the lower punch ejects the compact prior to repeating the cycle.

case, the compact green strength (strength after pressing) is improved, since there is no polymer film inhibiting powder–powder bonding. The penalty with die wall lubrication is slower die filling, since the lubricant also lowers the sliding friction between particles. Die wall lubrication is preferred in theory, but is not easy to incorporate into automatic compaction equipment, especially for complex shapes. Thus, lubricant particles are usually mixed with the iron powder as a final step before pressing. The typical concentration is near 0.5–0.6 wt%, but might range from 0.3 to 1.0 wt%.

5.2.2. Lubricant Types

Many lubricants are in use, including stearates, stearic acid, polytetrafluroethylene, polyvinyl fluoride, and even waxes. The lubricant molecules decompose during sintering, but some leave an undesirable residue termed *ash*. Common lubricants are identified in Table 5.2. They are polymer molecules ranging from 12 carbon atoms on up and melt at relatively low temperatures. Most of the lubricants have particle sizes in the 5–30-μm range.

Differences in compaction and sintered properties trace to the lubricant chemistry. An example is given in Figure 5.3, comparing the sintered density and strength for

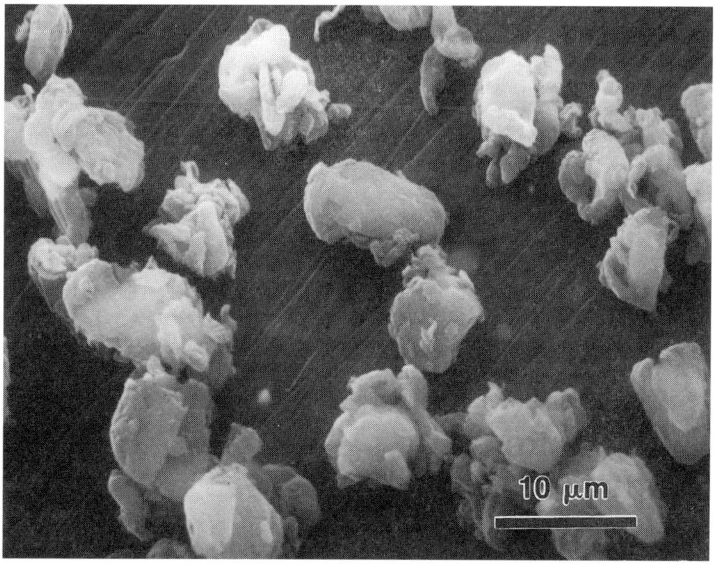

Figure 5.2. A scanning electron micrograph of lubricant powder prior to mixing with iron. (Photograph courtesy of Christine Kipphut-Furstoss.)

EQUATION BOX 5.1

The theoretical density of a lubricant–powder mixture is calculated by the inverse rule of mixtures

$$\frac{1}{\rho_T} = \frac{W_L}{\rho_L} + \frac{W_P}{\rho_P}$$

where ρ_T is the theoretical density of the mixture, consisting of W_L weight fraction of lubricant with density ρ_L and W_P is the weight fraction of powder with a theoretical density of ρ_P. As an example, using 1 wt% zinc stearate with 99 wt% iron (theoretical densities of 1.09 and 7.86 g/cm^3) gives a theoretical density for the mixture of 7.40 g/cm^3. The theoretical density sets an upper limit on die compaction.

TABLE 5.2. Lubricant Characteristics

Common Name	Oxide	Ash (%)	Softening Temperature (°C)	Melting Temperature (°C)	Density (g/cm^3)
Calcium stearate	CaO	9	115–120	160	1.03
Acrawax C Ethylenebisstearamide	None	2	120	144	1.05
Duramax B-1013 polymethylmethacrylate	None	9	105	160	1.1
Glycolube PM-100	ZnO	1	120	137	1.06
Lithium stearate	Li_2O	5	195	220	1.01
Paraffin wax	None	1	85	99	0.93
Stearic acid	None	2	45	55	0.85
Teflon Polytetrafluoroethylene	None	0	275	325	2.1
Zinc stearate	ZnO	14	110	120	1.09

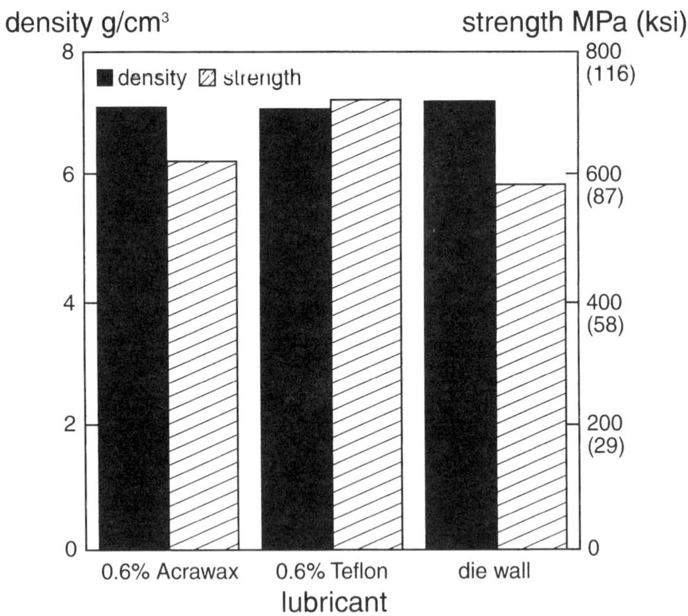

Figure 5.3. Comparison of lubricant effects on sintered density and strength for a steel alloy compacted at 690 MPa (50 tsi) and sintered at 1265°C (2310°F). The Acrawax and Teflon lubricants were mixed at the same 0.6 wt% level with the powder prior to compaction, while the die wall lubricant was applied only to the compact exterior. In spite of a small density difference, there is a major sintered strength difference. (Data courtesy of Anthony Griffo.)

an alloy steel with 0.6 wt% of two mixed lubricants (Acrawax, an ethylenebisstearamide and Teflon, a polytetrafluoroethylene) as compared to die wall lubrication. The density difference is small, but the sintered strength varies by 15% between lubricants. In some studies, sintered strength differences of up to 40% have been measured due to lubricant changes. More typically, a smaller variation occurs with shifts between lubricants, typically accounting for a 7% difference in ejection force, sintered hardness, or dimensional change in sintering. Although originally favored, the metal stearates (especially zinc stearate) are out of favor because of contamination problems. The low-cost and clean-burning character of ethylenebisstearamide makes it the most common lubricant for ferrous P/M.

5.2.3. Compaction Effects

Examples of lubricant effects on powder characteristics and compaction are demonstrated in Figures 5.4–5.7. The first two figures relate to a water atomized, −100-mesh stainless-steel powder. The powder was treated with three concentrations of lithium stearate for a range of mixing times. Figure 5.4 shows that the apparent density decreases with large concentrations of lubricant, because more relative volume is occupied by the low-density lubricant. The flow time (Hall flowmeter, 50 g-sample) is reduced when the amount of lubricant is increased, as shown in Figure

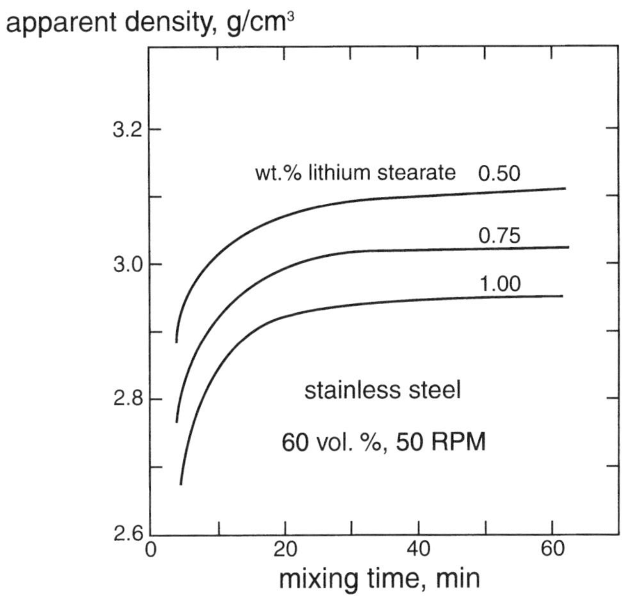

Figure 5.4. Apparent density versus mixing time for water-atomized −100-mesh stainless-steel powder treated with three concentrations of lithium stearate. Note the decrease in apparent density as the lubricant content increases. Typical mixing times in production are near 20–30 min. (Data from Michael Koczak.)

5.5. This reflects increased lubrication of the particles sliding past one another. Slower powder flow into the die cavity is a concern with die wall lubrication.

In the green state (the condition of the powder after pressing but before sintering), lubricants can lower the strength. Figure 5.6 plots the green density against lubricant content to illustrate a complex dependence on the compaction pressure for a reduced-iron powder. Figures 5.6 and 5.7 demonstrate the effect of lubricant content on the green density and the ejection force of a reduced-oxide-iron powder using zinc stearate as the lubricant.

Generally, the lubricant has a mixed effect on the pressure–density relation. For low-lubricant contents, the green density increases, because of easier particle sliding, that is, reduced friction. However, the optimal amount of lubricant decreases as the compaction pressure increases, because the lubricant occupies volume that inhibits particle compression. Alternatively, the ejection force continuously decreases as the amount of lubricant increases, and is typically in the range of 20–25 MPa (3–4 ksi) at the start of sliding, but falls to 10–15 MPa (1.5–2 ksi) once the compact starts to move. In practice, a balance is struck between reducing die wall friction, minimizing the radial load transfer, and maximizing the green properties.

Density is a primary determinant of strength. Hence, lower-green-density conditions correspond to a lower green strength. In Figure 5.6, the peak green density moves to a smaller lubricant concentration as the compaction pressure increases.

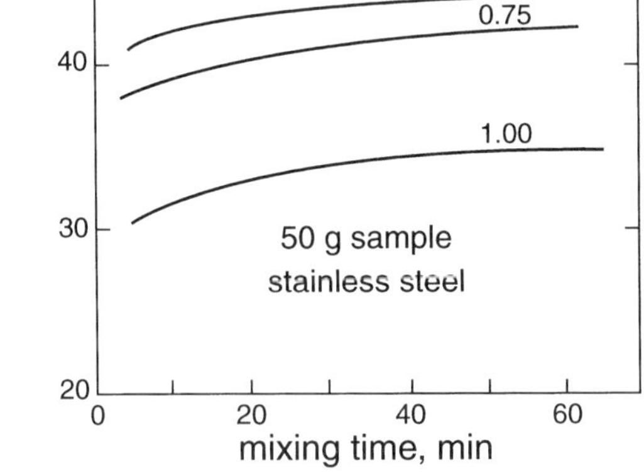

Figure 5.5. The flow times for 50-g samples of water-atomized −100-mesh stainless-steel powder treated with three concentrations of lithium stearate, versus mixing time. The powder flows faster (shorter flow time) for the higher lubricant content. (Data from Michael Koczak.)

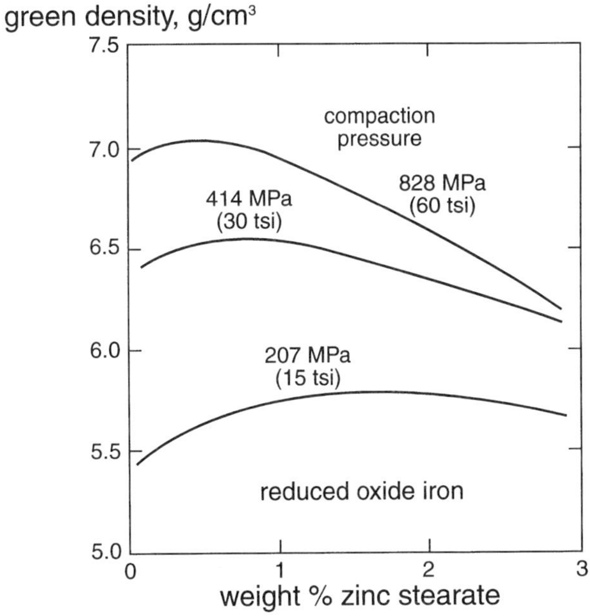

Figure 5.6. Lubricant content affects the green density. For a given compaction pressure there is a peak in green density, with the peak occurring at lower lubricant contents for higher compaction pressures. These data are for a reduced-oxide-iron powder.

That behavior results from better load transmission due to lubrication, but less efficient iron powder packing due to the volume occupied by the lubricant.

A major justification for lubricants arises after the compaction pressure is released. The tools are strong and hard, but the compact is weak. As the compact is pushed out of the tool cavity, stored strain in the compact causes it to press against the die wall, which wears the tooling. A beneficial effect of powder lubrication is to reduce the ejection force, the load needed to push the compact out of the die. As illustrated in Figure 5.7, the ejection force increases with higher compaction pressures. Accordingly, die wear increases, but lubrication counters the wear by lowering the ejection force. In practice, a balance is struck between the interparticle friction, green strength, green density, and the ejection force by using moderate lubricant levels. As the powder hardness increases, the required amount of lubricant increases.

Another role of the lubricant is evident in size growth on ejection. This feature is termed *springback*, meaning that the ejected compact is larger than the die. Typical springback expansions are between 0.01 and 0.2% of the die dimension, but can reach 0.5% with certain lubricants and powders. Such expansion can cause failure in the green compact, especially if the lubricant prevents cold welding of the particles. Lamination cracks form perpendicular to the pressing direction. As sketched in Figure 5.8, cracking is most likely to occur when the top of the compact is ejected

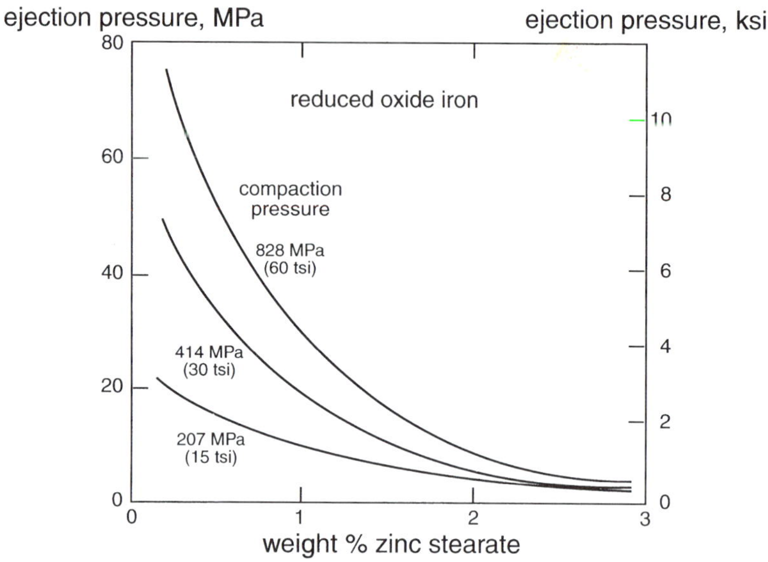

Figure 5.7. Ejection pressure needed to extract the compact after compaction, plotted against the zinc stearate lubricant content for three compaction pressures. Higher compaction pressures give higher ejection pressures, but more lubricant makes ejection easier.

and the bottom is still retained in the die cavity. The differential stress—which results because there is no constraint on the expanded portion, but a high stress on the confined portion—leads to cracking.

5.2.4. Mixing

Lubricants are usually mixed with the powder prior to compaction. Other alloying ingredients are added when the lubricant and iron powder are mixed. A common addition is graphite to form steel. Like the lubricant, the graphite particles are small to ensure homogeneous mixing with the iron. Figure 5.9 is a SEM of a typical graphite powder. It consists of a mixture of small flakes and nodules. Graphite has an ash content that contaminates the compact. Table 5.3 compares the attributes of a few common graphite additives. Although the particle size for the P/M grades are similar, the large differences are in surface area and purity.

Other common alloying additions introduced in the mixing stage include copper, nickel, and phosphorous compounds. Mixing is performed after these ingredients are weighed, using a rotating chamber, such as sketched in Figure 5.10. The interior is important in determining mixing efficiency. It might have baffles, spinning intensifier blades, or dividers to enhance mixing. Usually, only about 40% of the mixer is filled with powder to allow particle flow on each rotation. Two other important factors are the mixing time and rotation rate. Most desirable is for the mixer to rotate

EQUATION BOX 5.2

Green properties depend on the pressure more than any other variable. For the typical water-atomized iron, the green density ρ_G is an exponential function of the compaction pressure P

$$\rho_G = \rho_A + A \exp(-KP)$$

where ρ_A is the apparent density, and C and K are constants that change with each powder. These constants can be extracted by performing green density determination for a few compaction pressures. This is sometimes called a *constitutive equation*, since it relates the density to a process variable. Usually the green strength is directly proportional to the green density, at least to a first approximation. For a typical water-atomized iron powder, a green density of 6.8 g/cm^3 gives about 12 MPa green strength (1.7 ksi) and 7.2 g/cm^3 produces a green strength of 18 MPa (2.6 ksi). Thus, higher pressures give higher densities and strengths, but at a penalty in terms of tool wear and tool life.

Another conceptualization is to examine the energy imparted to the powder in compaction. Energy is calculated from the force–displacement curve, which for uniaxial compaction is just the density (constant area and mass)–pressure (constant area) curve. Two events consume the energy: particle rearrangement via sliding and particle deformation via contact flattening. The first gives a linear relation between density and work, and the second gives a square relation. Thus, green density is simply a function of the compaction work.

fast enough to lift the particles for cascading over one another. This is usually performed for about 30 min. If the powders are mixed for a longer time, they become more difficult to compact.

An alterative to mixing is to create composite particles with two or more phases clustered together on each particle. As conceptualized in Figure 5.11, the range of conditions goes from a simple mixture (with easy segregation) to agglomerates held together by polymers, to diffusion-bonded particles. Composite particles are important where homogenous products are required or where segregation is a problem. As compared to alloy particles, mixed powders have a lower cost and higher compressibility, but undergo segregation in handling that degrades final properties. In contrast, partially prealloyed or diffusion-bonded particles have higher cost, but give improved homogeneity. These are formed by heating the agglomerated powders to a high temperature where atomic diffusion welds the particles together. Diffusion-bonded and polymer-agglomerated powders are particularly important for high-performance P/M applications.

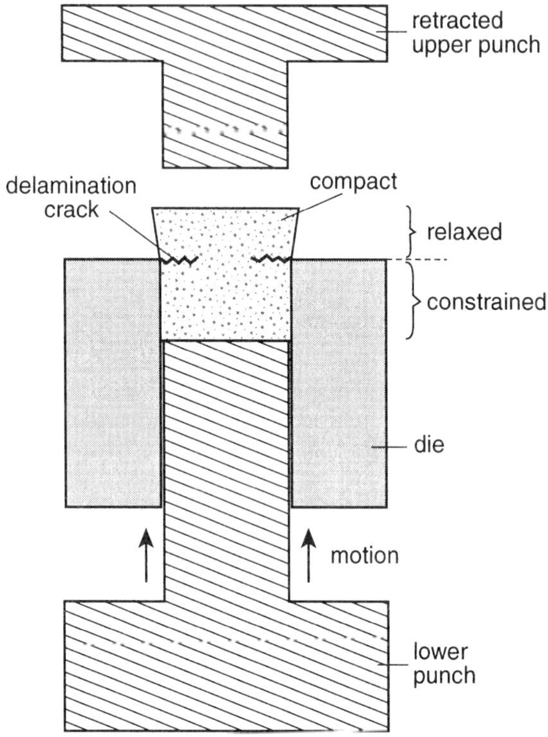

Figure 5.8. Springback is the elastic relaxation of a compact on ejection. The differences between constrained and relaxed regions can lead to cracking if the compact green strength is low or the springback is large.

The agglomerated powders are fabricated in fluidized-bed mixers, where heated air is used to lift and stir the powders with a polymer providing the glue between particles. Newer techniques apply electrical voltages to powder–water slurries to from agglomerates by an electrical-charge-induced dripping process. Another route is via spray-drying, where large agglomerates are formed from a slurry by spraying powder–binder–solvent droplets into a heated chamber.

5.3. PRESSURE EFFECTS ON FERROUS POWDERS

5.3.1. Particle Deformation

When a ferrous powder is compressed, the particles slide past one another, deform, bond, and harden. The greater the compaction pressure, the harder the particles become, thereby resisting further densification. Compaction initially deforms the particles at their contacting points and results in welding at those contacts. The higher the pressure, the higher the density. Examples of this behavior are given in

Figure 5.9. An important alloying addition for iron powder is graphite. This scanning electron micrograph shows an example of the graphite powder that is commonly mixed with iron to form steel during sintering. (Photograph courtesy of Anthony Griffo.)

Figure 5.12, where the green density is plotted against the applied pressure for three water-atomized iron powders and one reduced-oxide powder. Note the green density is expressed in grams per centimeter cubed (g/cm^3 or g/cc).

Powder characteristics have a large effect on compaction. As the pressure increases, the green density increases, but the gain in density with higher pressures progressively decays. This is because of a property known as *work hardening*. At the atomic level, the dislocations move and become entangled, thereby making further atomic motion difficult. Whenever a steel is deformed, it becomes harder and less ductile. Likewise, particle deformation during compaction causes particle hardening, and it is this increase in hardness that leads to progressive resistance to further compaction. At the higher pressures the three water atomized powders are essentially the same, but the reduced-oxide powder suffers a lower density. The main difference between the three water-atomized powders is in purity.

TABLE 5.3. Comparison of Graphite Powders Used in Powder Metallurgy

Screen Size	Particle Size (μm)	Specific Surface Area (m^2/g)	Carbon Content (%)	Ash Content (%)
−325 mesh	7	10	95.5	4
−325 mesh	4	20	99.0	0.02
−325 mesh	6	13	99.0	0.5

Figure 5.10. The lubricant, iron, alloying additions, and graphite are mixed by a tumbling action in a mixer. This double-cone mixer (top and bottom are in cone shapes) rotates on the horizontal axis with the ingredients inside. The tumbling action causes the powders to form a homogeneous mixture.

EQUATION BOX 5.3

The rotation rate for a mixer is determined by the velocity of the outer diameter of the mixing chamber. If the rotation is too fast, the powder is held by centrifugal force against the chamber wall and fails to mix. On the other hand, if the mixer is slow, the powder will simply slide on the wall and not mix. The best velocity is to have the chamber rotation rate in revolutions per minute (rpm) as follows:

$$N = \frac{32}{\sqrt{d}}$$

where N is the number of revolutions per minute and d is the outer diameter of the mixer in meters, remembering that about 3.25 feet (ft) are in each meter.

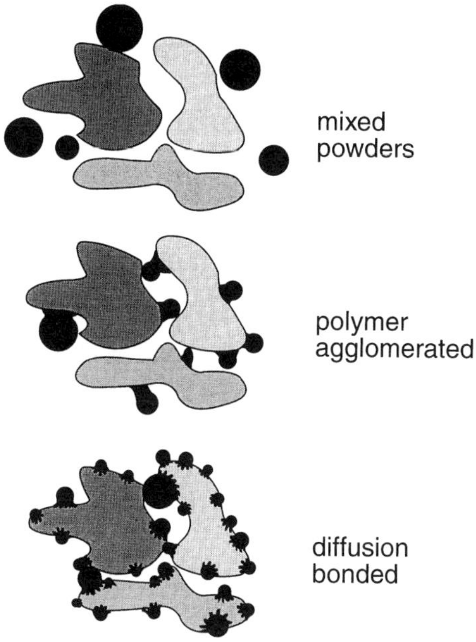

Figure 5.11. Newer technologies of polymer agglomeration and diffusion bonding have evolved to ensure that the alloying ingredients stick to the iron powders. Diffusion bonding heats the powders to a temperature where the additions metallurgically bond to the iron.

So why not more pressure? One reason is tool life. Buckling, fracture, wear, and fatigue of the punches and other tool components all become problems that cause premature tool failure with very high compaction pressures. Thus, in most instances sufficiently high pressures are applied to produce densities near 90% of theoretical or 7.0 g/cm^3. At higher pressures, trapping of contaminants in the compact decreases the sintered properties. Typically, iron and steel powders are compressed at pressures in the 550–700-MPa (30–50-tsi) range. A slight difference is apparent between powders in terms of green strength, as plotted in Figure 5.13. Green strength is the transverse rupture strength needed to break the compact.

Compaction pressure is defined as the load divided by the part cross-sectional area in the pressing direction. Pressure determines the degree of particle repacking and deformation. Generally, compaction pressure is not changed significantly with part height. However, for a constant pressure, any change in compact area requires an increase in load. The compression ratio relates the final density and height to the initial density (apparent density) and powder height. For ferrous powders it is typically about 2.5. Thus, if the apparent density is 2.9 g/cm^3 and the final density is 6.9 g/cm^3, the compression ratio is 2.36, meaning that the fill height needs to be 2.36 times the final height.

One major caution is to ensure that the compression ratio is never set to cause tool breakage. After all, the highest theoretical density is 100%, so if pressing conditions are set to exceed this value, the tools will be damaged. For example, if the theoretical density of iron is 7.86 g/cm^3 and the apparent density of the powder is 2.9 g/cm^3, the maximum theoretical compression ratio will be 2.71, and even this will cause tool damage. Accordingly, during setup it is necessary to start with low compression ratios to avoid tool damage. Since lubricants lower the theoretical density, care is needed to include their effects in setup calculations.

5.3.2. Green Properties

After pressing the particles are smooth, with flat faces where the contacts deformed. Figure 5.14 is a scanning electron micrograph of a water-atomized stainless-steel powder after compaction. The particles have been deformed and forced to conform to each other, with some interlocking. When the powder is pressed to just 5.5 g/cm^3, the diameter of the compressed faces are about 20% of the particle size, but when the powder is compressed to 7.1 g/cm^3, the faces are 33% of the particle size. Essentially, compaction pushes the particles into intimate contact as they form polygonal shapes. It is the particle interlocking that gives green strength.

Simply stated, the role of compaction pressure is to compress the particles into a shape, increase the density, and provide strength to the compact. Initially, when

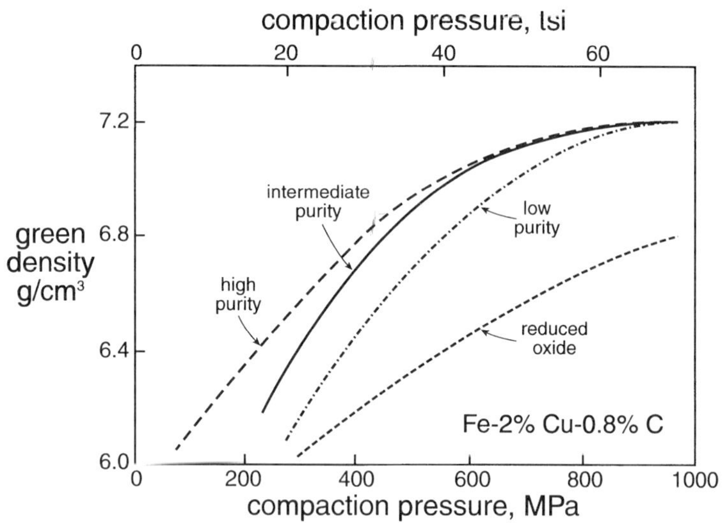

Figure 5.12. A comparison of the green density versus compaction pressure curves for four iron powders used to form an alloy of 2% copper and 0.8% carbon (FC-0208). Three water-atomized iron powders were selected with slight differences in purity level (99.0, 99.4, and 99.8% iron). Also, a reduced oxide powder is compared (98% pure). The plot shows the difference in water-atomized iron powders is most significant at lower compaction pressures. (Data from Stephen Claeys.)

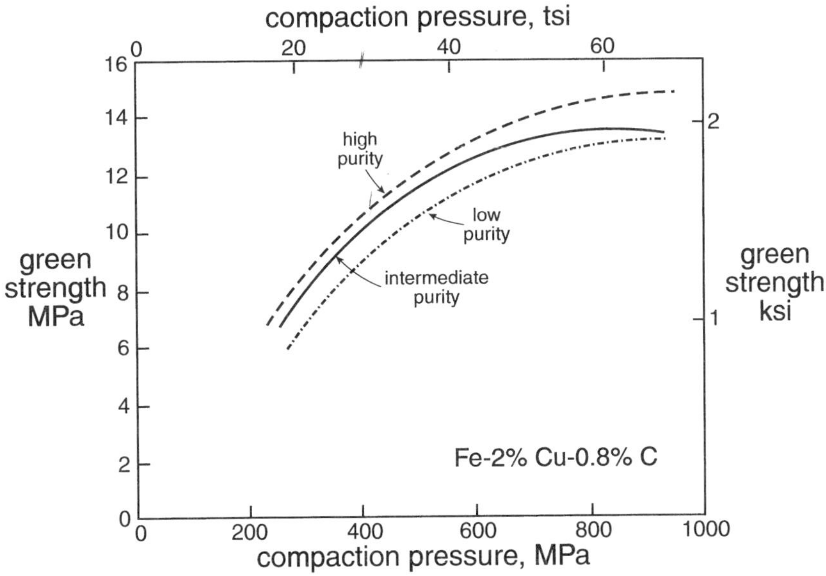

Figure 5.13. Green strength versus compaction pressure for the three water-atomized iron–copper–carbon powder mixtures used in Figure 5.12. The powder purity differences (99.0, 99.4, and 99.8% iron) translate into a green strength effect. Note these green strengths are about the same as that of chalk. (Data courtesy of Stephen Claeys.)

poured into the die cavity, each particle has four to six contacting particles. As pressure is applied and porosity is eliminated, the particles make new contacts, and can achieve an average of 14 contacts each. Thus, strengthening occurs for two reasons: first by the deformation and cold welding of particle contacts, and second by the formation of more contacts. Figure 5.15 plots the relation between the number of contacts per particle, a property known as the *coordination number*, and the density during die compaction. This simultaneous increase in coordination and contact flattening by working are the two primary changes that give higher densities.

5.3.3. Particle Hardness

One factor that greatly affects green density is the particle hardness. Soft particles deform more easily than hard particles. Thus, in examining various powders and their die compaction, the tendency is to use soft particles. Microhardness tests at various green densities clearly show particle hardening from compaction. Higher purity is one means of ensuring softer particles, as evident in Figure 5.12.

To form an alloy, it is a common practice to mix soft powders of differing compositions. Sintering then allows atoms to intermix and alloy. The alternative is to use prealloyed powders, but this makes compaction much more difficult. Purity is an important factor in determining compactability. For example, if an iron powder

EQUATION BOX 5.4

The compression ratio is determined by the green density and the apparent density. Each powder will exhibit subtle variations in compression ratio, reflecting shifts in moisture, temperature, particle size, and lubricant. The compression ratio depends on the punch travel, and fill height can be computed to determine the requirements for press setup conditions.

$$C_R = \frac{\rho_G}{\rho_A}$$

where C_R is the compression ratio, ρ_G is the green density, and ρ_A is the apparent density. For a final compact height of H, the fill height H_F is given as

$$H_F = H\, C_R$$

In other words, the compression ratio is the height dilation from fill to green conditions. The punch travel height H_T then is given as

$$H_T = H_F - H$$

Note that the ratio of H_T over H_F is the compression strain and often this is in the range from 2 to 3.

is prealloyed with carbon, the presence of just 0.5% carbon reduces the green density by 10% or nearly 0.8 g/cm^3. Carbon is desirable because it is a potent strengthener in steels. Yet, as an alloying addition it proves detrimental to compaction. In contrast, molybdenum has little detrimental effect on compressibility, so it is often prealloyed into the iron. The addition of 2% molybdenum in a prealloyed powder reduces the compressibility by just 2%.

To avoid a loss in compressibility, most steels are compacted from mixed elemental powders or iron alloys containing low alloying levels. An exception would be the stainless steels, where prealloyed powder is routine.

5.3.4. Die Wall Friction

The major difficulty with die compaction traces to die wall friction. This friction inhibits ejection, and more importantly causes density gradients in the green compact. Punch motion against the powder is similar to plowing snow. Close to the punch the packing is dense, but far removed from the punch, the powder is unaffected. This pressure decay with distance results because the powder spreads load to the die wall, in the form of friction.

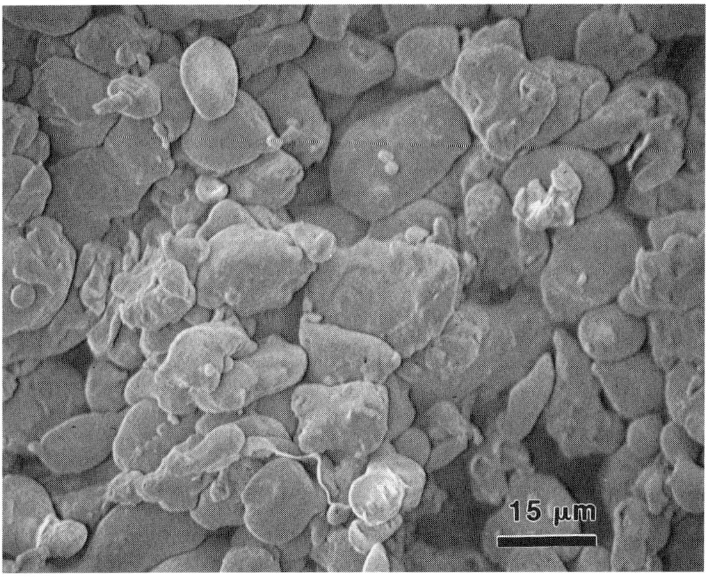

Figure 5.14. This scanning electron micrograph shows deformed and interlocked particles for a water-atomized stainless steel powder compacted to 280 MPa (20 tsi). With progressively higher pressures the particles deform to form polygonal shapes with an increased number of contacting particles for each grain.

As illustrated earlier, green density depends on compaction pressure. Consequently, density declines with distance from the punch. When the compact length is about 6 times the width, the powder is poorly compressed. This is one reason why most die-compacted parts are squat, to minimize density gradients. Also, density gradients are reduced when both the upper punch and lower punch move toward the compact center. *Double-action pressing* is a term used to describe the application of stress from both the top and bottom punches. In one variant, the die floats with respect to the lower punch, giving double action even though the lower punch remains stationary. Positive die motion ensures uniformity in green density. Inside the compact, the point at which the lowest density occurs is the neutral pressure plane, and in most instances the preference is for the neutral plane to be in the center of the compact.

As the compaction pressure and density increase, the die wall friction also increases. Figure 5.16 plots the measured ratio of the two pressures during compaction. The die wall friction is proportional to the radial pressure when the powder is compressed uniaxially. These data show that wall pressures climb to exceed 50% of the applied pressure. Thus, the higher the applied pressure, the higher the radial pressure and concomitant die wall friction.

If pressure decays with depth in a compact due to wall friction and green density depends on local pressure, then die compaction results in green density gradients.

Besides friction with the die wall, an additional factor is friction on the punches and core rods. Figure 5.17 shows the measured green density contours in a compact pressed near 700 MPa (50 tsi). These density gradients lead to nonuniform shrinkage or swelling in sintering, resulting in component distortion and loss of precision. Low density regions tend to change dimensions more in sintering as compared to high-density regions. Figure 5.18 illustrates the green density gradient using three micrographs. These were taken after sintering, and clearly show differences in the pore size and porosity between the top, side, and bottom surfaces.

Minimized density gradients are important to final dimension control. Accordingly, much study has gone into means of reducing density gradients. Unfortunately, most cures are ineffective, but a few concepts are important. These include keeping the compact thin in the pressing direction (thickness), lubricating the die wall, and double-action pressing. One alternative under study is to move the top punch in a circular motion, causing powder flow in a sideways motion to help eliminate low-density regions. The need for minimum density gradients increases as the component becomes more complex. Since die wall friction is the culprit, the more die wall per unit volume the greater the problem. This is one of the major barriers to the production of precise gears by P/M, since gear teeth have a large surface area leading to more die wall friction and larger density gradients.

Because of density gradients, problems occur with local weak spots, cracking on ejection, and subsequent component warpage in sintering. Thus, much effort is being

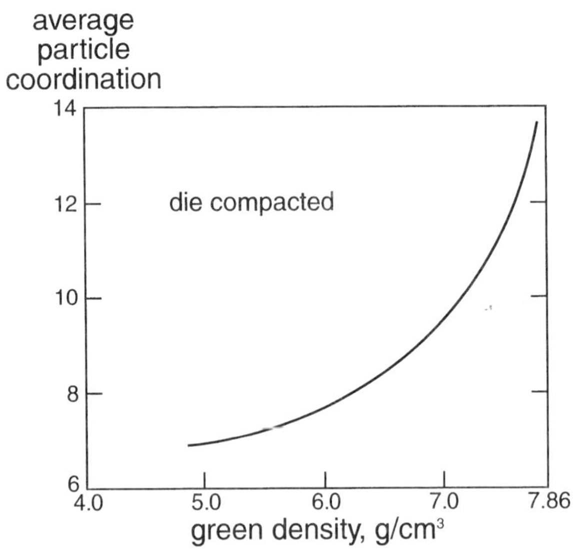

Figure 5.15. Particle coordination versus the green density. As compaction deforms and compresses the particles to increase the number of particle contacts, approaching a theoretical limit of 14 at full density. A variation exists, with some particles having more and others having less than this average.

EQUATION BOX 5.5

Pressure applied to a powder is distributed over a continuously increasing number of particles with distance into the powder compact. This results in a decaying pressure with depth below the punch. The analysis of pressure P at a distance X from the compact surface depends on two parameters: the applied pressure P_A and a collection of geometric and lubrication terms α

$$P = P_A \exp(-\alpha X)$$

where the parameter α is given by the compact diameter d, coefficient of friction with the tooling μ (which depends on lubrication at the die wall), and a powder parameter ξ that tells the radial stress as a function of the axial stress during compaction:

$$\alpha = \frac{\mu\xi}{d}$$

Essentially, pressure is lost to die wall friction since the powder converts the axial stress to a radial stress that gives sliding friction and lower pressure. The problem becomes larger with thick compacts and small compact diameters. Thus, most success occurs in producing uniform compacts with low heights and large diameters—the typical squat compact with a small length : diameter ratio.

applied to calculation of density gradients in pressing. Figure 5.19 shows the results of one set of calculations for a tapered component. For this geometry, pressure was applied on the top and bottom surfaces. Friction between the powder, die, and punches led to density contours. The calculations were performed using a concept of finite element analysis. In this approach, the component is numerically subdivided into hundreds of tiny computation cells. The placement of these calculation cells is visible on the component as a fine mesh. Each cell compresses and deforms according to the forces from the neighboring cells using equations that relate pressure to density. Computer calculation of what happens in each cell then determines the flow of mass and stress through the compact, resulting in the plotted density contours. Soon, most tooling design and compaction setups will be guided by such computer simulations, with the goal being reduction of density gradients and weak spots in the compact. The idea is to make the tooling correct with minimal trial and error.

5.3.5. Cold Forging

Very high compaction pressures have been used in pressing powders on a few occasions. The technique is best termed *cold forging*, since the pressures are far in excess of the normal pressing ranges and the rates of pressurization are usually high. In this regard the compaction rate is like forging. The powder is cold, but the pressures often exceed 1 GPa (over 72 tsi) and can climb to 3 GPa (225 tsi). Special tooling is required to withstand these pressures. Such compaction is performed largely for experimentation, but it is noteworthy that green densities approach the theoretical value. Lubricants cannot be mixed with the powder in this pressure range if such high densities are to be attained. Even 0.5% lubricant takes up about 4% of the volume, so densities over 96% of theoretical would be impossible. Also, powder purity must be high to avoid capture of contaminants inside the compact.

As an illustration of near-full-density cold forging, Figure 5.20 gives data for a water-atomized tool steel powder. Note that a density of 99% of theoretical is possible with a pressure near 3 GPa, which corresponds to about 225 tsi or 450 ksi. Don't try this experiment at home. It is very dangerous and not healthy for the tooling. More typical in ferrous powder metallurgy is a green density between 6.6 and 7.2 g/cm^3, which represents 84–92% of theoretical density. Higher green densities are possible with certain powders, but there are contamination problems that limit green densities to about 7.2 g/cm^3 (largely because the lubricant can no longer burn out, leading to cracks, blisters, and contamination). After pressing to these pressure ranges, the compact has a green strength between 10 and 20 MPa (typically between 1 and 3 ksi). As a point of reference, a green density of 6.2 g/cm^3 was typical in

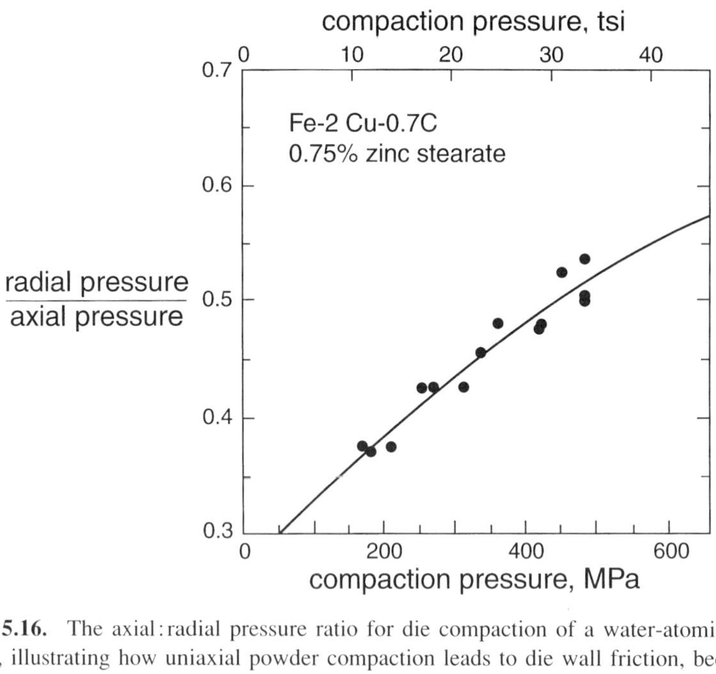

Figure 5.16. The axial : radial pressure ratio for die compaction of a water-atomized iron powder, illustrating how uniaxial powder compaction leads to die wall friction, because of the higher pressure and higher ratio transmitted in the radial direction. (Data from Hiroki Yanagawa.)

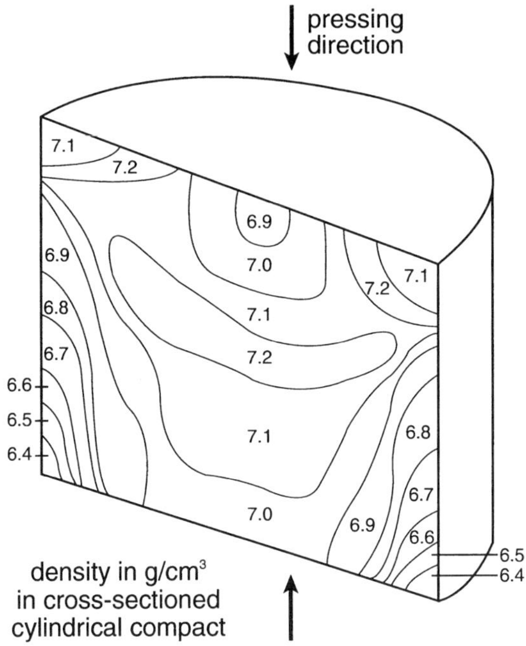

Figure 5.17. Experimental density distribution in a simple cylindrical compact. This mapping shows the density contours in grams per cubic meter traced out on a cross section of the compact. The friction with the die wall and punch faces gives this density profile. (Data from L. Edwards and M. Endeau.)

the early 1960s, while 7.0–7.5 g/cm^3 is a more typical production requirement for new applications.

Most of the studies on high-pressure compaction have been in research settings, although a few explorations of cold forging have been performed at the pilot production level. Early components produced by cold forging were simple in shape with low precision. Recent progress has extended the concept to the production of more complicated shapes, including gears. One benefit of cold forging is that sintering temperatures are low while still attaining target property and density goals. However, tool life is low, and this limits the applications.

5.4. POWDER EFFECTS ON COMPACTION

Compaction is influenced by the iron powder size, shape, and fabrication process. Figures 5.12 and 5.13 illustrate this with different iron powders used to make an alloy of 2% copper and 0.8% graphite (carbon) (FC-0208). These were all compacted under the same conditions, but the sponge powder formed by oxide reduction shows a lower green density. In contrast, the water-atomized iron powders are not signifi-

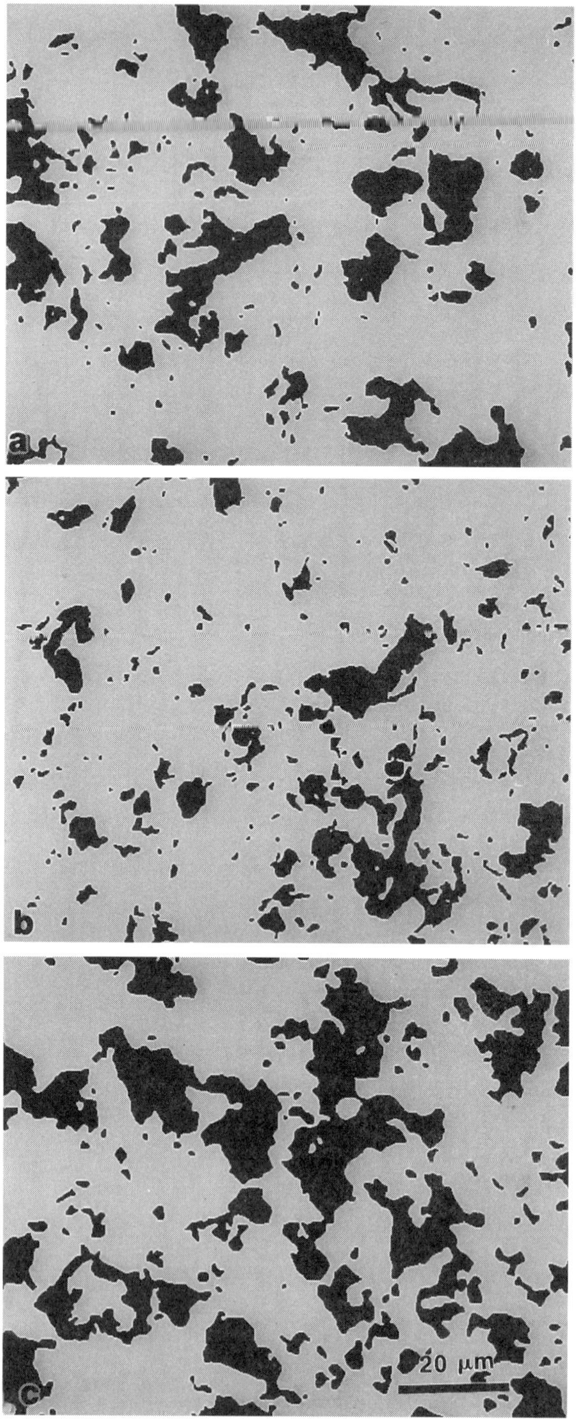
a
b
c
20 μm

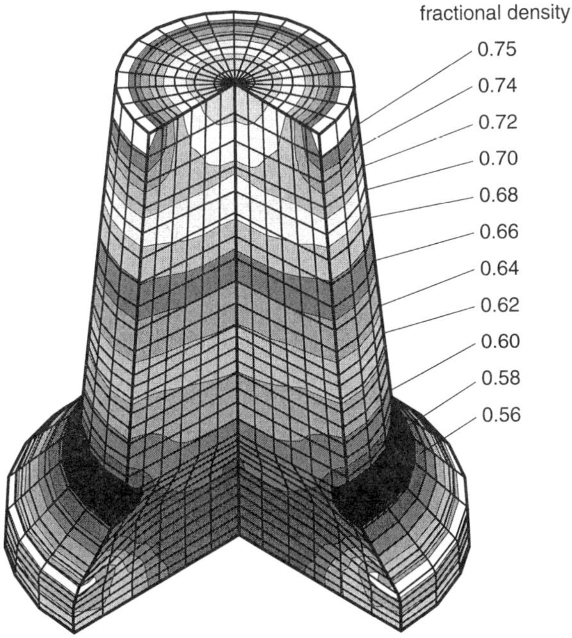

Figure 5.19. Computer-calculated density contours in a tapered compact with a flange on one end. The technique requires subdivision of the compact into many small cells or elements (shown by the grid). Each cell is used to calculate the impact of force and friction on the load transmission and density in the compact. (Calculation results courtesy of Tony Zahrah.)

cantly different when pressed at pressures of over 800 MPa (58 tsi), in part because they equally work harden to resist densification. Generally, the key factors affecting compressibility are the purity, particle size, particle shape, and internal porosity. Table 5.4 summarizes the variability in compressibility for pure iron powders by listing the pressure needed to obtain a density of 6.8 g/cm^3 along with the particle size, apparent density, and powder production process.

Larger particles without internal pores are most desirable for high densities. Smaller powders resist deformation. This is because the smaller particles harden at lower pressures during pressing since there are more contact points per unit volume. Small particles are also difficult to press because higher pressures are required to

Figure 5.18. Three micrographs showing the surface porosity (black phase) changes with location on a Fe–2Cu–0.8C compact after sintering: (a) top surface with average pore size of 12 μm; (b) side surface with average pore size of 10 μm; (c) bottom surface with average pore size over 12 μm. (Photographs courtesy of Anthony Griffo and Kate McKinley.)

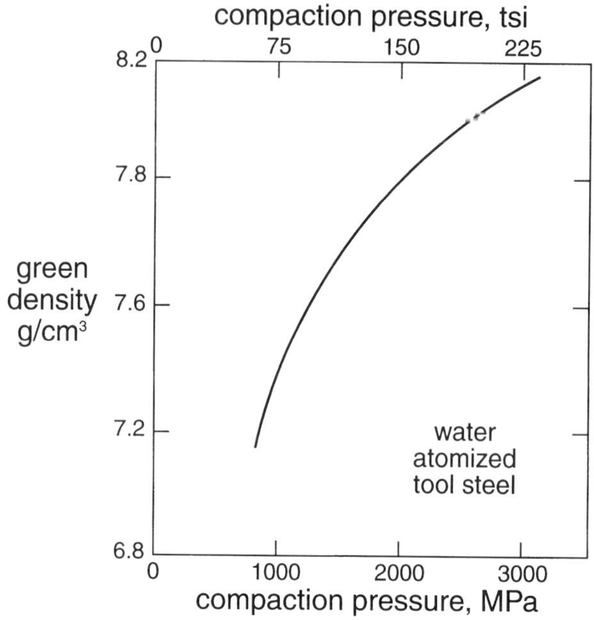

Figure 5.20. An example of ultrahigh compaction pressures and the corresponding green density that approaches the theoretical limit of 8.2 g/cm^3; these data are for a water-atomized tool steel. (Based on data from E. Y. Gutmanas.)

collapse small pores. Spherical particles pack to high densities, much higher than the typical water-atomized iron. However, during compaction the particles do not interlock. Lack of green strength is a problem. On the other hand, particles with internal pores prove difficult to compress. With such powders, the larger pores between particles collapse first. However, the smaller internal pores resist compac-

TABLE 5.4. Pressures Required to Obtain 6.8 g/cm^3 Green Density for Some Common Iron Powders

Powder Type	Median Particle Size (μm)	Apparent Density (g/cm^3)	Pressure for 6.8 g/cm^3 (MPa)
Water-atomized	104	2.92	392
Water-atomized	75	3.05	414
Water-atomized	75	2.94	450
Water-atomized	73	2.92	420
Water-atomized	63	2.90	500
Reduced-oxide	75	2.35	560
Reduced-oxide	53	2.50	1120
Reduced-oxide	50	2.30	700

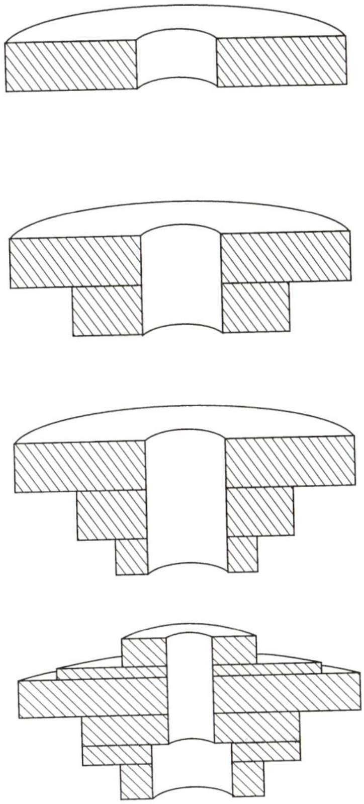

Figure 5.21. The horizontal lines in these compacts illustrate the concept of part levels. These four drawings show increasing complexity for die compaction.

tion. For that reason, sponge powders show a high initial compressibility, but resist compaction to high final densities. Small powders, hard powders, and sponge powders give more springback on ejection, and are more susceptible to cracking.

Factors that impair the compressibility also degrade the green strength. The higher the strength of the powder, the more difficulty expected in compaction. Prealloyed stainless steel exhibits rapid work hardening, so it is more difficult to press to high densities. Particle shape has a similar effect. An irregular shape gives a lower apparent density and hinders achievement of a high pressed density. However, an irregular powder gives a much higher green strength because of mechanical interlocking.

5.5. COMPACTION TOOLING

5.5.1. Definitions

Compaction involves several coordinated motions in a tool set to ensure proper density, shape, and properties in the compact. The pressing difficulty is assessed by

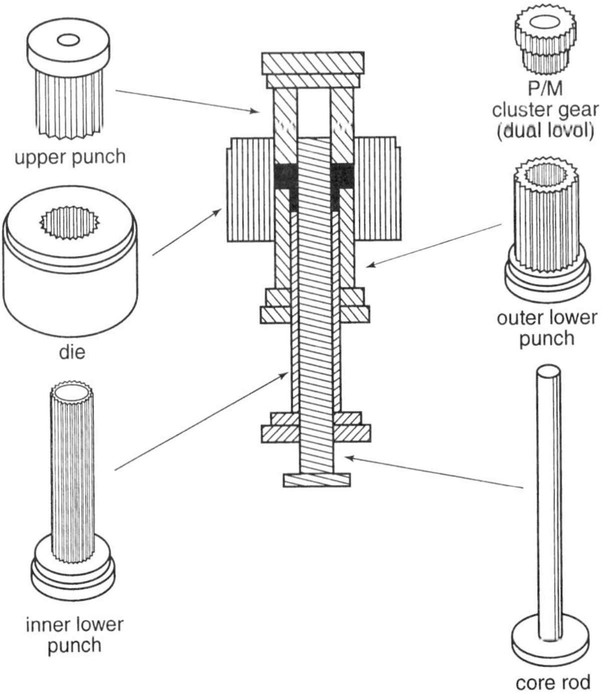

Figure 5.22. Tooling components and assembly for the fabrication of a dual-level cluster gear with a central core rod.

determining the number of levels in the component profile. Figure 5.21 sketches the concept of part levels. A simple two-level cluster gear is illustrated in Figure 5.22 along with components of the compaction tool set. The three main components are the upper punch, lower punch, and die body. Further, for this gear the lower punch is composed of three parts: a core rod at the center, an inner lower punch that forms the taller section, and the outer lower punch that forms the shorter section. The center drawing gives an idea on the assembly of these components.

With this first tool set we encounter the main components—die, punches, and core rods. There are two main punches, one the upper and the other the lower. These punches work inside the die. In turn, the die consists of a hard liner to prolong tool life and is contained in the die table. If multiple levels are required in the compact, then progressively more motions must be accommodated in the press and tooling. As a conceptual sketch, Figure 5.23 illustrates the platens and their arrangement for the production of a five-level hub component. It is the function of the press to drive each of these components into its proper location during compaction to ensure uniform pressing. Die set systems are preassembled tooling setups in which all the tool positions and motions were established prior to installation into a press. These are designed to maximize press operating time, since the die set is adjusted off-line and installed as a package, ready for production.

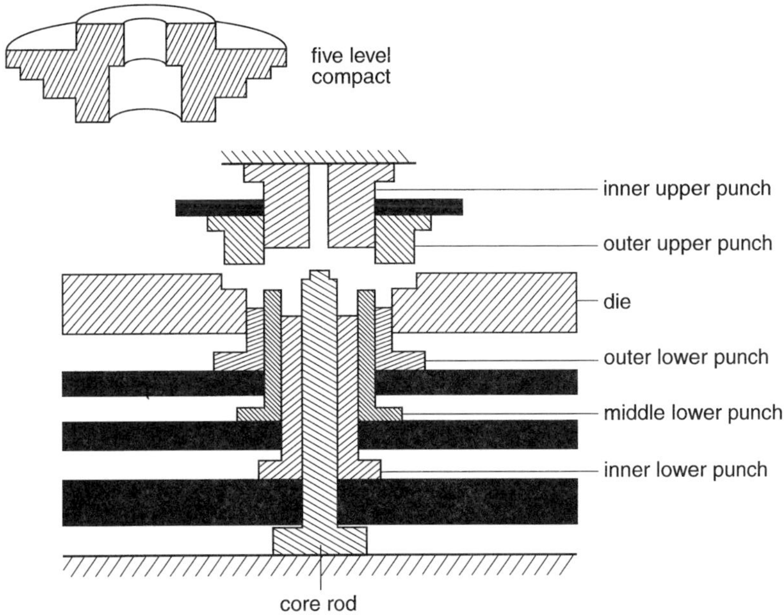

Figure 5.23. A five-level compact and the possible punch components needed to allow independent motions for the compaction of the sketched component.

5.5.2. Tooling Materials

Table 5.5 provides an overview of some common materials used in the construction of P/M compaction tooling. The list includes the most popular tool steels and cemented carbides. Most powder metallurgy die compaction is geared to larger production quantities and relies on tooling fabricated from tool steels or cemented carbides. The cemented carbides are sintered composites consisting of hard tungsten carbide

TABLE 5.5. Tooling Materials for Die Compaction

Composition[a]	Hardness (HRA)	Strength (MPa)	Applications
WC–3Co	92	1220	Bearing dies
WC–6Co	92	1585	Simple dies, simple cores
WC–10Co	91	1930	Hard particle punches
WC–12Co	89	2140	Complex cores, punches
WC–15Co	87	2500	Complex tooling components
WC–25Co	84	2500	Multi level dies
Tool steel CPM-10V	82	1500	Routine dies, punches, and cores
Tool steel D2	82	1960	General purpose

[a] WC–Co compositions are also known as *cemented carbides*.

grains in a solidified cobalt alloy matrix, WC–Co. Many of the carbides are modified by the addition of tantalum, titanium, or vanadium carbides. These are expensive materials, and machining of the carbide is difficult because they are very hard and strong.

Besides cemented carbides, tool steels are popular construction materials, especially for punches and shorter-production-run tooling. They are less expensive and are easier to machine, contributing to a lower cost. The cemented carbides have higher wear resistance, but less resistance to impact or damage as compared to tool steels. One of the curious developments over recent years has been the evolution of tool steels produced by powder metallurgy for use in P/M tooling.

5.5.3. Filling Operation

The complexity in P/M tooling comes from the desire to form multilevel parts in a single compaction stroke. Further, the inclusion of cores or holes increases the tool complexity. The chief problems come from the compression ratio and the need for customized filling. The normal sequence is sketched in Figure 5.24. The feed shoe drops powder from a storage reservoir into the open die cavity, with the lower punch in the fill position. It is possible to put too much powder into the cavity and then push some back out. This is called *overfilling* and is used for complicated shapes or where the powder exhibits poor flow. An alternative is to suck powder into the die cavity by putting the feed shoe over the tool cavity and then dropping the lower punch. Once filled, the powder mass is dropped to the pressing position to allow clearance for entry of the upper punch.

During compaction the two punches effectively work toward the compact center. After compaction, the upper punch retracts and the lower punch pushes the compact out of the die cavity. Now the cycle repeats with the feed shoe pushing the compact away from the die cavity.

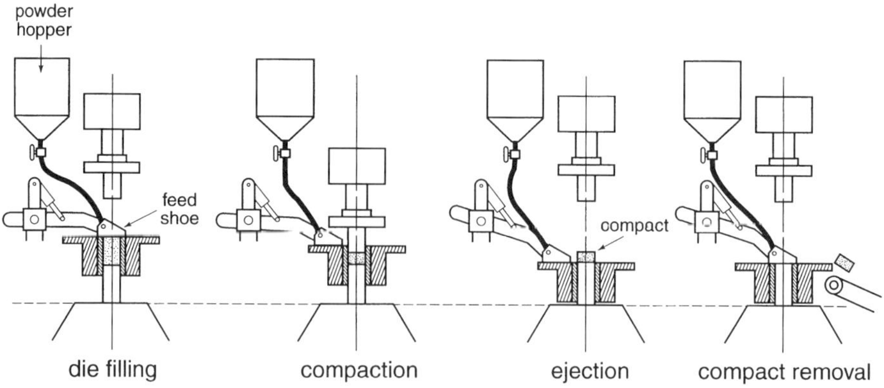

Figure 5.24. Powder filling by the feed shoe is coordinated with the die and punch motions. This sequence of four positions shows how the feed shoe can be used to push the compact away as it starts a new fill cycle.

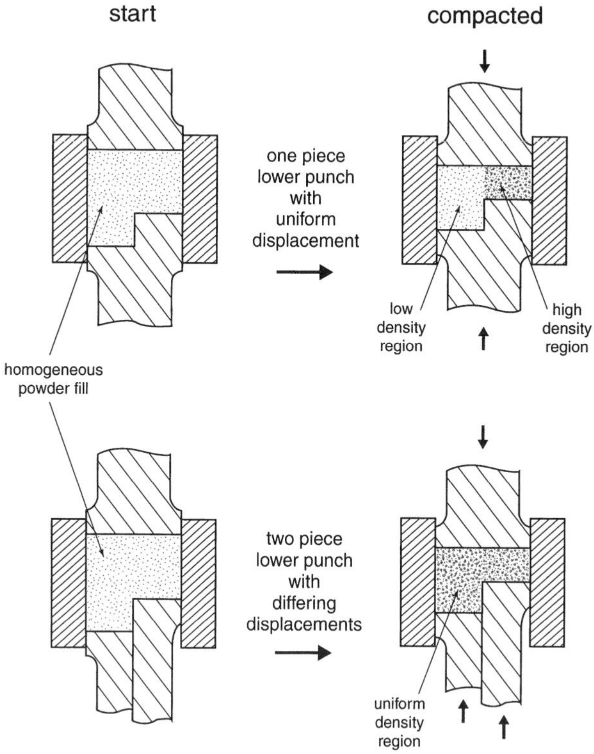

Figure 5.25. Differential motion in the lower punch is required to compensate for the different travels in compaction required to produce a constant green density. If the lower punch is simply notched, as illustrated on the top, then the thinner section will be much higher in final density. However, if the lower punch is segmented and programmed for differential motion, then uniform densification is possible.

Figure 5.25 illustrates the difficulty in forming a two-step component. The lower punch is split to give differential travel between the thick and thin section. Otherwise there is a conflict, with either the thick section remaining undercompacted or the thin section becoming overcompacted. Let us assume the thick and thin sections differ in final thickness by a factor of 2. Then for equal densification the lower punch for the thick section must travel twice as far as the lower punch in the thin section. This is not possible with a single lower punch, because the two halves require different fills. Thus, segmented punches are used, with different travels in compaction, to form multiple levels at the same densities.

Uniform powder fill becomes even more difficult for steps located near the compact center, such as in Figure 5.26. For example, if a stepped flange is located in the compact center, then an early press motion is required to shift the powder into proper location. The upper punch is segmented to perform this powder shifting or transfer after die filling. Punch segments forming thin sections are spring-loaded

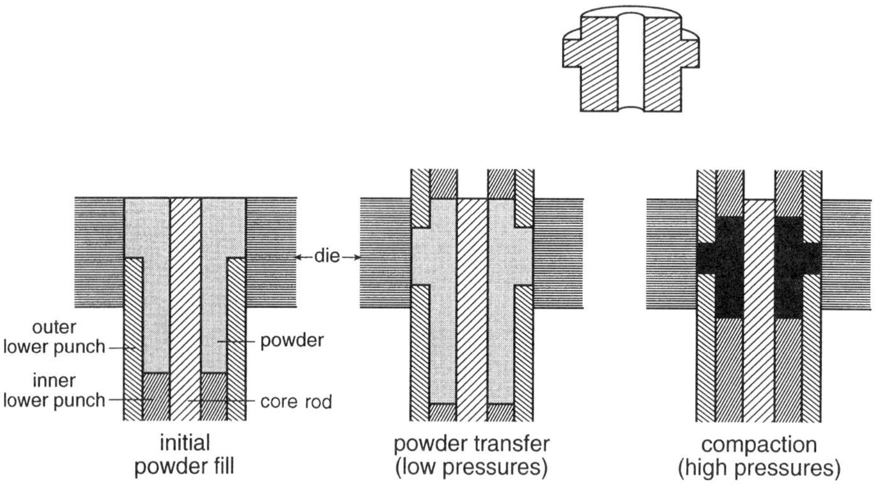

Figure 5.26. Use of an early outer punch motion to reposition powder for more uniform compaction. After filling, the top surface of the powder is flat, while the desired position is indicated in the center where powder transfer has occurred. High pressures are applied only after powder transfer is completed.

or otherwise controlled to push powder into the desired column height prior to compression. The option shown in Figure 5.26 is most effective in repositioning the powder prior to compaction.

Segmented punches need to be coordinated for motion when retracting from the compact to ensure that the compact does not stick to the upper punch. The actual motions are also coordinated to avoid cracks from differential stresses as the sleeves, core rods, and segmented punches are extracted.

It is common to find components with up to five levels in production. With this many steps, tooling becomes very complex to accommodate successful powder transfer and compaction without defects. Safety switches and shear pins are recommended in tool sets to avoid excessive stresses that might cause failure. It is easier to replace a shear pin as compared to a main crank, gear, or cam.

5.5.4. Tool Motions

The pressing sequence involves feeding powder into the cavity, pressurization of the powder, and ejection of the compact out of the tooling. The simplest case is for only one punch to move, with the die body and lower punch remaining fixed as sketched in Figure 5.27. Both the lower punch and die body remain fixed, as contrasted with double-action compaction.

As an alternative, the upper punch can be replaced by a fixed platen, leading to the concept of anvil pressing. A sketch of this variant is given in Figure 5.28. Here, the lower punch presses the powder against the die top, which is closed by a flat plate. Anvil pressing proves very rapid and is widely used for the production of simple shapes in high production volumes.

For complicated shapes these two options prove incapable of forming dense compacts with minimal density gradients. Accordingly, most pressing is performed with double action, meaning that the top and bottom punches move toward the die cavity center, although in many cases the lower punch remains fixed and the die body moves down.

Thus, the motions during pressing are sketched in Figure 5.29 for an opposed-ram press. Both punches move toward the die center during pressing, and ejection corresponds to the lower punch reaching the top of the die body. Withdrawal would achieve a similar situation, but now the die body moves while the lower punch remains fixed. This floating-die concept is illustrated in Figure 5.30. Differences in the form of ejection trace to press design decisions. After compaction, extraction of the compact from the die can be by moving the die down against the lower punch, termed *withdrawal*, or moving the lower punch into the die, termed *up-acting*.

5.5.5. Tool Design

A first consideration in designing tooling is the fill height needed to form the final compact height. The compression ratio and final height determine the needed fill.

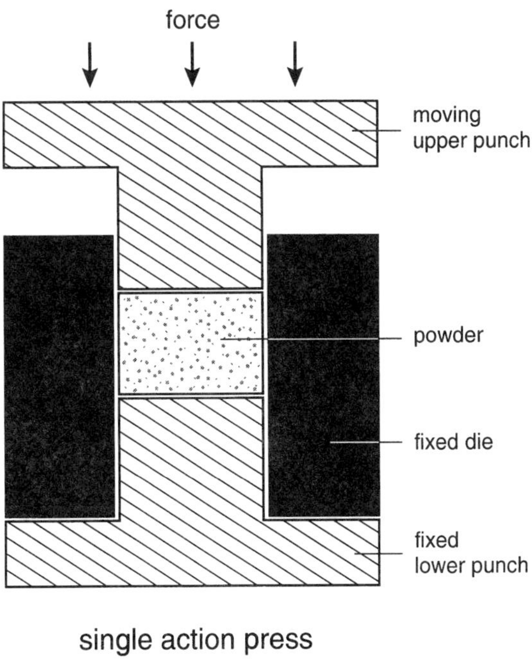

Figure 5.27. A simple uniaxial, single-action compaction cycle. The lower punch and die remain stationary while all the motion occurs in the upper punch.

The die space in the press determines the maximum fill height and the ejection stroke possibilities. Small presses can accept only small fill heights, so a marriage is needed between the tooling and press to ensure compatibility for thick components. Next is consideration of the compaction pressure, which determines the load for the cross-sectional area of the component. Care is needed to ensure that tool components are not stressed beyond their strength.

Depending on the actual component production scheme, some dilation of die dimensions will be needed to accommodate dimensional changes associated with ejection springback, thermal expansion in heating, shrinkage or swelling in sintering, and phase change in heat treating. For full-density products, these require that the die be 4 or 5% larger than the final component dimensions. However, more typical are low-dimensional-change systems where the die size is very close to the size of the final compact, usually differing by 0.1–0.3%.

Various retainers, adapters, and clamps are required to affix the tooling into a press. These members of the tool set must be sized to withstand the loads from

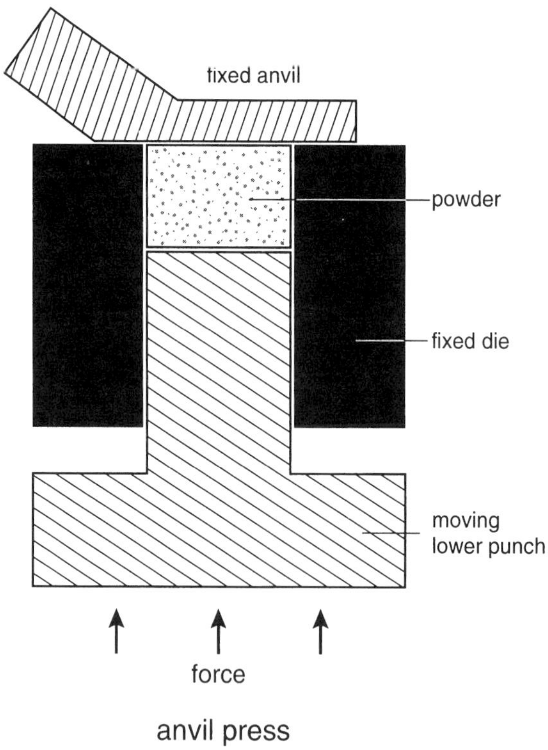

Figure 5.28. Anvil pressing is similar to single-action compaction, only the upper punch is replaced by an anvil that closes over the die cavity after powder fill, but without entry of an upper punch. This is suitable for simple compacts produced in high volume.

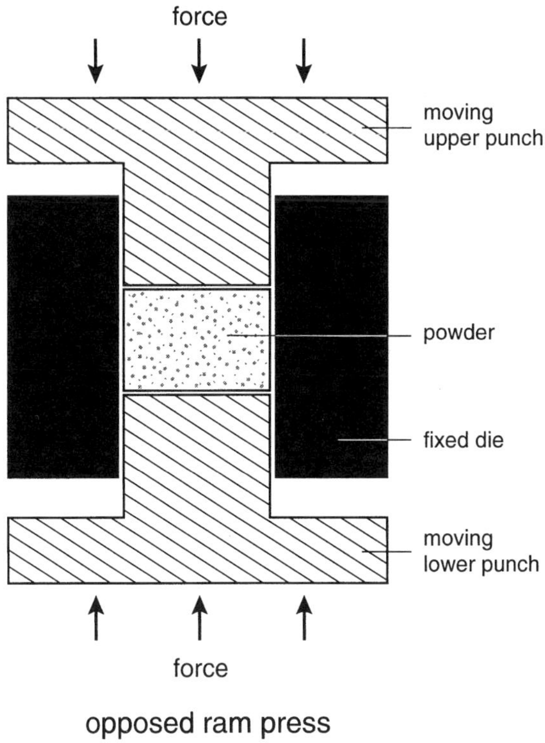

Figure 5.29. Opposed-ram compaction involves the motion of both punches toward the stationary die center. This is widely know as *double-action compaction*, indicating that the lowest density will occur in the compact center.

pressing and ejection. Additionally, sufficient bolts are needed to ensure that the components of the tooling remain fixed during service. The punches and die are clamped to the press and may need adapters to ensure proper fit. A sketch of such an arrangement is given in Figure 5.31. Most of the components of the die set are the same, independent of the final geometry. However, the die cavity and punches are custom-machined for the part specification.

5.5.6. Tool Life

One of the major fears in tool design is premature failure of a punch. Thus, stress calculations are performed to ensure that the punch size is sufficient to avoid yielding, bending, buckling, or fracture. If the punch buckles, then it is too long and thin. These calculations are based on standard strength of materials principles and are often supplemented by computer analysis of the tool stresses. Requirements for the tool material strength emerge from these calculations. Short-term failure indicates a low yield strength, while long-term failure indicates machining defects or tooling with a low fatigue strength. Sharp corners accelerate fatigue failure, so most tooling is polished to remove scratches, and any corner that can be rounded is given a radius to reduce stress concentration.

Besides prolonging tool life, many subtle design changes can greatly reduce the machining cost. Even so, tool construction is expensive, largely because of the close tolerances required for proper function. The tooling is usually toleranced for an allowed variation of 0.01% of the dimension. A tight fit is required to keep powder from working between sliding tool pieces. Since the particle size is usually over 20 μm, the fit between tool members must be at least this tight. A problem arises in ensuring close tolerances and tight fits in high-hardness tooling. Most steels warp slightly in heat treatment. Thus, cemented carbides machined using electrical discharge techniques are widely used in P/M.

When possible, a slight taper is desired to ease ejection of the compact out of the die. Avoiding thin sections and thin punches is important to prolong tool life and minimize defects on the compacts. Wall thicknesses below 1.5 mm (0.06 in) should be avoided. Likewise, the tool set will undergo deflection during loading, where punches compress and supports bend. These deflections are small, but can impact final dimensions and tolerances. Further, on multilevel compacts, deflection recovery on relief of the compaction load causes differential motion in the die cavity

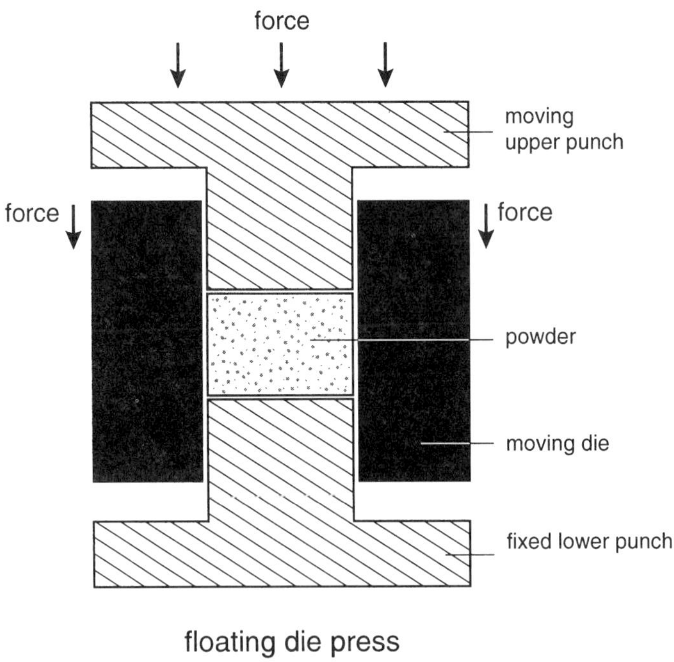

Figure 5.30. Floating-die compaction uses a fixed lower punch and motion of the die toward the lower punch in coordination with the upper punch motion, resulting in the same effect as double-action compaction.

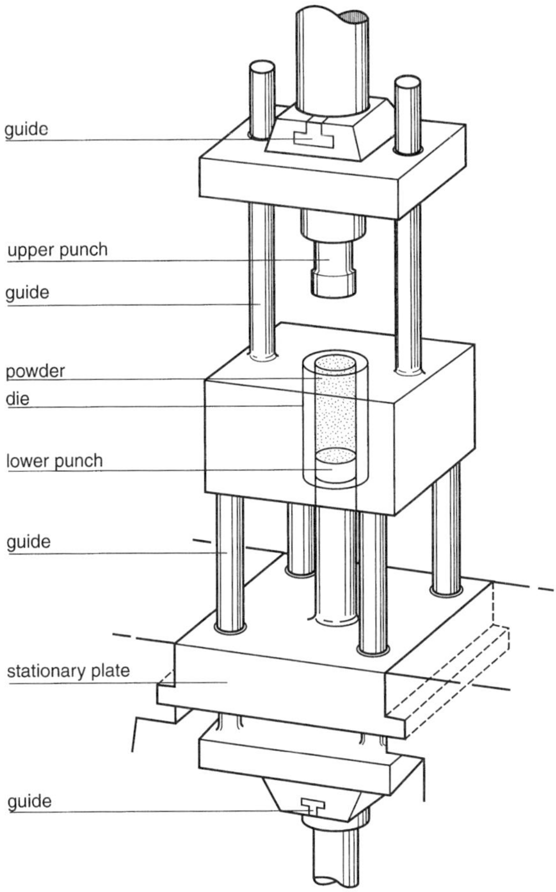

Figure 5.31. The die set is illustrated for a simple cylindrical compact. Die sets are often adjusted as packages that are inserted directly as an assembly into the press via a few guides.

with possible compact cracking. Accordingly, calculations are required to ensure that the relaxation is uniform, meaning that tooling must be sized to equate dimensional changes on unloading.

With experience, it is possible to produce long tapered sections and helical gears via die compaction. Internal tapers are formed using conical core rods, while external tapers require a conical die cavity. Tapers should meet at flats to ensure good tool matches. Helical components twist on ejection, with the lower punch also twisting to have the gear teeth follow grooves in the die cavity.

Most tool design is assisted by computer-aided engineering packages. These allow accurate drawings and analysis of form, fit, and function. Finite-element analysis is often used to assess the stresses in tool sets, and press manufacturers provide codes for analysis of press frame deflection. Of course, such efforts require knowledge on the part size and shape, as well as powder characteristics.

5.6. COMPACTION PRESSES

5.6.1. Press Types

Compaction presses are available in many sizes. The array of features is large and often confusing. In a simple analysis, the easiest way to distinguish between presses is based on *tonnage*. This is the force required to press a part, and is determined by the cross-sectional area of the part (or punches) perpendicular to the pressing direction and the required compaction pressure. The product of pressure times area gives the tonnage. Small components are fabricated on low-tonnage presses, and large components are fabricated on high-tonnage presses. Presses range up to nearly a 1000 tons [9 MN (meganewtons)]. However, the bulk of P/M compaction is performed on presses in the 60–250-ton range. Table 5.6 lists examples of feature combinations found in die-compaction presses. These examples include mechanical, hydraulic, anvil, and computer-numerical-controlled (CNC) presses.

A distinguishing factor on compaction presses is the number of part levels. Each distinct step in height (any change of 10% or more in height) in the pressing direction usually requires a separate punch motion. These levels might be on either the upper or lower punch. Great complexities in the press and tooling result from multiple level parts, so tonnage is only part of the equation. Fill depth is another limitation. For thick parts, the fill height must be large, but this also increases the press size. Indeed, the whole fill mechanism is important to press selection. Newer presses have die wall lubrication systems that coat the die prior to powder fill, while other

TABLE 5.6. Examples Press Configurations

Maximum Force [kN (ton)]	54 (6)	120 (13)	150 (17)	250 (28)	900 (100)	12000 (1300)
Type	Anvil	CNC hydraulic	Mechanical	Hydraulic	Mechanical	Hydraulic
Withdrawal force, kN (tons)	—	90 (10)	80 (9)	200 (22)	300 (33)	6000 (670)
Ejection force, kN (tons)	77 (3)	120 (14)	50 (6)	250 (28)	400 (44)	8400 (930)
Upper stroke, mm (in)	—	200 (8)	100 (4)	200 (8)	230 (9)	425 (17)
Lower stroke, mm (in)	28 (1.1)	100 (4)	35 (1.4)	100 (94)	75 (3)	200 (8)
Fill height, mm (in)	28 (1.1)	90 (3.5)	65 (2.6)	90 (3.6)	150 (6)	300 (12)
Maximum rate, strokes/min	150	30	50	30	24	2
Power, kW	1	18	3	33	200	130
Height, m (ft)	1.7 (5)	2.6 (8)	2.5 (8)	2.8 (9)	6 (19)	7 (22)
Weight, tons	0.7	4.6	2	6	45	46

presses time the fill to coincide with positioning of the lower punch to the fill position, thereby creating a vacuum to assist powder fill. Other controls on the powder feed process include the number of passes over the die cavity, dwell time, agitation velocity, and relative punch motion during fill.

The compact pressurization mechanism is another distinguishing feature of the presses. The variations include hydraulic, mechanical, pneumatic, and CNC systems. Hydraulic presses rely on a central pressure system and a network of servohydraulic valves to proportion hydraulic pressure to the pressing components. Mechanical presses have an electric-motor-driven flywheel. Cams and levers take power from the flywheel to move the press functions in a coordinated manner. Anvil presses form simpler shapes, but provide up to 150 parts per minute. A novel concept uses a circular motion to the top punch to give a rotary forging to the powder—a concept termed *rotoforging*. Newer presses address the need for independent motion of complex tool set components via electric motors, using concepts developed in computer controlled machining centers. Key to the success of computer numerical control is a closed-loop feedback system. In contrast to the hydraulic press, which achieves a preset pressure, or the mechanical press, which moves to a preset height, the CNC system allows constant assessment of the pressing situation to ensure that all motions are properly performed.

Hybrids exist that select certain features from each press type. In some systems the punches have pneumatic pressure buffers to control the peak pressure: The core technology is a mechanical press, but the upper punch has a pressure cylinder for controlling the peak pressure like a hydraulic press.

A major difference between mechanical and hydraulic compaction is in the response to subtle variations in powder fill. For the hydraulic press, the end of the compaction stroke is usually set by the peak pressure. This ensures uniform density in the components, but height variations depending on the uniformity of powder flow. In contrast, the mechanical press pushes the powder to a precise final dimension, but density fluctuates in proportion to the powder fill variations. CNC presses excel in this respect. All motions can be essentially programmed independently using the computer system, and for smaller presses the motions are achieved using servoelectric motors. Although more costly, these devices have achieved the greatest precision in production of complicated components, where complex positioning in the fill and pressurization cycles are required to avoid cracking.

Thus, in terms of press design, the partitioning of technology ranges from single level presses that use anvils instead of the upper punch, to computer numerical control, multilevel presses with die set systems for high-precision production of complicated multilevel components.

5.6.2. Pressing Cycles

The majority of iron and steel compaction is performed on mechanical presses. In this technology, the motions of the punches and die are coordinated to a single revolution of the main flywheel or drive shaft. All motions are referenced to a rotation, so common nomenclature is to relate pressing stages to the degrees of rotation, recognizing that 360° represents a full rotation of the crank. Figure 5.32 illustrates a cylindrical compact and plots the relative position of the die and punches during compaction as a function of the crank rotation. Further, this figure provides a schematic of the die and punch positions as related to press motions for this same cylindrical compact.

Figure 5.33 is a press in cross section to provide an internal view of the mechanical links. Main power is provided by an external electric motor that works the tool motions through a set of reduction gears. Powder is loaded into the die cavity by a feed shoe that sweeps over the opening right after compact ejection. Note the upper punch on the top crosshead is attached to the main drive shaft via an eccentric gear. Cams provide die support and coordinate compact withdrawal with the cycle.

Powder fill from the feed shoe is driven by a cam that coordinates ejection and filling with the rest of the compaction cycle. The upper punch position is determined

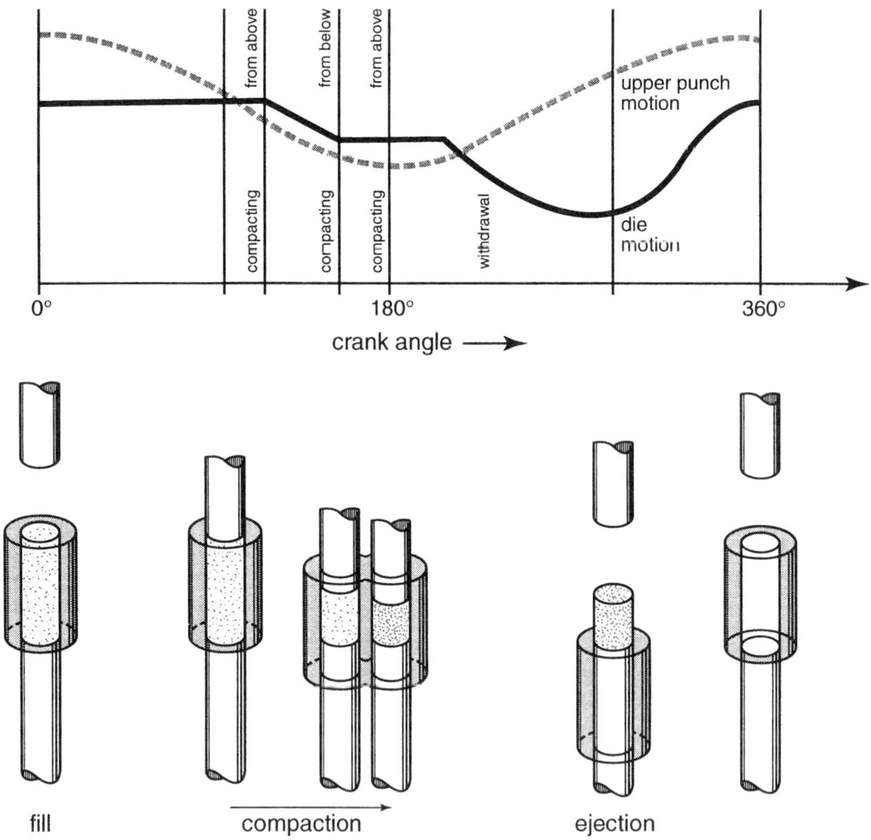

Figure 5.32. Press motions plotted as a function of the crank rotation angle. The situation drawn is for the double-action compaction of a simple cylinder with a fixed lower punch and floating die.

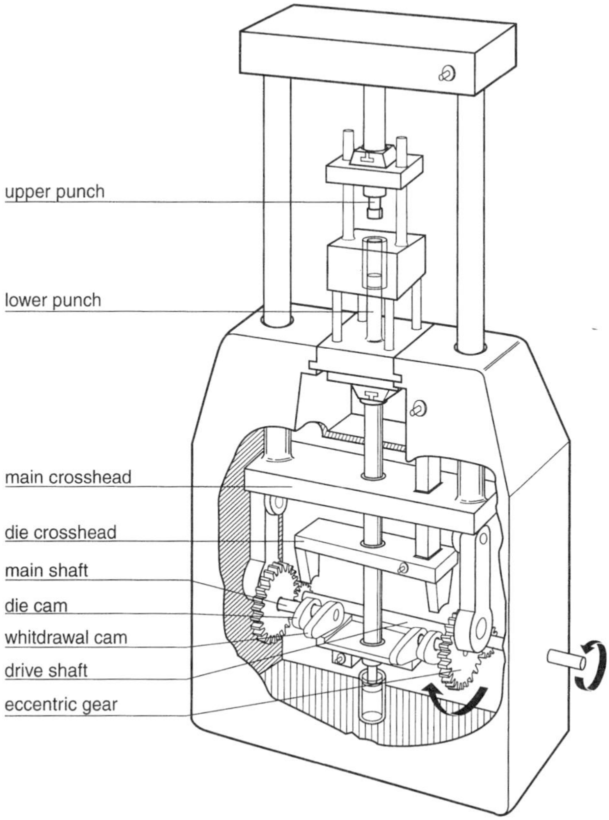

Figure 5.33. A simplified sketch of a die compaction press, showing an exposed view of the motions that drive the punch and die from a common power supply.

not by the eccentric, but by the relative top crosshead position with respect to the lower punch. For example, a longer punch is used to form a smaller compact, as illustrated in Figure 5.34. During compaction, the motion of the die from the fill position to the pressing position is controlled by its travel. While powder fill occurs, the die is at its upper position, but during compaction it floats past the lower punch. This is the source of double-action compaction. One of the main objectives in controlling tool motions is to ensure equal pressing from both the top and bottom of each level in multilevel tooling.

The end of the die float is controlled by the top position of the die support cams. Motion of the die with respect to the degree of crank rotation is controlled by adjustable stops, springs, and mechanical linkages. The ejection of the compact is controlled by the rotation of the withdrawal cam. In some complicated cases where the compact might experience cracking during ejection, the upper punch remains on the compact to provide a hold-down pressure. It is adjusted to ensure that the lower punch just reaches the top of the die cavity.

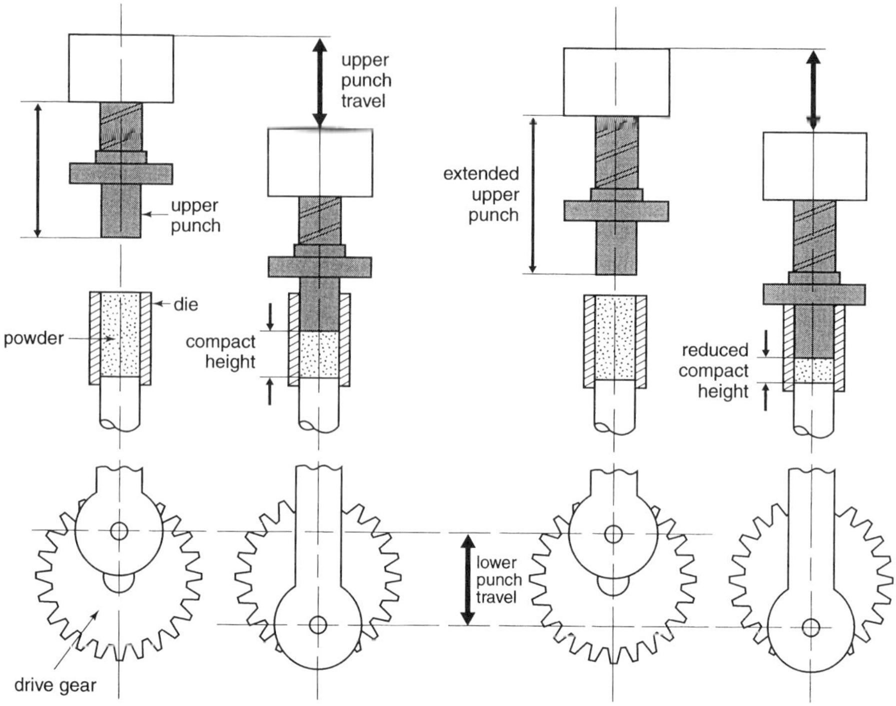

Figure 5.34. Movement of the top ram is coordinated by the crank position. To adjust the compact height (for a constant fill), the travel of the ram is adjusted by the effective position of the upper punch at the end of the compaction stroke.

The feed-shoe motion is usually coordinated to push the ejected compact away and deliver fresh powder into the cavity. The lower punch moves to the fill position. Since each P/M compact is different, the relative motions and forces are individually controlled and coordinated by cranks, motors, or valves. Only in a few instances are the setup functions computerized; thus, the bulk of P/M relies on experts to manually adjust each press startup to ensure proper powder flow and compaction.

5.6.3. Sensors and Controls

Most presses operate in an open-loop control mode. The position of the various tool members is adjusted at the beginning of a production run, but no monitor is provided for following operation conditions. Parts are selectively checked for weight and dimensions, but the preservation of the desired tool motions, pressures, fills, and other characteristics of compaction is not ensured.

A few advanced presses include pressure sensors, displacement transducers, strain gauges, temperature sensors, position sensors, and other monitors needed to watch for consistent production. Rarely does this advance to in situ control where instantaneous corrections are made, although a few hybrid computer controlled presses are now installed. If sensors are used, they usually measure the die temperature and strain, upper and lower punch loads (via in-line transducers), and relative displacements of the tool components. Hydraulic and CNC presses prove easier to instrument and control. For example, in-line load cells can be used to adjust pressure on the hydraulic cylinder via a servohydraulic valve. This results in one form of a hydraulic CNC press, best suited to complex, multilevel parts. The result is more consistent densities. Likewise, a few presses have multistep feed shoes. The first step weighs the powder charge to ensure that the proper amount is used in filling.

Computerization of die compaction has two objectives. One objective is to record the operating data, including setup parameters and production data. This allows initialization at the same conditions each time the tooling is installed. In one variant, the tool setup is fully automated using computerized electric motors. The second is to monitor trends in parts. Part-to-part variations and differences between production lots are both concerns. These require monitoring systems and logic systems. Robots might be coupled to the press for component extraction, placement on sintering trays, or inspection. These, too, are coupled into the logic and control system. Pick-and-place robots are used on large-volume production items, but much of the P/M industry relies on human intervention should there be a problem.

Monitoring is largely through load measurement in the pressing direction. Logic systems then decide what changes are appropriate when drifts in product quality are detected. Most important, these systems can stop production in cases where tool damage might occur. For example, if the powder fill changes, then excessive powder in the cavity might deform a punch. A load monitor during compaction would detect such a change and halt production. Generally only three monitors are used to detect product quality drifts at the press—changes in weight, compaction load, or dimensions. The latter parameter might be manually input to the control computer via an operator.

Only in a few instances have presses been equipped with robots and automatic inspection devices, so there is much opportunity to bring advanced sensors and controls into die compaction. This is an area where die compaction is significantly behind injection molding and many other manufacturing routes. For example, in situ ultrasonic inspection has been demonstrated using guided waves from transducers located on the die cavity. These transducers detect cracks, density variations, or other problems during compaction. Unfortunately, production operations are slow to embrace advanced inspection process.

5.7. DEFECT DETECTION AND CONTROL

One of the great barriers to expanded use of P/M for complex components has been poor understanding of defect avoidance in multilevel shapes. There are three main causes of green cracks: first is in the tooling materials and surface finish, second is in tool motions and powder flow during compaction, and third is in ejection. As noted earlier, the compact experiences considerable friction with the tooling, during

Figure 5.35. Optical micrograph of cracks at the junction of a straight wall and taper in a die-compacted component. (Photograph courtesy of James Marsden.)

both pressing and ejection. The strength of the green compact is low, yet the ejection stresses are high. Cracks, such as photographed in Figure 5.35, occur when the stresses exceed the low strength of the compact. Consequently, a rough tool finish, poor taper, misalignment, or other tooling defect that induces stress (friction) may cause cracks. The cure is to use slight tapers to the ejection portion of the die, smooth surface finishes, and immediately repair defects on the working surfaces of the tooling.

Assuming the tooling is not a problem, then the next most likely cause of green cracks is improper control of the tool motions during compaction. On multilevel components, the powder must be shifted to proportion for the same compression ratio at each level. When tool motions are not sequenced, then one region will be compressed early. When a neighboring region is compressed, it flows past an already densified region. The consequence is poor bonding along this interface. Characteristically, this defect will orient parallel to the press axis.

Cracks oriented perpendicular to the press axis (oriented in the die plane) usually occur during ejection. Figure 5.36 illustrates the bending of a cored compact during ejection. The friction between the core rod and die wall results in bending and cracking. One cause is a difference in stress relaxation of tool components. The elastic strain is proportional to the tool length. Longer tool components relax more than do shorter tool components; thus, a differential strain emerges after compaction. So even though proper ejection is expected, part of the compact might be unsupported. A solution would be to ensure that tool motions are known (or calculated) during ejection for positive contact with the compact.

What are the cures for cracks? One is to attain a high green density with a high green strength. This might mean reducing the internal lubricant and going to die

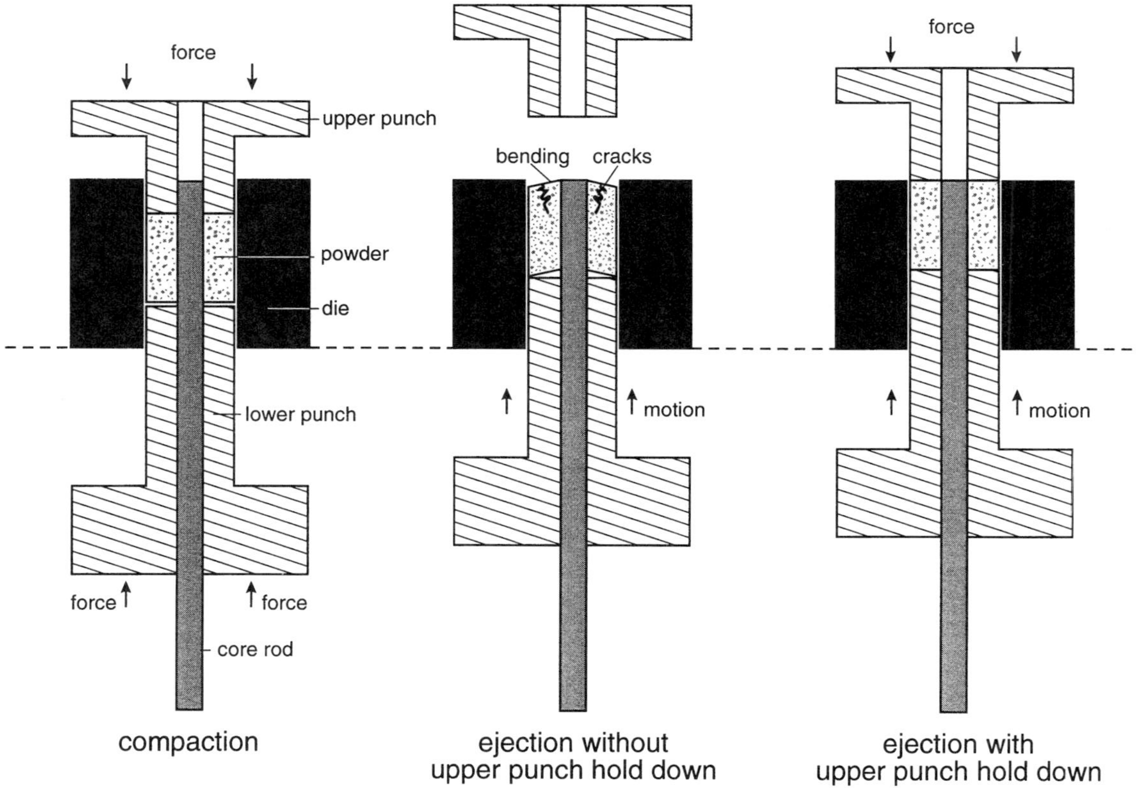

Figure 5.36. Compaction and ejection can lead to component cracking. This illustration shows how friction causes the compact to bend on ejection, leading to cracks. One remedy for this problem is to use upper punch hold-down, where pressure is kept on the compact by the upper punch during ejection.

wall lubrication, or compacting to a higher density, especially in the region that exhibits the crack. Another technique is to use external pressure from the upper punch during ejection. As illustrated in Figure 5.36, when the upper punch is used to pressurize the component during ejection, bending stresses are offset by the compression of the weak portions of the component. It is usually bending (induced by improper tool motion or die wall friction) that causes the problem, so eliminate bending by holding the shape rigid during ejection or eliminate the friction.

One other cause of cracks is the feed shoe. If it strikes the ejected part too aggressively, it might induce a crack or chip. This can be assessed by inspection of an ejected component that has been manually extracted from the press.

Usually, cracks originate at corners between part levels, especially when there are density differences or differences in support during ejection. They can be detected via microscopy, X-ray imaging, electrical conductivity, ultrasonic pulses, or other standard techniques. Green compact cracks do not heal in sintering. Hence, rejection early is important to avoid further wasted expense. With new computer modeling tools, most of the causes of cracking are evident in tool design and compaction simulations. With proper tool maintenance and closed-loop computer control of the press, it would appear that over time green component cracking will be eliminated. But until the bulk of the P/M compaction technology embraces these new tools, there will be a need to inspect for cracks during production setup.

5.8. WARM COMPACTION

One means to improve green density of die-compacted iron powder is to heat the powder prior to compaction. Usually this is combined with a polymer addition that glues the particles together. Various technologies known as *warm compaction* are emerging to perform simultaneous heating and bonding during pressing. Both the powder and the tooling are heated in warm compaction, giving a slight softening of the iron that produces a slight green density gain. A typical temperature for the powder and tooling might be 150°C (300°F), and compaction pressures are usually in the range of 700 MPa (50 tsi).

The polymer-coated powder is more costly than typical die-compaction grades, unless a simple lubricant is admixed with the powder. Various common stearates or other lubricants work, including Teflon. Heating of the die and punches requires modifications to the compaction press, and a heater is required in the powder feed mechanism. Both microwave and hot-oil heaters are available for bringing the powder up to a temperature near 150°C (300°F). Depending on which polymer is selected for coating the powder, ejection forces can be highly variable. Close temperature control is necessary, since product uniformity suffers if the polymer is too hot. In tests with various powders, the green density usually increases by 0.15 g/cm^3 over room-temperature compaction. After cooling to room temperature, the warm-compacted powder is stronger because chilling the polymer adds strength to the compact. However, there is no evidence of greater strength during ejection, which means that green cracks from ejection stresses are not reduced by warm compaction. Consequently, a hold-down pressure is required during ejection to avoid cracking.

Final properties of the sintered compacts are slightly improved over conventional compaction, due to the higher green density. When conventional compaction is performed to the same density, the sintered properties are equivalent. For example, compaction of an alloy consisting of Fe–2Ni–2Cu–0.9Mo–0.8C gives a sintered density [1300°C (2370°F) for 50 min] of 7.31 g/cm^3 following compaction at room temperature, and 7.32 g/cm^3 after warm compaction at 150°C (300°F). A comparison of properties shows no significant differences, although the room-temperature compact was generally harder (103 vs. 101 HRB), higher in impact toughness (23.5 vs. 23.1 J), and required less ejection force (31 vs. 35 kN).

The major role of warm compaction is in lowering the pressure required for attaining densities over 7.0 g/cm^3. Unfortunately, if warm compaction is performed at too high a temperature, then polymer degradation leads to lower sintered properties. Data from Fe–2Ni–0.9Mo–0.6C at 130 and 150°C (265 and 300°F) show green densities just below 7.3 g/cm^3 for warm compaction at 700 MPa (50 tsi), while conventional compaction gives 7.1 g/cm^3. After sintering, the steel warm-compacted at 130°C was the strongest, followed by the steel pressed at room temperature, with the steel warm-compacted at 150°C being lowest. Thus, precise temperature and process control are required to fully realize the benefits possible via warm compaction.

5.9. COLD ISOSTATIC COMPACTION

Although uniaxial die compaction is the dominant mode of powder consolidation, by no means is it the only approach. For complex shapes involving undercuts or large length : diameter ratios, cold isostatic pressing (CIP) is a common process. It relies on a flexible mold to apply pressure to the powder from all directions. The same rules apply for densification versus compaction pressure and green strength versus compaction pressure.

For the CIP process, powder is filled into the flexible tooling and the sealed tooling is pressurized in a vessel filled with a fluid such as oil or water. Lubricants are rarely used. However, for low-green-strength materials some wax can be added for increased green strength. Compaction pressures of up to 1400 MPa (200 ksi) are possible; however, cold isostatic pressing is usually performed at pressures below 420 MPa (60 ksi).

The tooling is rubber, latex, polyurethane, or polysilicone [room-temperature vulcanizing or (RTV)]. These are selected because they can be dipped or cast around a master mold. If the tooling is soft, it may become trapped in the pores between particles, leading to further problems, so hard and stiff polyurethanes are most popular. Otherwise, when pressure is released, the expanding tooling fractures the compact. For longer tool life and less sticking to the powder, it is best to use hard rubbers. The rubber mold can be complex in shape, but must be oversized to compensate for powder compression. During pressurization all seals must be under compression to prevent leakage of the pressurization fluid into the powder. Usually, an external support is required to help hold the bag shape during powder loading, and in some situations the bag is evacuated to remove air prior to compaction.

The two variants of cold isostatic pressing are the wet- and dry-bag techniques. With a wet bag, the filled and sealed mold is immersed into a fluid chamber that is pressurized by an external hydraulic system as illustrated in Figure 5.37. This particular conceptualization would be appropriate for forming a hollow tube by pressing the powder against a solid core. After pressing, the wet mold (or bag) is removed from the chamber and the compact extracted from the mold.

The dry-bag approach is favored in high volume production, because the bag is built directly into the pressure cavity. As diagramed in Figure 5.38, the flexible bag deforms but is not ejected. Two end plugs allow powder loading and component unloading on each cycle, without removing the bag assembly. Sealing is achieved by an upper punch that enters the bag before pressurization. In that situation, compaction stresses are generated by isostatic compression of the bag through a hydraulic fluid.

Isostatic pressurization is useful for making large, thin-wall, long, or homogeneous compacts. For example, one widely discussed application is to form steel cylinder liners for automobile engines. These are long and thin-walled, yet larger components

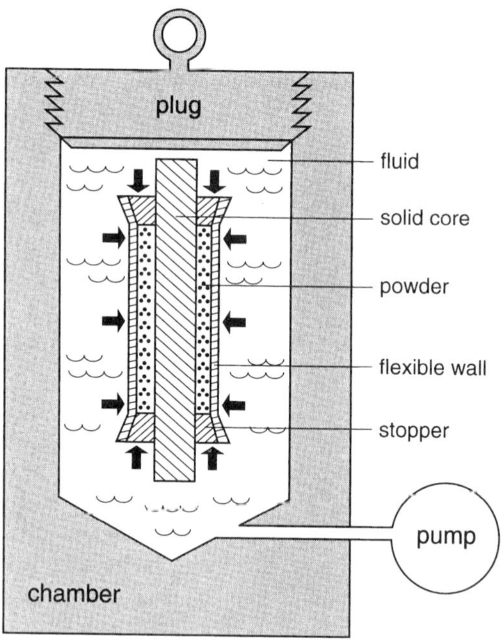

Figure 5.37. Cold isostatic compaction relies on a high-pressure chamber that is pressurized by an external pump. The powder is contained in a flexible bag, tube, or container that transmits pressure to the powder. It is a hydrostatic compaction process, since pressure is delivered from all directions. The process is best applied to thin-walled or long structures.

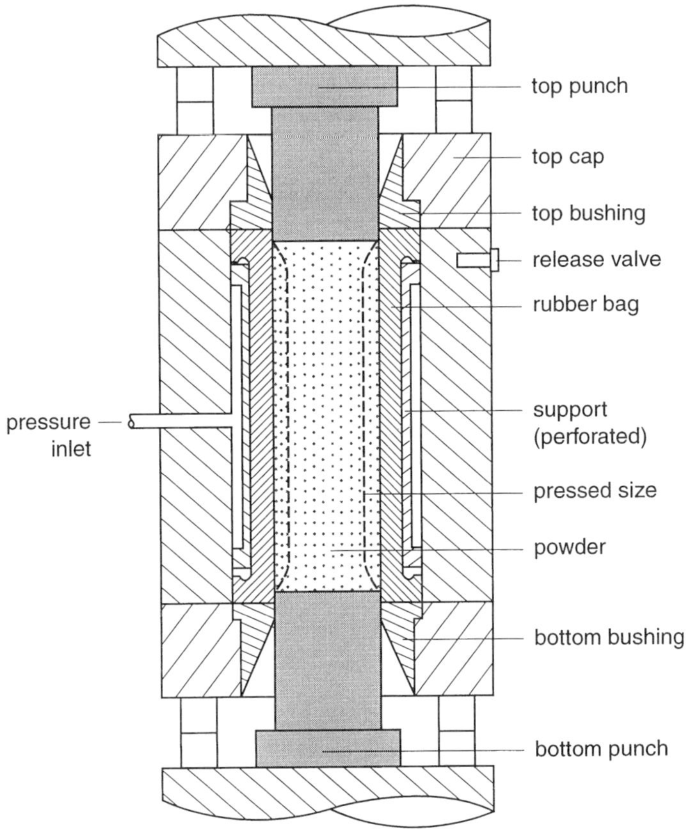

Figure 5.38. A cross-sectional view through a dry bag cold isostatic pressing operation designed to produce tubes. The flexible tooling is built into the chamber with powder loading and component unloading through the end caps.

that would be near impossible to fabricate by die compaction. Isostatic pressurization more efficiently reaches higher densities compared to die compaction. Also, density gradients are small in isostatically compacted components, allowing large sintering shrinkages without distortion. For this reason, it is widely applied to forming cutters from tool steels. Additionally, isostatic pressurization allows for more shape complexity, but sacrifices dimensional control and pressing speed. With isostatic compaction, the fluid provides uniform pressures and consequently, uniform density.

5.10. POWDER ROLLING

As should be clear, the initial compact shape can be generated by any of several methods. Although die compaction is the most common, powder rolling is applied

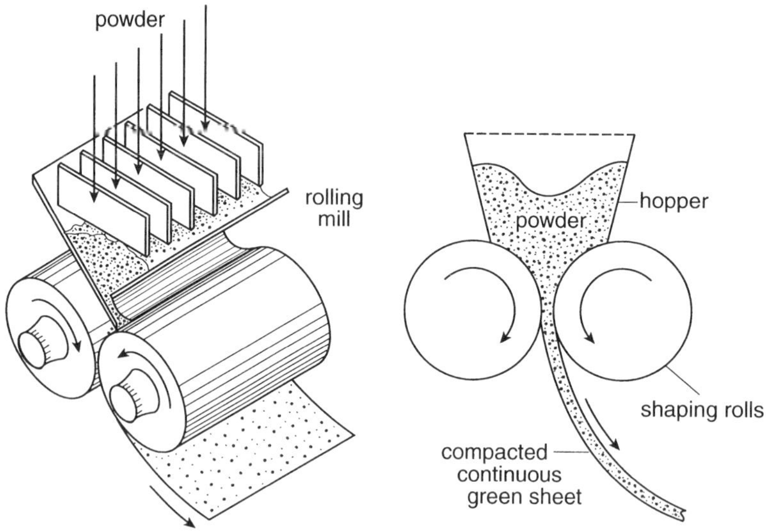

Figure 5.39. Powder rolling is an option for the continuous compaction of powder into a strip. A rolling mill is fed powder, and the compaction process generates a continuous green sheet that is a precursor to forming dense sheet products.

as a first step toward full-density steel sheet. Powder rolling is different from other compaction processes because it is a continuous process. After the first stage of compaction, in a rolling mill, the powder is not very dense. But subsequent rolling passes improve density.

Conventional rolling technology is used to continuously compact powder into a strip and is widely employed for the production of specialty alloy sheets. Loose powder is gravity-fed into the gap between two rolls, which generate the green sheet. Gravity feed is the only effective means to feed powder, since there is no strength to the powder prior to rolling and green densities are very low. Figure 5.39 illustrates a production operation. A binder can be added to enhance green strength, and water works for some powders. Two forces act on the powder as it is compressed. One is the normal force from the rolls that give compaction, but the other is a tangential force from friction between the powder and rolls. The compaction force must exceed the friction force in powder rolling. This is controlled by the roll diameter, gap size, and thickness of the powder feed. The allowed height reduction depends on the rotation angle from where the powder first contacts the roll to the point at which full compression occurs. For some powders this angle is as small as a few degrees.

The higher the initial powder density and the greater the reduction in thickness on rolling, the higher the final density. However, powders with high apparent densities have less friction and undergo only small thickness reductions in rolling. Increased friction, by use of irregular powder or roughened rolls, allows a larger angle of contact and more reduction in sheet thickness. Accordingly, annealed iron formed by water atomization is an ideal feed powder. Green densities after the first rolling pass range from 4.5 to 7.0 g/cm^3. The lower the particle hardness, the higher the green density. Common product thicknesses are in the 1–4 mm (0.04–0.16 in) range. The efficiency of densification initially increases with the application of a higher roll force; however, at very high green densities the process loses efficiency. With rolling mill gaps below 0.1 mm (0.004 in), the process will not work.

In a single-pass rolling operation, densities near theoretical have been achieved, but multiple-pass treatments are more common. Obviously, the final product is limited to a sheet form. In the green state, the rolled powder is too weak to be coiled, so immediate sintering is necessary. After the first sinter, repeated rolling and annealing allows fabrication of a full-density strip.

Currently, only a few sites practice powder rolling, and this is typically for specialty alloys. Related techniques involve powder extrusion through dies. Like powder rolling, the compact is formed in a continuous manner, but cold extrusion is rarely applied to steels because of short tool life.

5.11. EXPLOSIVE COMPACTION

Forging is a high-energy process that achieves rapid deformation of the powder compact. The concept of high-energy processing can be extended to include the momentary compaction from an explosive shock wave. Interest in explosive compaction of steels began in the 1950s, but has largely been the subject of research, with little effort to apply it in large-scale production. In early experiments powder was placed in a plastic bag; the bag was then submerged under water and subjected to the blast of a nearby explosive. More recently, the efforts have relied on closed systems that use a piston launched by an explosive. This is sometimes called *dynamic compaction*. The shock wave propagating through the powder both compacts and heats the powder. Under proper conditions the shock event results in full density. Always there is the problem with capturing the compact without causing it to fracture.

Interest in explosive compaction is usually associated with hard or unstable materials. By keeping the consolidation time short, new composites are possible. Two concepts of explosive and dynamic compaction are sketched in Figure 5.40. Depending on the explosive, detonation pressures of 190–350 kbar (about 3–5 million psi) are possible with detonation velocities of 6–8 km/s [13,000–17,000 miles per hour (mph)]. When the explosive is detonated, it launches a piston that produces a shock wave in the powder that gives compaction. Dynamic compaction uses an accelerated projectile that strikes the powder at a high velocity to supply the shock wave. This projectile can be launched by either an explosive or stored energy from pressurized gas. Typical compaction wave velocities are near 6 km/s (about 13,000 mph), with pressure pulses of 30 GPa (300,000 atms or 4.4 million psi), but the pulse duration is only in the microsecond range.

As the explosive mass : powder mass ratio increases, the compact density also increases. This explosive–powder mass ratio must be further increased to handle materials with high inherent strengths. Small particles are difficult to compact using shock waves. Product shapes are relatively simple and often cracked. Iron, steel,

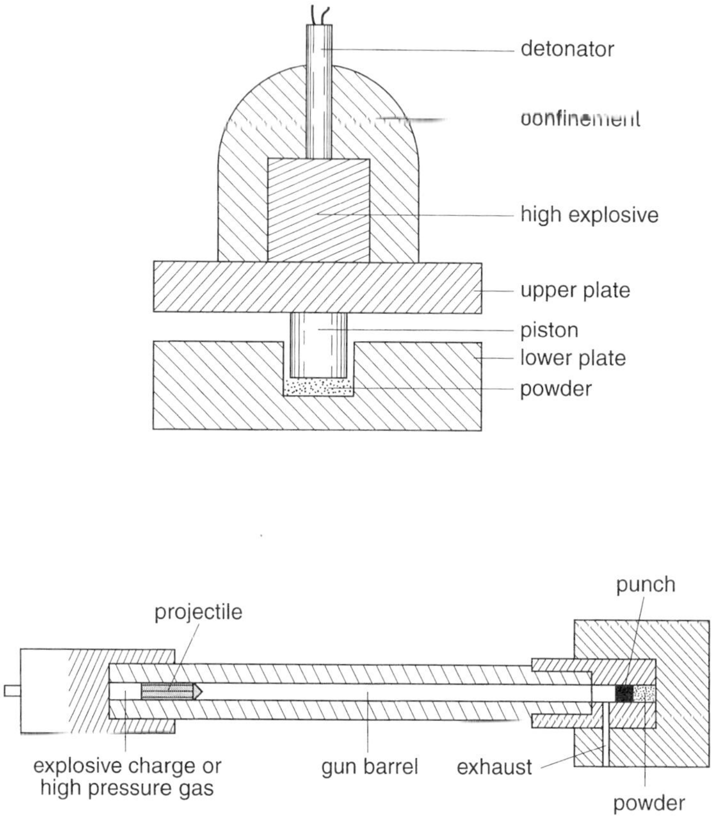

Figure 5.40. Explosive compaction uses a shock wave to consolidate powder. Two common means are shown here. The upper procedure drives a punch into the powder on the basis of direct application of the explosive energy to the punch. The lower procedure uses gas or explosive to accelerate a projectile that impacts the powder.

and stainless steel have been explosively compacted to densities from 7.4 to 7.8 g/cm^3. The best powders tend to be large, up to 1000 μm in size, and compacted strength is low, except when there is partial melting of the particles. Because of the processing difficulties, explosive compaction remains largely a laboratory curiosity.

5.12. SUMMARY COMMENTS

Powder compaction is an important means to shape and densify iron powder. Pressure causes deformation of the particles and induces cold welding to give handling strength. The higher the pressure, the higher the density and strength. However, tool life and press capacity limit the pressures applied in practice. Further, die compaction inherently produces density gradients. In turn, dimensional precision in the final component is impacted by these gradients.

Various efforts are in place to improve die compaction. First, iron and steel powders are being formed with higher purities to improve compressibility. Larger particle sizes, smoother particle shapes, and softer particles aid compaction, so these factors influence the decision on powder selection for high green densities. Also, warm powders and selected polymers aid compaction. This leads to a wave of interest in putting fluid beds, vibrators, and heaters on the powder feed shoe to bring the powder up to 120–150°C (250–300°F). The warm-compaction variant allows slightly higher green densities, in part by softening the iron and polymer. Unfortunately, ejection forces are higher, meaning that tool life is shortened. When room-temperature compaction is pushed to the same green density levels, the sintered properties are the same. Warm compaction simply lowers the compaction pressure required for a high density.

Two other developments should be noted. One is the use of green machining. Here the pressed compact is drilled, cut, or modified using traditional machining tools. Since the compact is weak, machining is relatively easy, but defects are also easy to form. No cutting fluids can be used in machining, so the cuts are relatively simple. For selected components where low-precision holes or undercuts are desired, this is a viable process. The second is repressing. Once a compact is formed, a higher density is possible by applying a second compaction stroke. Instead of ejecting the compact, reapply pressure to slightly increase the compact density. This is still in the research stage, but might be useful for components requiring higher densities.

SUGGESTIONS FOR FURTHER READING

Balazynski, T. Z., (ed.), *Explosive Welding, Forming and Compaction*, Applied Science, Essex, UK, 1983.

Dube, R. K., ''Metal Strip via Roll Compaction and Related Powder Metallurgy Routes,'' *Internat. Mater. Rev.* **35,** 253–291 (1990).

Engstrom, U., G. Veltl, and F. Petzoldt, ''High Density PM Steel by Warm Compaction,'' in *Competitive Advantages by Near-Net Shape Manufacturing*, H. D. Kunze (ed.), Deutsch Gesellschaft Verlag, Frankfurt, Germany, 1997, pp. 265–271.

German, R. M., *Powder Metallurgy Science*, 2nd ed., Metal Powder Industries Federation, Princeton, NJ, 1994, pp. 205–235.

Harnby, N., M. F. Edwards, and A. W. Nienow, *Mixing in the Process Industries*, Butterworths, London, UK, 1985.

James, P. J. (ed.), *Isostatic Pressing Technology*, Applied Science, London, UK, 1983.

Kulkarni, K. (ed.), *Powder Metallurgy for Full Density Products*, Metal Powder Industries Federation, Princeton, NJ, 1987.

Mosbah, P., and D. Bouvard, ''Finite Element Simulation of Die Compaction and Ejection of Iron Powder,'' in *Advances in Powder Metallurgy and Particulate Materials—1996*, Metal Powder Industries Federation, Princeton, NJ, 1996, pp. 7.23–7.37.

Pietsch, W., *Roll Pressing*, Heyden & Son, London, UK, 1976.

Popper, P., *Isostatic Pressing*, Heyden & Son, London, UK, 1976.

Salak, A., *Ferrous Powder Metallurgy*, Cambridge International Science, Cambridge, UK, 1996.

St. Laurent, S., and F. Chagnon, "Key Parameters for Warm Compaction of High Density Materials," in *Advances in Powder Metallurgy and Particulate Materials—1996*, Metal Powder Industries Federation, Princeton, NJ, 1996, pp. 5.125–5.138.

CHAPTER 6

Sintering Behavior

6.1. CHANGES DUE TO SINTERING

Sintering is the key step in transforming the green compact into a high-strength structure. It involves heating the compact to a temperature where the particles weld to each other, much like ice cubes in a freezer bond to one another. Because of strengthening during sintering, it is easy to follow the progress by measuring mechanical properties. Improved strength occurs by events at the atomic level. Weld bonds between particles form and grow through the motion of individual atoms.

Besides particle bonding, sintering contributes to changes in the structure as evident through a microscope, termed the *microstructure*. The grain size increases and can become much larger than the initial particle size. The transition from particle to grain occurs when the particles sinter-bond, forming a structure consisting of many grains. During sintering, the microstructure also exhibits changes in porosity, pore size, and pore shape. Sometimes the particles move relative to each other, giving a dimensional change to the compact. These dimensional shifts are important, since tooling is designed to ensure that final dimensions are to specification. Uncontrolled dimensional change is a major limitation in fabricating precision-sintered components. Further, surface area is consumed as pores become rounded. If the compact is pressed formed from mixed powders, then sintering also induces homogenization, leading to the formation of alloys by mixing atoms throughout the compact.

At the sintering temperature the high rate of atomic motion progressively leads to growth of bonds between the particles, termed *interparticle necks*. Atomic flow is highly dependent on temperature, with higher temperatures increasing atomic motion. Thus, up to a point, higher temperatures generally improve the sintered properties.

In this chapter we will first examine the fundamental sintering event, looking at the basic processing parameter effects with attention to the control of final dimensions and microstructure. In the next chapter, the practical aspect of sintering iron and steel will be examined by focusing on the furnaces, atmospheres, and common production cycles.

6.2. FUNDAMENTAL STEPS IN HEATING

6.2.1. Outline of the Sintering Cycle

The sintering furnace provides supervision to the time–temperature–atmosphere combination responsible for proper sintering. In a sintering furnace, the compact is heated to a peak temperature for a few minutes to allow heat to uniformly soak throughout the load. The first event during heating is the extraction of binders and lubricants. This is accomplished at temperatures below 550°C (1022°F). Subsequently, the compact is heated to a high temperature that induces atomic motion and sinter bonding, typically in the temperature range of 1100–1350°C (about 2000–2460°F).

Because iron will oxidize in air, the atmospheres for sintering must protect against oxygen attack. This is done by heating in a synthesized atmosphere, with nitrogen–hydrogen gas mixtures being most common. The nitrogen is neutral with respect to iron while hydrogen attacks any residual oxygen, forming water vapor that is removed by the atmosphere. Also, to attain the desired sintered mechanical properties, the carbon level in the sintering steel needs to be controlled. A usual route is via carbon doping of the process atmosphere with methane (CH_4) or a similar carbon source.

It is near the peak temperature in the sintering cycle that most of the particle bonding occurs. Although some time at the peak temperature is beneficial to ensure complete heating of the compact, extended times prove detrimental. *Oversintering* refers to the situation in which a steel exhibits a loss of properties with prolonged time at the peak temperature. Such events reflect microstructural coarsening and are evident by enlarged pores and large grains.

After the soak at the peak temperature, the sintered compact is cooled to room temperature and removed from the furnace. The phase transformations in a steel can be manipulated during cooling to induce high strengths. Thus, cooling rates may be controlled for some steels. This practice of heat treating during cooling is termed *sinter-hardening*. The alternative is to reheat the compact in a separate furnace after sintering. Postsintering heat treatments are treated in a later chapter.

One important aspect of sintering is control of the surface appearance of the compact. A reducing atmosphere is selected to remove oxides or rust. Hydrogen is a very potent reducing agent, and its use at low concentrations is widespread in P/M. Improper atmosphere control results in discoloration, or even worse, a loss of carbon and formation of a soft surface. The opposite is possible, where carbon is added to the surface, making it much harder.

Heating in sintering is accomplished by either placing the compacts in a furnace and applying heat via radiation from surrounding heating elements (a batch furnace) or by preheating a tunnel and passing the compacts through the tunnel (a continuous furnace). In the latter, a conveyor moves the work through a sequence of heating stages, allowing for polymer removal and oxide reduction before reaching the peak temperature. In the batch furnace, the temperature is controlled versus time to induce the desired sequence of reactions, while in a continuous furnace the temperatures and sizes of the heating zones are fixed, so the rate of travel for the compact determines the temperature–time combination. This is controlled by the belt or push mechanism speed.

Heat is delivered mainly by radiation, with contributions by convection (gas circulation) and conduction (heat flow) in the atmosphere. If sintering is performed in vacuum, then only radiation is active. This is ineffectual below about 500°C (932°F), so vacuum sintering is more difficult to control at low temperatures. Alterna-

EQUATION BOX 6.1

Heating in sintering occurs by conduction, convection, and radiation. Radiation from the furnace heating elements is usually dominant. The cooler compact absorbs heat in the same manner as the sun provides heat to the Earth's surface. Radiant heating depends on the temperatures of the heat source and compact

$$Q = 0.5\epsilon\sigma(T_f^4 - T^4)$$

where Q is the heat flow into the compact measured in J/(m^2 · s) (energy per unit area per unit time), σ is the Stefan–Boltzmann constant [about $5.7 \cdot 10^{-8}$ W/(m^2 · K)], T_f is the furnace temperature, and T is the compact surface temperature (both are absolute temperatures).

Convection and conduction require a gas phase, so they are not active in vacuum sintering. *Conduction* is heat passing through the atmosphere while convection involves hot atmosphere flow past the compact surface. In conduction, heat flow varies with the temperature difference between the furnace walls or heating elements and the P/M compact, and gas thermal conductivity κ

$$Q = \kappa \frac{\Delta T}{y}$$

where ΔT is the temperature difference between the heat source and the compact surface and y is the separation distance. Because of higher thermal conductivity, heating and cooling are much faster in hydrogen than in nitrogen, especially at lower temperatures where radiation is not a large factor. Flow of the heated atmosphere past the compacts enhances heating by convective heat flow

$$Q = h \, \Delta T$$

where h is the heat transfer coefficient, which depends on the gas and its velocity, and even varies with temperature. Now the ΔT parameter is the temperature difference between the compact and atmosphere. There is an optimal flow velocity to deliver the best convection heating. The total amount of energy deposited on a powder compact during sintering is the sum of the radiation, convection, and conduction contributions.

tively, atmospheres with high hydrogen contents have high thermal conductivities and give uniform heating, resulting in superior temperature control during sintering.

6.2.2. Lubricant Removal

A first function of the sintering furnace is to remove the lubricants and other polymer additions used for shaping, pressing, or bonding the particles. The process is termed *pyrolysis*, meaning the burning out of the polymer, but is often just termed the *burnout* or *delube stage*. Usually this means that the polymer is heated to a temperature where it becomes unstable and evaporates. Heat first melts the polymer and then breaks down its molecular bonds, forming small molecules that evaporate out of the compact.

Most polymers contain the same basic carbon–carbon, carbon–oxygen, and carbon–hydrogen bonds. Thus, they burn out over the same temperature ranges, in spite of apparent differences in bulk chemistry and lubrication properties. The evaporating molecules are mixtures of methane CH_4, carbon dioxide CO_2, carbon monoxide CO, water H_2O, and other combustion byproducts. Often polymer burnout is assisted by adding active agents to the atmosphere. A common procedure is to add a small amount of water or oxygen to help form carbon monoxide or carbon dioxide molecules that remove carbon. If the atmosphere is neutral, for example, argon, then it is possible to form molecules only via polymer cleavage. Unfortunately, this leads to a deficiency of hydrogen or oxygen, with the formation of soot. This black dust on the compacts is clear evidence of rapid heating and low oxidation potential in the delube atmosphere.

An example of the polymer loss versus temperature is given in Figure 6.1. This

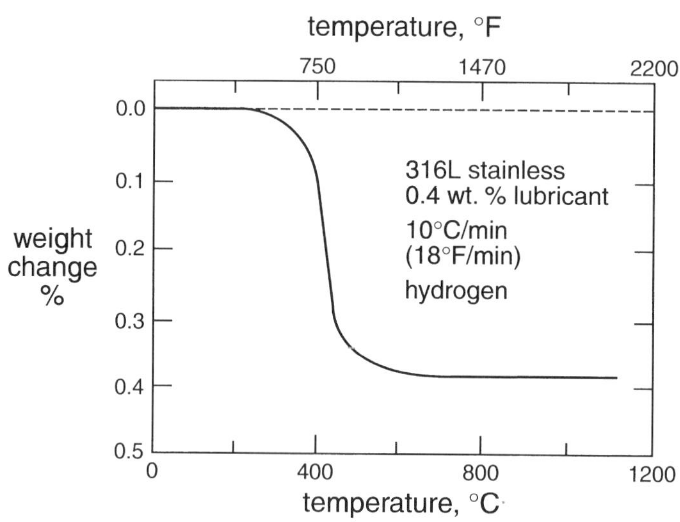

Figure 6.1. The weight loss during lubricant burnout for a stainless-steel powder compact containing 0.4 wt% lubricant heated at 10°C/min in hydrogen.

shows the progressive elimination of a common P/M lubricant from compacted stainless-steel powder during heating at the rate of 10°C/min (18°F/min). Most lubricants melt between 45 and 150°C (115 and 300°F), but do not evaporate until 300–500°C (570–930°F), so there is a temperature window during heating where the lubricant is liquid. For the case shown in Figure 6.1, little evaporation occurs up to 400°C (750°F), but essentially all the lubricant is gone by 550°C (1020°F). The heating rate must be slow during periods of rapid evaporation to avoid damage to the compact.

For very high green densities, the lack of open porosity inhibits polymer evaporation and slower heating rates are required. In advanced sintering furnaces, the atmosphere and gas flow are designed to help lubricant removal without slowing production. This requires a gently increasing temperature and often a long heating zone. Special gas inlets in the delube zone allow the addition of oxidizing species, such as water, to help in polymer burnout without sooting.

6.2.3. Atomic Motion

Curved surfaces are under stress. A balloon requires pressure to be inflated and increase its size or surface area. Likewise, particles have a stress that depends on their surface area. A flat surface has no curvature and has the least curvature stress. During sintering, any surface with a bump, or dip, will flatten over time to remove the stress. Since particles have a large surface area, they consist of many regions of high stress. It is the motion of atoms in response to this stress that produces sintering. The atoms have no knowledge of where to move to reduce stress, but random motion eventually produces a bond between particles. This is the same random motion that leads to water flowing downhill. There is no knowledge by the water of which way to move, but gravity causes all streams to flow downhill. Like gravity, the particle surface area contributes a sintering stress. For most powders encountered in P/M, the sintering stress is small, ranging from about 0.04 to 4 MPa (typically from 0.01 to 10 psi). These stresses are the same size as encountered in diving underwater in a swimming pool. These are weak stresses and one reason why sintering is somewhat slow.

Two factors contribute to produce sintering: sintering stress and random atomic motion. Since the stress is weak, atomic motion must be a dominant factor. For iron, the rate of atomic motion is very low at room temperature. Nevertheless, as iron is heated, atomic motion increases dramatically. This increase is described by an exponential equation termed the *Arrhenius equation*, named after a Swedish chemical engineer. Figure 6.2 plots the rate of atomic motion in austenitic iron versus temperature. At high temperatures the rates are very high. The effect is the same as seen by examining the flow of a liquid down a tube as sketched in Figure 6.3. The angle of the tube with respect to the horizontal is the stress gradient, which is analogous to the sintering stress. If the tube is filled with honey, the high viscosity (indicating low levels of atomic motion) results in slower flow, even with a steep gradient. However, with the same gradient or angle, the rate of flow is much higher if water

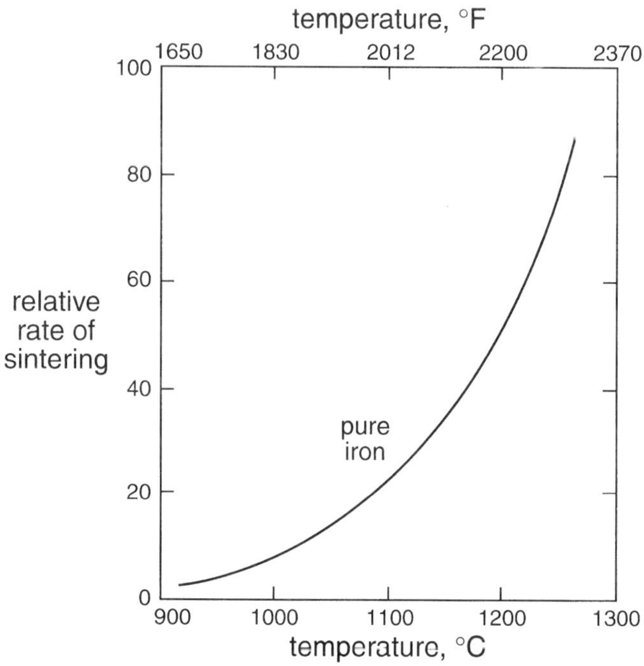

Figure 6.2. The relative sintering rate for pure gamma iron (γ-Fe) as a function of temperature, illustrating the dramatic effect of higher temperatures.

is the fluid, since it has much more atomic motion and a lower viscosity. Temperature controls the rate of atomic motion, and higher temperatures give more rapid sintering.

A reduction in both surface area and sintering stress occurs by growing bonds between contacting particles. This happens by atoms arriving via several paths. Two important distinctions in mass transport relate to the sources of the atoms that form the bonds. In *surface transport*, the atom comes from a pore surface in *bulk transport*, the atom originates from the interior of the particles. The importance of this distinction relates to shrinkage or dimensional change during sintering. Surface transport gives no dimensional change, while bulk transport produces shrinkage. Figure 6.4 sketches three particles undergoing sinter bonding and shows the several different paths the atoms can take in forming the sinter bond. The actual mechanisms include atoms skipping over the pore surface, passing along grain boundaries, and even evaporating across the pore.

The sinter bond between contacting particles is the critical region. It is the point where atoms are deposited to improve the compact strength. Neck growth occurs faster at higher temperatures. While both surface and bulk transport processes promote neck growth, the main difference is in shrinkage during sintering. Associated with shrinkage is an increase in density; alternatively, if the compact swells, then density decreases.

For densification to occur, the mass must originate from the particle interior with deposition at the neck. Most atomic motion is by diffusion. Sintering depends on the rate of diffusion (rate of atomic motion per unit time as measured by the atoms moving across an area per unit time) and the concentration or stress gradient (change in stress or change in concentration measured by atoms per unit volume and the distance over which the change occurs). In essence, diffusion is a measure of the inherent atomic motion, like a velocity times a distance. In Figure 6.3, the gradient is represented by the steepness of the tube, and atomic motion is represented by the fluid viscosity.

Bulk transport mechanisms give densification. They include volume diffusion, grain boundary diffusion, plastic flow, and viscous flow. Plastic flow is usually important during the heating period for high-green-density powders only. Grain boundary diffusion is important for densification for most steels. Surface transport

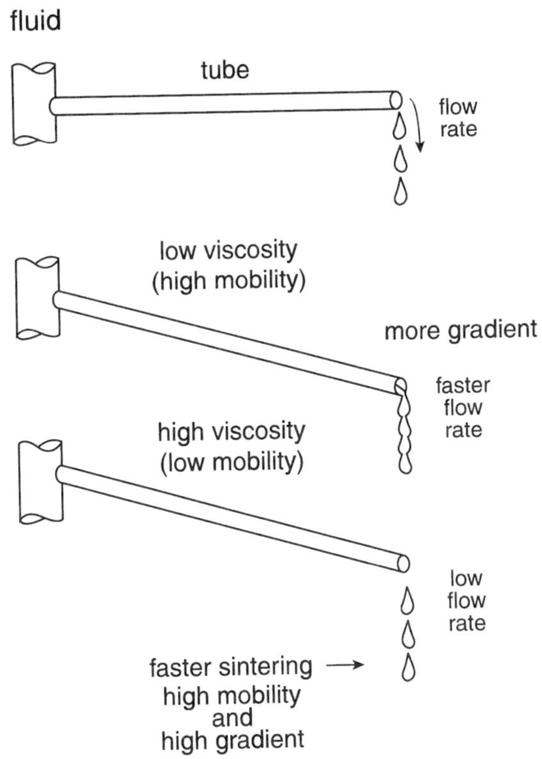

Figure 6.3. An analog description of sintering using a tube and liquid. The rate of fluid flow through the tube depends on both the tilt angle (the gradient) and the fluid viscosity (the motion). Low tilt angles and low mobility (high viscosity) make flow slow. In contrast, high tilt angles and low viscosities contribute to rapid flow. Likewise, in sintering, faster sintering occurs with smaller powders with high surface areas (large gradient) and high temperatures (high atomic mobility).

EQUATION BOX 6.2

The motion of atoms in sintering depends on at least two events. The first is that the atom have sufficient energy to break existing bonds with neighboring atoms Q_B, and the second is that a vacant site exists near the atom that requires a formation energy, too, Q_N. Probability arguments for atomic energy distributions in solids lead to the Boltzmann energy distribution. In this distribution, the probability of both events occurring simultaneously is described by an activation energy Q. Since probabilities are multiplied and the activation energies appear in exponential terms, the resulting Arrhenius equation gives the number of active atoms N_A as compared with the total number of atoms N_O as follows:

$$N_A = N_O \exp\left(-\frac{Q_B + Q_N}{RT}\right)$$

where R is the gas constant [8.32 J/(mol · K)] and T is the absolute temperature.

Most typically, the rate of atomic diffusion is considered as the diffusivity, which includes the frequency of atomic vibration, crystal class, and other factors. The resulting form is an Arrhenius equation

$$D = D_O \exp\left(-\frac{Q}{RT}\right)$$

where D is the diffusivity measured in m^2/s and D_O being the frequency factor. The activation energy Q is simply $Q_N + Q_B$. Many books tabulate diffusion data as D_O and Q, allowing calculation of D at any temperature, which is important in analysis of sintering behavior.

Besides diffusion, other material parameters follow a similar equation, including evaporation and high-temperature deformation. When assembled, the rate of sintering takes on a form that appears as follows:

$$\left(\frac{X}{G}\right)^n = \frac{gD\gamma\Omega t}{G^3RT}$$

Variations on this model were first proposed by Frenkel and Kuczynski in the 1940s. Here X is the size of the bond between particles, G is the grain size, g is a geometric constant, γ is the surface energy, Ω is the atomic volume, D is the diffusivity, t is the sintering time, and T is the sintering temperature. For many steels, n is probably near 6 and $g = 60$. Usually neck growth occurs by several transport mechanisms, and a simple equation of this sort is inaccurate. Hence, most of the calculations have relied on computer simulations to handle the complex problem of multiple sintering transport mechanisms.

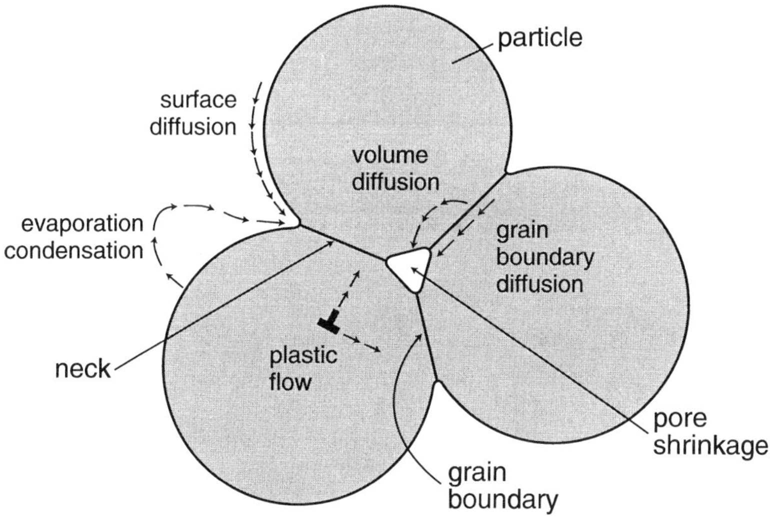

Figure 6.4. A three-particle sketch of sintering, showing the several possible paths of atomic motion involved with particle bonding (neck growth) and pore shrinkage (densification).

processes of surface diffusion and evaporation–condensation give no shrinkage. For iron, surface diffusion is the most important transport process at low temperatures. This means that compact strengthening occurs during the early portion of sintering without densification or dimensional change.

Because atoms move faster at high temperatures, sintering is greatly improved by use of temperatures higher than 1120°C (2050°F). However, warpage occurs because there is more bulk transport sintering and more dimensional change at higher sintering temperatures. Two other important changes occur at high temperatures. The first relates to the pore structure. Densification eliminates porosity, but at the same time the pores become smoother and can even enlarge, meaning that the total number of pores decreases while the average size increases. The other factor is grain growth. At elevated temperatures, the average grain size increases and the number of grains decreases. Both changes rely on atomic-scale events to move atoms around pores and across grain boundaries. Grain size is important to the sintered properties, including magnetic response, yield strength, hardness, and heat treatment response.

6.2.4. Liquid-Phase Formation

A liquid phase usually increases the sintering rate, but might also induce swelling when liquid forms. The liquids often form at the particle contacts, and this is the perfect site for improved particle bonding. Many ferrous alloys use a liquid essentially to braze the particles into a strong mass. Hence, liquid-phase sintering is less dependent on the basic rate of atomic motion for iron. Rather, sintering is rapid

EQUATION BOX 6.3

Grain growth is a natural consequence of sintering in both solid and liquid-phase variants. The number of grains decreases as the average grain becomes larger through coalescence (joining together of grains at their interface) or dissolution and reprecipitation (mass flow from one grain through the liquid to deposit on another grain). The generalized law for grain coarsening is as follows:

$$G^3 = G_0^3 + K\,t\,\exp\left(-\frac{Q_G}{RT}\right)$$

where G is the grain size after sintering at temperature T (absolute temperature) for time t with a starting grain size of G_0. The factor K depends on the way in which the grains grow and Q_G is an activation energy that depends on the ability of atoms to move.

The grain growth rate changes with the amount of second phase in the microstructure, so it is usually determined by experimentation. A small amount of liquid greatly increases the rate of grain growth. The rate constant K is modified to account for the effect of the shorter diffusion distance as the liquid content decreases. The model for grain growth during liquid-phase sintering is as follows:

$$K = K_0 + \frac{K_L}{V_L^{2/3}}$$

where $K_0 + K_L$ is the infinite dilution rate constant applicable to the no-solid condition, K_L is a parameter sensitive to the microstructure (factors such as contiguity), and V_L is the liquid fraction. This model agrees with the results of many studies.

once the liquid forms to provide bonding. Many important ferrous alloys rely on liquid phases, including those that use phosphorus, copper, or carbide additions. The effect of a liquid phase is similar to putting in a freeway, where everything moves faster on the freeway as compared with the slower city (surface) streets.

The surface tension from the liquid phase contributes a capillary force that pulls the particles together. Coupled with the capillary force is a higher rate of atomic transport in the liquid as compared with pure iron. Because of these two factors, fast sintering occurs when a liquid is formed during sintering. Consequently, about 70% of all ferrous components use a liquid phase during sintering. The rapid atomic motion makes it possible to use short sintering cycles as compared with sintering without the liquid. Because of the faster atomic motion, there is more dimensional change in liquid-phase sintering than with solid-state sintering. Thus, attention is needed to control the amount of liquid phase, heating rate, and hold times at the peak temperature to ensure proper sintered dimensions. Also, since atomic motion is rapid in a liquid, pores change shape and size rapidly, while grain growth is fast. Thus, it is common practice to use short sintering times for liquid-phase sintering systems.

6.2.5. Atmosphere Reactions

The atmosphere is important to successful sintering. Iron naturally forms an oxide, as is evident by its tendency to rust. The rate of oxidation increases as temperature increases. Hence, it is not possible to sinter iron in air without forming an oxide. The option is to protect the compact from oxidation by heating in a protective atmosphere. The atmosphere is important in determining the sintered properties. For iron and steel P/M, the atmosphere may be any of a variety of gases or even vacuum. In each case the atmosphere interacts with the alloy and can help or hinder sinter bonding.

The important role of the atmosphere in removing lubricant or binder constituents was discussed earlier. The polymers must be extracted at low temperatures to avoid undesired carburization reactions with iron. The properties of sintered steels are sensitive to the final carbon level. Thus, precise carbon control during sintering is required. This is possible via the atmosphere, where carbon can be added or subtracted from the composition based on the carbon content of the atmosphere. In turn, this requires attention to possible reactions with oxygen, water, methane, carbon monoxide, and carbon dioxide.

Figure 6.5 gives weight loss data associated with lubricant removal during sin-

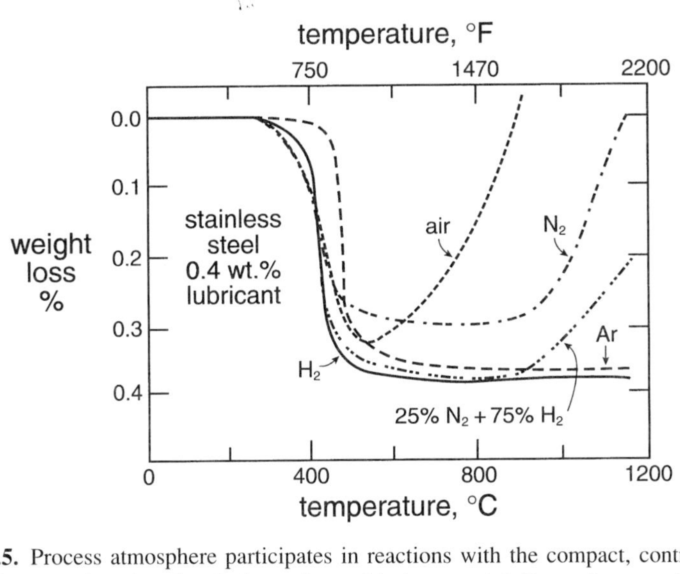

Figure 6.5. Process atmosphere participates in reactions with the compact, contributing to lubricant burnout, but also chemical reactions with the material. Depending on which atmosphere is used, wide swings are possible in the compact weight, as illustrated using thermogravimetric analysis results for a stainless-steel powder.

tering of a stainless-steel powder in several atmospheres. These data were obtained using a thermogravimetric analyzer (TGA) that measures the sample weight inside a furnace during constant heating. The results show that air is effective in removing the lubricant at low temperatures, but oxidation occurs at temperatures over 400°C (750°F), as is evident by the weight gain. In contrast, pure hydrogen gives excellent organic removal and no further reactions up to the sintering temperature. Nitrogen mixed with hydrogen leads to nitrogen adsorption at the higher temperatures, giving a weight gain. These nitrides are detrimental to corrosion resistance in stainless steels. Finally, sintering in argon is neutral, so it is less effective in removing the lubricant, but gives little adverse reaction at high temperatures. For each material there will be differing sensitivities to the sintering atmosphere.

The cost of the sintering atmosphere is an obvious consideration. High-purity atmospheres, such as hydrogen, produce some of the most repeatable sintered properties, but are also the most expensive. Alternatively, low-cost sintering is possible using a mixture of natural gas and air, but this is variable and gives less repeatable properties. Hence, the common compromise is to use a nitrogen atmosphere with hydrogen additions.

This introduction has presented the key aspects of sintering. These have included the extraction of the lubricant, heating to a high temperature to induce rapid atomic motion or form a liquid, and control of important chemical reactions with the sintering atmosphere. The balance of this chapter details the sintering of iron and steel.

6.3. SINTERING IRON AND STEEL POWDERS

6.3.1. Pure Iron

The atomic structure of iron affects the inherent sintering rate. Because of the phase transformation from alpha iron (α-Fe) or ferrite [the body-centered cubic crystal structure of iron stable below 910°C (1670°F)], to gamma iron (γ-Fe) or austenite (the face-centered cubic crystal structure of iron stable at most sintering temperatures), there is a dramatic shift in sintering rate on heating. Atoms diffuse faster in the body-centered cubic crystal structure, making ferrite the favored phase for sintering. Hence, iron exhibits faster sintering at temperatures near 900°C than at 1000°C (1652 and 1832°F). Consider that most ferrous sintering occurs between 1100 and 1300°C (2000 and 2375°F). Thus, on heating the sintering rate progressively increases, but the phase transformation from alpha to gamma slows sintering. As a demonstration of the behavior, Figure 6.6 that plots the sintering shrinkage during heating for a mixture of iron and nickel powders. The rate of sintering, as measured by the slope of the curve, shifts down when the transformation temperature is exceeded. Accordingly, much higher temperatures are required to regain the sintering rate possible in ferrite just before the transformation.

In sintering iron, there are two possible goals. For die-compacted, coarse, water-atomized powders with high green densities, the desire is to generate a strong sinter bond. Usually densification is not very desirable, since green density gradients in

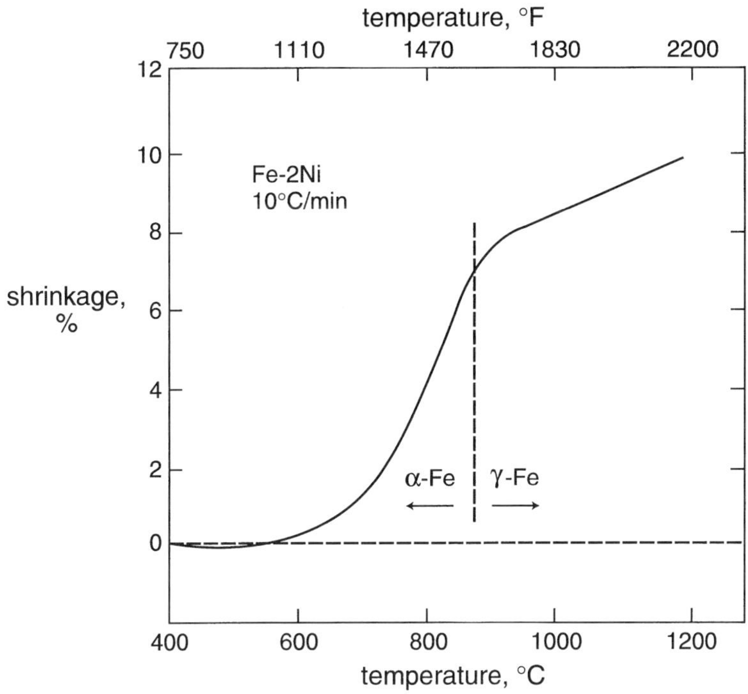

Figure 6.6. Shrinkage during heating of a Fe–2Ni compact heated at a constant rate of 10°C/min (18°F/min). At the phase transformation from alpha to gamma iron, the sintering rate declines due to a change in crystal structure. (Data courtesy of Haorong Zhang.)

the compact cause warpage, distortion, or loss of precision. On the other hand, for injection-molded structures the green density is low and small powders are used to encourage shrinkage in sintering. Densification is achieved during sintering in the latter case, while sintering only provides bonding in the former case. For larger powders, most of the densification is achieved in compaction. Alternatively, small powders are used when sintering is the basis for densification.

Table 6.1 provides a behavior contrast for these two sintering variants. Given here are two sintering cycles performed with iron powders. One used a small carbonyl iron powder that was injection-molded and sintered at 1300°C (2370°F) for one hour. The low initial green density, small particle size, and high sintering temperature produced 15.5% shrinkage. This shrinkage was isotropic, meaning that each dimension shrank by the same amount. The final density was close to 100%, meaning very few residual pores. For the water-atomized iron, it was compacted at a high pressure [near 700 MPa (50 tsi)] to a green density of 7.2 g/cm^3. As compared to the injection-molded steel, the particle size was larger, sintering temperature was lower, and that temperature was held for a shorter time. Consequently, only a small dimensional change occurred in sintering, just 0.25% shrinkage. Thus, the final

TABLE 6.1. Comparison of Densification and Bonding Behaviors in Sintering Steels

Attribute	Densification	Bonding
Shaping process	Injection molding	Die compaction
Iron powder size	5 μm	75 μm
Iron powder type	Carbonyl	Water atomized
Alloy	2% nickel–0.5% carbon	2% nickel–0.5% carbon
Green density	4.7 g/cm^3	7.2 g/cm^3
Heating rate to temperature	10°C/min (18°F/min)	15°C/min (27°F/min)
Maximum temperature	1300°C (2370°F)	1150°C (2100°F)
Hold time at temperature	60 min	30 min
Sintering atmosphere	Nitrogen–hydrogen	Nitrogen–hydrogen
Final density	7.8 g/cm^3	7.2 g/cm^3
Net dimensional change	15.5%	0.25%
Sintered strength	580 MPa (84 ksi)	420 MPa (59 ksi)
Elongation to fracture	15%	4%
Hardness	91 HRB	69 HRB

density was still close to 7.2 g/cm^3, which is about 8% porosity. Most important, since there was minimal sintering shrinkage, the component maintained size and shape. This is less an issue with injection molding since the forming pressure is *hydrostatic*, meaning the same in all directions. However, with die compaction the forming pressure is only in one direction, the pressing direction, resulting in pressure gradients that lead to density gradients. Since sintering shrinkage varies with green density, such green density gradients cause a die-compacted component to warp in sintering if shrinkage occurs.

As a comparison of these two components, consider the sintered hardness and strength. The injection-molded compact, sintered to a higher density, gave a higher hardness and strength. After heat treatment both compacts would be even stronger. A dramatic comparison is possible with the sintered ductility. The injection-molded compact has considerable ductility (15% elongation to failure) as compared with the die-compacted material (4% elongation to failure).

This comparison shows a fundamental aspect of iron sintering. Smaller powders, lower green densities, longer sintering times, and higher temperatures contribute to sintering shrinkage. Smaller iron powders have more surface area and respond to sintering more than do larger powders. Also, higher temperatures promote more atomic motion, and longer times cumulatively add more shrinkage. The net result is considerable sintering densification for the injection molded powder, but bonding without much shrinkage for the die-compacted powder. Two atomic transport mechanisms are important, grain boundary diffusion, and surface diffusion. The latter gives bonding without shrinkage, while grain boundary diffusion produces shrinkage as a byproduct of bonding. Various sintering cycles, powder sizes, and atmospheres can be used to manipulate the relative effect of each mechanism, providing control of the sintering response.

Thus, iron sinters with a rate that depends on the stable crystal phase. The ferrite structure is most desirable for a high sintering rate, but most steels are sintered at temperatures where austenite is stable. Iron particles bond via surface diffusion and grain boundary diffusion. The relative contribution of each mechanism depends on the specific details of particle size, composition, temperature, hold time, and green density.

6.3.2. Mixed-Powder Sintering

Most iron powder is mixed with alloying additions before compaction and sintering. Some common additives include graphite, phosphorus, nickel, copper, boron, silicon, chromium, manganese, and molybdenum. The initial structure is inhomogeneous. At one location in the microstructure is a pure-iron particle, while another is an alloying particle. Diffusion of the atoms is required to form a homogeneous alloy at all points. For example, Figure 6.7 shows the green microstructure of mixed iron and copper before sintering. This micrograph represents a random section through the compact, showing discrete islands of copper in the iron matrix. One requirement in sintering is to provide sufficient temperature and time to homogenize the alloy. Thus, the sintering cycle plays a dual role in forming an alloy by homogenizing

Figure 6.7. An optical micrograph of a mixed-powder green compact of iron and 2% copper, showing the discrete and localized condition for the copper before sintering. Homogenization requires that the copper diffuse uniformly into the iron between the copper particles. (Photograph courtesy of Anthony Griffo.)

while bonding the particles. Usually the homogenization process is slow as compared with the bonding event. This, dictates higher temperatures or longer times to fully distribute the additive. The situation is akin to mixing cookie dough. Initially, the ingredients are stratified in a bowl, but with repeated mixing the flour, sugar, milk, and flavoring take on a more consistent composition, a desirable attribute for the final cookies.

There are several recognized advantages to forming alloys from mixed powders. One benefit is easy composition adjustments—just add more or less when the powders are mixed. Also, the soft powders are easier to compact as compared with alloyed powders, and usually they are lower in cost as compared with prealloyed powders.

Homogenization during sintering is slow, since atoms of the additive must travel large distances to become equally present at all locations in the compact. This interdiffusion of atoms depends on the progressive random substitution of one atom with another. Figure 6.8 schematically illustrates the interdiffusion events, by which the

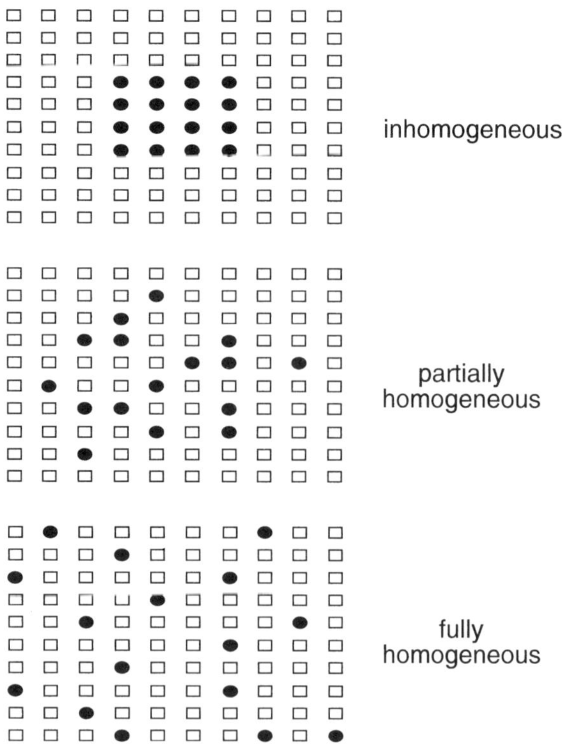

Figure 6.8. Schematic diagram of atomic level interdiffusion from a compact initially consisting of mixed particles heated to form a homogeneous alloy. The diffusion event depends on time and temperature, and is often slower than other aspects of sintering.

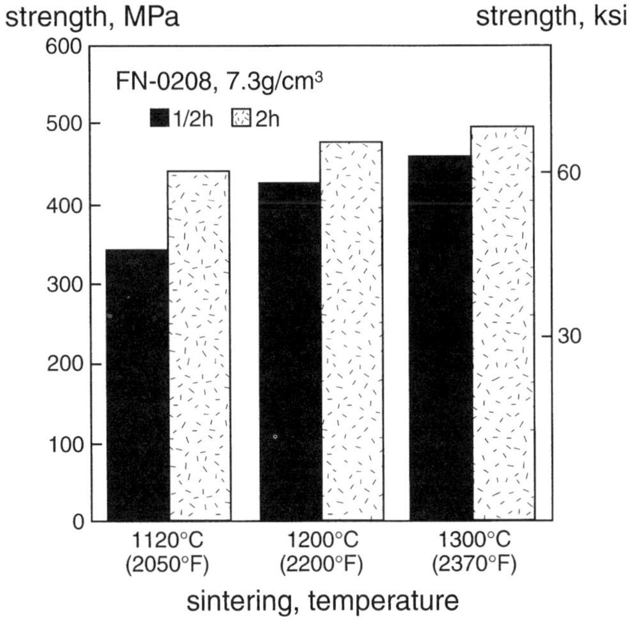

Figure 6.9. The combined temperature and time effects on tensile strength for an alloy of Fe–2Ni–0.8C (FN-0208) compacted to 7.3 g/cm^3. Note, in comparing temperatures, that 2 h at the lower temperature gives a higher strength than 0.5 h at the higher temperature. These data illustrate the interchangeable time–temperature combinations that contribute to the atomic motion required for sintering.

atoms intermix and eventually form a homogeneous alloy. Since atomic motion is faster at higher temperatures, it should be no surprise that mixed-powder sintering generally leads to better sintered properties with higher temperatures. Figure 6.9 is a bar-chart plot of the tensile strength of a Fe–2Ni–0.8C steel (FN-0208) formed from mixed iron, nickel, and graphite powders. The compacts were pressed to the same density of 7.3 g/cm^3 and sintered at 1120, 1200, or 1300°C (2050, 2190, or 2370°F) for either 0.5 or 2 h. Note the increase in strength with higher temperatures and longer times, both giving better homogeneity to the nickel distribution.

Diffusion is very dependent on temperature, as emphasized in Figure 6.2 for the sintering rate of iron. Figure 6.10 emphasizes the temperature effect on homogeneity by showing the diffusion rate versus temperature for nickel motion through iron. The change in crystal structure for iron on heating causes a change in the diffusion rate near 910°C (1670°F). After partial sintering, some regions are rich in nickel while nearby regions are low in nickel. A device for measuring the composition gradients in a sintered material is an electron microprobe. Figure 6.11 plots the microprobe-measured concentration of nickel with distance from the original nickel particle site after 1 h of sintering at various temperatures. High temperatures are needed to fully homogenize the nickel into the iron; this results in superior sintered

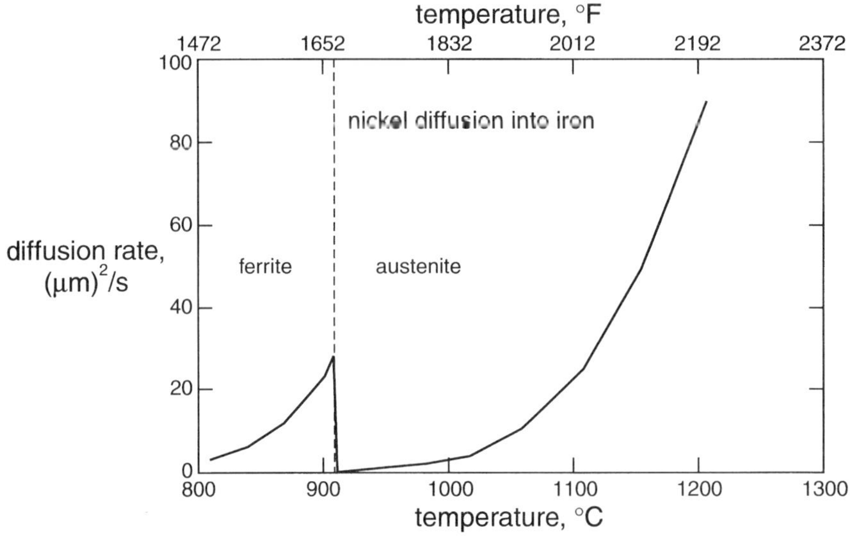

Figure 6.10. This plot gives the diffusion rate of nickel atoms spreading into iron. Note the dramatic decrease in rate when iron transforms crystal structures at 910°C (1670°F). That rate of diffusion is not recovered again until a sintering temperature near 1120°C (2050°F).

properties, both from the improved sinter bonding and alloy homogeneity. Other means to improve properties include selection of an optimal time–temperature pathway, use of smaller nickel powders, or increasing the green density via higher compaction pressures.

Sintering provides a means to homogenize mixed powders to form alloys with improved properties. Most ferrous alloys depend on sintering to create a more uniform distribution of alloying ingredients. Each element has a different rate of motion in iron, so the homogenization time depends on the composition. Usually, the rate of atomic diffusion is slow, so homogenization is slow compared with sinter bonding. A simple means to promote faster alloying is to use small additive particles. For a given composition, the smaller the additive particle size, the more uniformly dispersed it is in the green microstructure. Further, various additives are available to promote improved sintered properties. Table 6.2 illustrates the property ranges possible with sintering additives for iron. This gives the sintered strength for compacts pressed to approximately 6.8 g/cm^3, but sintered with different additives. The highest strength is not necessarily associated with the highest alloy level or highest sintering temperature.

6.3.3. Phase Changes during Sintering

As noted earlier, one complication in sintering ferrous alloys is the crystal structure change on heating. The low-temperature form of iron (ferrite) has a low solubility for carbon. On the other hand, the austenite phase is typically present at the sintering temperature and has a high solubility for carbon. The change in carbon solubility does not affect the sintering rate. However, since the phase transformation depends on carbon content, it is possible to affect the sintering rate by the level of carbon. High carbon levels slow sintering, since diffusion in austenite (stabilized by carbon) is slower than in ferrite. Nevertheless, a high carbon content and high temperatures produce a liquid phase that greatly increases the sintering rate. This is evident in sintered tool steels that sinter to full density with a small amount of carbon-rich liquid.

At low temperatures, the insolubility of carbon in ferrite means that a carbide exists with the ferrite. Second phases usually slow grain growth during sintering. In turn, since the rate of sintering varies inversely with the cube of the grain size, retarded grain growth helps sintering. Generally, the carbides dissolve once austenite forms. Thus, their effect disappears as the sintering temperature is attained.

6.3.4. Pore Structure Changes

The voids between particles initially contain lubricants, binders, and contaminants. As heating occurs, these species evaporate, leaving pores between the particles.

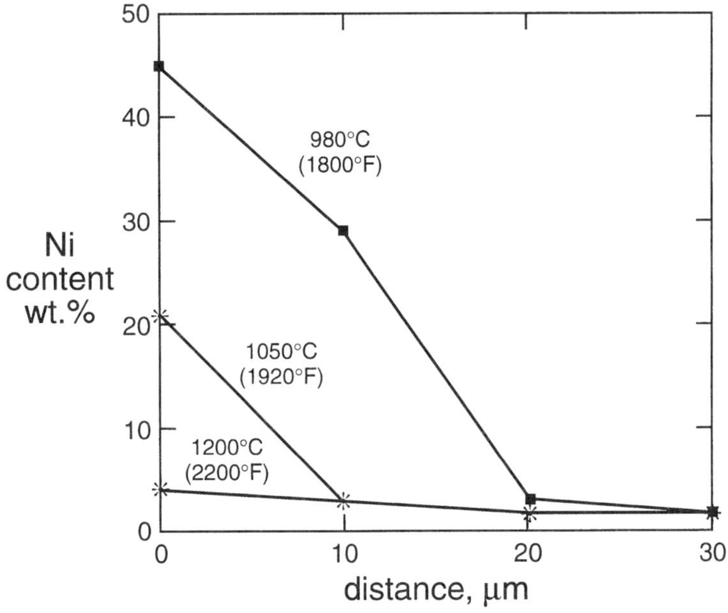

Figure 6.11. Plots of the nickel concentration profile in an alloy of Fe–2Ni after sintering at three temperatures for 1 h, showing that homogenization requires a temperature near 1200°C (2200°F) or longer times at lower temperatures. (Data courtesy of Haorong Zhang.)

Pronounced changes in the pore space take place as sintering progresses. If the powder is compacted to a high pressure, the pores are flattened. During sintering these flattened pores try to spheroidize. A consequence of pore spheroidization is pore shrinkage in the long direction and expansion in the short direction. Figure 6.12 schematically illustrates this change, with much exaggeration. Consequently, dimensional change in sintering is anisotropic. Sometimes, the compact height in the pressing direction expands, while the dimensions perpendicular to the pressing direction shrink. This anisotropic dimensional change creates a difficulty in reaching accurate final dimensions. The higher the compaction pressure, the greater this anisotropic dimensional change. Just the opposite behavior can occur with liquid-phase sintering. For example, in sintering Fe–2Cu–0.8C (FC-0208) at 1120°C (2050°F), the swell in the pressing direction was 0.03%, but the swell in the perpendicular direction was 0.62%.

Depending on the sintering time, temperature, and particle size, several other pore structure changes occur during sintering. The initially irregular pores become rounded and change size. Pore expansion results from a coarsening process and is called *Ostwald ripening*. It happens more with a wide pore size distribution, with the larger pores growing at the expense of the smaller pores. This is the same as the foam that appears on top of a soda or beer; the large bubbles grow and the small bubbles disappear. As a further aspect of sintering, the overall porosity might decrease. This is densification. Thus, changes in the pore structure depend on many details, including the initial pore size distribution and the sintering conditions. Figure

TABLE 6.2. Additive Effects on the Sintered Strength of Iron Powders[a]

Additive	Sintering cycle	Sintered Strength [MPa (ksi)]
None	60 min at 1250°C (2280°F)	175 (25)
3Cu–2Sn	60 min at 1000°C (1830°F)	310 (45)
2Cu–0.8C	30 min at 1120°C (2050°F)	495 (70)
2Ni–0.8C	30 min at 1120°C (2050°F)	385 (55)
4Mo–1.1B	60 min at 1200°C (2200°F)	500 (75)
8Ni–1.1B	60 min at 1100°C (2010°F)	400 (60)
3.5Ti	120 min at 1330°C (2425°F)	440 (65)
0.45P	30 min at 1120°C (2050°F)	400 (60)
2.2Cu_3P	60 min at 1100°C (2050°F)	480 (70)
1C	30 min at 1175°C (2150°F)	550 (80)
4Mn–0.4C	180 min at 1120°C (2050°F)	850 (125)
3VC	90 min at 1280°C (2335°F)	750 (110)
3Ni–3Mo–0.5P	60 min at 1250°C (2280°F)	450 (65)
1.5Cr–0.7C–0.5Cu	60 min at 1250°C (2280°F)	730 (105)
4Ni–1.5Cu–0.5Mo–0.5C	60 min at 1250°C (2280°F)	870 (130)
1.5Ni–1.5Cu–0.7C–0.5Mo	30 min at 1250°C (2280°F)	730 (105)
1.5Ni–1.5Cu–0.5Mo–0.5C	60 min at 1250°C (2280°F)	640 (90)

[a] Approximate green density 85% or about 6.8 g/cm^3.

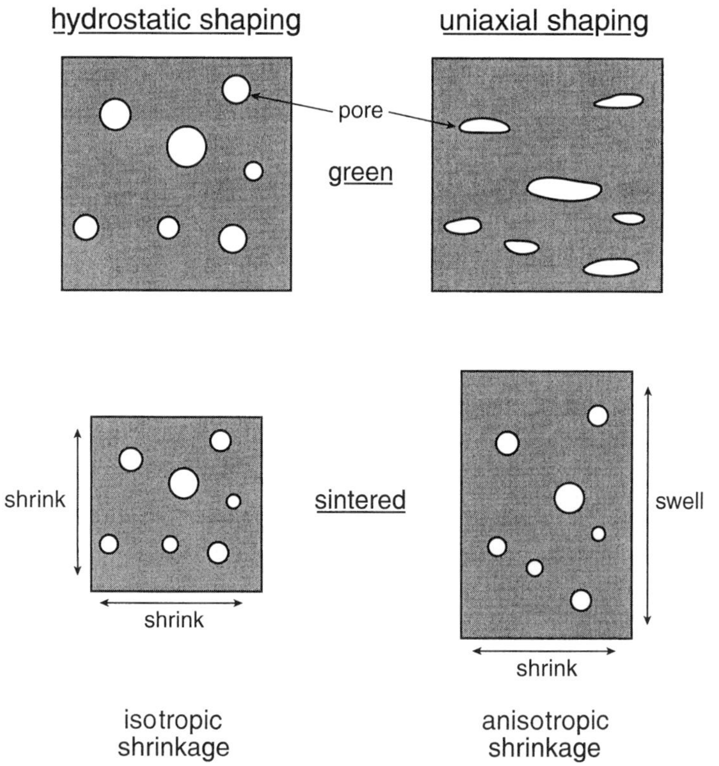

Figure 6.12. During sintering the pores attempt to form spheres. When a powder is compacted under hydrostatic conditions (such as with cold isostatic pressing), this spheroidization leads to isotropic dimensional change, but when a powder is compacted under uniaxial conditions (such as with die pressing), the dimensional change in sintering is anisotropic.

6.13 shows micrographs of two different sintered pore structures. Most typically, irregular pores become round and the number of pores declines over time. At high sintering temperatures and long sintering times, the pores are eliminated.

6.3.5. Formation of a Liquid Phase during Sintering

The formation of a liquid phase during sintering is a prevalent means of accelerating particle bonding in ferrous alloys. The liquid can originate from melting an admixed powder, such as in iron–copper alloys where the sintering temperature is over the melting temperature of copper. Liquid forms at the particle contacts, such as sketched in Figure 6.14. Another option is to generate a liquid by the interaction of two powders, such as when graphite and iron react to form a liquid at temperatures over 1130°C (2066°F). Liquid can form inside an alloy powder that is partially melted during sintering. This procedure is employed in sintering stainless steels and tool steels to full density. A final option is to form a transient liquid during heating by melting a powder that solidifies later. This is a temperamental process, used selectively in ferrous powder metallurgy. For example, mixtures of iron with about 2.5% titanium sinter via such a transient liquid phase.

A liquid is not necessarily beneficial to final properties. Some liquids attack the sintering material and cause swelling, while others lead to densification. Even for a given composition, the shift from densifying to swelling depends on the green density. In compacts with a high green density, the liquid pushes particles apart, while in compacts with a low green density the liquid pulls the particles together. Usually the effect of liquid-phase sintering is most evident at the contacts between particles, as seen in Figure 6.15. This micrograph shows the structure of an iron–carbon alloy after sintering at 1165°C (2130°F) for 1 h in hydrogen. The steel grains are outlined by a white solidified liquid phase, with rounded pores evident as the dark phase.

Coarsening is rapid during liquid-phase sintering, where the grains progressively enlarge over time. The growth of the larger grains occurs by the consumption of the smaller grains. Actually, the small grains preferentially coalesce into the larger grains or dissolve into the liquid while the larger grains continuously accept dissolved material out of the liquid. This leads to continual grain growth since the liquid is present.

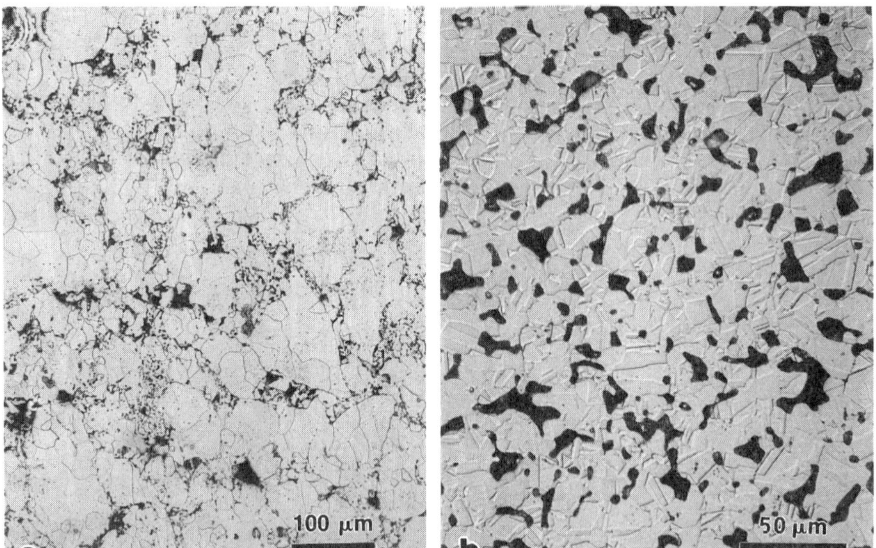

Figure 6.13. The pores (dark spots) in these two sintered steels are very different in shape and size, largely due to differences in sintering cycles. The arrangement of pores outlining the particles in (a) is detrimental to final properties, while the smooth pore structure evident in (b) is more desirable.

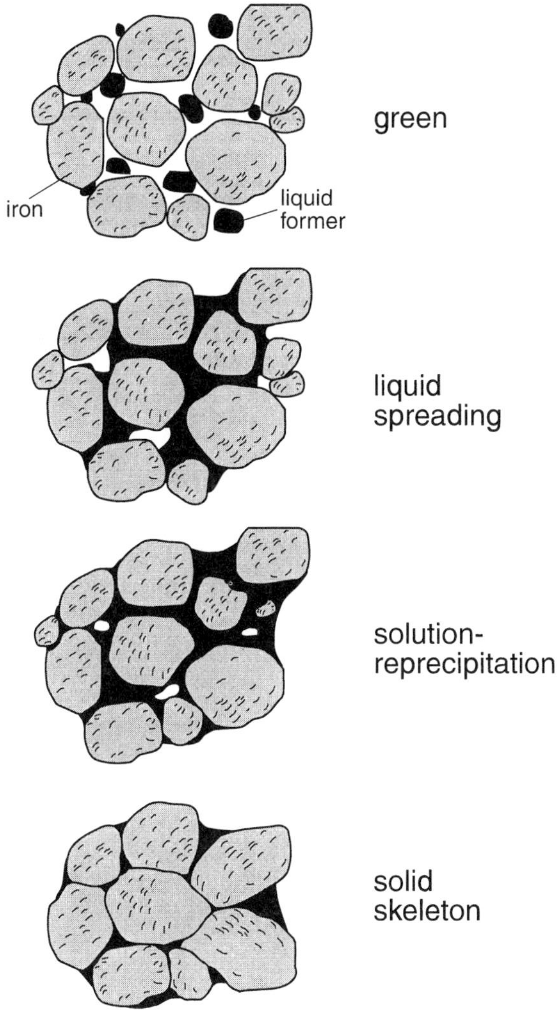

Figure 6.14. Liquid-phase sintering usually involves mixing an iron powder with a liquid forming powder (boride, carbide, phosphide, copper, tin) and heating to a temperature where the liquid forms, spreads, and contributes to particle bonding and densification.

The real benefit from liquid formation during sintering occurs when the liquid gives improved sinter bonding. For ferrous alloys, effective liquid phase sintering systems include those with copper, phosphorus, titanium, various carbides, and boron. Diffusion rates are high through the liquids, giving benefits in particle bonding. Further, the alloying from the additive usually promotes higher sintered strength. In cases where a high final density is desired, liquid-phase sintering is an economical means to densify the compact.

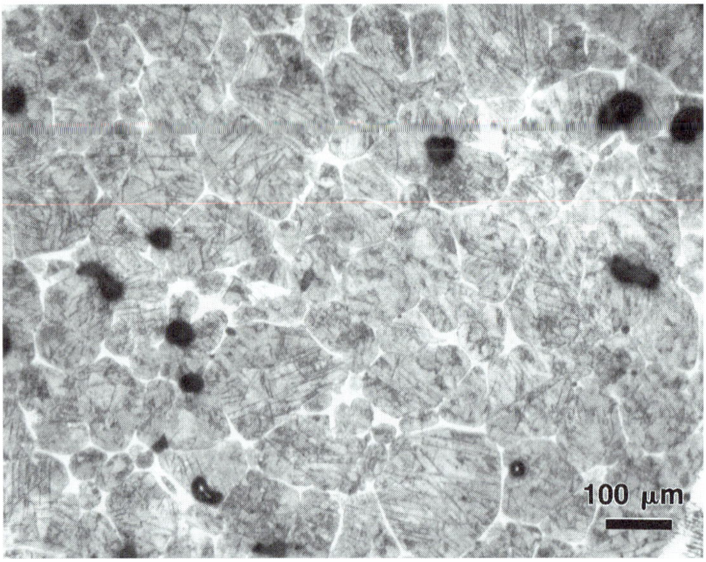

Figure 6.15. A characteristic microstructure associated with liquid phase sintering. This high density compact was formed by sintering Fe–2C (2% graphite mixed with iron) to 1165°C (2130°F) for 1 h in hydrogen. That portion of the structure that was liquid is evident as the white phase, with a small amount of porosity remaining as the dark rounded phase. (Photograph courtesy of James Dunlap.)

6.3.6. Loose-Powder Sintering

One of the easiest processes for sintering a compact uses powder poured into a crucible, or mold, and heated to sinter-bond the particles in place. The loose particles conform to the crucible shape, so no compaction is involved. Such a process is employed in sintering porous structures. Since there is no compaction pressure, the starting green density is at the apparent density, or tap density if the powder is vibrated. Crucibles are machined or cast from alumina-based ceramics to set the shape. Usually, loose-powder sintering is only applied to porous structures, since the shapes are simple and the sintered density is low. If very small particles are employed, then a high sintered density is possible. However, the geometry is no more complicated than the cavity in the ceramic crucible. Often this is not very precise, so the sintered size is not very precise.

As an illustration of loose powder sintering, consider the fabrication of simple disk filters. Holes are drilled through a ceramic plate and it is placed on a sheet of ceramic. The cavities in the ceramic plate are filled by pouring powder into the holes and leveled with a flat blade. Subsequently, the ceramic sheets are passed through a sintering cycle where the particles bond sufficiently for handling. Shrinkage is often in the 1% range, but variations in powder fill result in green density and concomitant final dimension fluctuations.

Loose-powder sintering is employed with stainless-steel and other high-alloy powders. The structures are used in applications where a high surface area is the primary concern. Besides applications as filters, these include sound attenuation surfaces, electrodes, gas distribution plates, bubblers, and polymer extrusion packs. Most typically, sintering temperatures are low and sintering shrinkage is small, leaving a porous structure. The production cost is low, but so are the sintered properties. Consequently, loose-powder sintering is limited in use. The technique is widely employed in sintering bronze filters for automotive applications, but sees less use in ferrous P/M.

6.3.7. High-Temperature Sintering

At this point we have established that sintering depends on atomic motion via diffusion, an exchange process at the atomic scale. In turn, diffusion increases dramatically with higher temperatures. Thus, temperature is the dominant parameter with respect to sintering. Because of this benefit, many high-performance ferrous alloys are processed at temperatures above the range historically employed in ferrous P/M. The higher temperatures induce more and faster atomic motion. The consequence is sintering densification, even with coarse water atomized powders. Instead of limiting the sintering temperature to 1120–1150°C (2050–2100°F), new furnaces operate at higher peak temperatures, more like 1250–1300°C (2280–2370°F).

The justification for high-temperature sintering is in the sintered properties. Example property gains are illustrated in Table 6.3. The sintered strength increases significantly. For example, in 316L stainless steel, the tensile strength and elongation to fracture increase from 310 MPa (45 ksi) and 7% when sintered at 1120°C (2050°F) to 380 MPa (55 ksi) and 25% when sintered at 1290°C (2350°F). Similar comparisons are possible for many ferrous alloys when considering magnetic, thermal, corrosion, or mechanical properties. Improvements of 15–50% are commonly observed for many engineering properties with higher sintering temperatures.

High-temperature sintering creates difficulties in process control. One problem is that traditional furnaces are not constructed from materials and conveyor mechanisms that can easily withstand the higher temperatures. Whereas stainless-steel belts

TABLE 6.3. Example Property Gains from High-Temperature Sintering

Alloy	Property	Units	Conventional Sintering	High-Temperature Sintering
Fe	Magnetic permeability	—	3300 at 1120°C	4600 at 1260°C
Fe–2Cu–0.8C	Yield strength	MPa	510 at 1120°C	615 at 1330°C
Fe–3Cu	Impact energy	J/cm^2	19 at 1120°C	32 at 1220°C
Fe–2Ni–2Cu	Fatigue strength	MPa	120 at 1120°C	135 at 1280°C
4640 steel	Fracture elongation	%	2 at 1120°C	4 at 1315°C
316L stainless steel	Fracture elongation	%	7 at 1120°C	25 at 1290°C
316L stainless steel	Impact energy	J/cm^2	28 at 1120°C	104 at 1315°C
316L stainless steel	Relative corrosion	% loss	0.06 at 1150°C	0.02 at 1350°C

EQUATION BOX 6.4

Sintering shrinkage changes the density. Density is the mass per volume. Assuming no loss of mass in sintering, which may not be true with lubricant burnout, the sintering shrinkage $\Delta L/L_0$ (or swelling) relates the sintered density ρ_S to the green density ρ_G as follows:

$$\rho_S = \frac{\rho_G}{[1 - (\Delta L/L_0)]^3}$$

If shrinkage occurs, then the sintered density is higher than the green density and $\Delta L/L_0$ is positive.

Another measure of sintering is the densification parameter Ψ measured by the change in density over the amount needed for full density

$$\Psi = \frac{\rho_S - \rho_G}{\rho_T - \rho_G}$$

where ρ_T is the theoretical density. For example, if a compact is 6.80 g/cm^3 green, with a theoretical density of 7.86 g/cm^3, and sinters to 7.00 g/cm^3, then the densification is 19% and the shrinkage is just under 1% in each dimension, actually about 0.96%.

are successful at the lower temperatures, silicon carbide belts are required for the higher temperatures. The sintering densification at the higher temperatures is due to grain boundary diffusion, a process that is not very active at traditional sintering temperatures. Final densities for high-temperature sintered materials are often over 7.2 g/cm^3. Thus, the improved properties allow P/M to be applied to new situations where failure might occur with lower densities. Figure 6.16 shows a microstructure after high-temperature sintering. Pores are still present. Nevertheless, the pores are highly rounded and exhibit much improved resistance to fatigue, impact, or corrosion situations.

For ferrous components formed from small powders with low green densities, such as with injection molding, high-temperature sintering is common. Sintering densification leads to shrinkages of up to 15% with final densities of 7.6 g/cm^3 or better. Figure 6.17 plots the effect of sintering temperature on the density of a small 316L stainless-steel powder. The structure was formed by injection molding. Densification is most evident when the sintering temperature approaches the melting temperature for the alloy. Note that the final density is essentially near theoretical, meaning no residual pores. Here is a case where high final densities and properties are attained not by high compaction pressures, but by high sintering temperatures. Conceptually, both routes (compaction at high pressures versus sintering at high

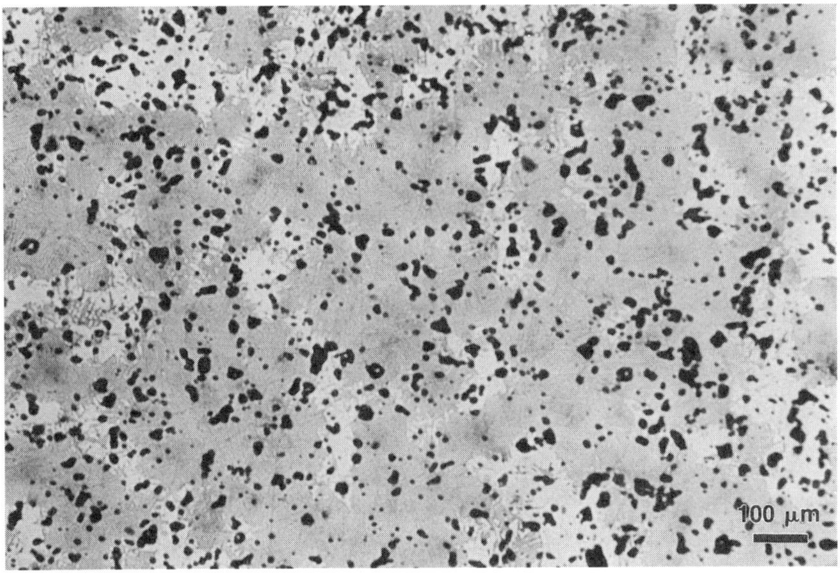

Figure 6.16. Microstructures after high-temperature sintering show very smooth pores and a homogeneous distribution of alloying elements. (Photograph courtesy of Christoph Degoix.)

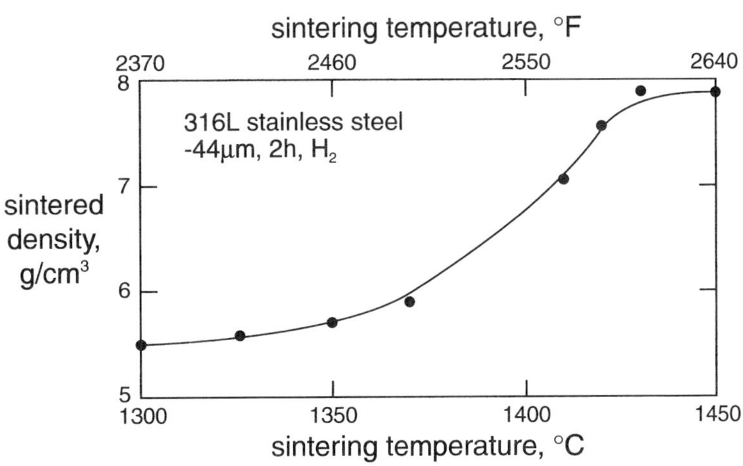

Figure 6.17. Temperature strongly affects the sintering densification of steels. This plot gives the sintered density for an injection-molded 316L stainless steel after sintering for 2 h in hydrogen at various temperatures.

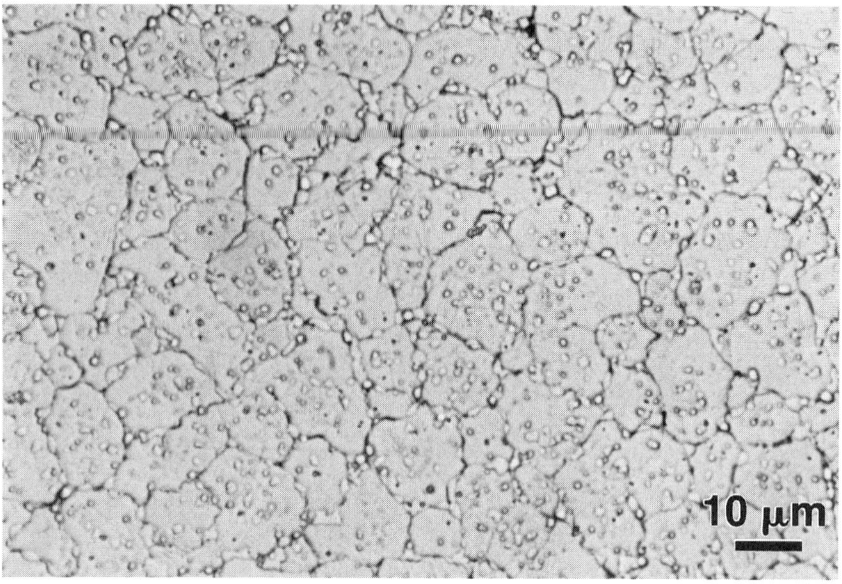

Figure 6.18. The microstructure of a full-density tool steel prepared by sintering in the liquid-phase region using prealloyed powders. (Photograph courtesy of Neal Myers.)

temperatures) can lead to the same final density, but high sintering temperatures distort die-compacted materials.

Other justifications for high-temperature sintering emerge when new alloying options are considered. Many additions do not diffuse into iron at the common 1120°C (2050°F) sintering temperature. By going to higher sintering temperatures, alloying additions such as manganese, chromium, titanium, and aluminum are viable for ferrous P/M. Another advantage is the easier reduction of oxides, making it possible to use alloying additions such as chromium. Indeed, the iron–chromium–carbon system forms a liquid phase at temperatures over 1184°C (2163°F), and the iron–molybdenum–chromium-carbon system forms a liquid at 1160°C (2120°F). Without high-temperature sintering, these novel liquid-phase systems are not viable for P/M. At higher temperatures, aluminum forms a corrosion resistant phase [1232°C (2249°F)] and titanium forms a liquid phase with iron at 1340°C (2445°F). Such reactions provide means to generate new property levels in steels, but only in the context of high-temperature sintering. These improvements are dramatic. For example, just 2% chromium doubles the hardness of a steel containing 0.35% C.

Finally, for high-performance ferrous alloys, only high-temperature sintering is employed, since pores are unacceptable to the sintered properties. Tool steels are examples where liquid-phase sintering is required to obtain full density and high hardness. Here sintering temperatures are often in the 1250–1350°C (2280–2460°F) range. Contrast the sintered microstructure shown in Figure 6.18 with the other steels depicted in this chapter. This is a full-density M2 tool steel sintered at 1245°C (2273°F) for one hour in hydrogen. There is no porosity, and the carbides exhibit a desirable small size and homogeneous distribution throughout the microstructure.

6.4. MICROSTRUCTURE CHANGES

The dominant microstructure change in sintering is growth of the interpaticle bond that provides strength to the compact. Changes also take place in the number of pores, the pore size, and the grain size. An example is given in Figure 6.19. These two plots show the porosity reduction during sintering and the associated changes in pore and grain sizes. The data are for a small (7-μm) iron powder compacted at 200 MPa (15 tsi) and sintered for various times at 870°C (1600°F) where ferrite is the stable phase. Although the porosity decreases with time, the pores and grain sizes both grow in average size. This means the small pores and small grains are preferentially consumed during sintering. Accordingly, the number of pores and

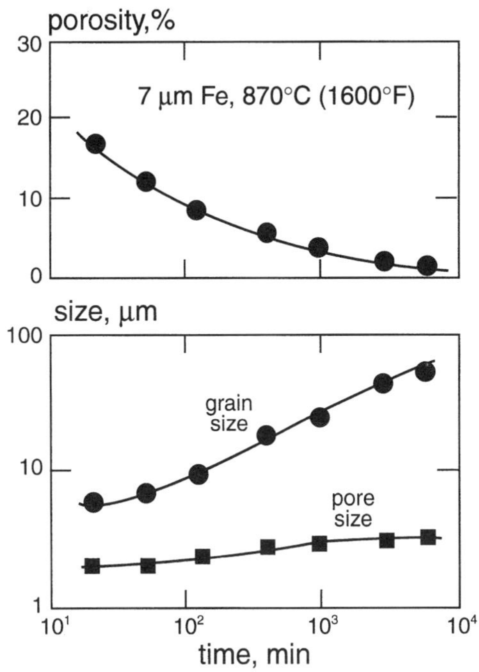

Figure 6.19. Sintering affects the microstructure, where grains grow, porosity declines, the number of pores decreases, but the average pore size increases. This behavior is illustrated here for a small iron powder compacted at 200 MPa and sintered in the ferrite temperature range for times up to 130 h. (Data from E. Aigeltinger and J. P. Drolet.)

the number of grains both decline. This example illustrates the multiple levels of microstructure change associated with sintering.

Microstructure coarsening is one negative aspect of sintering. Larger grain sizes result in decreased strength. So on one hand the strength increases as a result of improved sinter bonding, but on the other it decreases because of microstructure coarsening. Typically, a peak in sintered strength occurs at intermediate sintering times on the order of 30–60 minutes. Although pore growth is a common event during sintering, the simultaneous rounding of the pores is generally beneficial to mechanical properties. There is a curious interaction between the evolving grain structure and pore structure during high-temperature sintering. The number of pores decreases while the average pore size might increase. The growing grains sweep the pores with them. Initially, the pores slow grain growth, but lose effectiveness as they become rounder. Accordingly, rapid grain growth is observed late in the sintering cycle. If the final temperature is high, then the rate of grain growth dominants. One consequence is the isolation of pores inside the grains. When this happens, the pores are stable and will not be removed by prolonged sintering. Figure 6.20 is a micrograph of one such situation, where the gains have grown to a large size, but the pores remain dispersed inside the grains. Continued sintering is fruitless when this occurs, since most engineering properties decline.

If sintering is conducted under conditions where densification occurs, forming closed pores is possible. Closed pores are not connected to the compact surface. Compacts subjected to normal sintering at 1120°C (2050°F) show little pore rounding. However, at higher temperatures the pores become round and the compact densities. As the rounded pores collapse, they are no longer in communication with

EQUATION BOX 6.5

Many properties depend on the microstructure scale. One of the best documented examples in steels is known as the *Hall–Petch effect*. The yield strength σ_Y varies with the grain size G as follows:

$$\sigma_Y = \sigma_0 + \frac{\beta}{\sqrt{G}}$$

The parameter β is empirical, but σ_0 is known as the *friction stress*. It depends on the alloy and represents the inherent resistance to deformation and dislocation motion. Generally, the friction stress decreases as temperature increases. The Hall–Petch relation states that the yield strength is higher for smaller grain sizes. Thus, grain growth during sintering progressively reduces strength, although sinter bonding might simultaneously be increasing strength. One consequence is that steels become oversintered, where longer holds or higher temperatures result in microstructure coarsening and lower mechanical properties.

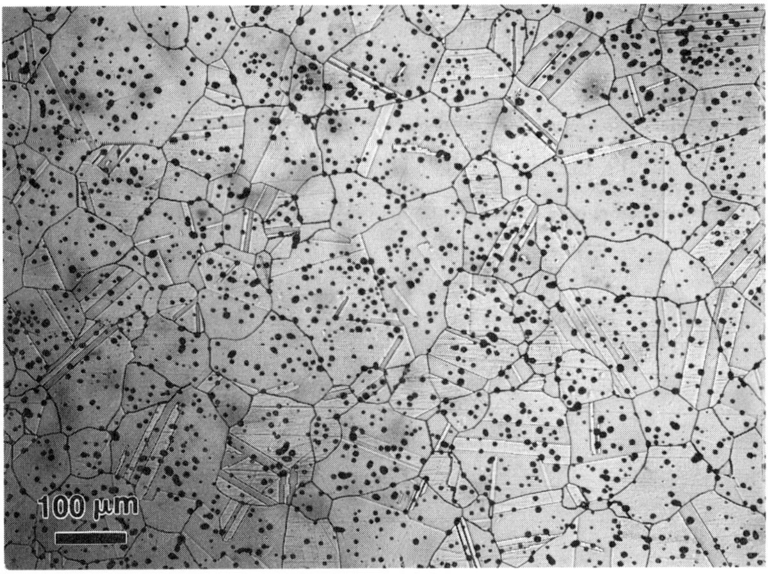

Figure 6.20. If heating is too rapid, the sintered microstructure forms isolated pores inside the grains, largely due to rapid grain growth at the high temperatures; generally this in an undesirable condition. (Photo courtesy of Haorong Zhang.)

the sintering atmosphere. Any contaminants trapped in the pores remain inside the compact and are not removed even with long holds at high temperatures. Pore closure is a natural event in sintering, especially when dealing with sintered densities over 7.0 g/cm^3. Once the pores close, grain growth accelerates as the pores no longer interfere with grain growth.

Various processing changes can be used to modify sintering and the evolution of the sintered microstructure. For example, smaller powders form smaller pores and increase the sintering rate, but often lead to more distortion. On the other hand, longer hold times at the peak temperature help densification, but add fabrication expense. Usually, high-temperature sintering should be done with shorter hold times at the peak temperature to reduce microstructural damage.

6.5. PROPERTY CHANGES

6.5.1. Property Evolution

The main justification for sintering is improved properties. Green compacts have some strength, but these are not competitive with other materials. Sinter bonding increases all typical engineering properties. Measuring strength and hardness is easy,

EQUATION BOX 6.6

During sintering the grain size distribution settles into a steady-state condition. Sometimes these are termed *self-similar distributions*, meaning that the distribution shape is always the same although the sizes are all enlarging. The bell-curve distribution of exam grades is a self-similar distribution, with the average being low for hard tests and high for easy tests, but the shape of the distribution is similar. The same holds true for microscope-measured grain size distributions measured on polished cross sections. The probability of sectioning a large grain is higher than that for sectioning a small grain. Additionally, the section is typically not through the middle of the grain; thus the apparent size is not the true size. Thus, the usual measure is based on the random test line intercept length. A test line is randomly placed on the microstructure, and the line lengths are measured to reflect the grain size.

An intercept size distribution function has been isolated for such data in sintered alloys. The cumulative normalized grain intercept size distribution is given as

$$F(L) = 1 - \exp(-0.7L^2) \tag{6.47}$$

where $F(L)$ is the cumulative fraction of grains having a size L, where $L = G/G_{50}$ is the normalized intercept size. Here G is the grain size and G_{50} is the median grain size, corresponding to the grain size at the 50% point on the cumulative distribution.

and both improve dramatically during sintering. Just in heating to the sintering temperature there is significant gain. As the hold time at the sintering temperature increases, there is further benefit, to a point, beyond which over sintering occurs and properties decline. After proper sintering, the mechanical properties of P/M steels are competitive with alternative production practices.

During heating to the sintering temperature, there is progressively more atomic motion and particle bonding, giving a higher strength. Figure 6.21 plots the strength of a 304L stainless steel versus temperature during heating from room temperature at a rate of 10°C/min (18°F/min). Note how the starting point is the green strength, which is 12 MPa (1.7 ksi). By 500°C (930°F), all the lubricant has been removed and the strength is starting to improve. Finally, when the compact is held at the peak temperature for a half hour the sintered strength is quite high, reaching 350 MPa or (50 ksi), about 30 times the green strength.

6.5.2. Optimized Properties

In the context of sintering, if some is good, will more be better? That leads to a mixed response. As previously illustrated, many properties improve as sintering temperature increases. Temperature is not the only variable, density, composition, and microstructure are also important. Higher densities improve properties. For example, Figure 6.22 plots the sintered tensile strength versus density for 4% nickel steels with three carbon levels. These compacts were sintered at 1120°C (2050°F) using mixed powders. With no carbon and a low density, the sintered strength is about the same as in some plastics, near 250 MPa (35 ksi). With higher sintered densities the strength increases. Strength is greatly improved from the combined density gains and carbon additions. In the best condition, the strength is over 3 times that of the low density condition. Proper sintering and productivity in the furnace dictate short hold times at the peak temperature for most ferrous sintering, typically less than one hour.

6.5.3. Processing Effects

Density goes hand in hand with sintering to improve sintered properties. Another demonstration of that fact is given in Figure 6.23, which plots both the yield and fatigue endurance strengths for a common iron–copper alloy, iron with 2% copper and 0.8% C, known as FC-0208. The plot gives both strengths (fatigue strength corresponds to 10 million cycles) for pressed and sintered compacts processed to

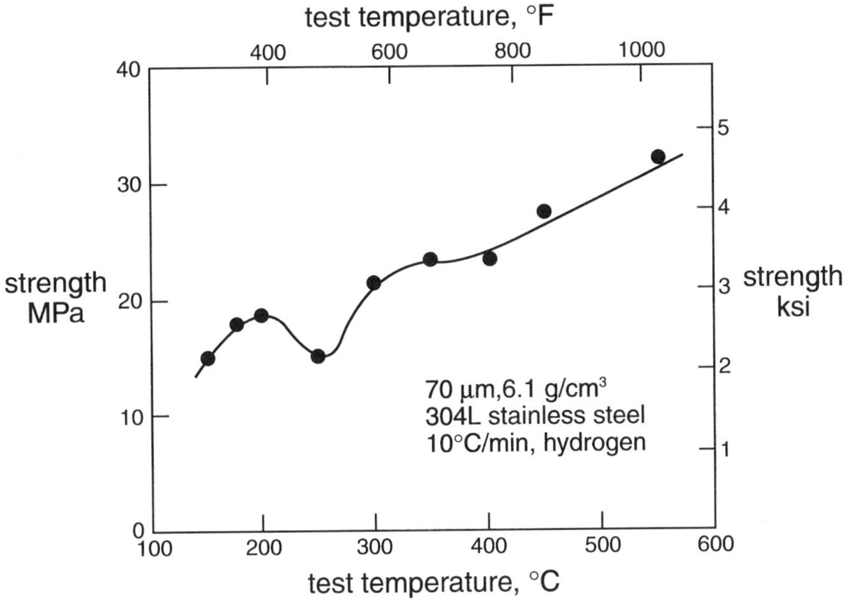

Figure 6.21. Strength at temperature during heating for a 70-μm water-atomized 304L stainless steel heated at 10°C/min (18°F/min) in pure hydrogen. Most of the strengthening occurs at temperatures over 1000°C (1832°F). (Data courtesy of Gregory Shoales.)

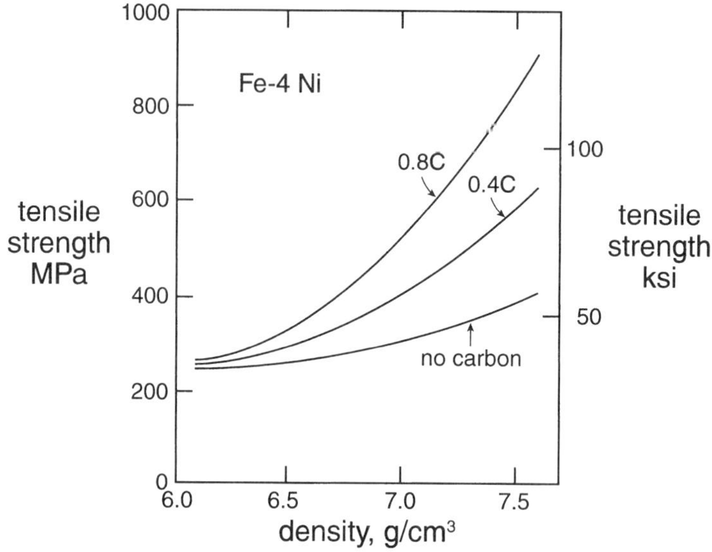

Figure 6.22. Sintered tensile strength for an alloy of iron with 4% nickel and three carbon contents. Note the dual effects of carbon and density on strength. (Data from Howard Sanderow.)

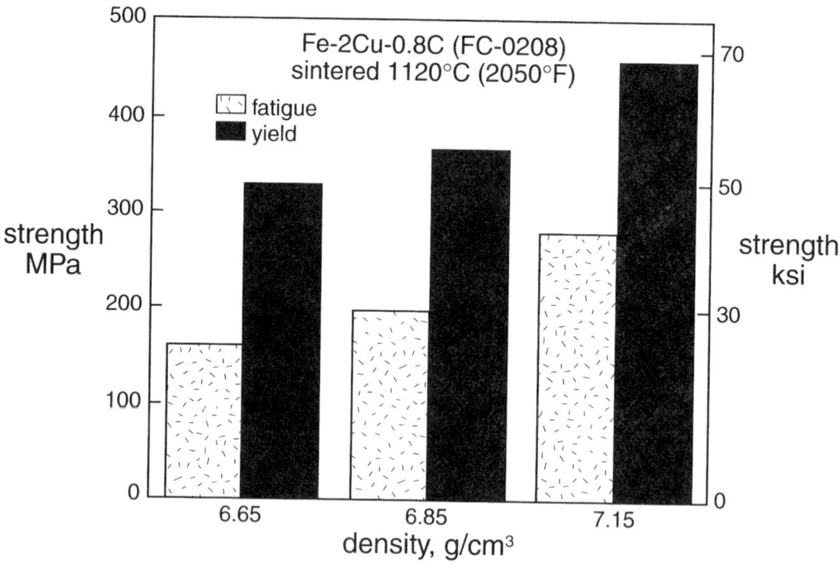

Figure 6.23. Yield and fatigue strength of sintered Fe–2Cu–0.8C (FC-0208) shown as a function of the green density, for a constant sintering cycle of 1120°C (2050°F) for 25 min. (Data from Kevin Christian.)

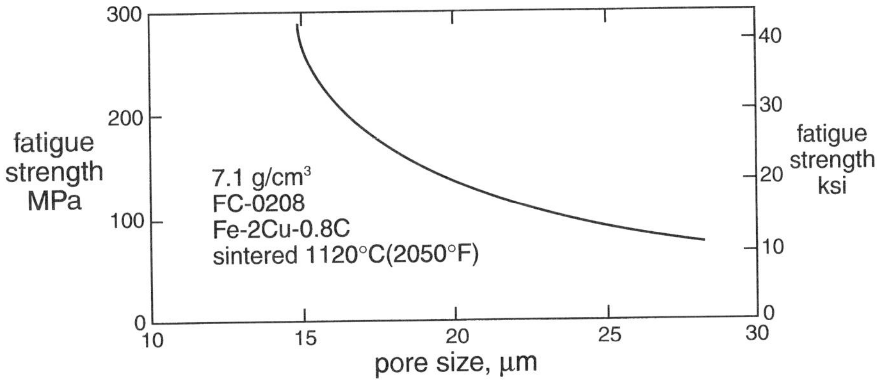

Figure 6.24. Fatigue strength is dependent on the pore size as evident from these data taken from Fe–2Cu–0.8C compacts sintered from a variety of powders. A large pore size lowers fatigue strength. (Data from Kevin Christian.)

three density levels; 6.65, 6.85, and 7.15 g/cm^3. Note that with a constant sintering cycle of 25 min at 1120°C (2050°F), there is still a possible 40% gain in yield strength and fatigue strength by compacting to a higher density. The highest fatigue strength is competitive with common steels formed by traditional metalworking routes. Fatigue strength is improved by specific pore structure changes, namely, smaller pores that are rounded. This is illustrated in Figure 6.24, where the fatigue strength is shown as a function of the pore size. At a fixed density, the number of pores increases as the pore size decreases. Small pores tend to be rounded and have less stress concentration effect. Pores that are highly curved, meaning irregular pores, lack resistance to fatigue. When sintering is adjusted to create rounded, smooth, small pores, the fatigue strength is much higher. Initially, longer sintering times or higher sintering temperatures achieve this goal, but oversintering leads to pore growth and a loss of fatigue strength.

A previous plot (Figure 6.9) illustrated the strength levels attainable in sintering with various time and temperature combinations. The combined effect of sintering temperature and time were given for an iron–nickel steel, FN-0208, which corresponds to Fe-2Ni-0.8C. The compacts were processed to 7.3 g/cm^3 or about 92% of theoretical. Sintering was performed at two times and three temperatures, leading to the six tensile strengths in this plot. Note that for each temperature, sintering for two hours gives better strength than sintering for 0.5 h at the higher temperature. Necessarily, when compared on just a constant time basis, higher sintering temperatures give higher strengths. However, when higher temperatures are not available, longer hold times at the peak temperature give similar levels of bonding. This leads to a concept of integral work as a rational basis for sintering cycles. In simple form, this says that the amount of sintering, as measured by the time–temperature combination, can be the same with many different sintering cycles.

Ferrous alloys exhibit improved properties when sintered. Accordingly, one may

EQUATION BOX 6.7

Integral work in sintering is a new concept arising out of research performed by Professor D. Lynn Johnson at Northwestern University. He has discovered that essentially the same degree of sintering is possible with many different combinations of heating rate, peak temperature, and hold time. Assuming no changes in mechanism, like phase transformation or melting, he shows that the integral work of sintering $\Theta(T,t)$ can be represented by the following equation:

$$\Phi(T,t) = \int \frac{1}{T} \exp\left(-\frac{Q}{RT}\right) dt$$

where T is the absolute temperature, t is the time, R is the gas constant [8.32 J/(mol · K)], and Q is the activation energy for sintering. Integration is performed from the start to end of the sintering cycle for the desired combination of heating rates, holds, and cooling rates. Much of the early analysis was applied to ceramic densification with varying combinations of heating rate, temperature, and hold time. However, this integral work term is a unifying principle that allows intelligent assessment of tradeoffs in cycles or furnaces for equivalent levels of sintering. In this form the units of $\Theta(T,t)$ are seconds per kelvin (s/K), reinforcing the interplay between longer times at lower temperatures giving equivalent levels of sintering.

ask why all components are not sintered at the highest temperatures to obtain the highest properties. The answer lies in two basic factors. First, higher temperatures cause dimensional changes. Sintering densification is inversely dependent on the green density. Since die compaction induces density gradients in the powder, then high-temperature sintering causes nonuniform shrinkage and warpage. Second, higher temperatures are more costly to apply. Not only is there more energy consumed in heating to the higher temperatures; more significant is the added equipment cost and complexity. Furnaces that operate at higher temperatures are more expensive to build, require more exotic construction materials, and often have shorter lives and lower load capacities. Each of these factors increases cost. Thus, an element of sufficiency is involved in determining the sintering temperature—what combination of sintering temperature and time is sufficient to meet the property demands for the application? The answer to that question varies considerably between P/M operations and is usually highly influenced by the equipment that is available.

6.6. DIMENSIONAL CONTROL

6.6.1. Importance of Dimensional Change

A major concern associated with ferrous sintering is sustaining predictable dimensional changes. The atomic motion events associated with sintering, coupled with thermal expansion and phase transformation events, swing dimensions over a wide range. Universal interest is in obtaining precise final dimensions. So, for some steels, sintering conditions are selected for zero dimensional change, meaning that the final component will be the same size as the die size, although the strength will be significantly improved. Other ferrous components are sintered at high temperatures where significant grain boundary diffusion occurs, leading to large sintering shrinkages. If the green body is uniform in initial structure, then shrinkage will not cause warpage and final dimensions will be repeatable and predictable. Dimensional uniformity is the counterbalance to desires for high final properties. Usually, as density levels are driven higher there is a loss of product dimensional uniformity. Repressing or machining after sintering is frequently used to correct these problems.

Dimensional control in sintering is one of the largest barriers to expanded use of P/M. Analysis of various materials shows that for a final product size variation, 30–40% of that variation is from powder sources (lot-to-lot variation in powder, mixing, or lubricant homogeneity), 15–25% is from compaction sources, but 40–50% is from sintering. Similar results are attained with both injection-molded and die-compacted components.

Many steel components could be processed by P/M, if the tolerance spread could be reduced, especially in automotive transmissions and other moving mechanical systems. The need is for more complicated shapes, with smoother surfaces and tighter tolerances. Unfortunately, a gear that is slightly out of round generates noise and wears faster than does one that is symmetric. Thus, subtle production factors, including powder fill in the die cavity, heat transfer in sintering, and lubricant distribution in the mix, become barriers to expanded P/M production. Yet modest tolerance improvements, meaning more predictable final dimensions, immediately open opportunities for 50% growth in current P/M materials.

When carefully examined, literally every process parameter affects final dimensions. The composition is very influential, with some ferrous alloys exhibiting swelling during sintering and others exhibiting shrinkage. Table 6.4 lists the dimensional size change from the die size for several common P/M alloys die compacted to

TABLE 6.4. Dimensional Changes from Die Size for Various Ferrous Alloys, Die Compacted to 6.8 g/cm^3 Green Density and Sintered at 1120°C (2050°F) for 30 min

Alloy	Designation	Size Change (%)
Fe–2Cu	FC-0200	0.41
Fe–2Cu–0.5C	FC-0205	0.26
Fe–2Cu–0.8C	FC-0208	0.03
Fe–0.8C	F-0008	−0.05
Fe–0.5C	F-0005	−0.08
Fe–2Ni–0.8C	FN-0208	−0.18
Fe–2Ni–0.5C	FN-0205	−0.21
Fe	F-0000	−0.25
Fe–2Ni	FN-0200	−0.37

Source: Data from Thomas Jesberger.

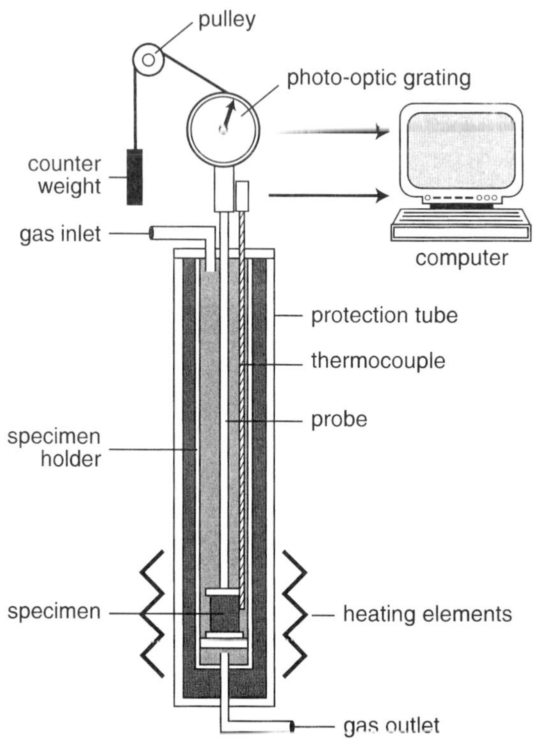

Figure 6.25. Dilatometry allows measurement of the compact dimensions in a furnace during sintering, thereby providing insight into the temperatures and possible causes of dimensional change during sintering. This is a typical design based on a vertical pushrod coupled to a photooptic grating and computer data acquisition system.

a density of 6.8 g/cm^3. Negative values indicate shrinkage, while positive values correspond to swelling. A few of the alloys exhibit shrinkage, such as iron–nickel alloys, but others produce swelling, such as iron–copper. One unique attribute of the Fe–2Cu–0.8C composition is its near-zero dimensional change when sintered at 1120°C (2050°F) for 30 min. This is useful in reducing distortion of complicated shapes.

Inhomogeneously distributed additives, improperly sized powder, and even changes in powder shape impact sintered dimensions. In die-compacted systems, anisotropic behavior is observed between the pressing direction and perpendicular to the pressing direction. This is traced to how the particles deform in pressing and the resultant pore shape. High green densities behave differently from low green densities, again reflecting pore shape differences with compaction conditions. Finally, the key sintering parameters of burnout cycle, heating rate, atmosphere, peak temperature, time at the peak temperature, and cooling rate all have a significant and measurable impact on final dimensions.

6.6.2. Dilatometry

Dilatometry is a useful means for following dimensional change during sintering. These dilation meters continuously measure the size of a component in a furnace during sintering. The measurement is performed by contacting the compact with an external measuring transducer that uses electrical, magnetic, or optical signals to determine the instantaneous dimensions. Figure 6.25 shows a cross section through such a device, here with a vertical, counterbalanced probe connected to a photooptic grating. As the test material shrinks or expands, the probe motion sends a signal to the computer along with temperature measurements from the thermocouple.

An example of the insight possible using dilatometry is plotted in Figure 6.26. The test alloys were iron with 8% copper, in one case without added graphite while in the other with 0.8% added graphite. This figure plots the net expansion versus cycle time for a sintering run of 10°C/min (18°F/min) to 1120°C (2050°F) with a soak of 60 min, followed by cooling to room temperature. First note the complex size changes. During heating there is thermal expansion, reaching about 1% expansion by 725°C (1340°F). At this temperature the iron phase starts the transformation from alpha to gamma (ferrite to austenite), giving a net contraction. The transformation is completed by slightly over 900°C (1650°F).

On continued heating, swelling starts when the copper melts at 1083°C (1981°F). Once the copper has melted, there is little dimensional change during the soak at

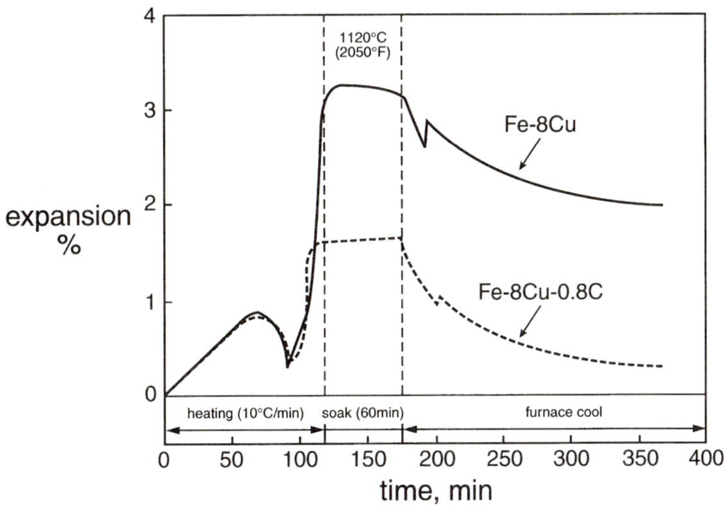

Figure 6.26. Example dilation information during sintering of Fe–8Cu and Fe–8Cu–0.8C alloy compacts. The compacts expand from thermal expansion during heating, shrink as a result of the ferrite-to-austenite phase transformation, and expand because of dissolution of copper into the iron when copper melts. The net dimensional change differs by a factor of 6 between these two alloys. (Data courtesy of Anthony Griffo.)

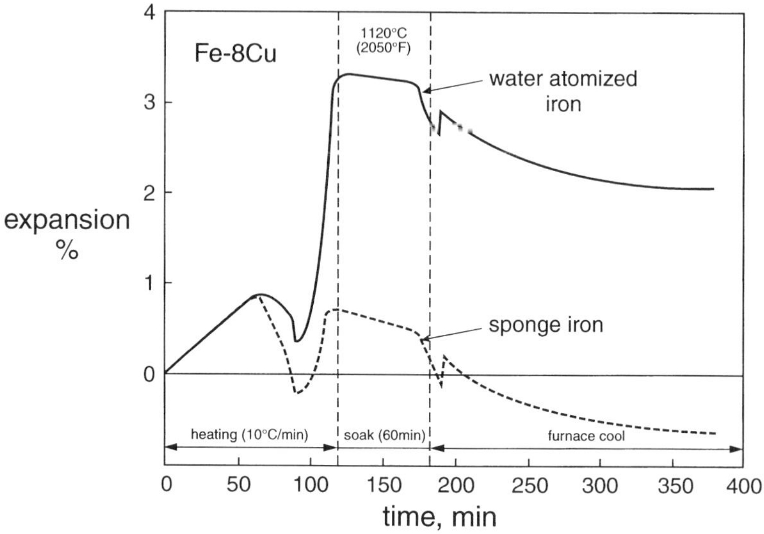

Figure 6.27. Comparison of the dilatometry for two iron powders using a composition of Fe–8Cu. The water-atomized iron exhibits a large net expansion, while the sponge iron exhibits a net shrinkage, since there are pores for molten copper absorption in this porous powder. (Data courtesy of Anthony Griffo.)

the peak temperature. Finally, on cooling the copper solidifies, giving a slight swelling peak, the compact shrinks as a result of thermal contraction, but then undergoes a swelling burst with the gamma-to-alpha transformation. The bottom line on the sintering run is a net swell of 0.3% for the Fe–8Cu–0.8C composition, but 2.2% swell for the Fe–8Cu composition.

Melting copper causes iron to swell, but the addition of carbon (graphite) reduces this swelling reaction. The event traces to how carbon dissolves preferentially into the iron during heating, thereby reducing the swelling attack by molten copper. This fact is most evident at the point where copper melts on the dilatometer traces.

In a similar manner, dilatometry can be used to measure dimensions in the pressing direction to compare with those with the perpendicular direction. Differences in powder, mixing, compaction, alloying, heating rate, maximum temperature, hold time, and atmosphere can be quantified using dilatometry. Another example is given in Figure 6.27, which compares the sintering of Fe–8Cu formed from two different iron powders. One was a water-atomized powder, and the other was a reduced-oxide sponge powder. Note the difference in final dimensions. Considerable difficulty in meeting a dimensional specification would result if one powder were substituted for the other in production.

Many comments that follow are based on insight gained by dilatometry. Another tool for following dimensional change in sintering is high-temperature video imaging. Here the aperture on a video camera is coordinated with a flash to illuminate the compact inside a furnace, allowing digitization of all dimensions and determination of sources of warpage. Such advanced tools are currently available only in research settings.

6.6.3. Green Structure Effects

If viewed closely enough, every processing change impacts dimensions in sintering. Some examples of this were evident in the dilatometry plots. Sometimes, processing factors have a small effect, but in others the effects are dramatic. For example, the impurity content in the powder, especially the oxygen level, acts to remove carbon. Carbon monoxide and carbon dioxide are formed during heating, leading to a net reduction in carbon and dimensional change problems. Usually decarburization is directly proportional to the starting oxygen level. There are many examples of components going out of dimensional specification because of improperly controlled oxygen levels in the starting powders.

Green density is another important factor. Higher green densities give improved strength, largely because the sintered density is higher. Still, there is a change in the dimensions after sintering. Figure 6.28 plots example strength variations for an

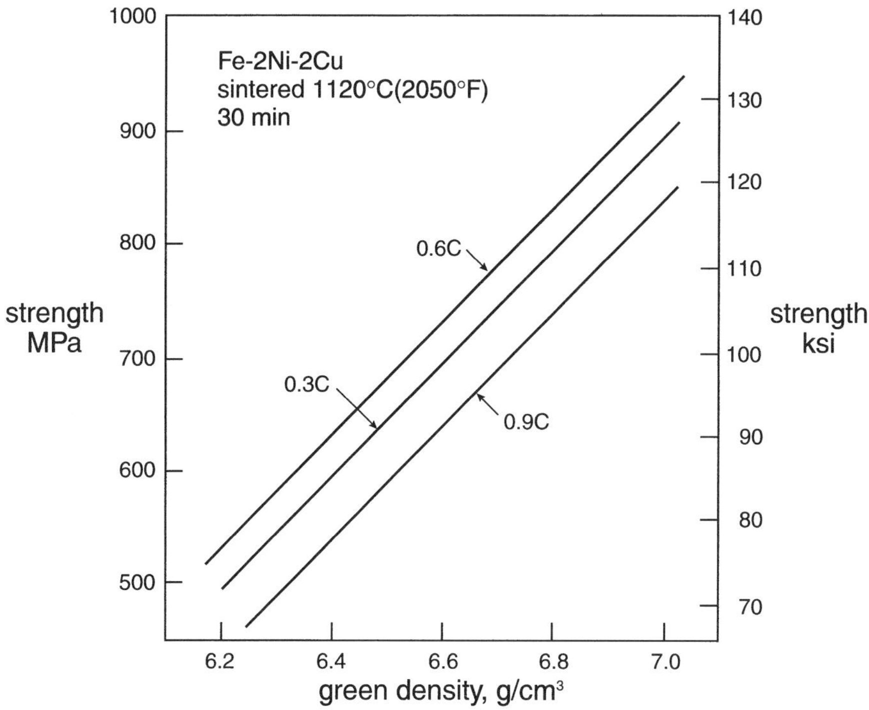

Figure 6.28. Green density effect on the sintered transverse rupture strength for Fe–2Ni–2Cu compacts containing various carbon levels. The highest strength occurs for 0.6% carbon.

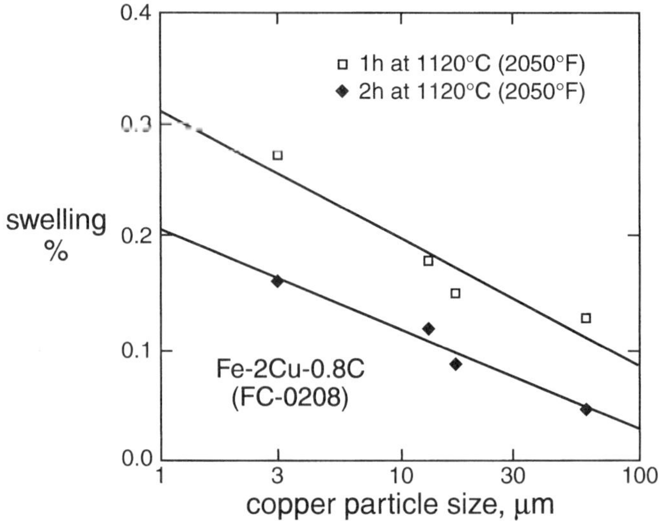

Figure 6.29. Dimensional change in sintering Fe–2Cu–0.8C using water-atomized iron powder (heat at 10°C/min to 1120°C and hold for 1 or 2 h, which equals 18°F/min to 2050°F) showing reduced swelling with a larger copper particle size and longer hold time. (Data courtesy of Anthony Griffo.)

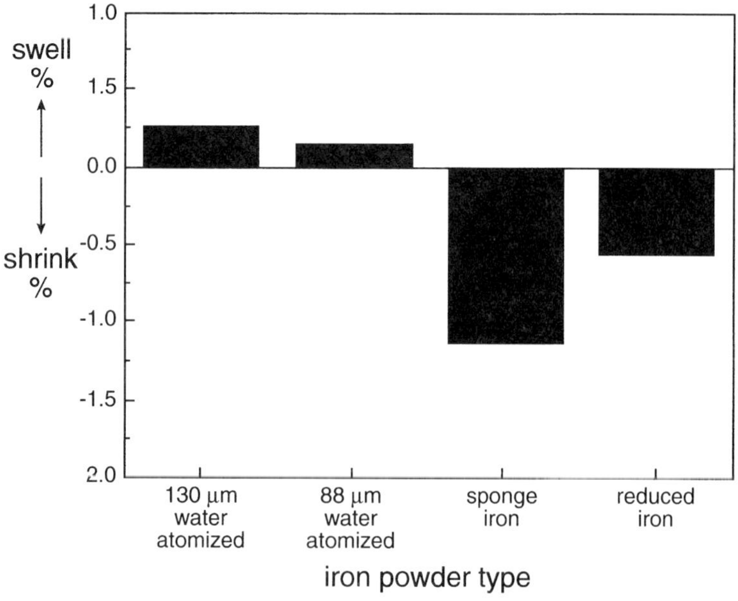

Figure 6.30. A demonstration of the iron-powder-type effect on net dimensional change in sintering an alloy of Fe–2Cu–0.8C at 1120°C (2050°F). It is possible to custom blend iron powders for a zero net dimensional change. (Data from Anthony Griffo.)

iron–nickel–copper steel at three carbon levels. The highest strength occurs at the 0.6% carbon level. Interestingly, at the lower green densities the compacts shrink, but at densities over 6.3 g/cm^3 for 0.9% C there is swelling, and the shift from shrink to swell is 6.5 g/cm^3 for 0.3% C.

Recognizc that any change in the green microstructure affects dimensional change. A few plots help illustrate the character of the effects. The iron–copper–carbon system is the best studied, so the following comments emphasize this system. Figure 6.29 plots the swelling versus sintering time and copper particle size (logarithmic scale) for an alloy of 2% copper and 0.8% carbon, FC-0208. The compacts were fabricated with all other parameters being held constant. There is less swelling with the larger copper particles and longer sintering times.

The iron particles also effect dimensional change. Figure 6.30 demonstrates this behavior for four iron powders. For the FC-0208 composition, note that two of the powders give swelling and two give shrinkage. Smaller iron powders produce shrinkage, especially at lower green densities. Full density is possible with use of a subsieve iron powder. With all else being equal, a blend of the two powders would form a zero-dimensional-change composition. This is widely practiced, leading to many proprietary powder blends. The difficulty is that each blend must be matched to the other processing factors to achieve zero dimensional change. The dimensional change data in Figure 6.31 illustrate two other factors, namely, amount of added

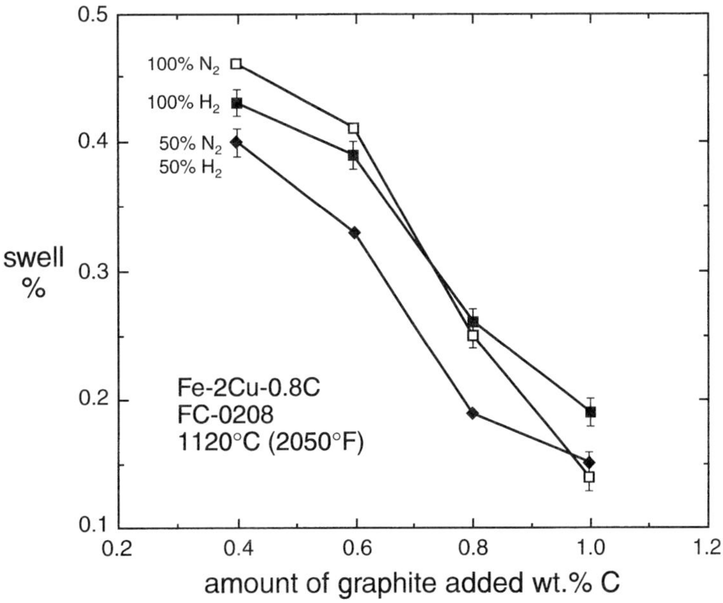

Figure 6.31. The amount of graphite mixed into a Fe–2Cu alloy influences the net sintered dimensional change as illustrated using data with three sintering atmospheres for a green density of 6.8 g/cm^3, heated at 10°C/min (18°F/min) to 1120°C (2050°F), which was held for 1 h.

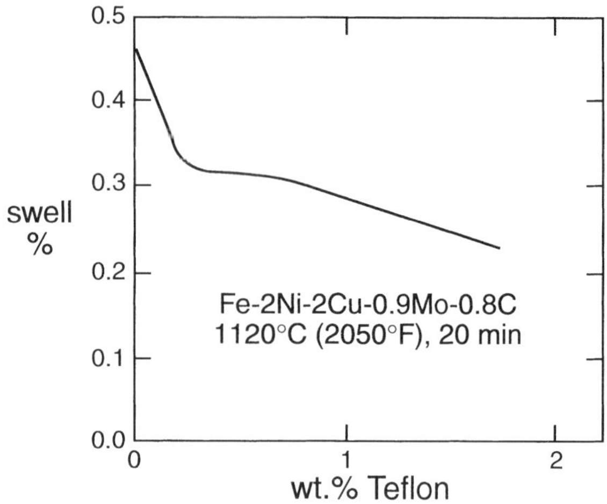

Figure 6.32. Swelling in sintering a Fe–Ni–Cu–Mo–C steel in 20% hydrogen–80% nitrogen at 1120°C (2050°F) for 20 min after compaction at 690 MPa (120 ksi or 60 tsi). Swelling decreases as more lubricant is added. (Data from Anthony Griffo and Christoph Degoix.)

graphite and sintering atmosphere. Here again the composition consists of 2% copper. For this iron powder, copper powder, and sintering cycle, the least swelling occurs with the highest carbon content. For most of the compositions the lowest swell is in an atmosphere consisting of mixed nitrogen and hydrogen. Pure nitrogen produces the largest swelling at low graphite contents, but pure hydrogen gives the largest swelling at high graphite contents.

Other factors trace to the lubricant type, amount, and even the particle size of the lubricant. Figure 6.32 plots the swelling of Fe-Ni-Cu-Mo-C compacts sintered in 80% nitrogen–20% hydrogen for 20 min at 1120°C (2050°F). As the amount of lubricant increases, the swelling decreases, for a constant compaction pressure of 690 MPa (50 tsi). Likewise, slight changes in impurities impact final dimensions. For iron with 2% copper, Figure 6.33 plots the sintering shrinkage versus silicon content in the iron for three carbon levels; 0.6, 0.8, and 1.0%. Note that silicon causes expansion, largely because of its retardation of carbon dissolution into the iron. All the compacts for this study were processed under equivalent conditions to emphasize the role of a trace element in the 100–900-ppm concentration range. Unfortunately, the silicon concentration in the iron powder is seldom specified.

As a final demonstration, consider the possible effect of furnace design. Depending on the component size, shape, and placement in the furnace, there will be changes in heating. For example, in one study on sintering an alloy of Fe–2.5Ni, two furnaces operating at the same temperature for the same time gave nearly threefold differences in dimensional change.

There is no simple behavior. Many process variables interact with each other to affect dimension change in sintering. How can we rationalize such process sensitiv-

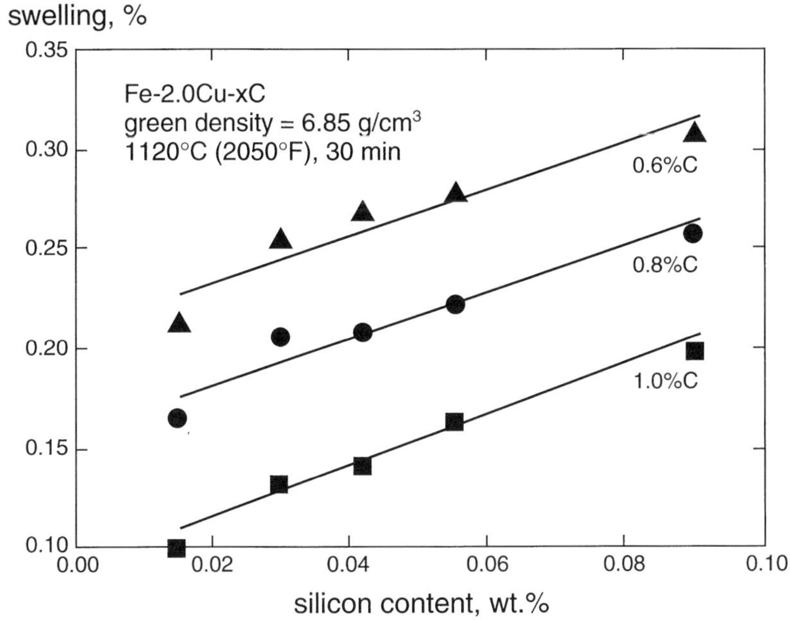

Figure 6.33. Swelling in sintering as a function of the trace silicon content in the iron powder and carbon level for iron–2% copper alloys pressed to 6.85 g/cm^3 and sintered under typical conditions. (Data courtesy of Satoshi Uenosono.)

ity? The sintering cycle plays a role in dimensional change because of bonding, pore rounding, pore coarsening, and grain growth. Yet another part of the story is in the composition and the green microstructure. Composition shifts change the chemical reactions during heating. Accordingly, the phase transformations and reaction kinetics are affected. These prove difficult to predict and must largely be isolated by experimentation. Likewise, the green microstructure influences dimensional change. Compensation for shifts in the green microstructure are possible via adjustments in process variables or composition. However, empirical studies are needed to know which variable to shift. Generally, sintering temperature–time combinations are the final means to adjust dimensions, but possibly at the expense of sintered properties.

6.6.4. Effects from Sintering Variables

There are several recognized and important sintering variables. Temperature clearly is the most important. The effect of temperature is most dramatic in examining liquid-phase sintered alloys. For example, Figure 6.34 presents micrographs of an iron–copper alloy quenched from 1075 and 1110°C (1967 and 2030°F). The electron contrast between iron and copper shows the copper particle as a discrete island in the iron matrix. Copper melts at 1083°C (1981°F), between the two quench temperatures. Once it melts, the prior site for the copper particle becomes a void as liquid

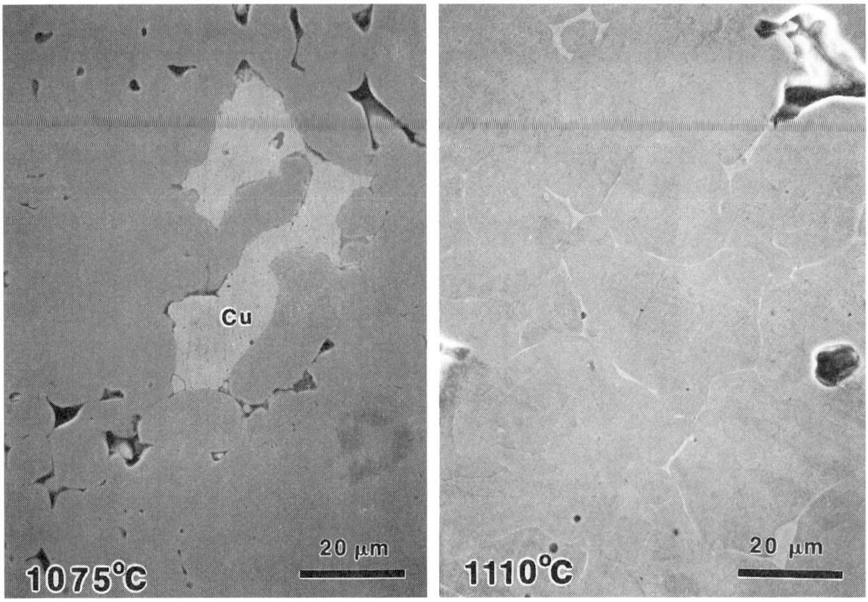

Figure 6.34. These two SEM micrographs were obtained by quenching Fe–Cu compacts from just before copper melting (a) and just after copper melting (b). The dramatic change in the copper placement, from isolated particles to thin liquid films, shows why there is a large bust of swelling in the dilatometry traces during copper melting. (Photographs courtesy of Anthony Griffo.)

copper spreads between the iron particles. This spreading causes swelling. The cause is very apparent, and is dependent on temperature. If the proper temperature is not achieved in sintering this alloy, there are dramatic differences in final dimensions.

As emphasized in Figure 6.35 for die-compacted powders, higher sintering temperatures give more shrinkage. Here, there is swelling up to 1150°C (2100°F), but shrinkage at higher temperatures. Longer times would increase shrinkage, but higher green densities lower shrinkage. The data plotted in Figure 6.36 emphasize this point for die-compacted stainless steel. Higher green densities reduce shrinkage, and as green density increases, there is less difference between sintering temperatures. The green density effect simply reflects the reduced pore space. Less shrinkage occurs with low initial porosities. Accordingly, with injection-molded steels there is much larger shrinkage because the green structure starts with 40% porosity.

Time is another processing parameter that affects properties and dimensional change, especially at shorter times. As the steel consumes the initial surface energy, the rate of consumption also declines. This is similar to the rate at which a fire burns. Once started, it burns rapidly, but as the fuel is consumed, the burn rate decreases to a smolder. An example of this behavior is given in Figure 6.37, which shows the sintering shrinkage for 1200°C (2200°F) sintering of an iron–cop-

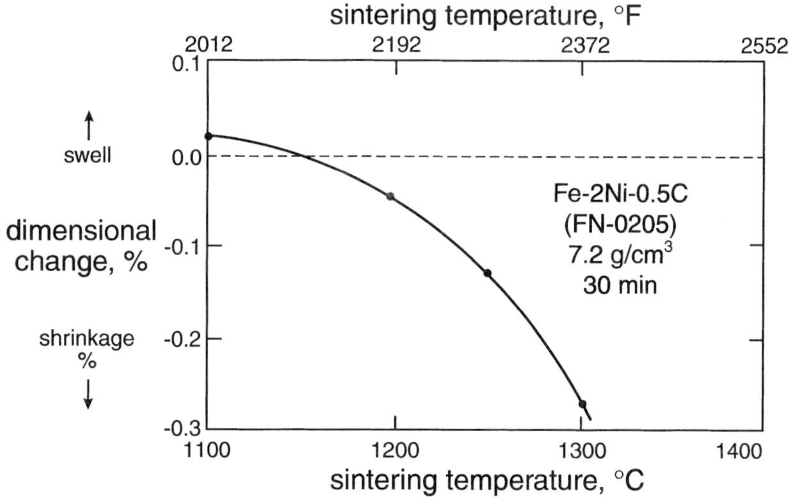

Figure 6.35. An example of the temperature effect on net dimensional change for a Fe–2Ni–0.5C (FN-0205) pressed to 7.2 g/cm^3 and sintered for 30 min at various temperatures. By 1150°C (2100°F), there is essentially no dimensional change, but swelling is evident at lower temperatures, and shrinkage dominates at higher temperatures. Longer times would also increase the shrinkage. (Data from Howard Sanderow.)

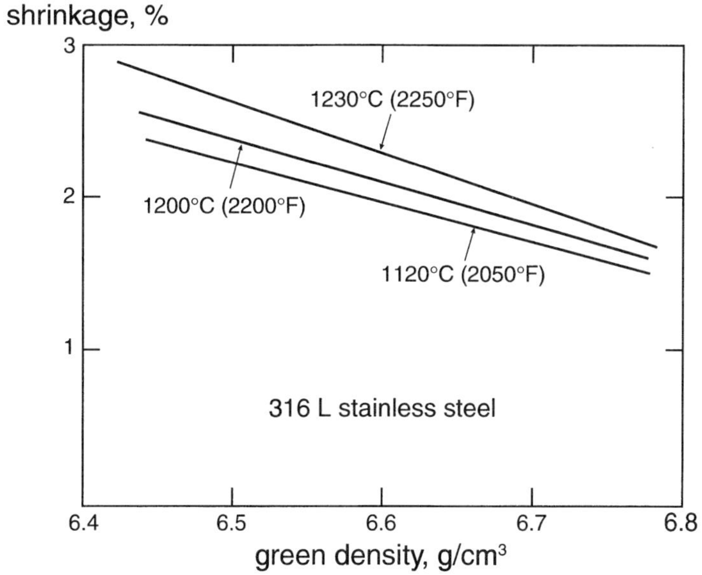

Figure 6.36. Sintering shrinkage versus green density for 316L stainless-steel compacts sintered at three temperatures. Note the benefit of a higher green density in reducing the net dimensional change.

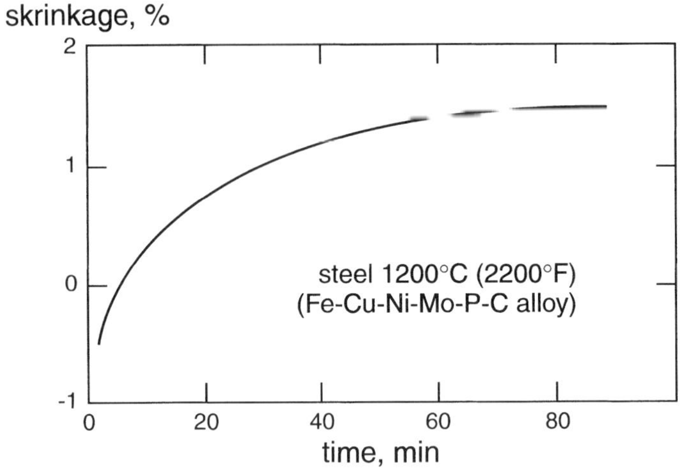

Figure 6.37. Compared with temperature, time affects dimensional change as evident by these data for a steel sintered at 1200°C (2200°F). In practice, most ferrous alloys are sintered for times ranging from 20 to 60 min. (Data courtesy of Deepak Madan.)

per–nickel–molybdenum alloy. In the first 20 min the dimensional changes is relatively large, over the next 20 min the relative shrinkage is much less, and so on.

Related to time is the heating rate. Usually, industrial furnaces heat too slowly to affect final dimensions. However, this is not always true. Fast heating generates stresses that interfere with uniform dimensional changes. Most problems arise at lower temperatures where heat transport is slow and lubricants or binders remain in the compact. Heat is delivered only to the outside of the compact while the core is cold. The temperature difference causes a stress that warps or even cracks the component.

6.6.5. Dimensional Precision

In sintering, a problem is not the absolute dimensional change, but the scatter between parts. Figure 6.38 compares the standard deviation in a dimension perpendicular to the pressing direction for a die-compacted iron–nickel–carbon steel (FN-0208) at a 7.1 g/cm^3 density. In the as-pressed condition, the dimensional scatter is just over 0.01%, while it climbs to 0.017% after sintering at 1120°C (2050°F), but nearly doubles on sintering at 1220°C (2230°F). Thus, the common problem with higher sintering temperatures is not in the net dimensional change, but in the dimensional scatter. Consequently, high sintering shrinkages mandate homogeneous green microstructures. As suggested by the drawings in Figure 6.39, this is why injection-molded steels are superior for high-sintered-density applications; they sinter to higher densities with less dimensional scatter because of homogeneous green structures.

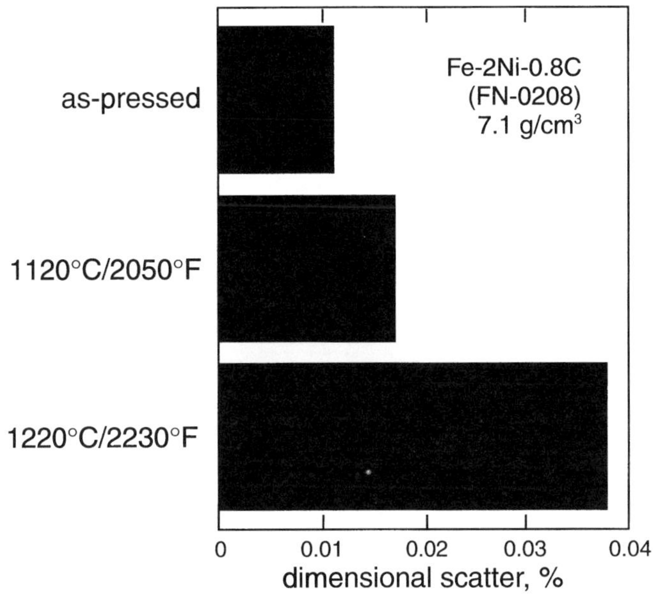

Figure 6.38. Dimensional scatter in sintering Fe–2Ni–0.8C (FN-0208) compacts at a density of 7.1 g/cm^3 at three temperatures. The higher sintering shrinkage at higher temperatures contributes to more dimensional scatter. (Data from Howard Sanderow.)

Sintering atmosphere is another factor. Usually cost and material compatibility dictate the sintering atmosphere. However, there are some secondary effects on dimensional change. Many metals react at high temperature, forming hydrides or nitrides with the atmosphere. However, iron is relatively impervious to these reactions. The difference in heat transport between atmospheres impacts dimensional precision. Vacuum has no thermal conductivity, while hydrogen has a high thermal conductivity. Nitrogen is intermediate in thermal conductivity. Differences in these atmospheres are evident as a dimensional change sensitivity to heating rate.

Concern with dimensional control in sintering is inherent to net-shaping by P/M. Many factors influence the total dimensional change, the uniformity of shrinkage in a compact, and the variations between compacts. Sintering practice focuses on controlling these factors to produce high-precision components. However, it is not an easy process. For example, Figure 6.40 illustrates a classic problem by plotting the axial (in the pressing direction) and radial (perpendicular to the pressing direction) dimensional changes for Fe–2Cu–0.8C sintered at 1120°C (2050°C) with different graphite addition levels. Zero dimensional change in one dimension corresponds to 0.2–0.4% dimensional change in the perpendicular dimension. So, tooling design and compaction conditions must account for this difference. As illustrated in this chapter, many such interactions bear on efforts to control the sintering behavior of ferrous systems.

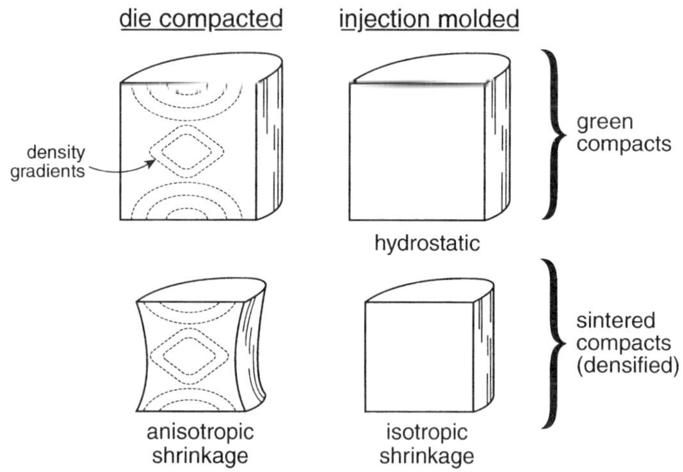

Figure 6.39. A suggestion of the green density gradient effect on warpage in sintering. Compacts formed with isotropic green densities by hydrostatic compaction show more uniform sintered dimensions. Unfortunately, most compacts are die-pressed, leading to density gradients. During sintering these density gradients produce nonuniform dimensional changes and compact warpage. It is this loss of precision that limits the sintering temperature for most ferrous alloys, with a corresponding penalty in properties.

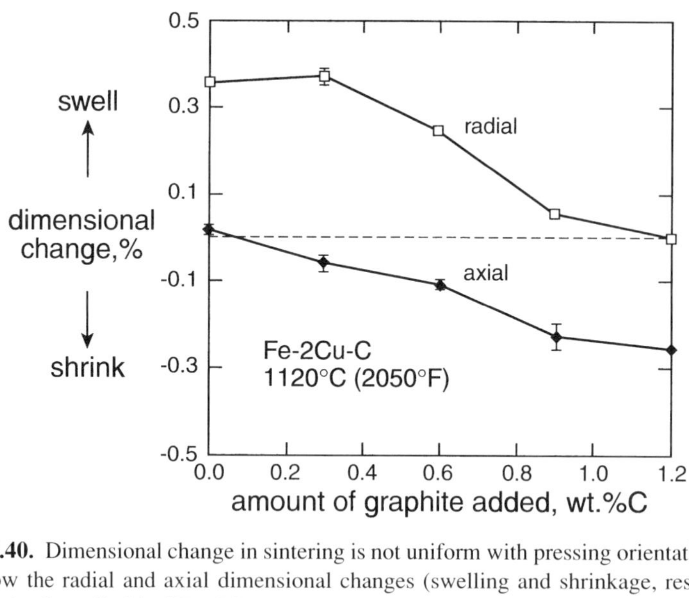

Figure 6.40. Dimensional change in sintering is not uniform with pressing orientation. These plots show the radial and axial dimensional changes (swelling and shrinkage, respectively) from die size for cylindrical Fe–2Cu compacts containing various amounts of added graphite. This behavior is a reflection of the green pore shape effect. (Data from Anthony Griffo.)

SUGGESTIONS FOR FURTHER READING

Chaklader, A. C. K., and J. A. Lund (eds.), *Sintering '91*, Trans Tech Publications, Brookfield, VT, 1992.

Exner, H. E., ''Principles of Single Phase Sintering,'' *Rev. Powder Metallurgy Phys. Ceramics* **1,** 7–251 (1979).

German, R. M., *Sintering Theory and Practice*, Wiley-Interscience, New York, NY, 1996.

German, R. M., G. L. Messing, and R. G. Cornwall (eds.), *Sintering Technology,* Marcel Dekker, New York, NY, 1996.

Griffo, A., and R. M. German, ''Dimensional Control in the Sintering of Iron–Copper–Carbon via Particle Surface Area,'' *Internat. J. Powder Metallurgy* **30,** 399–407 (1994).

Lenel, F. V., *Powder Metallurgy Principles and Applications*, Metal Powder Industries Federation, Princeton, NJ, 1980, pp. 211–267.

Nayar, H., ''Production Sintering Atmospheres,'' in *Powder Metallurgy*, vol. 7, Metals Handbook, E. Klar (ed.), American Society for Metals, Metals Park, OH, 1984, pp. 339–350.

Sanderow, H. I. (ed.), *High Temperature Sintering*, Metal Powder Industries Federation, Princeton, NJ, 1990.

Uskokovic, D. P., H. Palmour, and R. M. Spriggs (eds.), *Science of Sintering*, Plenum Press, New York, NY, 1989.

CHAPTER 7

Production Sintering Operations

7.1. SINTERING ATMOSPHERES

The last chapter focused on the behavior of ferrous powder compacts during sintering. The treatment looked inside the component and assessed the effects of processing variables on dimensional change and sintered strength. In this chapter, we change the perspective to look at the means for performing production sintering operations. Two attributes of concern are the furnace and the atmosphere in the furnace. This combination sets many practical limits on sintering. First attention is given to the atmosphere, the gaseous chemical environment surrounding the component during heating.

7.1.1. Atmosphere Types

The sintering atmosphere performs several roles; the most important one is to protect the compact from attack by air at high temperatures. Therefore, leaking furnaces are not acceptable. Further, the atmosphere helps in removing lubricants or binders, transferring heat to the compacts, extracting surface oxides, controlling the final carbon content, and adjusting impurity levels such as the nitrogen content.

Several atmosphere conditions are possible inside a sintering furnace, ranging from oxidizing (carbon dioxide or oxygen), neutral (argon, helium, or vacuum), reducing (hydrogen or carbon monoxide), nitriding (nitrogen or ammonia), carburizing (methane or propane), and decarburizing (carbon dioxide or steam). These options are not necessarily mutually exclusive. For example, it is possible to have a combination such as hydrogen and methane that is carburizing and reducing at the same time. In the overall practice of sintering, there are many possibilities; however, for ferrous P/M, there are a few common sintering atmospheres. These are categorized as follows:

- Hydrogen, essentially pure hydrogen with slight contaminants of water or oxygen
- Nitrogen–hydrogen, formed from a mixture of the two gases over a wide ratio
- Dissociated ammonia, a mixture of 75% hydrogen and 25% nitrogen formed from NH_3

TABLE 7.1. Compositions of Sintering Atmospheres

Constituent	Hydrogen	Nitrogen–Hydrogen	Dissociated Ammonia	Endothermic
% N_2	0	75–98	25	39
% H_2	>99.9	2–20	75	39
% H_2O	<0.01	0.001	0.004	0.8
% CO	0.0	0.25	—	21
% CO_2	0.0	0.05	—	0.2
% CH_4	0.0	0	0	0.5
ppm O_2	10–25	5	10 to 35	10–150
dew point, °C	−70 to −20	−75 to −50	−50 to −30	−16 to 10

- Endothermic and exothermic gas, generated from natural gas and air, a mixture of several species
- Vacuum, a very low pressure of air about one-millionth of atmospheric pressure

Table 7.1 tabulates the composition of the first four atmospheres. Since a vacuum is essentially the absence of an atmosphere, it is not included in this tabulation.

Pure hydrogen is explosive, is very light, and exhibits a high thermal conductivity. Its reactivity with oxygen helps in the sintering of alloys containing elements that form stable oxides. The most common examples are stainless steels, where chromium is usually present in the alloy at concentrations from 12 to 20% to prevent rusting in service. Chromium provides protection by forming a stable surface oxide, too thin to see, but tenacious enough to prevent corrosive attack. However, this attribute makes sintering difficult. Instead of sintering iron or chromium in the stainless steel, if the oxide is not removed, we are faced with the more difficult task of sintering the oxide of chromium Cr_2O_3, which requires temperatures over the melting point of stainless steel. Thus, we must reduce the oxide, meaning reacting hydrogen with the oxide to form water vapor that is swept away in the furnace. In situations of this sort, pure hydrogen is the preferred sintering atmosphere, in spite of concerns over reactivity, flammability, and cost.

Nitrogen is essentially neutral or even beneficial for many ferrous alloys, since residual nitrogen in the steel increases strength. Many sintering atmospheres are based on nitrogen, with additions of hydrocarbons, alcohol, moisture, hydrogen, carbon monoxide, or methane to control the final carbon level. Some common compositions are between 80 and 97% nitrogen. The higher the nitrogen content, the lower the cost and flammability. A nitrogen–hydrogen mixture is not flammable below 5% hydrogen. Since 80% of all air is nitrogen, it is possible to construct on-site generators that inexpensively take it from air via cryogenic or molecular sieve techniques. For a few ferrous alloys, nitrogen forms compounds that degrade properties. Special concern is with nitrides of titanium, aluminum, chromium, and tantalum, so use of an atmosphere containing nitrogen is avoided if the steel contains these alloying elements.

Dissociated ammonia is a reducing atmosphere used to sinter many steels. It is delivered as a liquid under pressure. The ammonia molecule is catalytically broken into its hydrogen and nitrogen components in the dissociation step

$$2\ NH_3\ (g) \rightarrow 3\ H_2\ (g) + N_2\ (g)$$

where the nomenclature (g) indicates a gas. The dissociation reaction occurs on a heated catalyst [near 1000°C (1830°F)] to give a nearly pure atmosphere of hydrogen and nitrogen. It is a reducing atmosphere, since the hydrogen reduces oxides, meaning that hydrogen and oxygen react to form water vapor.

Endothermic atmospheres are low-cost mixtures of nitrogen and hydrogen, with water, carbon monoxide, and carbon dioxide. They are formed via a catalytic reaction of natural gas and air. Natural gas is about 93% methane (CH_4). Some unconverted methane may remain. The use of endothermic atmospheres is on the decline because of poor control, since daily variations originate from fluctuations in the amount of water, methane, butane, and other hydrocarbons in natural gas. Exothermic gas is a related mixture of air and natural gas, but is not used for most ferrous sintering because of carbon control problems.

As for reducing character, pure hydrogen is most favorable if water vapor is removed to maintain a pure atmosphere. Liquid hydrogen supplies maintain the inlet atmospheres to less than 1 ppm oxygen and 8 ppm water. However, moisture generated inside the furnace lowers the ability to reduce oxides and makes the atmosphere decarburizing with respect to most materials. Nitrogen is considered neutral in most situations, except where nitriding can occur. Dissociated ammonia is both reducing and nitriding at the same time.

Sintering in a vacuum is really processing at a reduced pressure, since it is not practical to remove all gaseous molecules from a process environment, even in space. Under controlled conditions, vacuum sintering provides a clean, reproducible, and nonreactive environment. This is achieved by heating in a sealed chamber with a pumping mechanism continuously extracting evolved vapor. The pressure in the sintering chamber might be between 0.01 and 0.0001% of normal atmospheric pressure. Most steels can be sintered under a vacuum, but this may not necessarily give the best composition control nor the best sintered properties. One advantage is that it allows use of novel alloying additions in steels, like titanium, tantalum, and niobium, and it can be used for high-temperature sintering of tool steels and stainless steels. Vacuum sintering is a batch process that is best suited to lower production quantities. Many industrial vacuum furnaces employ graphite heating elements. Any residual oxygen evolving from the steel component reacts with the graphite to form carbon monoxide that is reducing.

7.1.2. Delubing and Binder Burnout

In heating a P/M component to the sintering temperature, the first step is to extract the processing aids (lubricants) and contaminants (oxide, moisture). If carried to high temperatures, these impurities will alter the compact chemistry, properties, and dimensions.

Essentially all P/M products use polymer processing aids for shaping, molding, compaction, lubrication, flow, or powder agglomeration. Polymers are the most common and these are decomposed and extracted in the first stage of heating. Most polymers decompose into a complex array of molecules during heating—methane, ethane, butane, propane, carbon monoxide, carbon dioxide, water, and so on. This decomposition starts as low as 150°C (300°F), peaks near 450°C (840°F), and is usually completed by 550°C (1020°F). The atmosphere has a role in polymer burnout since it can provide reaction species to help break apart the polymer molecules. For example, oxygen, carbon dioxide, or water provide sources of oxygen to form carbon monoxide or carbon dioxide with carbon in the polymer. Further, the atmosphere flow sweeps away the burnout products, preventing high-temperature reactions with the component. Unfortunately, if the steels see the oxidants at high temperature, there will be oxidation and decarburization of the steel. So the use of atmospheres with oxidants is restricted to low temperatures.

When burnout processes are done improperly, sooting occurs along with surface blemishes or discoloration of the P/M component. In the worst case, defects are generated in the component. Sooting occurs because the compacts were heated too rapidly, without sufficient oxidant (water or oxygen). During heating the polymer directly forms graphite as a smoke that falls on the compacts. It is more of a problem with high loading in the furnace, high heating rate, low atmosphere flow rates, and large or high-density components.

Best practice in removal of the polymer phases is to heat slowly over the range of temperatures where burnout occurs, but ensure the compacts are to 650°C (1200°F) before entry into the high-temperature portion of sintering. Atmosphere turnover is important to removing the burnout products. Unfortunately, few production operations adjust gas flow to balance changes in furnace loading. Recent tests determined that the best polymer extraction occurs in an atmosphere based on mixtures of hydrogen and nitrogen. As illustrated in Figure 7.1, the higher the hydrogen content, the greater the weight loss, meaning less carbon residue from the polymer. Indeed, the best procedure is to use pure hydrogen or hydrogen with water vapor below 200°C (390°F) and nearly pure nitrogen above this temperature. This combination proves difficult to manage in some furnaces, and the usual situation is to use a high-nitrogen-content atmosphere for the whole cycle.

The indications of improper burnout are sooting, low strength, high carbon level, blistering, and property or dimension variations with location in the furnace. For most cases the cure is slower heating, more oxidant in the atmosphere, and a faster atmosphere flow over the components. Computer simulations of furnace performance, including polymer burnout, help identify problems. Further, closed-loop control systems have been implemented that take readings from the furnace and use constant adjustments to the composition and atmosphere flow rate to improve burnout.

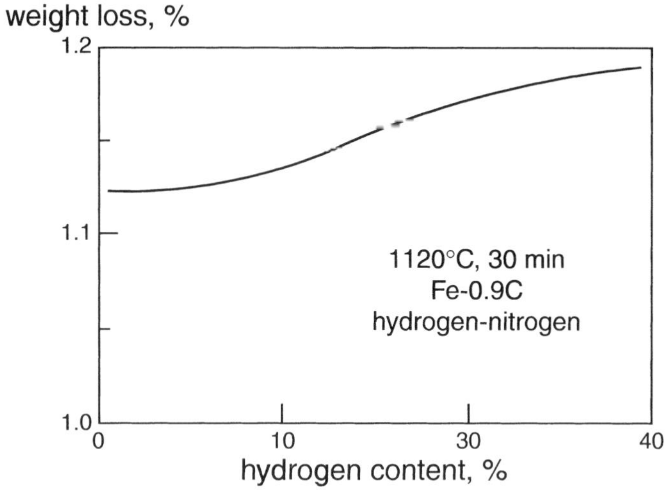

Figure 7.1. The weight loss in sintering a steel powder versus the hydrogen content in the atmosphere (balance nitrogen). High hydrogen levels result in loss of oxygen and reduced nitrogen absorption. (Data from Eric Streicher.)

7.1.3. Oxidation–Reduction

Thermochemical reactions are important in selecting the atmosphere for sintering ferrous P/M alloys. The dominant reactions are oxidation–reduction and carburization–decarburization. The reactions involving iron (Fe) are well understood. An equilibrium between oxygen and an oxide such as FeO can be written in reaction form as follows:

$$Fe\ (s) + \tfrac{1}{2}O_2\ (g) \leftrightarrows FeO\ (s)$$

Iron (Fe) and iron oxide (FeO) are solids as indicated by the (s), and oxygen is a gas (g). Reaction equations with a symbol $\leftrightarrows$ indicate that the reaction can go both ways. A high concentration of iron and oxygen leads to iron oxide production. Alternatively, a high concentration of iron oxide favors reduction and the formation of oxygen and pure iron. It is like a balanced playground seesaw—push down on one side, and the other side goes up; add oxygen, and iron oxide production is favored. Figure 7.2 plots the oxygen partial-pressure effect on this reaction. This is a plot of the balance between the oxygen pressure and temperature, showing the balance between oxidation (iron and oxygen going to iron oxide) and reduction (iron oxide going to iron and oxygen). At a typical sintering temperature, reduction requires that the atmosphere have less than 10^{-13} parts of oxygen to avoid oxidation. Alternatively, oxidation occurs when the oxygen partial pressure in the sintering atmosphere is above the line. Note that higher sintering temperatures make reduction easier.

On a relative scale, iron oxide is easy to reduce using hydrogen. The reaction is described by the following symbols;

$$FeO\ (s) + H_2\ (g) \leftrightarrows Fe\ (s) + H_2O\ (g)$$

If iron (Fe) is heated in the presence of steam (water vapor, H_2O), then iron oxide (FeO) and hydrogen (H_2) form. Alternatively, if iron oxide and hydrogen are heated together, the reaction produces iron and water vapor. A constant supply of hydrogen is required for total oxide reduction, meaning that the water vapor product must be constantly replaced by fresh hydrogen. Note that the water vapor and hydrogen contents appear on opposite sides of the reaction. This means that the direction of the reaction depends on the concentration ratio of water and hydrogen, a parameter measured by the dew point. The dew point tells the temperature at which water vapor will condense and is a measure of the moisture content in the sintering atmosphere. A dew point of 7°C corresponds to 1 vol% water content and is oxidizing for most metals, while −42°C corresponds to 0.01 vol% water vapor and is reducing for most metals.

Thus, atmosphere purity is a concern in sintering. The difference in ease of reducing various oxides is categorized in Figure 7.3, where the dew point for reducing the oxide is given for most common metals. For example, ferrous oxides are more

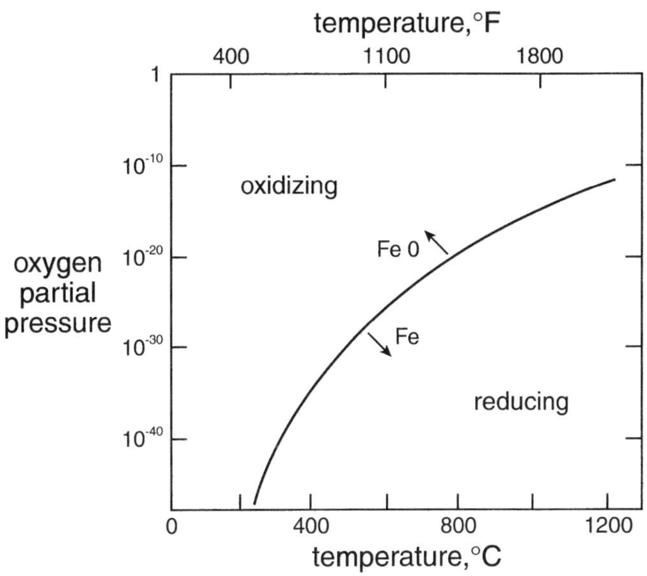

Figure 7.2. A plot of the oxygen partial pressure (oxygen as a part of the total pressure) versus temperature for the iron–iron oxide equilibrium. Conditions represented above this line cause iron to oxidize; below this line, low-oxygen partial-pressure conditions lead to oxide reduction.

EQUATION BOX 7.1

Oxide reduction occurs during sintering as controlled by thermochemical reactions. During the sintering of iron, reduction is easy, but chromium oxide reduction proves more difficult. Assume the solids are of fixed composition, such that a metal M is in equilibrium with oxygen gas O_2 and the oxide M_xO_2, where the subscript x represents the stoichiometry of the oxide:

$$x\,\mathrm{M}\ (s) + \mathrm{O}_2\ (g) \leftrightarrows \mathrm{M}_x\mathrm{O}_2\ (s)$$

The metal and oxide are solids, as indicated by the notation of (s), and the oxygen is a gas (g). For this reaction, there is an equilibrium constant K defined as follows:

$$K = \frac{a_{\mathrm{M}_x\mathrm{O}_2}}{a_{\mathrm{M}}^x P_{\mathrm{O}_2}}$$

where a designates a thermodynamic quantity known as the *activity*. For the solid phase, the activity is unity (meaning that there is plenty of each solid available for reaction) and the P_{O_2} oxygen partial pressure is the only factor that determines which way the reaction progresses. Thus, the equilibrium constant for the oxidation–reduction reaction depends only on the inverse partial pressure of oxygen, and lower partial pressures favor oxide reduction. In turn, the equilibrium constant reflects the standard free energy ΔG for the reaction

$$\Delta G = -RT\ln(K) = RT\ln(P_{\mathrm{O}_2})$$

where R is the gas constant and T is the absolute temperature. The free energy for such reactions can be determined from tables or charts available in standard metallurgical reference books, or these days via computer routines. The key concept is that low oxygen and water vapor pressures are required to reduce oxides and to keep the P/M compact from oxidizing.

easily reduced than chromia (chromium oxide = Cr_2O_3). Thus, atmosphere conditions that prove reducing for iron might be oxidizing to chromium. For good sintering, reducing all the oxide is usually necessary before reaching a high temperature. The higher stability of chromium oxide, not iron oxide, dominates the atmosphere selection.

Two common reducing species that are effective in sintering iron are carbon monoxide (CO) and hydrogen (H_2). The oxidation–reduction reaction involving carbon monoxide is similar to that shown above for hydrogen:

reducing dew point at 1120°C in hydrogen

dew point

over 0°C (easy)
: gold, silver, cobalt, copper nickel, palladium, platinum, molybdenum, tungsten

-20 to 40°C (reasonable)
: tantalum, manganese

-60 to 80°C (very difficult)
: niobium, silicon

below -80°C (impossible)
: aluminum, titanium, vanadium, magnesium, beryllium

Figure 7.3. At a typical sintering temperature for steels, 1120°C (2050°F), this ranking gives the relative stability of each oxide. Iron oxides are easily reduced, even in high-dew-point atmospheres, while many oxides prove very stable and difficult to reduce in any practical sintering atmosphere.

$$\mathrm{FeO\ (s)} + \mathrm{CO\ (g)} \leftrightarrows \mathrm{Fe\ (s)} + \mathrm{CO_2\ (g)}$$

Like the hydrogen–water effect, the partial-pressure ratio of carbon monoxide CO and carbon dioxide CO_2 is important. Here a high partial pressure of carbon dioxide leads to oxidation (it pushes the reaction to the left) and a high partial pressure of

EQUATION BOX 7.2

One of the measures of impurity effects is the dew point. It tells the temperature at which water vapor will condense; accordingly, it is a measure of the relative moisture content and the atmosphere oxidation–reduction potential. The following equation allows calculation of the volume percent of water in an atmosphere V_{H_2O} if the dew point is known as T_D (in °C):

$$\log_{10}(V_{H_2O}) = -0.237 + 0.0336\,T_D - 1.74 \cdot 10^{-4}\,T_D^2 + 5.05 \cdot 10^{-7}\,T_D^3$$

Oxide reduction depends on attaining a low dew point, and this becomes easier as temperature increases.

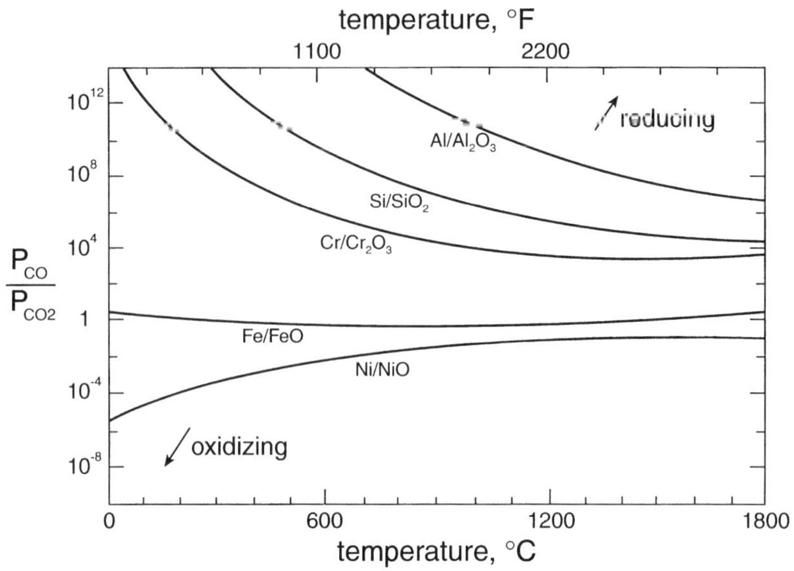

Figure 7.4. A plot of the balances between oxidation and reduction for five metals as a function of the temperature, expressed as the carbon monoxide to carbon dioxide pressure ratio.

carbon monoxide leads to reduction, since it pushes the reaction to the right. Figure 7.4 is a plot of the reactions versus the partial pressure ratio over a range of temperatures for iron, nickel, chromium, silicon, and aluminum. It is easier to use carbon monoxide to reduce iron and nickel versus aluminum, silicon, and chromium.

7.1.4. Carburization and Decarburization

As described above for oxidation, there are similar concerns with atmosphere-controlled carbon reactions. Carbon is a potent and necessary strengthening agent in steels. Precise control of the final carbon content mandates an understanding of carburization and decarburization events during sintering. Because of the reactions between carbon and oxygen and hydrogen, it is possible to add or subtract carbon from the compact on the basis of specific atmosphere attributes.

As a first example, the general reaction between iron (Fe) and its carbide (Fe_3C) is considered in terms of the carbon monoxide (CO) or methane (CH_4) reactions,

$$3\ Fe\ (s) + 2\ CO\ (g) \leftrightarrows Fe_3C\ (s) + CO_2\ (g)$$

$$3\ Fe\ (s) + CH_4\ (g) \leftrightarrows Fe_3C\ (s) + 2\ H_2\ (g)$$

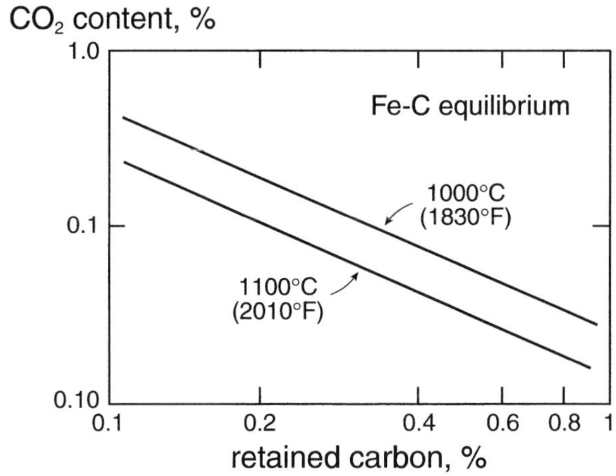

Figure 7.5. Retained carbon depends on the carbon dioxide content in the atmosphere. A high final carbon level requires a low carbon dioxide level, especially as the sintering temperature increases.

As an example, Figure 7.5 plots the relation between the retained carbon level in steel and the carbon dioxide content in the sintering atmosphere at two temperatures. Note that higher retained carbon levels mandate less carbon dioxide. If the carbon dioxide content is high, then the upper carbide reaction is driven to the left, producing decarburized iron and carbon monoxide. Sintering in a vacuum with graphite heating elements leads to a reaction between residual oxygen and graphite. Carbon monoxide forms via the reduction of iron oxide at temperatures over 800°C. On the other hand, carbon-based reduction of chromium oxide does not occur until nearly 1250°C. Graphite heating elements also react with carbon dioxide or hydrogen to form carbon monoxide or methane (usually the former reaction is favored), both of which are carburizing.

Besides heating elements, carbon can be introduced from the decomposed polymers or by intentional additions of graphite to the powder mixture. As evident in Figure 7.6, carbon additions go into solution early in the heating process, usually in the 700–900°C (1300–1650°F) range. In this micrograph, carbon is evident as pearlite, but the particles are still poorly bonded. Consequently, decarburization in the high-heat zone is more of a concern as compared with carburization reactions.

Consider the reaction involving the carbide in steel with water vapor (steam) during sintering:

$$Fe_3C\ (s) + H_2O\ (g) \leftrightarrows 3\ Fe\ (s) + CO\ (g) + H_2\ (g)$$

As the water content in the atmosphere increases, the steel is decarburized. Thus, atmospheres with low water vapor levels (dry or low dew points) are most effective in controlling carbon during sintering. Figure 7.7 plots the relation between dew

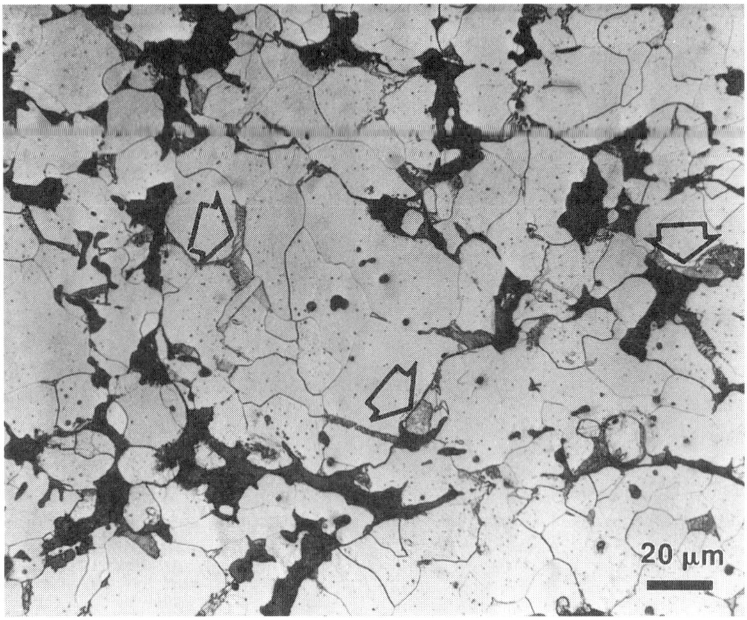

Figure 7.6. A mixture of iron and graphite heated to 925°C (1700°F) to show that carbon is dissolved into solution prior to significant interparticle bonding as marked by the arrows. (Photograph courtesy of James Marsden.)

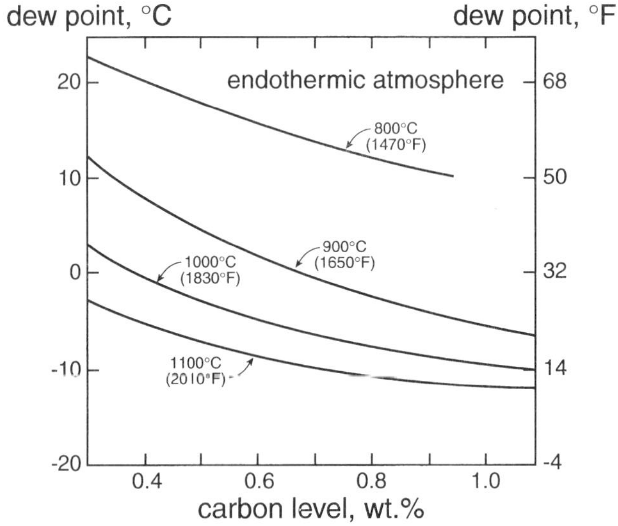

Figure 7.7. Dew point and temperature effects on the stabilization of various carbon levels during sintering in an endothermic atmosphere. The final carbon level can be controlled by selection of the proper dew point. (Data from Fritz Lenel.)

point, a measure of the atmosphere water content, and the final carbon level for endothermic atmosphere sintering. At the typical sintering temperature, high final carbon levels require low dew points. Similar plots can be developed for any sintering atmosphere, alloy, and temperature.

To a lesser degree, there are also decarburization events possible in sintering in dry hydrogen. A reaction between carbon and hydrogen to form methane is decarburizing, but it is slower and often ignored. Therefore, sintering in high-purity hydrogen is a desirable option, since it gives both oxide reduction and carbon control.

In the preceding reactions, FeO and Fe_3C were used to illustrate the oxygen and carbon roles in iron reactions with the atmosphere. Actually, the oxygen and carbon do not have to form compounds. Instead, oxygen and carbon dissolved in the steel will participate in these same reactions, even if they do not form the oxides and carbides.

7.1.5. Key Reactions

Since there are several possible reactions in steels, it is necessary to consider the five main gas species involved in the reactions: CO, CO_2, H_2, H_2O, and CH_4 (carbon monoxide, carbon dioxide, hydrogen, water, and methane). Two key atmosphere reactions are as follows:

$$CO_2\ (g) + H_2\ (g) \rightleftarrows CO\ (g) + H_2O\ (g)$$

$$CH_4\ (g) + H_2O\ (g) \rightleftarrows CO\ (g) + 3\ H_2\ (g)$$

Full control of the carbon reactions involves measuring the concentrations of carbon dioxide, carbon monoxide, hydrogen, water, and methane. However, in each of the preceding reactions the carbon monoxide and carbon dioxide species are always on opposite sides of the reaction. Thus, quantitative measure of the carbon potential is possible by measuring the ratio of these two gases, just like the dew point.

Most steels are sintered using continuous furnaces with nitrogen–hydrogen atmospheres. If oxygen is present, then either the compact or the atmosphere is doped with carbon (graphite or methane) to compensate for decarburization. To obtain target sintered properties, oxide reduction and controlled carbon levels are required. A map can be constructed for iron and steels based on the atomic fraction of oxygen (f_O = atoms of oxygen divided by total atoms in atmosphere) and carbon (f_C = atoms of carbon divided by total atoms in atmosphere) showing the balance between oxidation and reduction. Figure 7.8 is such a map covering several temperatures. The horizontal axis is the atomic fraction of carbon (e.g., pure methane is 20% carbon), while the vertical axis is the atomic fraction of oxygen (water is 33% oxygen). For each temperature the indicated line represents the balance between oxidation and reduction. Reduction is achieved by processing in an atmosphere below the equilibrium line at low oxygen contents.

Figure 7.9 uses the same plotting scheme, but now shows the temperature profiles for carburization and decarburization. It is possible to measure the atmosphere qual-

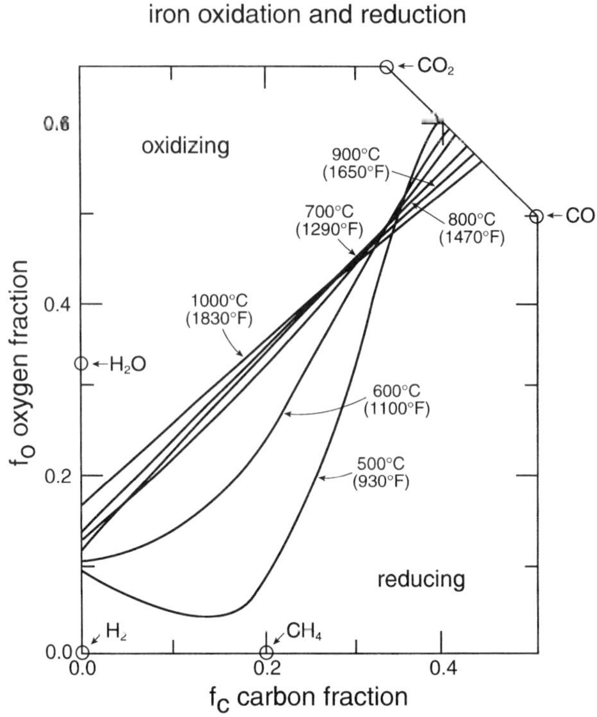

Figure 7.8. Iron oxidation and reduction as a function of temperature and atmosphere composition. Here atmosphere composition is measured in terms of the fraction of active carbon and oxygen levels, where nitrogen and hydrogen are considered neutral. The reduction of iron oxide requires a low oxygen fraction. (Data from Larry Darkin.)

ity and from these plots to determine the conditions where oxides will be reduced, while avoiding decarburization or carburization. Typically, this requires continuous adjustments of the atmosphere composition during heating via computer control.

7.1.6. Impurity Effects

Many chemical reactions occur in sintering. This is evident in Figure 7.10, which traces the change in carbon, oxygen, and nitrogen contents in an iron powder compact during heating in hydrogen. All three contaminants are removed. The total weight loss during heating is nearly 2% through the formation of carbon monoxide, carbon dioxide, steam, and nitrogen vapors.

Control of the process atmosphere provides an opportunity to improve sintering and to adjust the steel chemistry. However, the atmosphere is not constant. The compacts carry contaminants (air, moisture, oxides, carbon, and polymers) into the furnace. The evolution of these contaminants during heating changes the atmosphere composition. Further, the support materials and heating elements may interact with the sintering material via the atmosphere. Consequently, impurity control in the sintering environment is not simple.

Oxide reduction occurs at progressively higher dew points as temperature increases. Higher sintering temperatures shift the metal–oxide reaction toward oxide reduction. Figure 7.11 diagrams the oxide stability versus temperature, showing two regions; oxidizing and reducing. The powder compact enters the furnace under oxidizing conditions, especially during polymer burnout. During heating it crosses into a reducing condition. If excessive oxidation occurs during heating, then subsequent reduction leaves a high-porosity surface region such as pictured in Figure 7.12. In contrast, the desired surface structure is shown in Figure 7.13, where there are no gradients in pore structure or carbon content.

However, on cooling the compact recrosses the oxidation–reduction boundary, allowing oxidization during cooling. If that occurs at a high temperature, the compact will be discolored. Most important is the atmosphere quality after attaining the peak

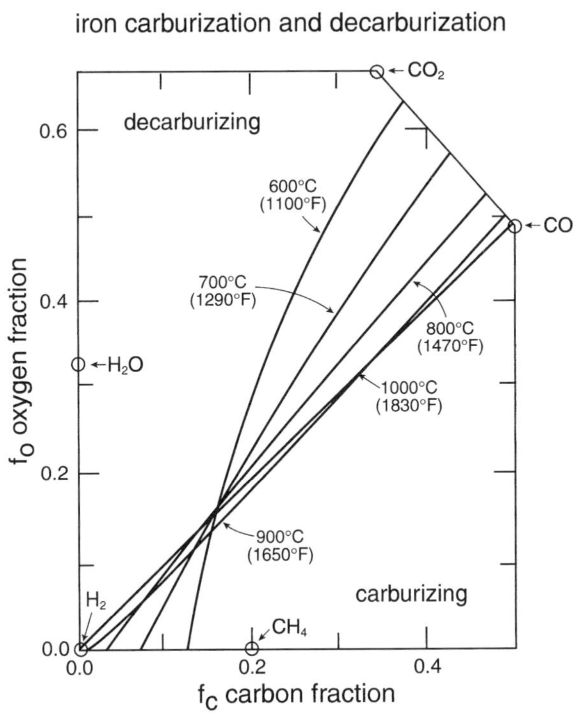

Figure 7.9. Atmosphere carbon and oxygen effects on decarburization and carburization of iron at various temperatures. Above the line represents decarburization, while below the line represents carburization. Mixtures of atmosphere ingredients and proper selection of the process temperature provide a means to control the final carbon level. (Data from Larry Darkin.)

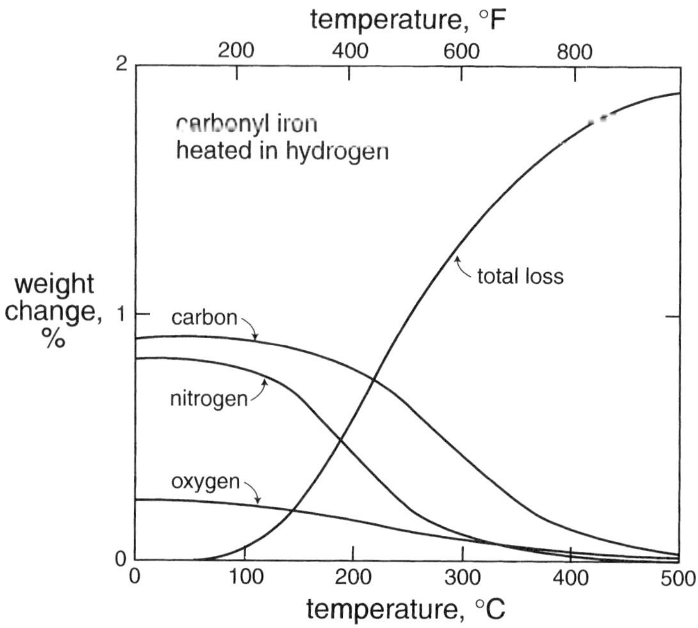

Figure 7.10. The weight change during heating in hydrogen for a high-carbon-content carbonyl iron powder. The three impurities—oxygen, nitrogen, and carbon—are progressively extracted at temperatures up to about 500°C (930°F). (Data from John Warren.)

temperature. Consequently, most furnaces inject fresh gas into the cooling zone. Because the dew point reflects the reactions in a furnace, it and temperature are important monitors of a sintering furnace. A high dew point arises from furnace leaks, drafts at the entrance or exit of the furnace, or improper balance between the feed of material and the atmosphere flow.

Steel powders exhibit a weight loss during sintering that can easily reach 1.5 wt%, resulting from both reduction and evaporation events. For example, consider the high temperature reduction of solid Fe_2O_3 by gaseous hydrogen:

$$Fe_2O_3\ (s) + 3\ H_2\ (g) \rightleftarrows 2\ Fe\ (s) + 3\ H_2O\ (g)$$

The water is in the form of steam; thus, it is swept along by the atmosphere through the furnace. For most steels it is imperative to fully reduce surface oxides during heating to the sintering temperature. However, the reduction of surface oxides is easier (requires less purity in the atmosphere) at higher temperatures. This is evident by the plot showing the balance between reduction and oxidation for stainless steel versus temperature and atmosphere in Figure 7.14. Less hydrogen is required in the process atmosphere at a given dew point for the higher sintering temperature. However, use of a high nitrogen content is not necessarily acceptable. Nitrides of chro-

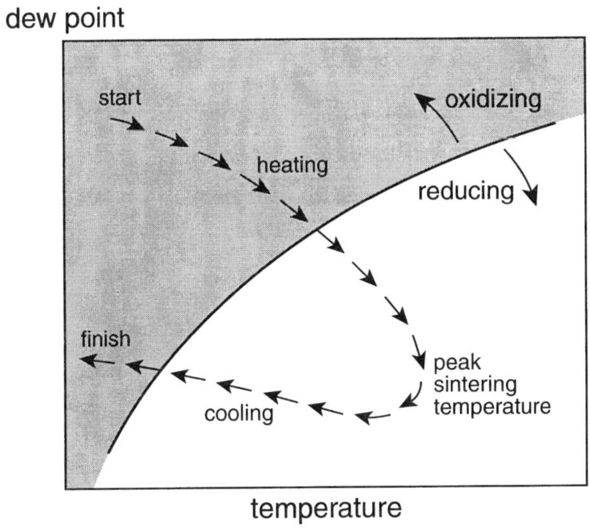

Figure 7.11. During sintering the powder compact starts in air under oxidizing conditions. Before reaching peak temperature, the condition needs to be reducing, and this must be maintained to a low temperature on cooling to avoid discoloration.

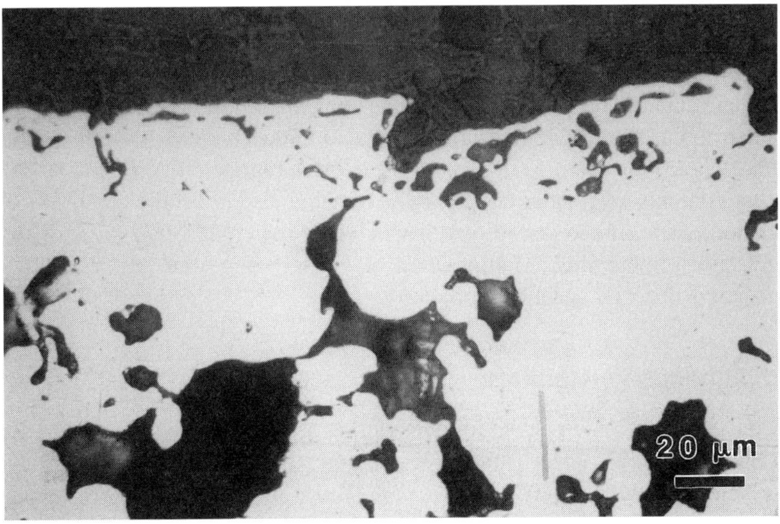

Figure 7.12. During the early portion of sintering this Fe–2Cu–0.9C compact was surface-oxidized. During subsequent sintering that surface oxide was reduced, leaving behind the layer of small surface pores as evident in this cross-sectional optical micrograph. (Photograph courtesy of James Marsden.)

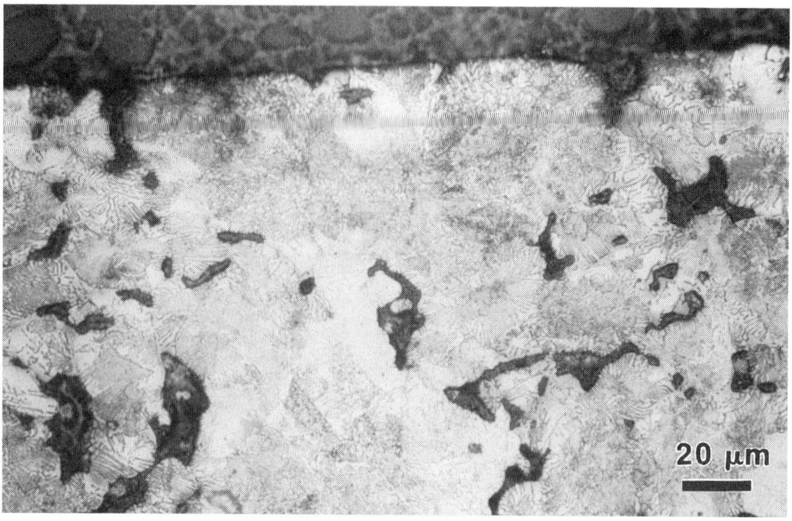

Figure 7.13. In contrast to Figure 7.12, this microstructure shows the desired homogeneous distribution of surface pores and surface carbon associated with proper sintering. (Photograph courtesy of James Marsden.)

mium reduce corrosion resistance, giving a dramatic shift in corrosive attack as illustrated in Figure 7.15. Thus, care is required in specifying a sintering atmosphere, since major property shifts occur.

The effects from impurities in the sintering atmosphere carry over to dimensional change. For some alloys, swelling peaks at 10–15% hydrogen (balance nitrogen) and low mechanical properties are reported for this atmosphere. Figure 7.16 plots the dimensional change in sintering a 6.8-g/cm^3 iron compact with 0.9% carbon. The atmosphere was intentionally varied from low to high hydrogen contents. Higher hydrogen contents give better oxygen reduction, but also affect the carbon level and distribution. Atmosphere conditions that maximize strength often give larger size changes and low ductility. Manipulation of the sintering atmosphere to attain an optimized condition is a delicate process in P/M.

7.1.7. Atmosphere Monitors

Two concerns exist with respect to the sintering atmosphere composition. The first is the nominal or bulk composition; for example, the nominal hydrogen content in a nitrogen atmosphere at 10% versus 20%. But a more important concern is the impurity level, the few parts per million of oxygen, water, methane, or other contaminants. Common monitors of these impurities are through the dew point, carbon level, and oxygen level. This requires three separate monitors. Various infrared and solid-state sensors allow detection of contaminants, both at the atmosphere inlet and at various position in the furnace. It is difficult to place the sensors in the hot portion

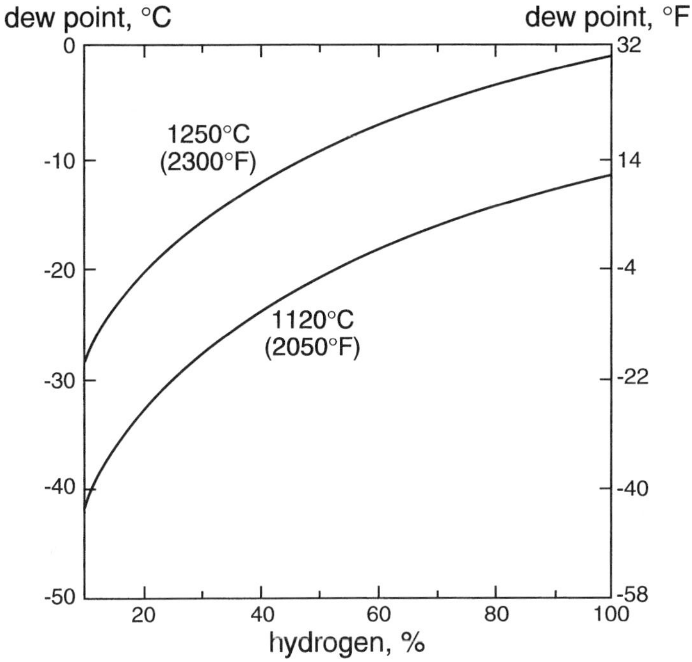

Figure 7.14. The temperature, dew point, and atmosphere combinations for sintering stainless steels under reducing conditions. Less hydrogen is required for higher sintering temperatures and lower dew point atmospheres.

of the furnace, so the normal situation is to pull samples of the atmosphere out through small ports and pass that atmosphere over sensors.

Common examples of the sensors include ceramics (alumina and zirconia) for oxygen and moisture, semiconductors and infrared detectors for carbon species, thermal conductivity bridges for hydrogen and nitrogen, electrolytic cells for moisture, and chilled mirrors for moisture. Newer concepts rely on infrared and laser light absorption to detect gas composition, especially the carbon monoxide : carbon dioxide ratio and the water : hydrogen ratio. A recent development is a semiconductor hot-plate sensor that exhibits a conductivity change dependent on the atmosphere composition and can be tuned for specific species. When heated by passage of an electric current, various chemical species can be monitored with a single sensor.

A typical sensor system is illustrated in Figure 7.17. A pump pulls a sample of the atmosphere from the furnace hot zone. If multiple positions are to be monitored, then sequential valving is used to rotate between sample ports. However, the atmosphere composition in the high-heat portion of the sintering cycle is most important. Calibration gases are used to ensure that the readings remain accurate.

Although typically used in a periodic mode, these systems have graduated to a level where constant monitoring is possible. The signals from the monitors are used

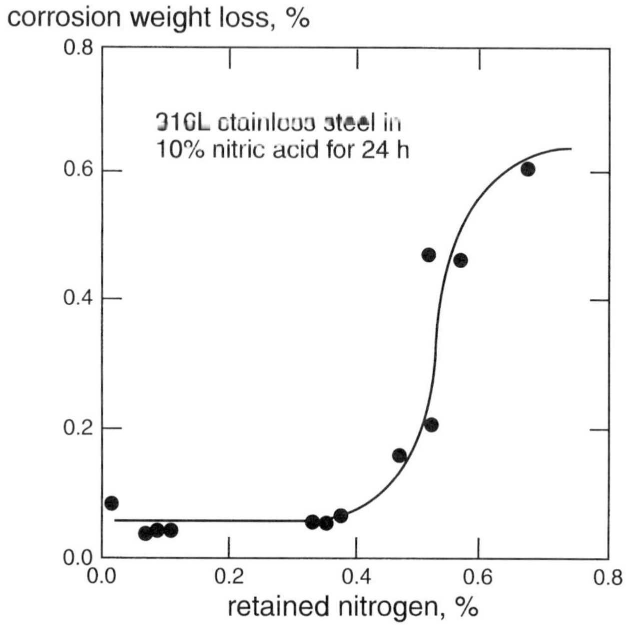

Figure 7.15. Corrosion attack of a sintered 316L stainless steel as a function of the final nitrogen content. The corrosion test was a 24-h exposure to a 10% nitric acid solution. Residual nitrogen from the sintering atmosphere negatively affects corrosion when it is present at concentrations over 0.4%. (Data from Gunha Li.)

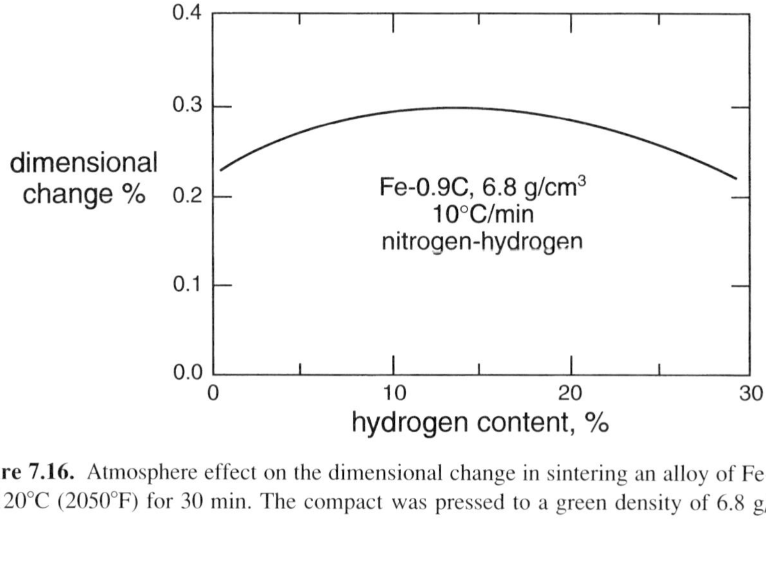

Figure 7.16. Atmosphere effect on the dimensional change in sintering an alloy of Fe–0.9C at 1120°C (2050°F) for 30 min. The compact was pressed to a green density of 6.8 g/cm^3.

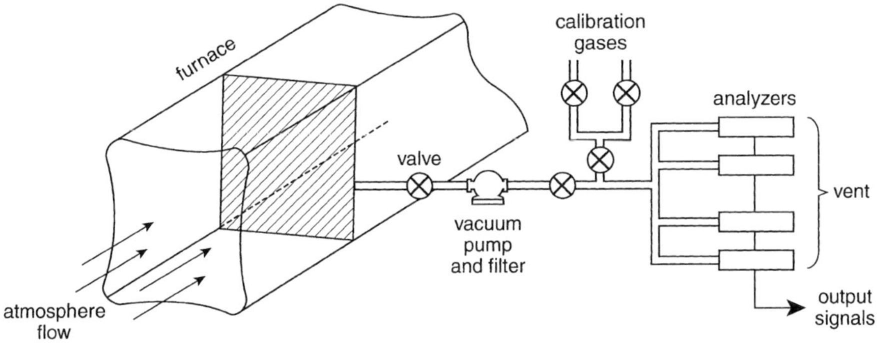

Figure 7.17. Schematic layout for a gas analysis system for a production sintering furnace. The atmosphere is pulled at selected points in the furnace via sampling ports and passes over a range of detectors for moisture, oxygen, and carbon determination.

to correct the atmosphere composition and flow. Inherent in this form of monitoring is the development of a model for the atmosphere interaction with the compact. These vary between furnaces and alloys; thus, the models used in the control systems are empirical. For example, Figure 7.18 plots one such control curve showing the required dew point for various target carbon levels in sintering a low-alloy steel at 1120°C (2050°F) for 30 min. Generally such algorithms control dew point, dopants

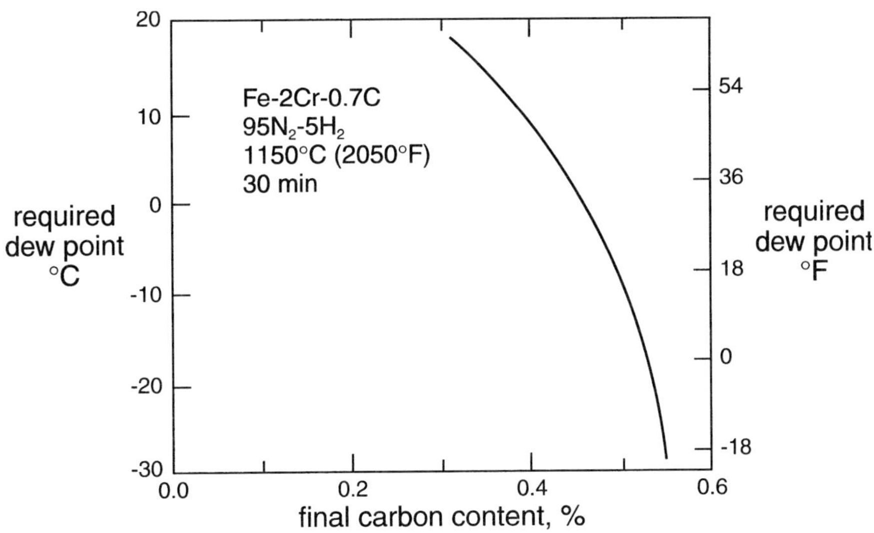

Figure 7.18. A demonstration control curve for obtaining a desired final carbon level from a Fe–2Cr–0.7C initial composition sintered at 1120°C (2050°F) for 30 min in a hydrogen–nitrogen atmosphere.

(such as methane), and flow rate to correct any drifts in atmosphere quality. Further, such systems can be programmed to set off alarms if conditions indicate leaks, open doors, or other abnormal conditions. Experience with closed-loop feedback control systems on production sintering furnaces show carbon control of $+0.05\%$ is possible. This produces consistent tolerances, properties, and predictable heat treatment response.

7.1.8. Atmosphere Sources

Atmospheres are generated at the sintering plant or delivered in bulk from outside generation facilities. The usual case is for a sintering operation to start with bulk delivery and shift to local generation as consumption grows. However, very high consumption levels are usually required to justify the expense of a generation plant. Consequently, most P/M production is performed using atmospheres taken from large storage facilities.

Local generation of an atmosphere is always the case with vacuum sintering. The pumps required to evacuate the furnace are attached to the furnace. This is because conductance through a vacuum is poor, so short distances are necessary from the generation point (the pump) and the furnace. Likewise, dissociated ammonia and natural-gas-based atmospheres require some form of conversion close to the furnace.

For many steel sintering operations, the choice is a nitrogen–hydrogen blend formed from liquids. Purified bulk liquid nitrogen and liquid hydrogen are stored near the sintering facility. The liquids are heated to generate a gas, and these gases are mixed at the furnace. Because of the high purity available in the purified liquids, these atmospheres prove consistent and high quality. They also contribute to the cost of the atmosphere. On-site generation of these gases is possible. For larger-production operations, there are adsorption–deadsorption membrane concepts that allow extraction of the nitrogen directly from air.

Many technologies are involved in atmosphere generation, purification, and safe delivery. Most desirable are liquid delivery systems. The generation facilities are centrally located, and delivery occurs by trucking the liquids to storage tanks. This is the same as oil and gasoline delivery. Long-term contracts determine the terms of delivery, quality, price, and ownership of the storage tanks.

7.2. SINTERING FURNACES

7.2.1. General Features

The sintering furnace provides time–temperature control for the sintering cycle. It also contains the sintering atmosphere. There are two forms—batch (discontinuous) or continuous. A batch furnace is loaded with the material to be sintered and is heated to temperature. Batch furnaces are less productive, but have a high level of flexibility for low production volumes. Further, vacuum sintering and pressure-assisted sintering are performed only in batch furnaces. They also provide options for special cycles and small production quantities. In many batch sintering treatments, particularly for sintering, the furnace walls are kept cold via reflective heat shields around the working zone. An alternative design is a hot wall where the heating elements are outside a retort that contains the sintering material and protective atmosphere. The work can be placed in the hot zone from the front, top, or bottom, and various elevators, hoists, or lift-trucks might be employed to place large loads into the furnace.

A continuous furnace provides thermal treatment by controlling the position of the compact in a preheated furnace. In continuous furnaces, the parts are moved through multiple zones by a conveyor, such as a belt, pusher, lifting beam, or roller assembly. Usually, the transport mechanism limits the furnace operating temperature. Often the conveyor is fabricated from a wire mesh that restricts the maximum temperature to 1200°C (2200°F). A few systems have been converted to ceramic or refractory metal belts, allowing maximum temperatures of 2000°C (3632°F), but these are easily damaged. Alternatively, pusher mechanisms move work through a furnace at higher temperatures, where trays are pushed one against another.

The first zone in a continuous furnace initiates compact heating; removes lubricants, binders, and contaminants from the pores; and starts atmosphere reactions with the powder. Binder burnout and outgassing products are at high concentrations in the preheat zones, so atmosphere flow is adjusted to push burnout products to the furnace entrance. After a transition region, the next zone is the high-heat region, where the actual time, temperature, and atmosphere conditions are maintained. Cooling takes place in the last zone, where the compact is subjected to a high gas flow. By selective placement of gas inlets and outlets, each furnace zone may contain a different atmosphere, allowing tailoring of the chemical reactions during sintering. If necessary, the furnace heating elements can be external to the atmosphere, with heat radiating through a furnace muffle. A circulating fan may be employed to ensure homogeneous heating, cooling, or lubricant removal. A uniform temperature distribution helps maintain dimensional repeatability.

The optimal furnace type and dimensions depend on the intended production quantity, material to be sintered, operating costs, type of atmosphere, and postsintering cooling rate. Production furnaces exist for loads ranging from less than 1 kg to 250,000 kg. Routine production sintering handles about 100 kg/h, but large furnaces exist with capacities near 1 ton per hour. The author has seen production of over 1 million compacts a month in a laboratory furnace and over 20 million compacts a day in a batch vacuum furnace, so there are no clear rules on the sintering furnace design for various production quantities. Large components require slower heating and larger furnaces, thereby mandating a higher expense for sintering. As a first estimate, the cost of sintering (atmosphere, energy, and maintenance) is probably between $0.10 and $1.00 (USA) per kilogram. This is a rough estimate since many variables (original furnace cost, installation expenses, labor rates, gas usage, energy consumption and costs, depreciation schedules, and maximum temperature) are involved in these calculations. In ferrous powder metallurgy, even this small cost can account for 20–30% of the total component fabrication expense.

7.2.2. Furnace Types

7.2.2.1. Batch. As many different designs exist for the sintering furnace as there are furnace manufactures. Vacuum sintering is always performed in a batch mode. The door of the furnace is opened to allow components to be loaded into the cold furnace; then the door is sealed before heating and not reopened until the completion of the cycle. Figure 7.19 is a sketch of a batch furnace, which normally runs under a vacuum, but can be used with a flowing process atmosphere. A few semicontinuous vacuum sintering furnaces exist where three chambers are put in sequence. The work dwells in each chamber for a soak time and then is moved into the next chamber, so the cycle actually is sequential sintering in batch furnaces. Vacuum sintering is mostly applied to tool steels and other alloys with high nickel or cobalt levels, including stainless steels, and many magnetic alloys. One problem is contamination from polymers. Consequently, special traps are required to capture vapors during compact heating. These traps need to be cleaned frequently to avoid contamination. Batch furnaces can range from very small chambers to a few that can hold an automobile.

Often vacuum furnaces have the option of plumbing argon, nitrogen, or hydrogen into the chamber. Retort and tube furnaces also operate as batch furnaces. Here a support structure is inserted into a furnace, and the work is contained in a chamber to control the atmosphere. Batch sintering furnaces are used mostly in low-volume production, research, and pilot production.

7.2.2.2. Belt. A large variety of continuous furnace systems are available. However, the most widely used option is a mesh belt furnace. As sketched in Figure 7.20, it consists of a constantly moving belt and fixed-temperature profile furnace.

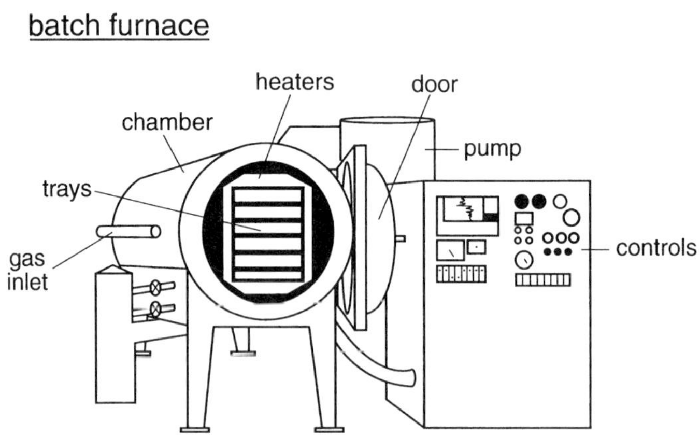

Figure 7.19. Batch sintering involves loading the P/M compacts into a cold furnace with a door that is closed prior to establishing the process atmosphere (vacuum, inert, reducing, etc.). The cycle is performed with stationary compacts and might require a day for completion.

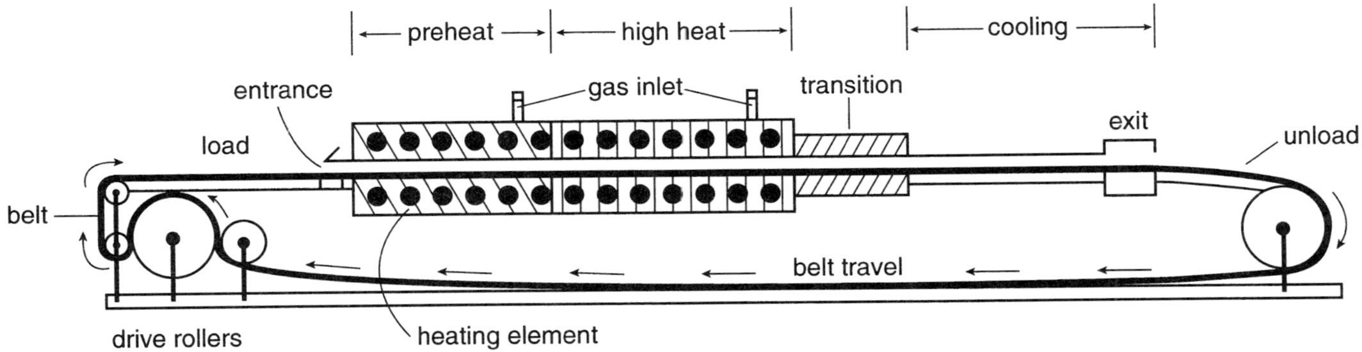

Figure 7.20. The mesh belt furnace is the most productive means of continuous sintering, as long as the peak temperature is low enough to ensure long belt life. The mesh belt is continuously pulled through the furnace with P/M compacts loaded directly onto the belt.

TABLE 7.2. Examples of Standard 1120°C (2050°F) Belt Sintering Furnaces

Belt width, cm	30	45	60
Hot-zone length, m	2.0	2.5	2.5
Hot-zone height, cm	15	15	20
Preheat zone length, m	2.0	2.5	2.5
Cooling zone length, m	5.5	7.4	7.4
Overall length, m	15	19	19
Overall width, m	1.5	1.6	1.7
Belt speed range, m/h	6–18	9–27	9–27
Belt load, kg/m	10	15	20
Power, kW	95	160	225
Atmosphere, m^3/h	55	120	145
Water, m^3/h	2.2	4.3	5.0

These constitute over 80% of all installed ferrous sintering capacity. In its simplest form, a horizontal belt passes through the hot zone with supports at the ends of the furnace from two rotating drums. The belt passes below the furnace on its return. Widths usually range from 15 to 60 cm (6 to 24 in) and belt velocities range near 5–20 cm/min (2–8 in/min). As the belt passes through the furnace, it carries the work through progressive stages of preheat, high heat, and cooling.

Most belts are constructed from high-temperature stainless-steel alloys that limit the belt to a maximum sintering temperature near 1150°C (2100°F). Compacts to be sintered are loaded onto the belt at the entry where a flame curtain helps prevent air entry. The attributes of three sizes of belt furnace are summarized in Table 7.2. Belt speed and hot-zone length determine the time at temperature, with up to a maximum of 6 m (about 20 ft) of hot zone. Typical combinations of belt speed and hot-zone length give times at the peak temperature between 10 and 40 min. Usually a series of preheat zones burn out the polymers and bring the work up to the sintering temperature on its way to the high-heat zone. A final zone is used to cool the components before exit from the furnace. The maximum loading is about 50 kg per m^2 of belt (about 12 lb/ft^2 of belt area), giving typical production rates about 70–300 kg/h (150–650 lb/h). Lower loads are necessary at higher temperatures, and lower loads extend the belt life. Occasionally, the floor of the furnace is inclined from both ends to form a humpback, where the highest point is also the hottest.

For higher sintering temperatures it is necessary to use ceramic belts or other conveyor mechanisms. Usually, the ceramic is silicon carbide (SiC), but may be alumina (Al_2O_3) or silicon nitride (Si_3N_4). On a comparative basis, silicon carbide represents the best performance per unit cost. The ceramic belts work well for smaller loads and smaller furnaces, but as the temperature increases and the furnace size increases, there is a cost penalty for ceramic belts. Maximum operating temperatures are probably 1500°C (2730°F), which is too high for most steels, so the ceramic is not a limitation in terms of maximum temperature. Some ceramic belt furnaces use molybdenum as a reinforcement for high temperature strength.

7.2.2.3. Alternatives. An alternative conveyor mechanism is to load the compacts onto trays and to push the trays through the furnace, one after another. As illustrated in Figure 7.21, the pusher mechanism is outside the furnace, but the train of trays continuously provides output. These furnaces have very high maximum temperatures [over 1800°C (3300°F)], and provide good support for the sintering material. For this reason they are widely used in sintering injection-molded steels. The larger pusher furnaces handle up to 400 kg/h (nearly 900 lb/h) with floor loadings that approach 450 kg/m^2 (90 lb/ft^2) and hot zones of 2.4 m (8 ft) length.

Several other designs of high-temperature continuous furnace are in use. These are known by the means of moving the work through the high-heat regions. One is the walking beam. In this design trays with the sintering components are periodically lifted and advanced by a beam in the furnace floor that is tied to two revolving drums. A sketch of the high-heat region is given in Figure 7.22. Walking-beam furnaces operate up to 1370°C (2500°F). Related is a design termed the roller hearth, where rollers in the furnace floor convey the trays containing the work. Both designs are used for high-temperature sintering of large production quantities, especially for larger components.

7.2.3. Heat Sources

One of the requirements in sintering is uniform heating to ensure repeatable dimensions and properties. Most P/M sintering furnaces rely on electrical heating. The power from a high electric current passing through the heating elements generates radiant energy that heats the work. Although several materials are available for the heating elements, silicon carbide and the refractory metals (tungsten and molybdenum) are most common. Other elements might include alloys of nickel–chromium, iron–chromium, iron–chromium–aluminum, and graphite. The selection of heating

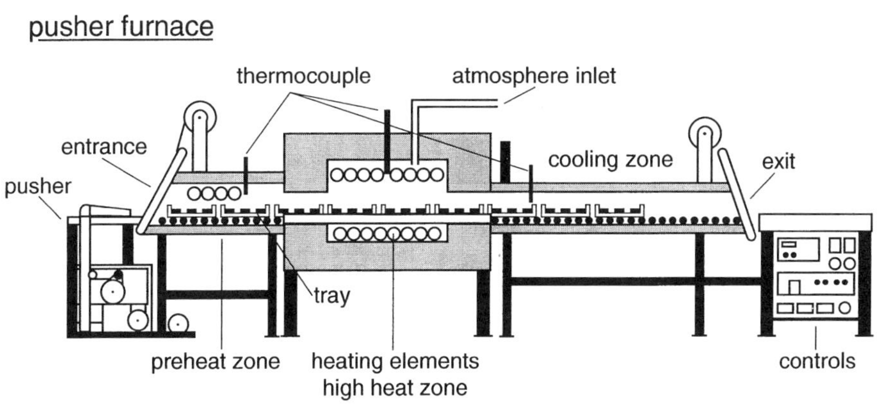

Figure 7.21. A sketch of a continuous pusher furnace, where trays are stoked one against another to create a constant flow of work through the preheat and high-heat zones.

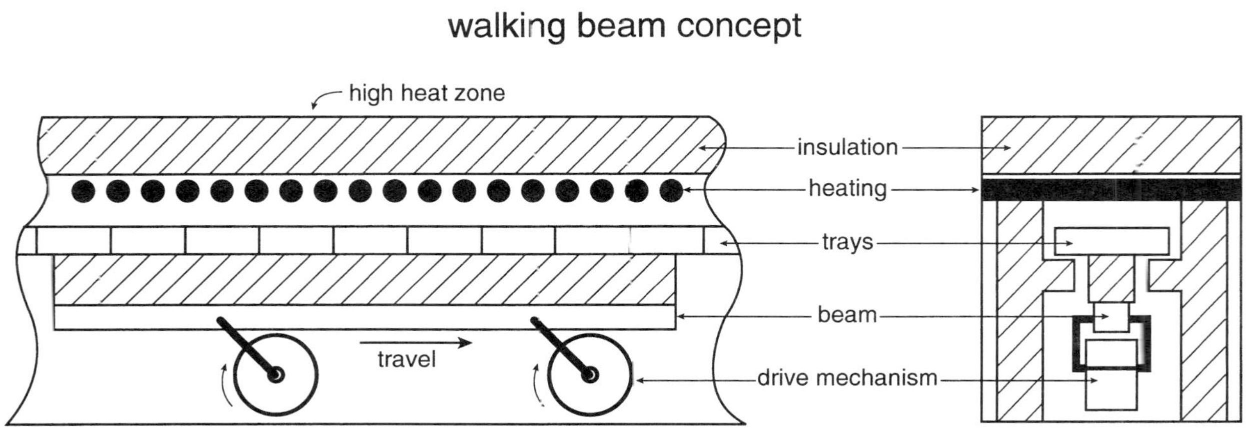

Figure 7.22. Walking-beam furnaces move the sintering trays through the high-temperature portion of the furnace via a sequence of steps, as controlled by a lifting beam in the floor of the furnace. On each rotation of the lifting mechanism the trays are moved forward.

elements depends on the furnace design, maximum temperature, and atmosphere. As the process temperature climbs, there are only a few candidate construction materials, especially when the heating elements are inside the working zone.

For vacuum sintering of steels, tool steels, and various high-alloy steels (including some stainless steels), graphite is effective because it can operate at high temperatures. In the presence of oxygen, graphite generates a reducing condition, especially when used in sintering materials that give off trace amounts of water vapor:

$$H_2O\ (g) + C\ (s) \rightarrow CO\ (g) + H_2\ (g)$$

where both carbon monoxide and hydrogen provide reduction. Consequently, vacuum furnaces constructed from graphite are useful for batch sintering steels, stainless steels, and tool steels at elevated temperatures.

Silicon carbide is effective as a heating element for sintering temperatures up to 1250°C (about 2300°F) in inert or reducing conditions, but can go much hotter in air. At very high temperatures [around 1500 to 1600°C (near 2800°F)] it is attacked by oxygen. Molybdenum is a heating element for high-temperature furnaces, especially the walking-beam and pusher designs. It must be protected from air at high temperatures, so usually the furnace construction consists of two chambers—one for the work and process atmosphere and a second outer chamber to protect the heating elements.

New furnaces are being constructed to accelerate low-temperature heating. A microwave supplement to traditional heating is effective in bringing the P/M compact up to temperature rapidly. The microwave units are mounted in the furnace wall and provide surface heating only. In traditional furnaces heat transfer is slow at lower temperatures, so the direct heat deposit from the microwave proves useful in controlling temperature and heating rate. These supplemental heating modifications are installed in the burnout and preheat zones.

EQUATION BOX 7.3

At low temperatures, microwaves are being used to improve normally slow heating rates. Microwave heat is absorbed only at the surface of a metal, yet that can help remove binders or lubricants at low temperatures. The depth of microwave penetration X varies with the inverse square root of the microwave frequency ν

$$X = \frac{1}{\sqrt{\pi \nu \mu \sigma}}$$

where σ is the conductivity and μ is the magnetic permittivity. Usually the microwave frequency is selected to couple with water. The depth of penetration in steels is very shallow, and only surface heating occurs.

TABLE 7.3. Belt Alloys and Relative Performance in Sintering Furnaces

Alloy	Nominal Composition	Relative Performance
304 stainless steel	Fe–8Ni–18Cr–1Si–1Mn	Poor
314 stainless steel	Fe–20Ni–26Cr–2Si–1Mn	Poor
330 alloy	Fe–35Ni–19Cr–1Si–2Mn	Moderate
520 alloy	Fe–35Ni–21Cr–2Si–1Nb	Moderate
600 alloy	Ni–7Fe–16Cr	Excellent
601 alloy	Ni–14Fe–23Cr	Good
800 alloy	Fe–32Ni–21Cr–1Mn	Good

7.2.4. Supports

Most sintering is performed by placing the compacts directly on the sintering belts, or on trays or plates that may be stacked. Trays are used exclusively in batch furnaces. For delicate components, the supports are contoured to maintain compact shape during sintering and prevent components from welding to one another. These supports might be fabricated from steel, stainless steel, molybdenum, silica, graphite, alumina, or other high-temperature materials.

The support material must remain rigid at high temperature without contaminating the compact. For high temperature, alumina-based materials are popular, because of the combination of low cost, high melting temperature, low vapor pressure, and resistance to attack during sintering. They are followed by graphite, silica, and various iron or nickel alloys. For graphite supports, a surface barrier is required to avoid iron-carbon reactions during sintering. These are applied as slurries or paints of small ceramic powders on the graphite. Alumina (Al_2O_3) is probably the best barrier coating because of its low cost, but alumina–silica (Al_2O_3–SiO_2), zirconia (ZrO_2), magnesia (MgO), and calcia (CaO) are inert with respect to most steels.

The wire belts for continuous furnaces are usually fabricated from iron–nickel–chromium alloys, with 20–60% nickel and 16–26% chromium. Table 7.3 lists some belt alloys. Under cyclic heating and stress conditions typical for sintering furnaces, the belts undergo stretching and abrasion, and eventually fail. The higher the silicon and iron content, the shorter the belt life. Certain atmospheres accelerate failure, especially those containing methanol, methane, and high hydrogen contents. Figure 7.23 plots the effect of temperature on the relative life of a stainless-steel belt. For these tests the belt was loaded to a high stress and the time to significantly deform the belt was measured and plotted the vertical axis on a logarithmic basis. Note the near catastrophic fall in belt life at temperatures over the usual 1120°C (2050°F) sintering temperature.

7.2.5. Process Measurement

The P/M sintering furnace has several process monitors. Depending on the sophistication of the facility, these might include temperature, dew point, carbon potential,

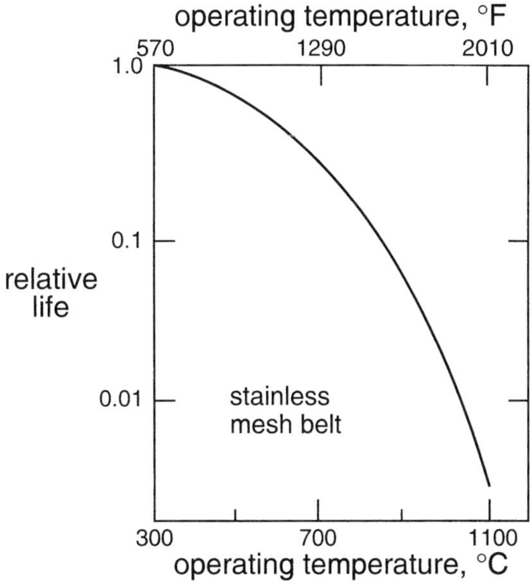

Figure 7.23. A plot of the relative life for a stainless-steel mesh belt versus operating temperature. This plot illustrates the rapid decline in belt life at temperatures over 1120°C (2050°F). With normal loading rates, temperatures, and atmospheres, the belt will last on the order of a year.

oxygen level, and atmosphere flow rate. New sensors are available for recording temperature inside a furnace using optical probes, infrared imaging, laser ultrasonics, and radio telemetry. Also, laser and semiconductor sensors have been developed to report atmosphere composition versus position in the furnace.

Thermocouples are the most common temperature measurement devices. These consist of two wires joined together at a tip. Voltage is generated by heating the tip, and the voltage is proportional to the temperature. In production, this temperature is at the heating elements. However, the sintering components may not be at the same temperature, leading to a difference of 100°C (180°F) between the reported and actual sintering temperatures. Many P/M operations will cross-calibrate the heating element control thermocouple with the work temperature by performing periodic temperature profiles of the furnace. An alternative is to use optical or infrared pyrometers that are sighted into the furnace to measure component temperature.

Several control options exist for hitting target sintering cycles. Most sintering is performed using open-loop controls. Here the furnace is set to a desired temperature–time–atmosphere profile, but there is no assurance that this cycle is actually attained. Indeed, large differences often occur. Adaptive control measures the quality of the furnace output to detect problems; for example, a low hardness indicates decarburization or insufficient temperature. Corrections are made for subsequent parts, but the lost quality is corrected only after possibly producing more defective material. The use of statistical process control charts is an integrated aspect of adaptive process control. With the availability of inexpensive computer systems, process monitoring is automated with data loggers. These continuously record the status of the furnace and atmosphere to assess process uniformity and pinpoint problems. Although statistical process control and data logging detect changes in quality, they do not avoid occasional defective parts.

Closed-loop feedback control is the newer concept. Here the sintering environment is instrumented and monitored. Corrective actions are made instantaneously so that any drifts in part quality are sensed and corrected. Early uses were in the processing of defense and aerospace components, using eddy-current sensors and gas analysis sensors in the hot zone. As the technology advances, these closed-loop feedback control concepts are becoming more widespread. Ultrasonic probes via laser pulse techniques are available to test for the level of sintering inside a furnace. This provides a nondestructive test of component quality that could be used for decisions on the sintering furnace operation.

A target for sintering monitors and controls is to properly adjust the sintering atmosphere. Once heat is deposited at the compact surface, it is necessary to distribute the heat into the compact. This depends on the component's thermal conductivity. Heat conduction through the compact is not a large problem under normal heating rates. Only with large compacts, rapid heating, or high initial porosities will there be surface stresses sufficient to cause cracking. In such cases it is appropriate to include fans or jets to induce turbulence over the compact.

For many steels, there is much to be gained from managing the cooling rates. An excellent example is in a process termed *sinter-hardening*. Most steels are sintered in the austenitic (face-centered cubic crystal structure) phase. To maximize the strength and hardness, on cooling it is necessary to extract heat rapidly. Manipulation of cooling has led to considerable redesign of furnace cooling systems and computer analysis of gas flow paths. Turbulent flow aids in cooling.

7.3. COMMON SINTERING CYCLES

Each P/M manufacturing operation has its own set of sintering cycles. These differences result from decisions on powder vendors, alloy compositions, polymer additions, compaction conditions, and the furnace (design, operation, and atmosphere). Thus, there is no universal sintering cycle. However, there are similarities between operations. In this section a variety of common sintering cycles are detailed to help understand how the principles discussed above are applied in practice.

Table 7.4 reflects the variety in common sintering cycles. This table lists some of the important parameters associated with sintering ferrous alloys. It gives details such as the particle size, heating rate, maximum temperature, time at the peak temperature, process atmosphere, support material, and furnace type. Beyond these tabular data, the following comments explain the sintering cycles for few example ferrous systems.

EQUATION BOX 7.4

A simple model for sintering was derived in the 1940s by Frenkel. It assumes that all densification is by a process of viscous flow, which is the same process by which honey flows. This model is suitable when some liquid forms or when sintering is controlled by atomic diffusion. According to this concept, shrinkage during sintering $\Delta L/L_0$ is given as

$$\frac{\Delta L}{L_0} = \frac{3\,\gamma\, t}{4\, D\, \eta}$$

where γ is the surface energy, η is the viscosity, t is the sintering time, and D is the particle diameter.

As a first approximation, this model is valid for P/M sintering, although it has little atomic-level accuracy. The predicted sintered density ρ_S for a compact starting at a fractional green density ρ_G is given as

$$\rho_S = \frac{\rho_G}{[1 - (\Delta L/L_0)]^3}$$

A problem with this approach is determination of the viscosity. Usually, a few experiments can be performed to make an estimate. Viscosity, like diffusion, changes with temperature, becoming much lower at higher temperatures, producing more shrinkage.

7.3.1. Iron

Since pure iron is relatively weak, it is rarely used for structural applications. Instead, various alloying additions are added for strengthening. Except for carbon, the diffusional homogenization of these additions is often slower than the rate of interparticle bonding. The exception is in sintering iron for magnetic applications. Here the presence of alloying additions often degrades the magnetic response. Pure iron is easy to sinter. A high sintering rate exists prior to the ferrite to austenite phase change at 910°C (1670°F). Alternatively, because of the slower diffusion rate in the austenite phase region, much higher temperatures are required to regain the same rate of sintering. Usually, temperatures near 900°C (1652°F) or over 1120°C (2050°F) are used for soak times of 20 min.

For pure iron, the sintering atmosphere is of minor concern. It exhibits excellent sintering in essentially pure nitrogen with just enough hydrogen to ensure complete oxide reduction. Sometimes adsorbed nitrogen is a problem, so pure hydrogen or vacuum is employed for ultimate performance in sintered magnets.

At lower temperatures, iron sinters by surface diffusion control, meaning that

TABLE 7.4. Common Sintering Cycles as Applied to Ferrous Alloys

Alloy	Particle Size (μm)	Green Density (g/cm^3)	Furnace Type	Atmosphere	Dew Point (°C)	Heating Rate (°C/min)	Maximum Temperature (°C)	Time at Peak Temperature (min)	Shrinkage (%)
Fe–1C	5	2.9	Batch	Vacuum	—	10	1200	60	15
Fe–2Cu–0.8C	100	7.0	Belt	N_2–H_2	−20	10	1120	30	0.0
Fe–2Cu–0.8C	120	6.8	Belt	N_2–H_2	−40	10	1120	10	0.4
Fe–2Ni	5	5.1	Pusher	H_2	−20	15	1250	60	15
Fe–2Ni–0.5C	75	7.3	Batch	Vacuum	—	10	1260	120	0.7
Fe–2Ni–0.5C	80	7.2	Belt[a]	N_2–H_2	−40	5	1200	60	0.04
Fe–2Ni–0.8C	75	7.3	Belt	N_2–H_2	−30	15	1150	30	0.01
Fe–50Ni	5	5.0	Pusher	N_2–H_2	−30	10	1250	60	14
Fe–6Ni–2Mo	80	7.1	Pusher	N_2–H_2	−40	15	1315	45	3.4
Fe–0.8P	66	6.7	Belt	N_2–H_2	−40	25	1120	25	0.4
Fe–6Si	100	6.5	Batch	H_2	−45	10	1200	120	3.7
4600 steel	80	7.0	Belt	N_2–H_2	−40	5	1120	30	−0.05
316L stainless steel	15	5.0	Pusher	H_2	−40	5	1325	90	16
316L stainless steel	22	6.1	Batch	H_2	−40	20	1300	240	2.4
316L stainless steel	45	5.2	Pusher	H_2	−40	12	1250	45	2.4
316L stainless steel	80	6.9	Batch	Vacuum	—	8	1300	60	1.1
M2 tool steel	10	4.9	Batch	H_2	−40	10	1215	45	15
T15 tool steel	58	6.3	Batch	Vacuum	—	5	1270	75	9

[a] Ceramic belt.

particle bonding is active, but densification is low. At the higher temperatures and higher densities the predominant sintering mechanism is grain boundary diffusion. Grain growth during sintering slows densification. For die-compacted water-atomized iron, there is little dimensional change in sintering, while for small carbonyl iron powder there is extensive densification.

7.3.2. Iron–Carbon Steels

The most inexpensive and effective alloying addition for iron is carbon. It is the mainstay of all steels. Carbon is added as mixed graphite and dissolves into iron to stabilize the austenite phase during heating to the sintering temperature. Carbon diffusion rates are high, so it homogenizes easily during heating, unlike many other alloying ingredients. The iron-carbon steels can be sintered at 1100°C (2000°F) in high-nitrogen atmospheres using times on the order of 30 min at the peak temperature. Depending on the initial oxygen content in the powder, enough graphite is added for the desired final carbon content and decarburization due to oxygen. For a first approximation, the carbon loss equals the oxygen contamination. Hence, if an iron powder contains 0.2% oxygen and the target final carbon level is 0.6%, then a first guess would be to add 0.8% graphite. The final carbon level can be assessed from the amount of cementite in the sintered microstructure. Micrographs of four final carbon levels are given in Figure 7.24. These range from a low carbon level (0.25%) to a high carbide content (1.2% carbon), and include one example close to the eutectoid composition.

7.3.4. Alloy Steels

Sintered nickel or copper steels are used in large quantities by the automotive industry, where strength and cost are most important. Accordingly, much attention has been directed to these steels in this chapter. Usually, a coarse iron powder is mixed with alloying powders and pressed in uniaxial tooling. Dimensional control during sintering is very important and usually size can be held to close tolerances.

Because of the high densities (90% of theoretical) attained in the green state, little sintering densification is required for the final application. Thus, sintering is at temperatures in the 1120–1150°C (2050–2100°F) range for short times (20–30 min) to avoid dimensional change. Higher temperatures induce sintering shrinkage, leading to less dimensional precision, but higher properties.

Most of the alloys are formed from mixed elemental powders that homogenize during sintering. As evident by the differential etching iron particles in Figure 7.25, copper does not uniformly diffuse into the iron, unlike carbon. Longer times are required to attain a homogeneous microstructure. Consequently, after normal sintering cycles the core of the iron powder is soft while the surface is hardened by the copper addition. Since diffusional homogenization is slow, interest is growing in partially prealloyed and fully prealloyed powders. The process atmosphere must be reducing to remove oxides, and hydrogen–nitrogen mixtures (up to 97% nitrogen) are effective in this regard. Solid-state sintering is by a mixture of surface diffusion

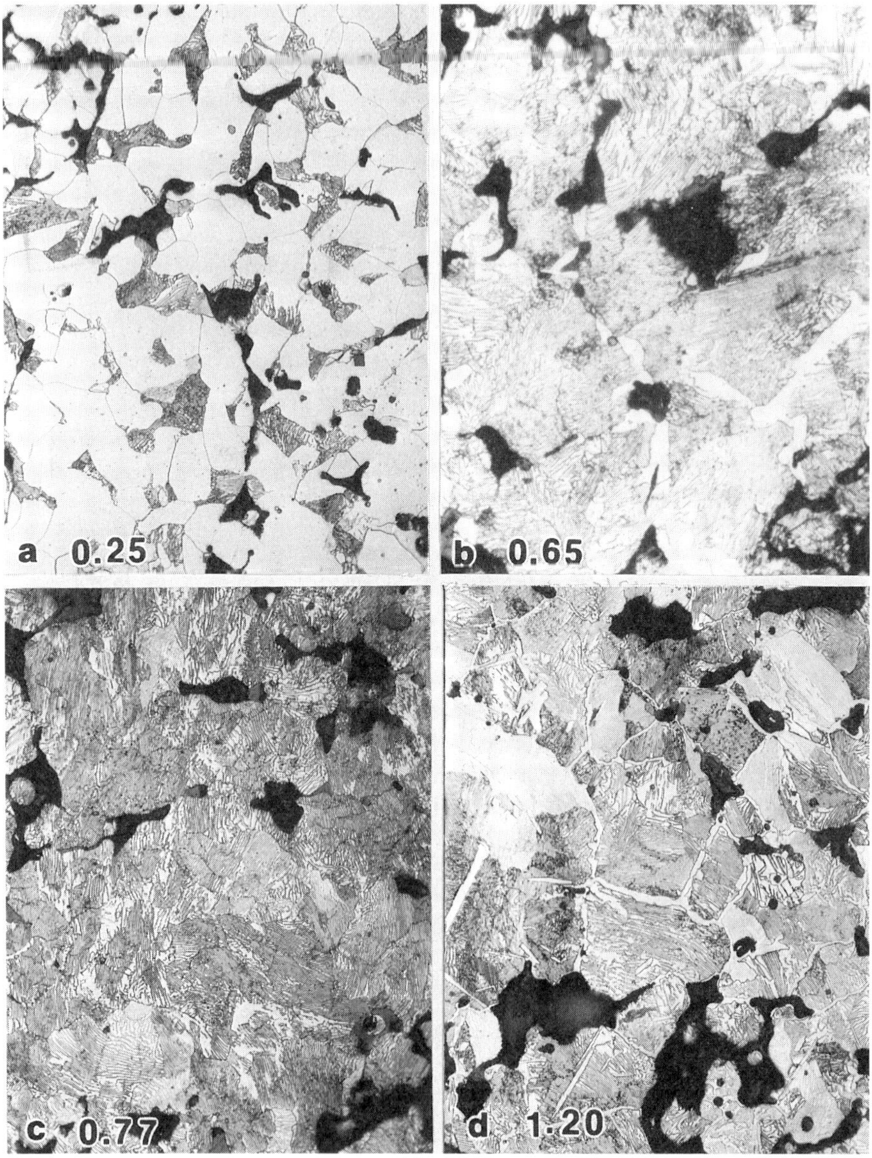

Figure 7.24. These four photographs illustrate the microstructure changes associated with carbon uptake in sintering at 1120°C (2050°F) for compact densities of 6.8 g/cm^3: (a) 0.25% carbon; (b) 0.65% carbon; (c) eutectoid steel with about 0.77% carbon; (d) 1.2% carbon. (Photographs courtesy of James Marsden.)

and grain boundary diffusion processes, although many alloys form a liquid phase during sintering.

In most sintered structural-steel components, over 90% of the composition is iron. Iron–copper–carbon compositions are the most common alloy, where copper melts during sintering. It is common to infiltrate these compacts with molten copper to eliminate residual pores. Figure 7.26 is a micrograph of one such steel, with carbon dissolved into the iron and a white phase that is copper from the infiltration. Nickel is another common addition for improved strength. In low concentrations, phosphorus is used because of its potent hardening of iron. Phosphorus forms a liquid phase with iron at temperatures above 1050°C (1920°F). The most popular compositions contain 0.45–0.80% phosphorus.

As an alternative, for low-green-density steel structures, possibly those formed by injection molding, densification is sought during sintering. This is aided by a liquid, possibly induced by the addition of copper, phosphorous, or silicon. In contrast with the die-pressed materials, smaller particles, higher sintering temperatures, and longer sintering times are employed to increase the sintered density. The densification rate is high at temperatures near 900°C (1650°F). At higher temperatures, sintering is slow and grain growth is rapid. For example, the change in sintering temperature from 750 to 1000°C (1380 to 1830°F) gives a 20-fold increase in grain size for a low-carbon steel (Fe–2Ni–0.8C), while maximum densification occurs near 850°C (1560°F). For small mixed elemental powders encountered in injection molding and slip casting, the sintering temperature is near 1250°C (2280°F). These

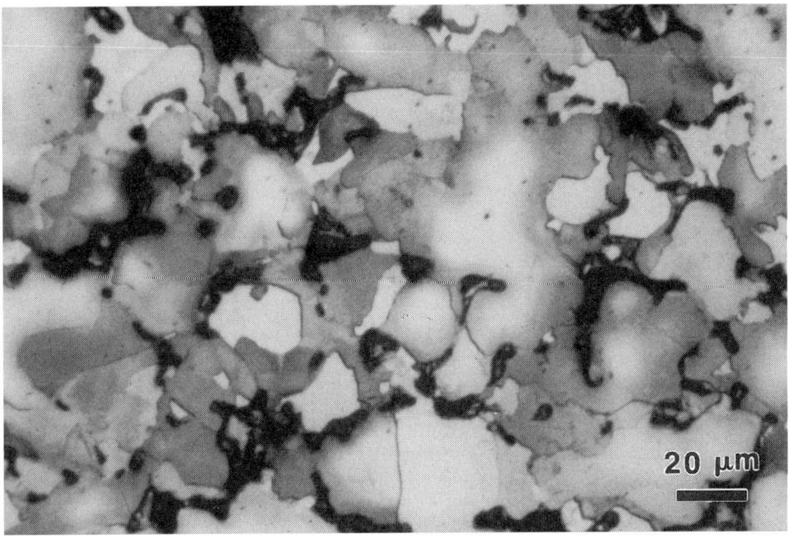

Figure 7.25. This etched microstructure shows the pure iron cores (white regions) and regions of copper diffusion (shaded regions) associated with homogenization in sintering. The particle centers are soft, but the surfaces are hard as a result of copper alloying. (Photograph courtesy of James Marsden.)

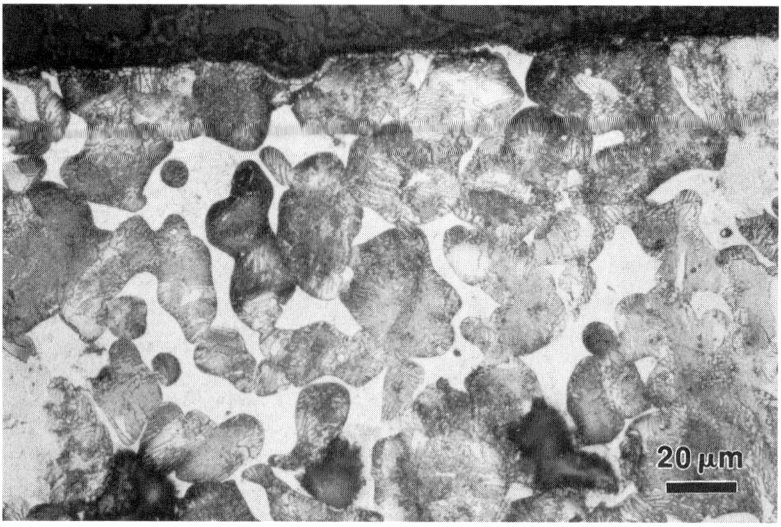

Figure 7.26. Copper-infiltrated iron–copper–carbon steel. The light phase is the solidified copper, while the iron grains containing carbon are of varying shades. This is a properly sintered and infiltrated microstructure with little residual porosity. (Photograph courtesy of James Marsden.)

high temperatures help form the alloy. Subsequent postsintering heat treatment response is improved by the higher sintering temperature. Incomplete densification often traces to impurities or reaction products trapped in the pores. Typical sintering atmospheres are hydrogen–nitrogen mixtures and vacuum, and the support materials are commonly stainless steel for low-temperature sintering and alumina-coated graphite for high-temperature sintering.

7.3.4. Stainless Steels

Stainless steels are selected for their corrosion resistance. So, preservation of the protective chromium is an important aspect of sintering. For die-compacted, water-atomized powder, shrinkage increases with temperature. It is modest at temperatures near 1100°C (2000°F), but often these low temperatures do not produce satisfactory strength or corrosion properties. For improved strength, ductility, toughness, and corrosion resistance, the sintering temperature must be higher and 2–3% shrinkage occurs, with increased dimensional control problems. Alternatively, in slip casting, injection molding, or loose gravity sintering, there is no compaction, so smaller powders are used to induce uniform densification during sintering. Here sintering temperatures are in the 1250–1365°C (roughly 2280–2490°F) range.

Improper control of the atmosphere during sintering significantly degrades corrosion resistance. Carbon, nitrogen, or oxygen contaminants lead to faster corrosion. Accordingly, care must be taken to reduce contamination during sintering. A typical cycle involves heating in hydrogen to temperatures at which any polymers decompose [approximately 450°C (840°F)] and holding until decomposition is complete. Subsequent heating is at approximately 5–10 °C/min (9–18 °F/min) to the sintering temperature. A hold at approximately 1000°C (1830°F) helps remove contaminants. The need for such a hold can be measured via atmosphere quality. For die-pressed, coarse stainless-steel powders, net shrinkage is probably in the range of 0.2% at 1100°C (2012°F) and 2.5% at 1250°C (2280°F).

For the smaller particle sizes and lower green densities (as encountered in injection molding), shrinkage is near 14% with a sintering temperature of 1325°C (2420°F). Faster sintering can be induced by adding liquid forming agents. Boron proves very effective, and the sintered properties are competitive with wrought stainless steel. Copper, silicon, and aluminum also form liquid phases.

Vacuum sintering is viable, as long as the pressure is not so low that chromium evaporation occurs. A maximum sintering temperature of up to 1360°C (2480°F) is used, depending on the alloy and desired level of densification. Rapid cooling, possibly by forced convection in the furnace, is desirable to improve corrosion resistance. The total time of the sintering cycle is approximately 18 h.

Sintered stainless steels are capable of a wide range of final properties, depending on the composition and microstructure. For precipitation hardenable alloys sintered to near full density, even with oxides in the microstructure such as shown in Figure 7.27, final yield strengths of 1100 MPa (155 ksi) with 12% elongation are possible. Alternatively, for the austenitic stainless steels, a sintered yield strength of 250 MPa (35 ksi) and considerable ductility (30% or more) are common.

7.3.5. Tool Steels

Tool steels are a hybrid between a cemented carbide and steel. The final sintered microstructure consists of a dispersion of carbides in a steel matrix. A typical composition might range from 0.6 to 1.5% carbon, with 15% refractory metal (Cr, W, Mo, or V depending on the grade). Low oxygen levels are needed to prevent decarburization in sintering.

Supersolidus liquid phase sintering is the most prevalent technique in sintering the high-alloy tool steels. The powder is heated to a temperature where a carbide–metal eutectic forms. The liquid disintegrates the particles and allows rapid densification. Sintering is best performed in vacuum near 1250°C (near 2280°F). Precise temperature control is required to form sufficient liquid for densification, with minimized carbide coarsening. The quality of the final product is very sensitive to the elimination of porosity while sustaining a small carbide size. A form of pressure-assisted sintering is employed in some furnaces, where the vacuum furnace is backfilled with inert gas while the compact is still at high temperature, thereby collapsing any residual porosity.

7.3.6. High-Alloy Ferrous Systems

Beyond the steels, other ferrous systems are sintered for use in electronic or magnetic applications. Here the concern is not with the strength, but microwave absorption,

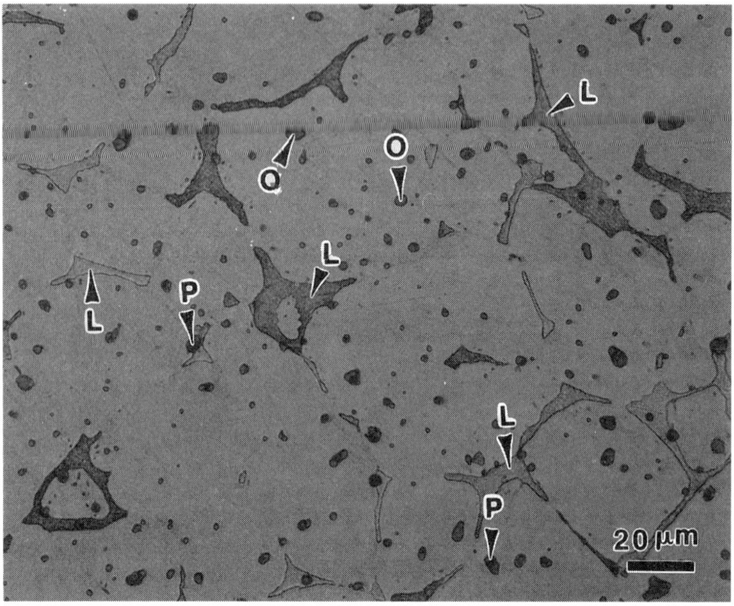

Figure 7.27. A 17-4 PH precipitation hardened-stainless steel, sintered to full density using supersolidus liquid-phase sintering of a water-atomized powder. The microstructure consists of four phases, a small amount of porosity marked P, the bulk stainless grains, the solidified liquid marked L, and oxide inclusions marked O. The corrosion and tensile properties were very high because of the near full-density product. (Photograph courtesy of Donald Kubish.)

thermal expansion, oxidation resistance, magnetic saturation, wetting by molten glass, or other special features associated with these compositions. Many new alloys are being developed, including steel composites containing borides, silicides, aluminides, and nitrides. High-alloy steels usually require high-temperature sintering. This push is dictated by the need for a high density in the final product. Figure 7.28 illustrates two of the novel high-alloy ferrous systems under current development. Both alloys use liquid phases for densification.

Pores not only lower the sintered strength but also degrade magnetic and electrical characteristics. Thus, high final densities are needed to increase the sintered properties. Therefore, high-temperature sintering and small powders can be used to induce densification in sintering. The alternative is double press and double sinter using a high-purity water-atomized iron powder.

Table 7.5 lists several materials and sintering cycles encountered in the specialized high-alloy ferrous systems. A very difficult alloy to sinter is the FeCrAlY composition, commonly pronounced "fee-crawl-ee." These alloys are associated with high-temperature exposures in oxidizing atmospheres, for use in heat treating equipment, furnace heating elements, and catalytic converters. Many useful magnetic alloys contain high nickel contents (up to 50% Ni). These are usually formed from mixed

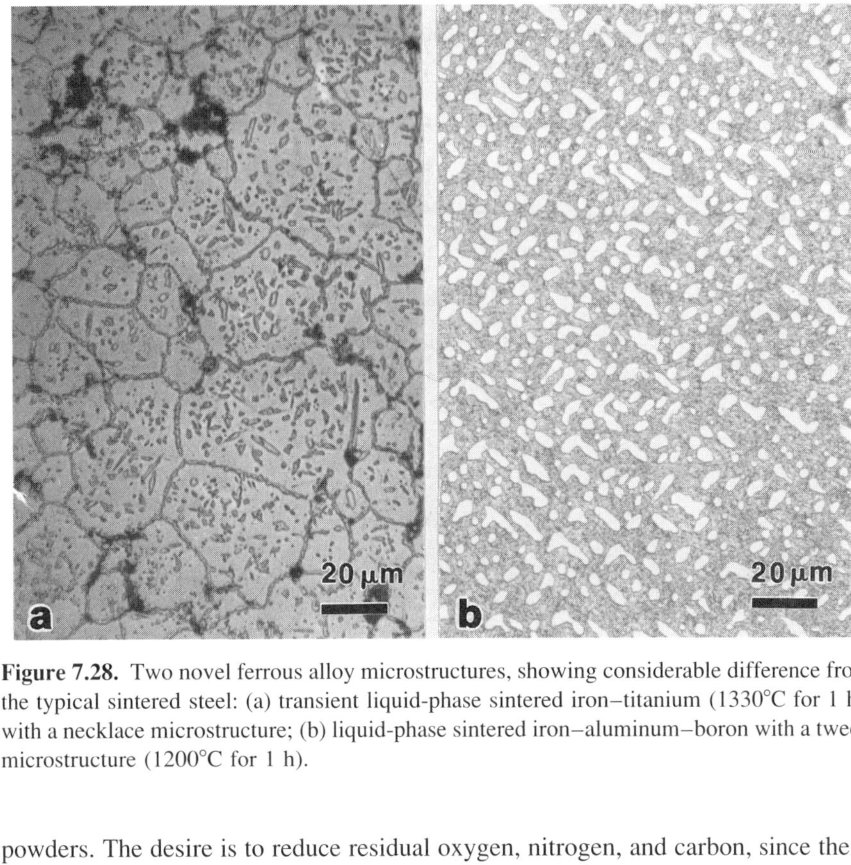

Figure 7.28. Two novel ferrous alloy microstructures, showing considerable difference from the typical sintered steel: (a) transient liquid-phase sintered iron–titanium (1330°C for 1 h), with a necklace microstructure; (b) liquid-phase sintered iron–aluminum–boron with a tweed microstructure (1200°C for 1 h).

powders. The desire is to reduce residual oxygen, nitrogen, and carbon, since these elements degrade magnetic properties. Hence, sintering is performed in pure hydrogen.

Another class of special alloys is formed by mixing two insoluble metals or a steel and ceramic. One use is to tailor properties: wear, thermal, magnetic, or hardness. A current example is a mixture of invar and silver shown in Figure 7.29 that is used to attain a high thermal expansion coefficient in a high-thermal-conductivity alloy. These products are composites, and common metal–ceramic composites are termed *cermets*. Figure 7.30 is a cermet microstructure for a sintered steel–titanium carbide. It has a mixture of metallic and ceramic characteristics. Several useful compositions exist with up to 50 vol% ceramic phase, with stainless-steel or tool steel matrices being a favorite for corrosion resistance or strength. The ceramics might be alumina, chromium boride, titanium carbide, or tungsten carbide.

7.4. SPECIALIZED SINTERING CYCLES

Most of the specialized sintering cycles involve high temperatures. These are for improved sintered properties, such as fatigue life. High temperatures are not always

TABLE 7.5. Some Special Sintering Materials and Cycles

Composition	Particle Size (μm)	Green Density (g/cm^3)	Heating Rate (°C/min)	Peak Temperature (°C)	Hold Time (min)	Atmosphere	Support Material	Sintered Density (g/cm^3)	Application
Fe–4Al	70	6.5	18	1250	30	N_2–H_2	Stainless steel	6.5	Corrosion
Fe–0.8P	66	6.6	25	1120	25	H_2	Stainless steel	6.7	Magnetic
Fe–50Co	5	4.2	50	1250	120	H_2	Alumina	8.2	Magnetic
FeCrAlY	6	3.9	10	1350	120	H_2	Molybdenum	7.3	Oxidation
Fe–50Ni	5	5.0	10	1250	60	N_2–H_2	Alumina	8.0	Magnetic
Fe–3Si	70	6.5	10	1230	60	Vacuum	Stainless steel	7.0	Magnetic
Fe–5Ti	37	6.1	200	1400	120	Argon	Alumina	7.4	Corrosion
Stainless steel + alumina	100	5.8	10	1350	180	H_2	Alumina	7.1	Wear
Tool steel + alumina	5	5.6	5	1300	60	H_2	Alumina	6.7	Cutting

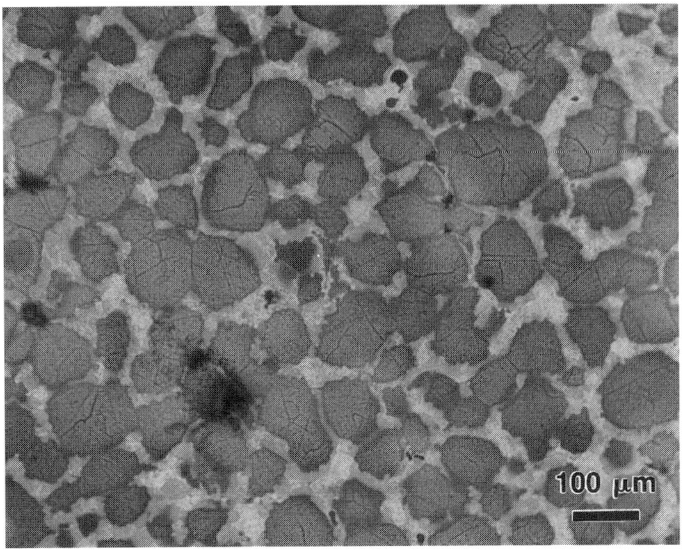

Figure 7.29. Microstructure of a metal–metal composite used for thermal management applications, consisting of invar (Fe–36Ni, dark phase) and silver (light phase). (Photograph courtesy of Jean Murray Stewart.)

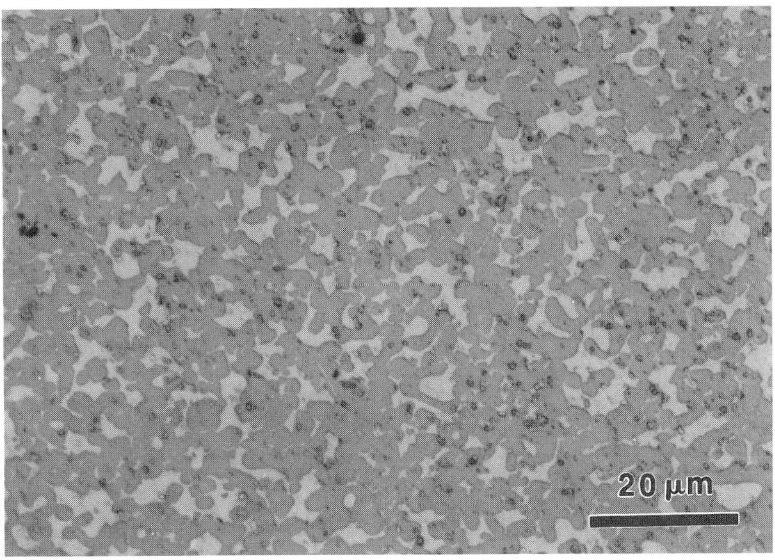

Figure 7.30. This is a microstructure of a cermet processed to near full density by liquid-phase sintering at 1400°C (2550°F) for 2 h, producing a mixture of hard, ceramic grains in a steel matrix. (Photograph courtesy of Barbara Lograsso.)

beneficial. Figure 7.31 illustrates the oversintering of a high-alloy steel. Both the strength and ductility were higher with lower sintering temperatures. Without proper control, there can be considerable microstructural coarsening, distortion, and component warpage. Further, evaporative loss of crucial elements is a problem at high temperatures. The classic example is chromium evaporation from stainless steel. High sintering temperatures improve particle bonding. However, a high sintering temperature can evaporate chromium, leading to a loss of corrosion resistance as sintering temperature increases. The cure for this problem is to increase the atmosphere pressure during sintering to suppress chromium evaporation, typically using hydrogen. But this requires a special furnace and is seldom used; thus, porous stainless steels processed by powder metallurgy usually have inferior corrosion resistance when compared with other metal fabrication techniques. The opposite is true when the stainless steel is sintered to full density.

Tool steels, stainless steels, and many composites have specialized sintering cycles that use liquid phases. These products are a relatively small portion of all of sintered steels. Boron is an effective sintering aid, and might be used at a concentration of just 0.015%. Depending on the alloy, the sintering temperature will be between 1200 and 1300°C (roughly 2200 and 2400°F), and sintering is usually performed in vacuum or hydrogen.

Another novel technique is pressure-assisted sintering. This evolved largely in the sintering of cemented carbides used for cutting tools. The vacuum sintering furnace is programmed to the high-heat temperature, but just before the start of the

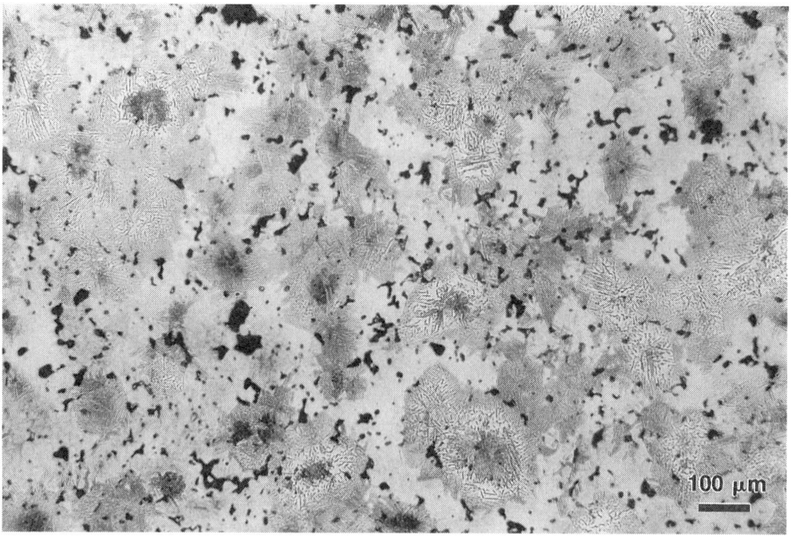

Figure 7.31. An oversintered microstructure of an alloyed steel (Fe–2Cu–2Ni–0.9Mo–0.8C) processed at 1350°C (2460°F) for 50 min in a nitrogen–hydrogen atmosphere. The mechanical properties were degraded by the microstructure coarsening from the excess heating. (Photograph courtesy of Christoph Degoix.)

cooling cycle, the furnace is filled with high-pressure gas. This pressure effectively squeezes the compact and gives a higher final density. Obviously, for the pressure to be contained, the vacuum furnace must be of special construction. These expensive furnaces may operate at pressures up to 50 atm [5 MPa (750 psi)] to push the final pores out of the compact.

SUGGESTIONS FOR FURTHER READING

Bocchini, G. F., ''Influence of Process Parameters on Precision of PM Parts,'' *Powder Metallurgy* **28,** 155–165 (1985).

Fagan, R. (ed.), *Vacuum Technology, Practical Heat Treating and Brazing*, ASM International, Materials Park, OH, 1993.

German, R. M., *Sintering Theory and Practice*, Wiley, New York, NY, 1996.

Jenkins, I., and J. V. Wood (eds.), *Powder Metallurgy: an Overview*, The Institute of Metals, London, UK, 1991.

Klar, E. (ed.), *Powder Metallurgy*, Vol. 7 of *Metals Handbook*, E. Klar (ed.), American Society for Metals, Metals Park, OH, 1984.

Lall, C., ''Principles and Applications of High Temperature Sintering,'' *Rev. Particulate Mater.* **1,** 75–107 (1993).

Lei, G., R. M. German, and H. S. Nayar, ''Corrosion Control in Sintered Austenitic Stainless Steels,'' *Progr. Powder Metallurgy* **39,** 391–410 (1983).

Sanderow, H. (ed.), *High Temperature Sintering,* Metal Powder Industries Federation, Princeton, NJ, 1990.

Smith, G. D., and R. A. Smith, ''Characteristics of Current and Advanced Wire Mesh Belt Alloys for Sintering Furnaces,'' *Industrial Heating* 61–63 (Oct. 1995).

CHAPTER 8

Full-Density Processing

8.1. CONCEPTUAL BASIS FOR FULL-DENSITY PROCESSES

8.1.1. Introduction

Except bearings and filters, densification is a necessary aspect of attaining high performance levels in P/M products. Pressing and sintering to densities near 7.0 g/cm^3 is suitable for many applications, but residual pores still limit the property levels. The justifications for full density usually derive from the higher strength, toughness, and fatigue resistance, resulting in higher reliability and longer service life. Full densification can be achieved by several routes—for example, by repeated die compaction and sintering. However, most full-density production processes involve the simultaneous application of pressure and temperature, as covered in this chapter. If the pressure is low, then the densification rate is paced by atomic diffusion, and that is normally slow. On the other hand, if the pressure is very high, then the particles instantly deform to give full density.

It is noteworthy that performance levels from full-density P/M products can exceed those obtained in wrought steels. These property gains require control over defects, impurities, and microstructure. Since full-density processing is expensive, it is justified only by the property gains. Important concerns need to be addressed before commitment to full density; these relate to the need for the higher properties, assessment of alternative alloys, and justification of the added processing cost. In several instances, it is more cost-effective to increase alloying versus resorting to a full-density P/M option on a low-alloy steel.

8.1.2. Prior Particle Boundaries

One difficulty in full-density processing is interfacial contamination. Even with the elimination of all pores, the degree of particle bonding might be poor, leading to premature fracture. Contaminants often locate at a critical area, the zone of interparticle contact. Consequently, fracture resistance can be significantly degraded by contamination. Figure 8.1 shows scanning electron micrographs of the fracture surfaces of a high-alloy steel consolidated by hot isostatic pressing. One shows the outlines of the flattened, but poorly bonded, particles, while the other exhibits considerable tearing as an indication of toughness and ductility.

Segregation leads to particle boundary contamination as illustrated in Figure 8.2, which includes optical and transmission electron micrographs of the particle bounda-

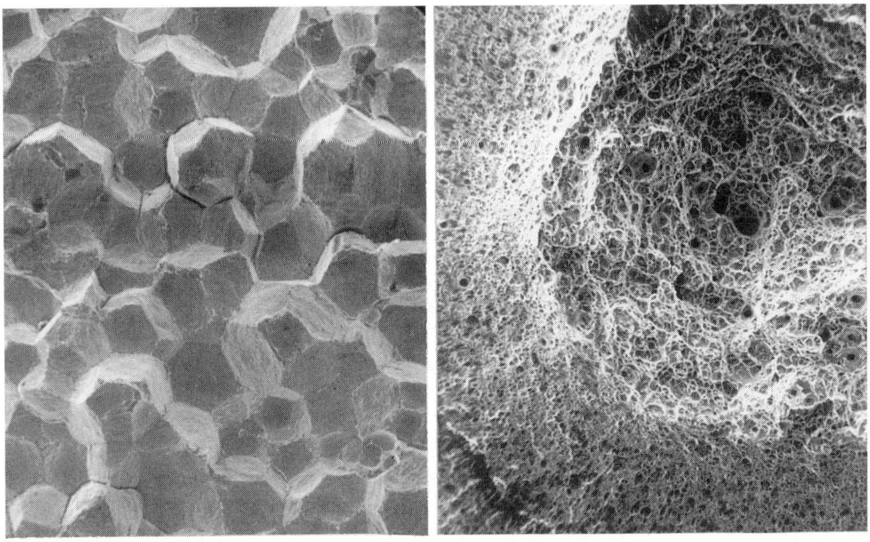

Figure 8.1. Scanning electron micrographs of fracture surfaces from hot isostatically pressed compacts of maraging steel. Contamination of the powders led to preferred failure between the flattened particles in one case, while good metallurgical bonding forced considerable pulling and flow before fracture in the other case.

ries in a full-density high-alloy steel. This localized second phase allows easy fracture along the particle contacts. The method of powder fabrication contributes to the problem. Most typically the contamination is an oxide or similar brittle film (possibly a nitride or carbide). Therefore, full density is not a total assurance of competitive properties, unless steps are taken to reduce particle boundary contamination. Because of purity advantages, gas-atomized powders are usually preferred. Furthermore, avoiding contamination before consolidation may require inert handling. Several sites have installed cleanrooms for powder handling and inert-gas gloveboxes for assurances of the powder quality.

8.1.3. Options

Most often, full-density techniques emphasize the simultaneous use of pressure and temperature to eliminate pores. Full-density processing is performed using a combination of heating and pressurization. Three material characteristics contribute to the improved densification. First, steels become weaker at higher temperatures. Second, they have higher ductilities, meaning less chance of cracking during deformation. Finally, at high temperatures a steel will not work-harden or undergo the strengthening observed in room-temperature deformation. Consequently, heating allows easier densification. This behavior is illustrated in Figure 8.3 with respect to the density

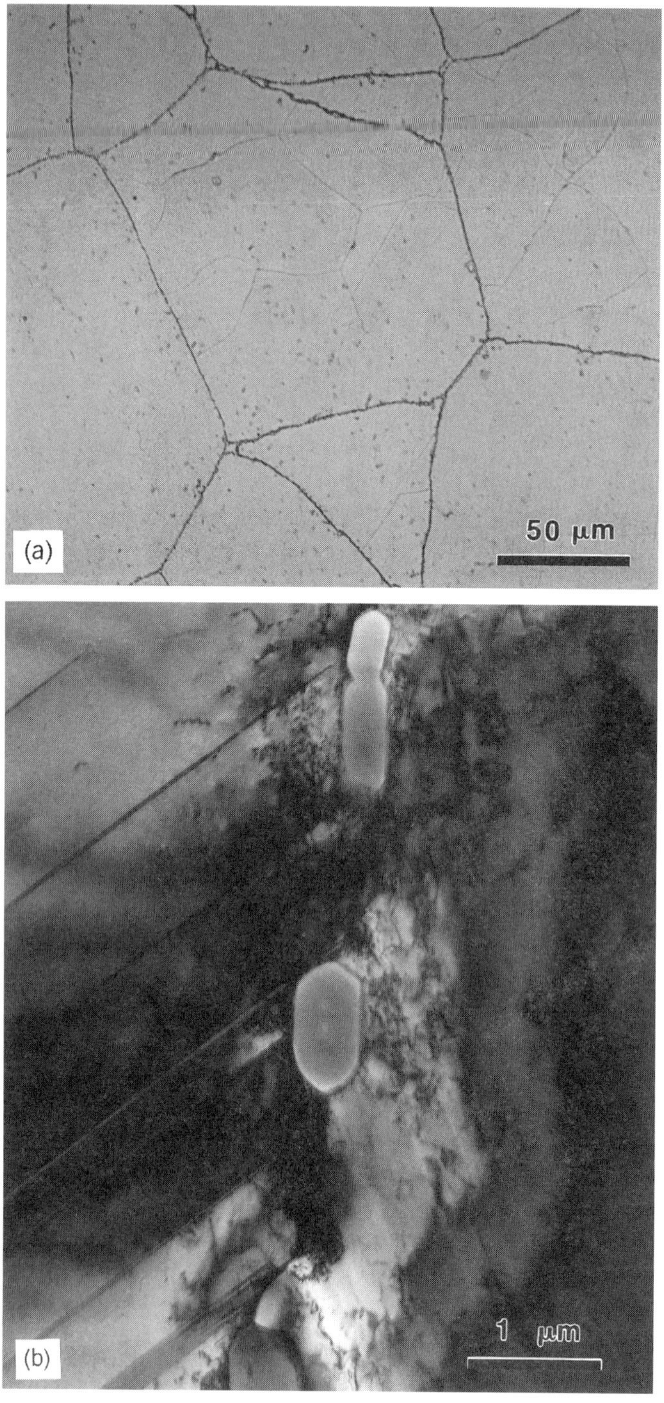
50 μm
(a)
1 μm
(b)

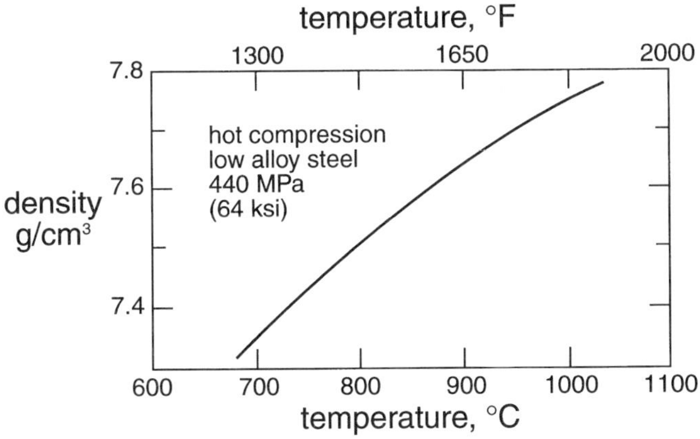

Figure 8.3. Density as a function of the hot compression temperature for a low-alloy steel. Compression was performed at a constant pressure of 440 MPa (64 ksi). Densification is easier at higher temperatures, since the alloy softens.

of a hot-compressed low-alloy steel powder. The pressure was held constant at 440 MPa (64 ksi) to measure the relative effect of temperature on final density. Higher temperatures soften the steel, giving a higher density. Other factors, such as the powder type, are less important; however, powder cleanliness (low-impurity content, especially low oxygen levels) is crucial to high final properties.

Conceptually, the effect of temperature is the same as with polymers that soften on heating. Because of this similarity, many densification processes are treated with models similar to those used for polymer forming. The simultaneous application of temperature and pressure is characteristic of the full-density processes treated in this chapter.

It should be noted that small powders can be sintered to full density, especially at high temperatures. No pressure is required for this form of densification. In these cases, the compact can be formed by any of several methods. This is the basis for attaining high properties in injection-molded steels.

As an option, additives are available to increase the sintering rate. Similarly, liquid-phase sintering is a potent technique for densification. The combination of a capillary force, lubricating liquid, and rapid atomic motion results in high sintered densities. Figure 8.4 is a microstructure of a liquid-phase sintered martensitic stainless steel. It was doped with a small quantity of boron to form a liquid during sintering, apparent as the light phase between the rounded grains. Liquid-phase

Figure 8.2. Micrographs of a steel processed to full density by hot isostatic compaction, but with a residual carbide film decorating the boundaries between particles: (a) optical micrograph; (b) transmission electron micrograph of the carbides. (Photographs courtesy of John Smugeresky.)

EQUATION BOX 8.1.

Viscous flow applies to the motion of polymers, but at high temperatures powder densification is similar. Accordingly, the densification rate can be modeled using a simple viscous flow model

$$\frac{d\rho}{dt} = \frac{3P_E\epsilon}{4\eta}$$

where ρ is the fractional density, t is the time, so $d\rho/dt$ is the densification rate, and P_E is the effective pressure, ϵ is the fractional porosity, and η is the viscosity. Usually viscosity decreases as either temperature or porosity increases. Note that porosity and fractional density are related. Thus, integration is possible, leading to a simple relation between fractional porosity and time

$$\ln\left(\frac{\epsilon}{\epsilon_0}\right) = -\frac{3P_E t}{4\eta}$$

which states that the porosity decays from an initial value of ϵ_0 with time. However, this assumes that the effective pressure is constant. During densification the contacts between particles grow, so the effective pressure decays, making densification slower than anticipated from this model. Even so, such a model provides a first basis for describing hot consolidation of powders.

sintering is used for full-density products such as tool steels, stainless steels, and iron-based magnets. A related technique is to infiltrate the pore structure after sintering. As with liquid-phase sintering, the microstructure is a composite with solidified liquid in the pores between the solid particles. On the other hand, for larger powders there are several techniques that rely on the simultaneous application of temperature and pressure to attain full density.

8.2. PROPERTY GAINS

Usually, performance improves with density. Therefore, full-density processing is coupled with high alloying levels to boost the properties. Both factors (densification and alloying) work together, as illustrated in Figure 8.5 for varying carbon levels in a 4% Ni steel. All the mechanical properties improve with density. Further, as the carbon content is adjusted, strength increases, but toughness and ductility decrease. Densification improves all properties, while alloying targets specific properties. For applications involving impact or cyclic loading, alloying and densification are simultaneously applied to maximize properties.

On one hand, full density provides an avenue to improved properties; on the other hand, it represents a greater expense. Specialized equipment is required, and prealloyed powders are necessary to reach highest property levels. The type of powder most responsive to full-density processing is a clean, prealloyed powder, typically formed by gas atomization.

For certain applications requiring high reliability, the full-density P/M steels are cost-effective when compared with other metalworking technologies. This is especially true in high-alloy steels (such as tool steels), where the raw material is expensive. Microstructure control gives better properties as compared with wrought technology. An example with 4640 steel is diagrammed in Figure 8.6. All three tensile properties (yield strength, ultimate strength, and elongation) are improved by higher densities. Although strength and ductility exceed wrought values, fatigue properties are sometimes less certain. Again, the difficulty usually traces to contaminants on the particle boundaries. Sufficient deformation during densification eliminates any embrittlement associated with contamination. Therefore, some full-density P/M products are forged after densification.

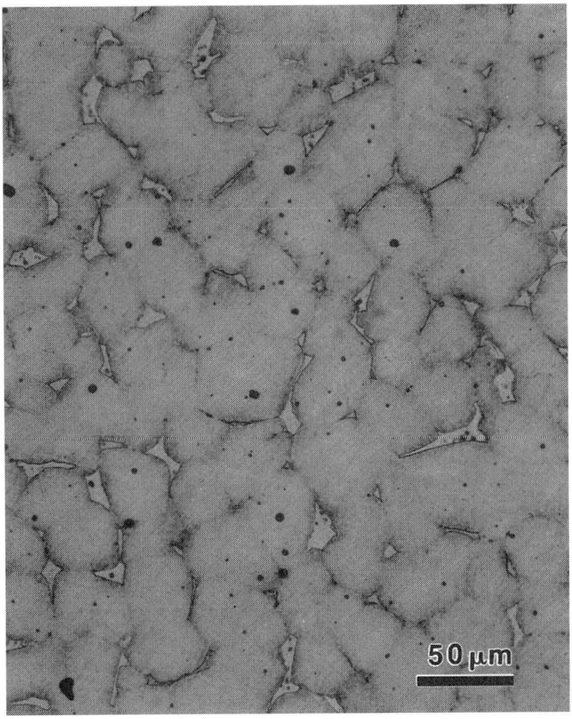

Figure 8.4. Optical micrograph of a full-density 17–4 PH sintered stainless steel consolidated by doping with boron to allow liquid-phase sintering. (Photograph courtesy of Anthony Griffo.)

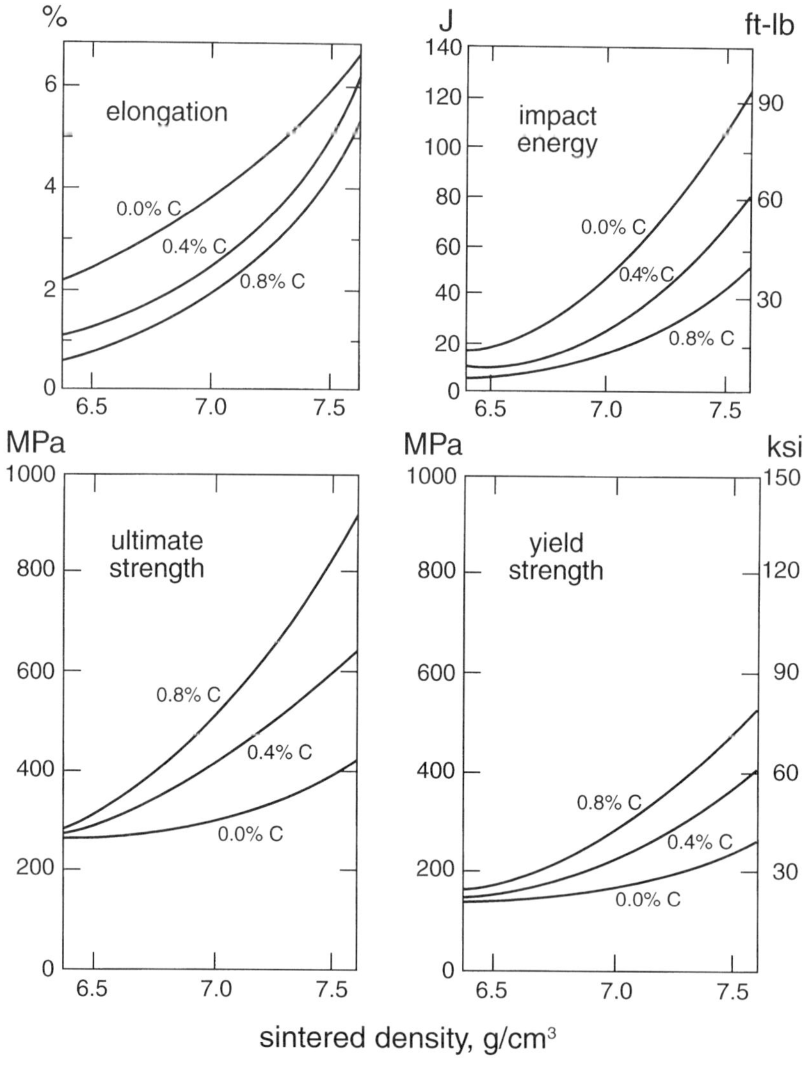

Figure 8.5. Mechanical property variations with density for Fe–4Ni alloys with varying carbon levels: 0.0, 0.4, and 0.8% C. These plots show the benefits from densification, in that all properties improve, while alloying increases strength at the expense of toughness and ductility. (Data from Howard Sanderow.)

8.3. POWDER FORGING

8.3.1. Preforms

In powder forging, a sintered preform with 10–25% porosity (roughly 6–7 g/cm^3) is hot-densified in a single blow. Closed dies are used, so density is controlled by the preform mass (volume is set by the die cavity). Consequently, the preform weight is closely monitored to provide full density in the final shape. Forging is a fast deformation process, conducted at elevated temperatures where the steel has a low strength and high ductility. Fundamentally, forging occurs under a combination of temperature and stress that causes the dislocations to move. Densification is immediate, since the forging stress exceeds the material strength, forcing collapse of any pores. P/M forging is competitive for selected shapes where a high fatigue strength is important.

As illustrated by the process schematic in Figure 8.7, traditional press and sinter powder metallurgy is used to fabricate precursor material for powder forging. Most preform designs are created using computer programs. The preform shape and density have a large influence on the success of the forging operation. After compaction, the preform is sintered at 1120°C (2050°F) for 20 min in an atmosphere of hydrogen and nitrogen. At the induction heater the compact is taken to temperatures near 1000°C (1830°F) for a few seconds before transfer to the tooling. The temperature where each alloy is weakest depends on the carbon content—for example, with no carbon, hot forging is best done at 800°C (1470°F) and requires only 160 MPa (23

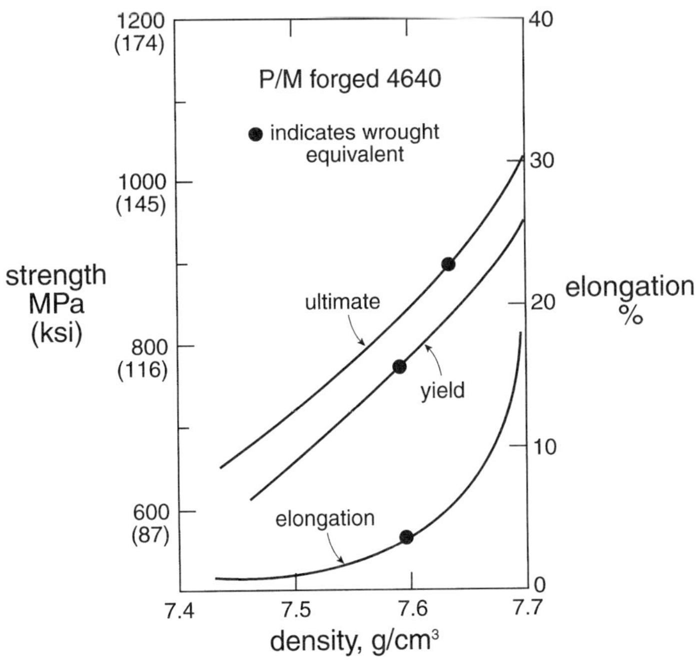

Figure 8.6. The strength and ductility for a 4640 steel hot-compressed to illustrate the density effect on tensile properties. When the last 2% porosity was eliminated, the compacts exceed wrought properties, as indicated by the solid marks on each curve. (Data from Andrew Crowson.)

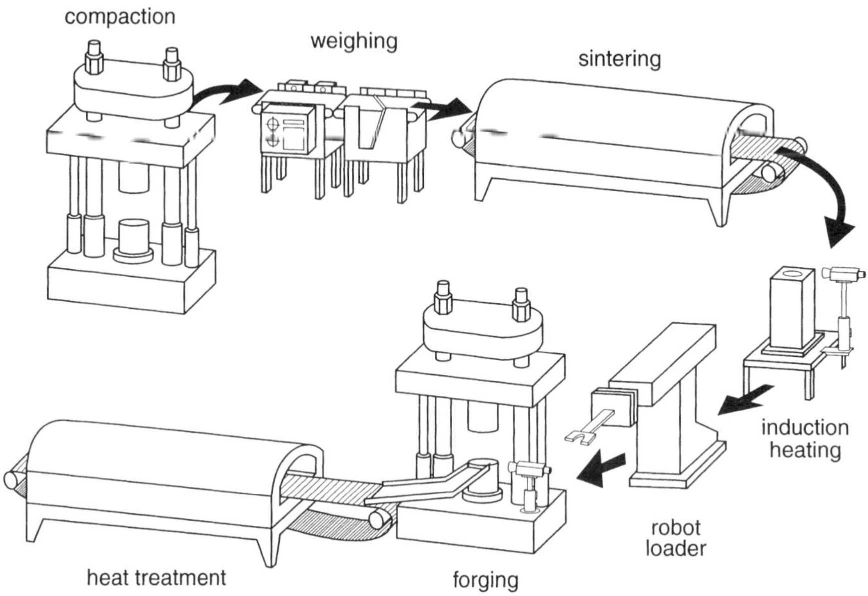

Figure 8.7. Powder forging begins with die compaction of the preform. This compact is inspected for weight, sintered, and hot-forged, with a final postforging heat treatment. Possible secondary steps might include sizing, machining or grinding, and shot peening.

ksi); but with 0.7% carbon, the best temperature is 950°C (1740°F) with 265 MPa (38 ksi). Most operations run near 10 forgings per minute.

8.3.2. Processing Options

In one version of powder forging, the flow in the die cavity is from one direction. Here, the preform profile is the same as the profile of the final compact; all densification is in the pressing direction, often termed hot *repressing*.

The alternative is to make the preform smaller in profile and larger in height than the final compact. This is called *upset forging*, since compaction causes the compact to flow outward to contact the die wall during densification. The two processes are contrasted schematically in Figure 8.8. Upset forging has an advantage in obtaining more uniform densities in the compact, especially in the corners. A disadvantage is cracking if too much flow occurs. The difference is whether there is contact with the die wall at the beginning of the forging stroke. Die wall lubricants, typically mixtures of water and graphite, keep the steel compact from sticking to the die and punches.

Since room-temperature deformation to full density requires stresses far higher than practical, powder forging relies on heated compacts for easier densification. The heating makes it easier for the steel to flow to fill out the die cavity. The preform mass must be sufficient to fill the part shape at full density. If there is too little mass, then the compact is porous, while too much mass damages the tooling. A fracture limit occurs with respect to total strain, dependent on lubrication. Consequently, as illustrated in Figure 8.9, a forging window results that defines the change in dimensions from the preform to final compact. If the initial compact height is too high, failure will occur from buckling or fracture, while a low initial compact height gives insufficient densification.

Temperature determines the pressure necessary to achieve densification. Usually, the die is at 200–400°C (400–800°F). Pressures for densification are not routinely measured; most forging is for full-density products, and the press goes to a final compact height instead of a specific pressure. Thus, an important need exists to control the pressed weight. Tool life is usually in the 50,000–100,000 cycle range.

In one variant of the process (STAMP), billets of steel, tool steel, or stainless steel are produced from gas-atomized powders. A flowchart for the process steps is given in Figure 8.10. The gas-atomized powders are poured into steel containers. These containers are heated and evacuated to remove contaminants. For initial densification, the sealed container is heated to 1100°C (2110°F) and densified in a hot isostatic press. The dense compact is hot-die-forged to ensure the highest properties. These billets are subsequently extruded, rolled, or otherwise processed into mill

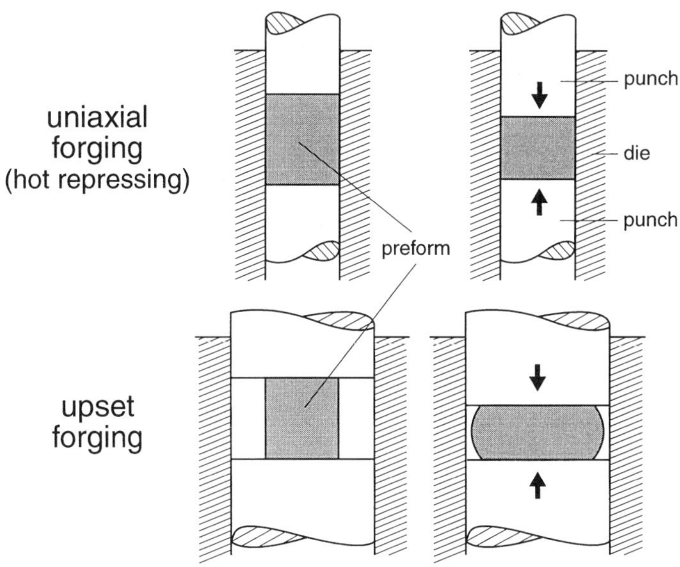

Figure 8.8. Two variants of powder forging. In the upper repressing case, the compact is always in contact with the die walls, so densification is by height change only. In the lower case of upset forging, the compact flows outward during initial forging to eventually meet the die wall after undergoing lateral flow. This latter process gives more uniform densification and higher mechanical properties.

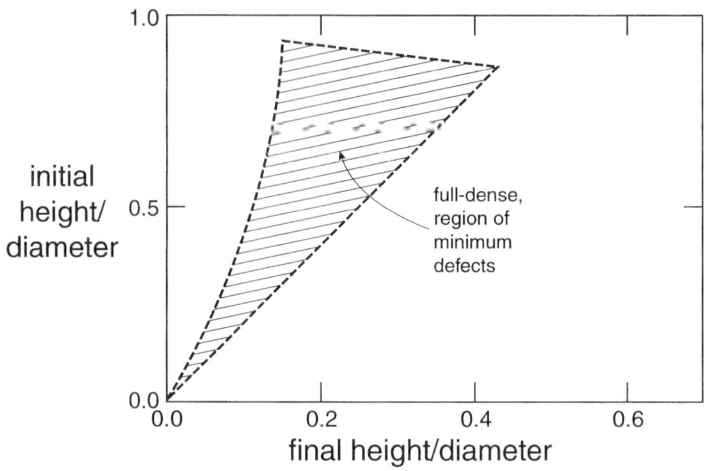

Figure 8.9. A schematic representation of the forging window concept. This plot shows the final and starting height : diameter ratios for a simple cylindrical compact. Sufficient deformation is needed to densify the compact without causing fractures, resulting in a definition of the needed initial compact size to satisfy the final dimensions.

shapes. Main uses are in very demanding applications, where defects are very costly, such as in oil-well drilling equipment.

One new notion related to powder forging is known as *ausforming*. Although not fully developed, the idea is to hot-deform a P/M steel preform while it is cooling from the austenite-phase region [usually from 900°C (1650°F)]. To form gears, a sintered preform is induction heated, and while cooling, it is deformed by rolling against precision gears that impress grooves into the compact. This technique provides a high-hardness, dense surface layer that is useful for bearings and gears. However, the interior of the compact is not fully dense, only the surface regions subjected to rolling.

8.3.3. Property Gains

The mechanical properties from powder forging are attractive. A major advantage is a higher strength, and there are no large defects. Consequently, the dynamic properties (fatigue and fracture toughness) are superior to wrought products. The larger the change in compact height during forging, the higher the final properties. Figure 8.11 illustrates this effect showing the impact energy for a forged steel–height change in forging. Larger height changes put more work into the compact and break apart any contamination films between particles. For comparison, this alloy has a toughness of just 10 J (7 ft · lb) at a density of 7.0 g/cm^3.

Contaminants seriously degrade properties, in spite of full densification. Thus,

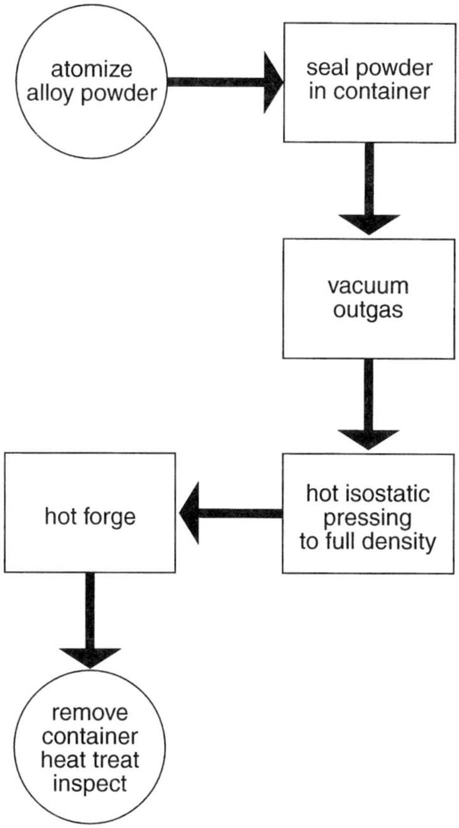

Figure 8.10. Schematic of a P/M process for the production of large billets of high-alloy steels. Hot isostatic pressing is applied to atomized powders that are subsequently forged to form high-quality ingots for final deformation.

the preform must be protected against environmental contaminants. Currently, many automotive connecting rods are forged from powder preforms. They are fabricated to 100% density, with less than 0.2 vol% inclusions and less than 300 ppm of oxygen. If the oxygen level reaches 1500 ppm, the impact toughness falls to one-third of that at 300 ppm of oxygen. Likewise, toughness depends on both factors. With just 0.2% oxides, the impact toughness drops by a factor of 3; with 1% oxides, it drops by a factor of 6. A small inclusion size and low oxygen content are most important to the fatigue life.

Besides the performance gain, powder forging offers a lower processing cost as compared with conventional fabrication. For example, on the automotive connecting rod shown in Figure 8.12, the production cost decreased by 10% by going to powder forging. This savings is significant, considering that more than 100 million connecting rods are fabricated in the United States each year.

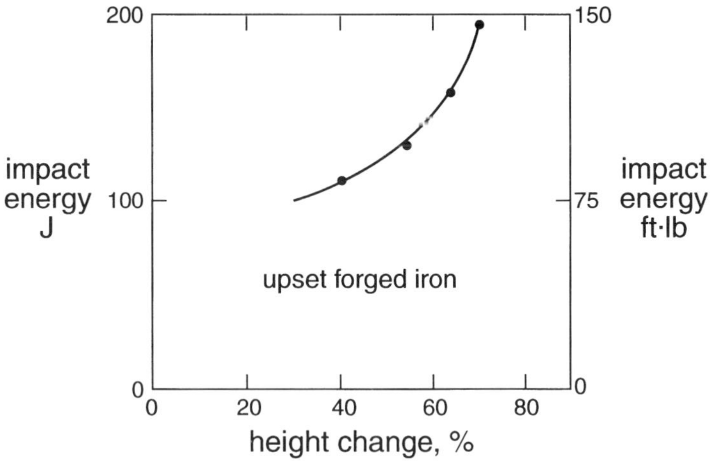

Figure 8.11. Charpy impact energy for powder-forged steels versus the height change in forging. (Data from W. Brian James.)

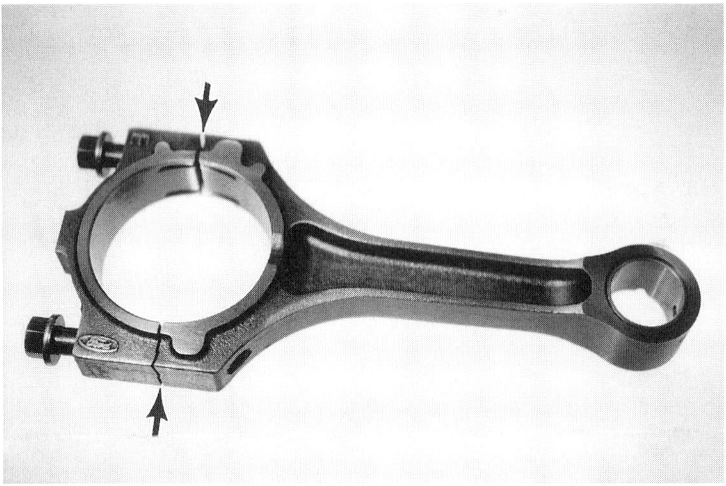

Figure 8.12. An automotive connecting rod formed by hot-forging an alloy of Fe–2Cu–0.8C. The rod is subsequently cracked (at the arrows) for assembly and bolted back together on the cracked surface. (Component courtesy of Charles Webster and photograph by Julian Thomas.)

8.4. HOT PRESSING

8.4.1. Uniaxial

Hot pressing with pressing along onc axis is a form of pressure-enhanced densification. It is very similar to die compaction, but the tooling and powder are heated. A sketch of the typical arrangement is given in Figure 8.13. Compaction is along the vertical axis, so the compact profile remains constant during densification, but the compact height decreases. Just as with die compaction, pressing can be single-action (only one punch moves) or double-action. The die and punches are usually made from graphite to allow resistance or induction heating. The strength of graphite limits the peak pressures. Since graphite burns in air, it is common to do hot pressing in a vacuum chamber. Also, a vacuum helps prevent oxidation of the powder. Other common die materials include refractory metals (usually alloys of molybdenum) and sometimes ceramics such as alumina or silicon carbide. After cooling, the compact resists ejection, so it is appropriate to eject the compact while it is hot.

Uniaxial hot pressing is slow and has poor control over heating and cooling, because the tooling is slow to heat and cool. The maximum operating temperatures are much higher than required for steels, and most operations use pressures below 50 MPa (7 ksi). Because the pressures are much lower than those encountered in hot forging, densification is slow. Atomic motion is required to fill the pores, unlike

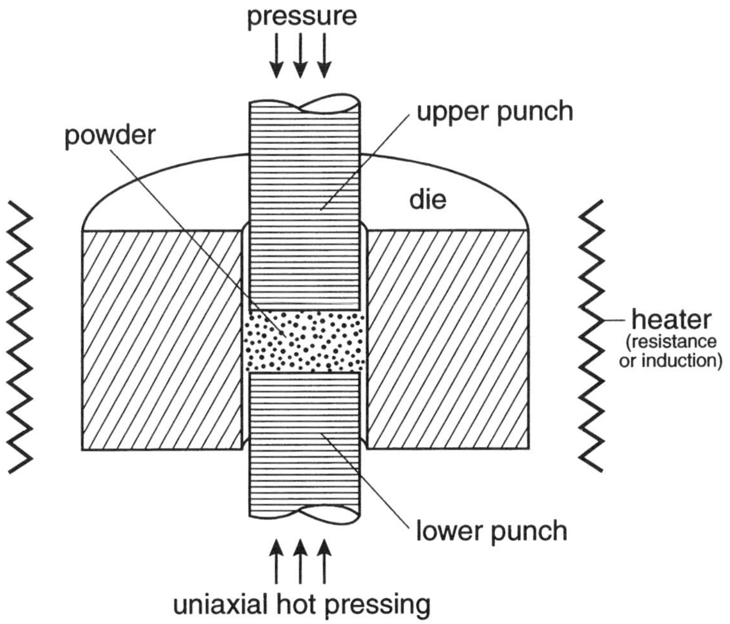

Figure 8.13. A cross-sectional view of uniaxial hot pressing, where a powder in the die cavity is simultaneously heated and compacted.

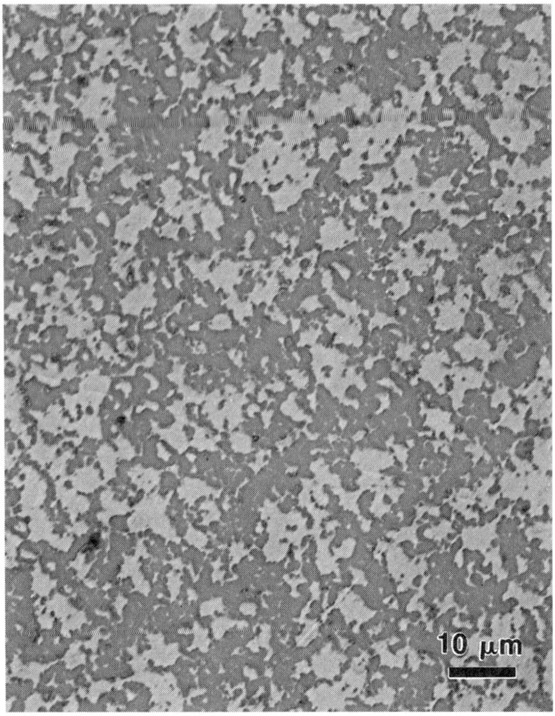

Figure 8.14. Microstructure of a composite consisting of tool steel with 50 vol% titanium carbide. This composite was densified by uniaxial hot-pressing a powder mixture at 1200°C (2200°F). (Photograph courtesy of Barbara Lograsso.)

hot forging, which stresses the powder to a point where the particles deform by dislocation motion. Thus, hot-pressing cycles might involve 15–60 min at the peak temperature, while hot-forging cycles give densification instantly.

The dies and apparatus are expensive, especially if pressing is done under a vacuum. Contamination of the compact from the die is a perpetual problem. However, uniaxial hot pressing is widely used to fabricate unique compositions in research settings. One of the largest production uses is in the fabrication of superabrasive cutting tools. Some of these diamond–metal composites use an iron alloy matrix to hold the diamonds in place. Another common use is in the fabrication of steel–ceramic composites. Examples of such composites are the steel–titanium carbide (ferrotic) alloys. These alloys are nearly impossible to sinter because of the high TiC ceramic content, typically 45 vol%. Figure 8.14 is a microstructure of such a composite where the hard carbide constitutes 50 vol% of the material.

Two heating techniques are used in hot pressing. A resistance heater outside the die provides heat at the surface, which eventually penetrates into the die cavity and powder. More efficient heating is obtained via induction. Here a water-cooled copper coil surrounds the die, and oscillating current through the coil induces currents in

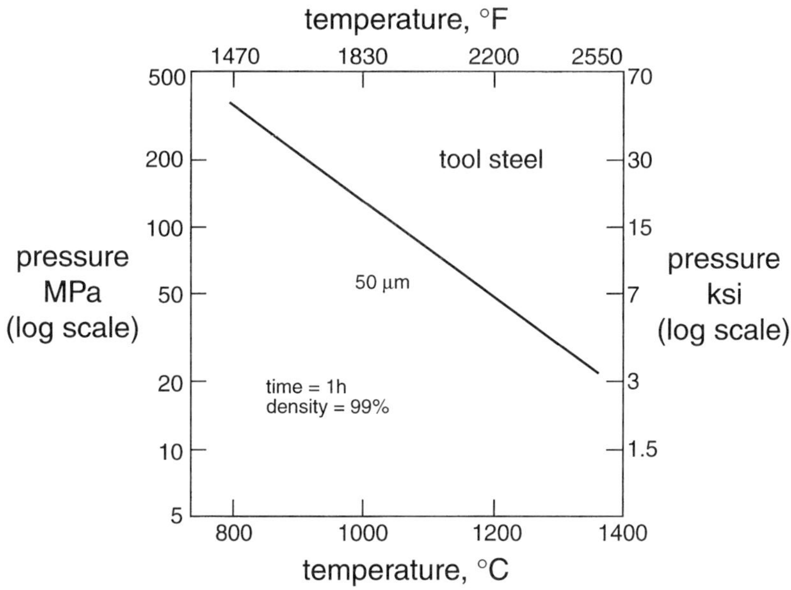

Figure 8.15. Pressure and temperature combinations required to densify tool steel. For these calculations the powder is a 50-μm tool steel and the hold time is 1 h at temperature and pressure. (Data from E. Arzt, M. F. Ashby, and K. E. Easterling.)

the die and powder. Induction heating is very effective for graphite tooling, since the high resistance of the graphite gives rapid heating from the induced currents.

The full-density techniques interplay different combinations of the controlling parameters of pressure, temperature, and time. Plots of these interplays provide guides to the best processing conditions. Figure 8.15 is one such plot showing the combinations of temperature and pressure needed to give full density with a 50-μm tool steel powder in 1 h. Note that the pressure scale is logarithmic, meaning that much lower pressures can be applied with small temperature increases.

8.4.2. Isostatic

Over the past 30 years, a newer form of hot consolidation has become very popular for many high-performance alloys. It is a form of gas pressure bonding that was developed in the 1950s, best known today as *hot isostatic pressing* (HIP). Similar to hot pressing (in contrast to hot forging), HIP usually uses lower pressures to densify a powder by atomic movement. The actual atomistic events include atomic diffusion, plastic flow, and creep. Computer calculations allow projections of the combinations of temperature, pressure, and time needed to attain full density. Figures 8.16 and 8.17 plot the results from two such calculations, one for a mild steel and the other for an austenitic stainless steel. The pressure is less than the yield strength of the powder at the consolidation temperature, as suggested by the nearly vertical

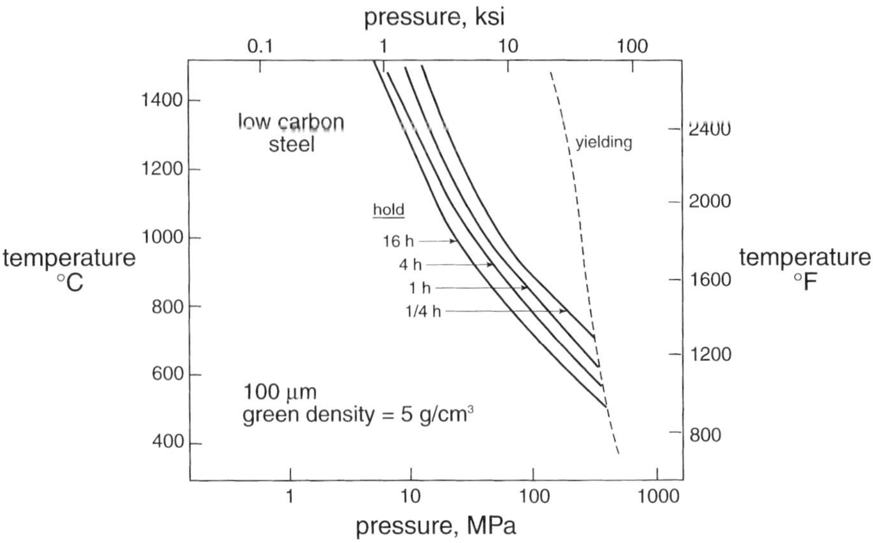

Figure 8.16. The computer-calculated time, temperature, and pressure combinations needed for full densification of a 100-μm low-carbon steel.

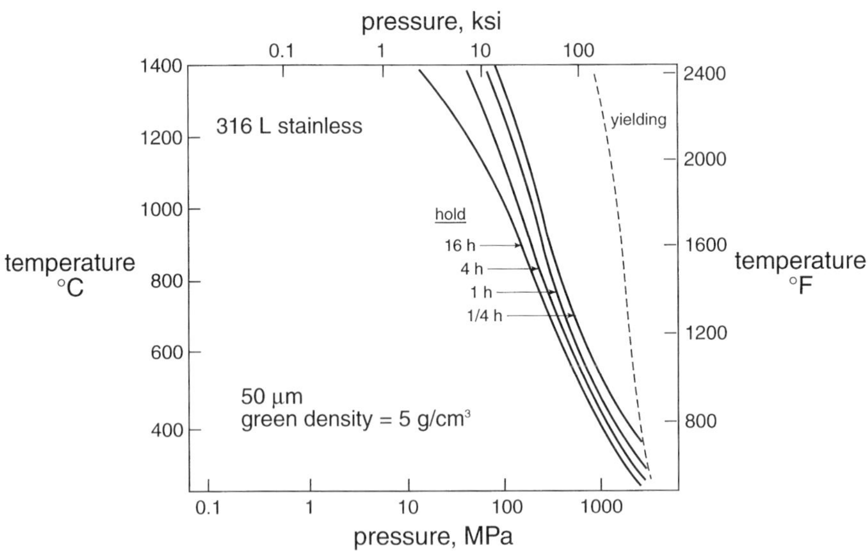

Figure 8.17. Computer calculation of the time, temperature, and pressure combinations needed to densify a 50-μm austenitic stainless steel (316L) powder to full density.

EQUATION BOX 8.2.

Creep-controlled densification is well modeled and extends the concepts from viscous flow to include details of the diffusion events. The densification rate (change in density per unit time), $d\rho/dt$ depends on the creep strain rate de/dt (change in strain per unit time) as follows:

$$\frac{d\rho}{dt} = A\,\frac{de}{dt}\,\frac{\rho(1-\rho)}{[1-(1-\rho)^{1/M}]^{M}}$$

where ρ is the fractional density, A is a geometric constant, and M is an empirical factor that depends on the powder characteristics. For a first pass, the creep strain rate varies with the absolute temperature T and stress σ with a phenomenological equation as follows:

$$\frac{de}{dt} = C\,\sigma^{m}\exp\left(-\frac{Q}{RT}\right)$$

where R is the gas constant [8.3 J/(mol · K)] and C and m depend on the actual mass flow events, microstructure, and material. The activation energy Q reflects the controlling atomic movement step, such as grain boundary or volume diffusion. Computer programs estimate all the material constants to make predictions of the densification behavior from these models.

lines marking the onset of yielding. In each case, the time required for densification is shorter with higher pressure or higher temperature.

The combinations of pressures, temperatures, and times are similar for both hot isostatic and uniaxial hot pressing. A key difference is in the pressure medium. Pressure is applied by a heated gas in hot isostatic compaction. Operating temperatures range beyond the melting point of all steels—up to 2000°C (3630°F)—and pressures over 200 MPa (30 ksi) are possible. The chamber diameters are as large as 1.5 m (about 5 ft). The pressure vessel is formed from a large block of steel or from wire windings on a tubular pressure vessel. The wire is wound under tension. Thus, the vessel is under compression even with a high-pressure gas on the inside. As illustrated in Figure 8.18, the wire winding is both on the outer diameter and around the yoke that holds in the end plugs. The powder in a container is at the very center of the device. Should the containment vessel fail, the wire winding will prevent the pressure chamber from exploding. A few catastrophic failures of HIP vessels occurred in the past, leading to wire winding designs geared to prevent human injury.

To prevent gas penetration into the pores, the powder must be contained in a pressuretight container. These HIP containers are fabricated from materials that are deformable at the consolidation temperature, so glass, iron, steel, stainless steel, and tantalum are common choices, depending on the maximum temperature. The container filled with powder is heated and vacuum-degassed to remove volatile contaminants. After prolonged evacuation, the container is sealed, possibly by welding. Once the powder-filled container is in place, the HIP vessel is heated and pressurized to force densification. Pressure is generated by argon or nitrogen that is flooded into the vessel. During the HIP cycle, the pressure vessel is simultaneously pumped and heated, but the pressure vessel walls are kept cold to minimize loss of strength. After the HIP cycle finishes, the compact is cooled and the container stripped from the densified compact.

An overall schematic of the workflow is provided in Figure 8.19. The shape of the container that holds the powder is oversized to allow for densification. Computer techniques are used to predict distortion, so most HIP products are close to final

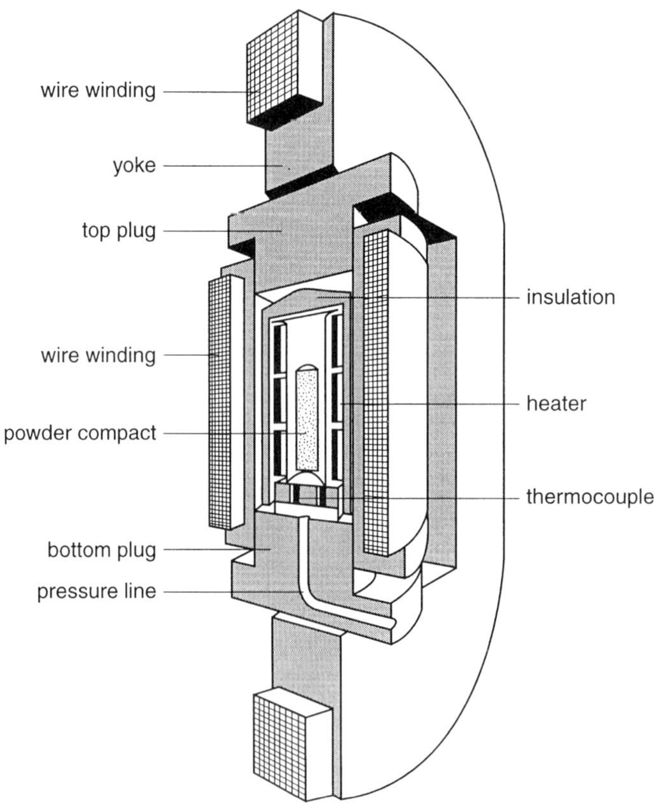

Figure 8.18. A cutaway illustration of a wire wound HIP pressure vessel used for hot consolidation of powder compacts. The wire winding is placed around both the cylindrical portion of the chamber and around the yoke, restraining the top and bottom plugs.

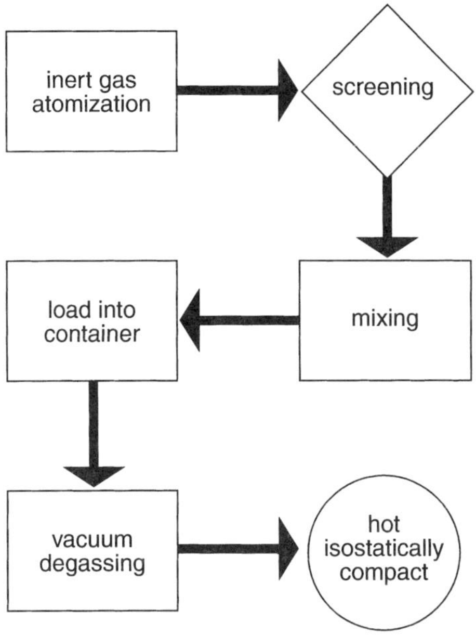

Figure 8.19. A schematic diagram of the production sequence for hot-forming a billet from a gas atomized powder, where the powder is first screened, blended, canned, vacuum-outgassed, and finally densified by the simultaneous action of pressure and temperature in a hot isostatic press.

shape. In the aerospace community the close approximation to final dimensions is called *near-net shape*, indicating that some machining is necessary to produce the component. Common HIP steel products are high-quality tool steels, aircraft components, cutting tools, metalworking and plastic processing tools, oil-well drilling equipment, and computer printer parts.

Figure 8.20 shows a schematic of the hot isostatic pressing cycle in terms of the simultaneous heating and pressurization. Table 8.1 lists some HIP cycle variants for full densification of ferrous powders. Because the pressure rise is slow, HIP cycles are long and can range up to several hours each. If properly performed, the material is fully dense without contaminants in the microstructure such as shown in Figure 8.21. Although pore-free, another significant feature is the homogeneous microstructure, free of contamination and segregation.

One variant is to use sintered compacts with densities more than 92% of theoretical. This density approximately corresponds to the point of pore closure. Although there is residual porosity, the pores cannot be filled by the pressurization gas. Then, the component already has the desired shape and the closed pores allow for hot consolidation without a container. Figure 8.22 is an optical micrograph of an

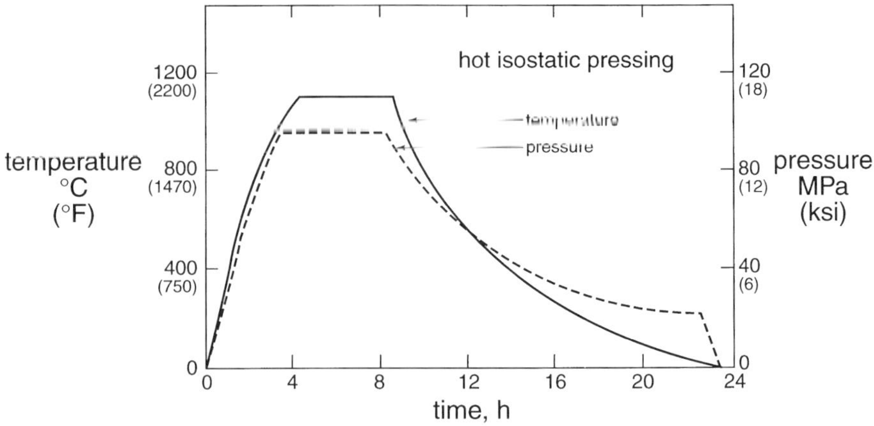

Figure 8.20. Example of the temperature and pressure curves during a production HIP run, showing a cycle time of nearly one day. Today, much effort is going into shortening these cycles to 6–8 h, largely by preheating the compact and rapid pressurization of the chamber.

iron–nickel magnetic alloy consolidated by such a containerless HIP process. The left panel shows the microstructure at 95% density after sintering, and the right microstructure shows the hot isostatically compacted microstructure after 30 min at 103 MPa (15 ksi) and 1200°C (2200°F). Note that all the small, spherical pores present after sintering have been eliminated but the grain size is much larger after HIP.

Typically, the surfaces of the compact are contaminated by the container and need to be removed by chemical dissolution, machining, or abrasion. Because consolidation is by hydrostatic pressure, there is little shear on the particle surfaces, so particle boundaries may be evident after HIP. Those are invariably weak aspects of the microstructure and limit mechanical properties. Accordingly, for some applica-

TABLE 8.1. Hot Isostatic Pressing Conditions for Full-Density Iron and Steel

Material	Pressure (MPa)	Temperature (°C)	Time (h)
4-µm carbonyl iron			
75-µm sponge iron	200	805	1
70-µm low-alloy steel	98	1000	1
190-µm maraging steel	150	800	1
100-µm austenitic stainless steel	210	1200	3
120-µm martensitic stainless	160	1150	3
steel	150	1150	3
65-µm Fe–10Al–5Si	200	1000	1
80-µm tool steel	100	1100	1
165-µm iron superalloy	69	1200	3

tions, subjecting the HIP compact to postconsolidation deformation is necessary. Alternatively, microstructural contamination can be avoided by use of gas-atomized powders handled under inert conditions. When properly executed, HIP is ideal for large, complex components intended for high-performance applications.

8.5. SPECIALIZED TECHNIQUES

Invention abounds in full-density processing. Every year new techniques are announced. Unfortunately, processing costs are high with most techniques, leading to modest use. Reduced processing cost with dimensional precision is the perpetual goal of full-density P/M. The following four sections help in the appreciation of some newer technologies.

8.5.1. Granular Forging

Because densification by hot isostatic pressing requires expensive pressurization and containment equipment, much effort has been expended on alternatives. Some popu-

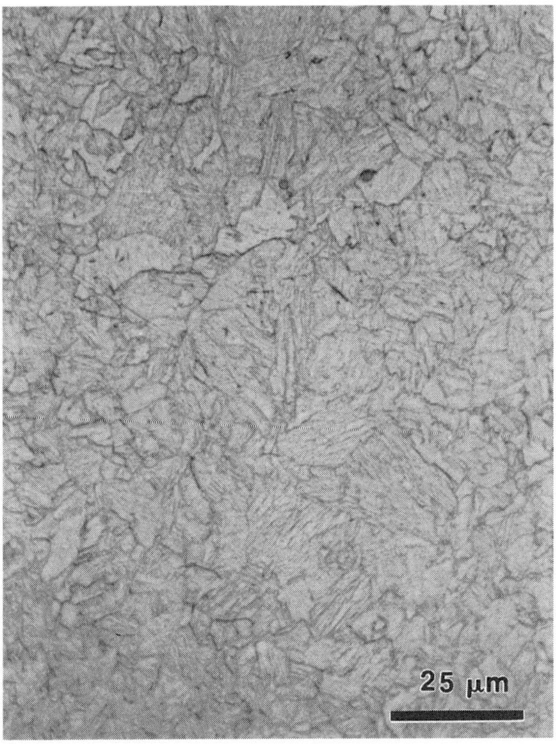

Figure 8.21. Optical micrograph of a full-density maraging steel consolidated by HIP at 1200°C (2200°F) in a 3-h cycle at 105 MPa (15 ksi).

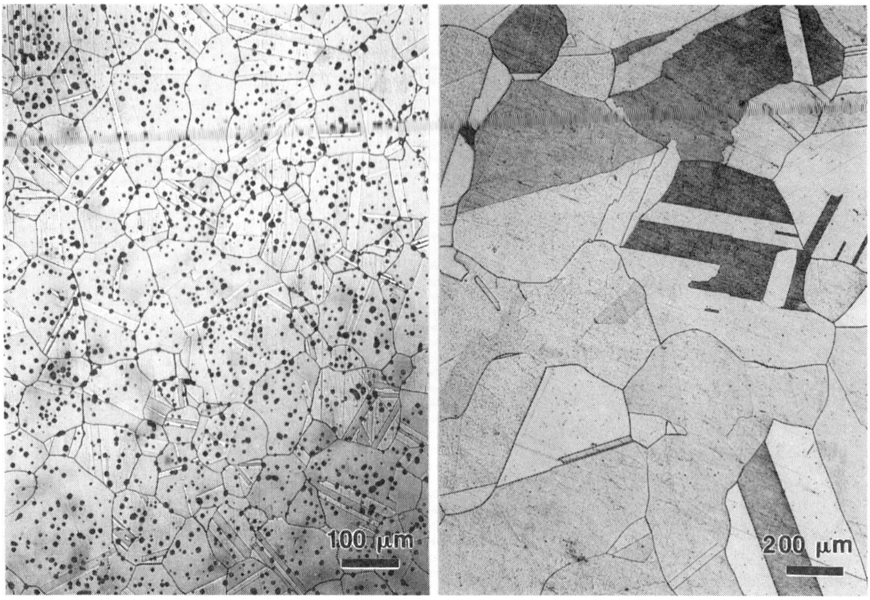

Figure 8.22. Two nickel–iron alloy compacts fabricated by first sintering carbonyl powders to 95% density [1410°C (2570°F) for 1 h in hydrogen] as shown on the left. These compacts were then containerless-densified by hot isostatic compaction at 103 MPa (15 ksi) and 1200°C (2200°F) for 30 min in argon, giving the microstructure shown on the right. (Photographs courtesy of Animesh Bose and Gerald Camus.)

lar options use uniaxial pressures acting on a low-viscosity medium containing a powder preform. Force transmission is through either a low-melting-point liquid (grease), soft metal (copper), or granular solid (alumina or graphite). The medium transforms the uniaxial force into a pressure on all surfaces of the compact. Figure 8.23 illustrates one such approach that uses pressed granules to densify a powder preform. This compact is embedded in the granules, which are heated before the consolidation cycle. The original granules were ceramics, leading to the name *ceramic consolidation* or *Ceracon*.

When graphite pellets are used, they can be heated by direct electrical conduction using a low-voltage (10-V) and high-current (2000-A) discharge for a few minutes. Figure 8.24 sketches a cross section of an electrically heated granular compaction scheme, which is also called the *electroconsolidation process*. If a high-frequency alternating current is used, there is reason to suspect that a plasma might be induced in the pores. The plasma would clean surface contaminants from the particles, and external pressure would compress the heated and cleaned particles. Plasma-activated sintering is used to describe this process, although it is also known as *spark sintering*, but both involve pressurization by hot-pressing powder undergoing direct heating from electric current passage.

Such hybrid compaction routes are faster than hot isostatic compaction. Another advantage is that consolidation can be achieved using a compact with open pores, because the granules will not penetrate the pores. Further, the consolidation resembles forging more than hot pressing, since the pressures exceed the yield strength. However, because the pressure transmission is not uniform in all directions, the height change is about twice that of the perpendicular dimensions. Typical consolidation conditions for steels are temperatures over 1000°C (1830°F) and pressures over 560 MPa (80 ksi). In one experimental variant, a very short-duration electric pulse is applied from a capacitor bank, resulting in full densification with a 30,000-V electric discharge that lasts only 300 μs, but achieves a current density of 10,000 A/cm^2.

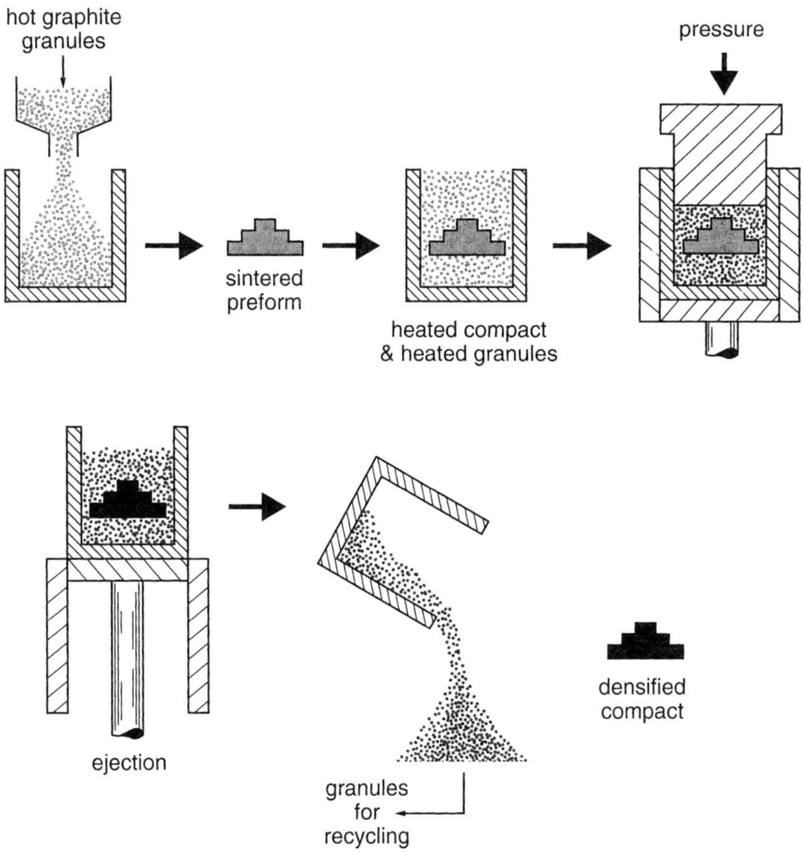

Figure 8.23. One approach to generating hydrostatic pressures without resorting to gas pressure is to embed the porous compact in hot ceramic or graphite granules and transmit pressure to the compact by pressing on the granules. Such an approach works without the necessity of sealing the pores and is a blend of forging and hot isostatic compaction.

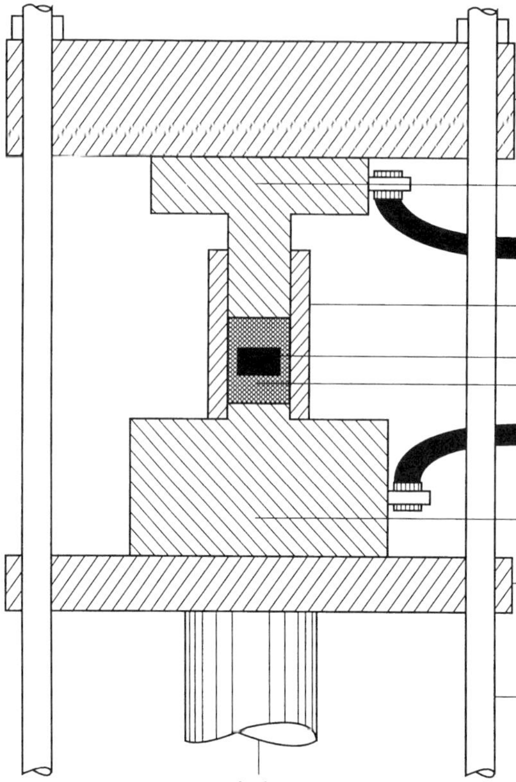

Figure 8.24. Granular pressing with simultaneous heating is performed using graphite granules heated by the direct passage of electric current. The powder compact is embedded in the granules that heat because of their high electrical resistance. Pressure is applied by the upper and lower rams acting on the graphite granules.

8.5.2. Gas Forging

A newer full-density approach starts with a sintered compact that is inserted hot into a heated pressure vessel for consolidation. Pressure is created by introduction of a liquid that vaporizes to generate a pressure pulse. Liquid nitrogen or liquid argon produces rapid pressurization when injected into a hot pressure vessel. Such a quick-HIP or pneumatic forging approach eliminates the pressurization time associated with conventional HIP. Densification is nearly instantaneous, because the pressure is far greater than the steel's strength. Peak pressures of up to 1000 MPa (145 ksi) have been recorded, effectively forging the compact to full density, but without the use of a container if the compact is over 92% dense.

A related, but lower-cost (and slower) process applies a low pressure, but increases sintering densification rates using an activating agent added to the powder.

Pressure is generated by encapsulating the powder under a vacuum in a glass container. At elevated temperatures, the glass is soft, but leakproof, and transmits atmospheric pressure against the powder. Hence, it is known by the term *consolidation at atmospheric pressure* (CAP). Combined with that low atmospheric pressure is a thin coating of boron on the particles. Boron forms a low-melting liquid phase that promotes densification. Thus, the combination of a liquid phase between the particles and the low stress leads to slow densification from the atmospheric pressure. As an example, a 16-h cycle at 1200°C (2200°F) proves effective in consolidating a tool steel doped with 0.1% boron.

For faster cycles, pressure-assisted sintering is available. Here sintering at high temperatures is first performed in a vacuum furnace. Once closed pores are formed, an external pressure is applied by filling the vacuum furnace with gas. The gas pressure might reach 3 MPa (450 psi). A related approach relies on heating an alloy powder to a temperature where a liquid phase forms inside the powders. Usually, the compact is first solid-state-sintered in a vacuum to a density typically near 95% of theoretical (about 7.5 g/cm^3 for most steels). Then the compact is heated to a temperature where a liquid phase forms. At this point the furnace is pressurized. This combination of liquid and external pressure gives rapid densification. Typically the compact size is small, but the final mechanical properties are similar to wrought alloys. Although widely demonstrated with steels, this technique is still in the development stage.

8.5.3. Hot Powder Extrusion

The hot extrusion of a steel powder is another method to achieve full density. This delivers a high level of particle deformation, so particle boundary contamination is not a problem. Prealloyed powders are used to obtain high densities and mechanical properties. Long shapes with constant cross sections are the main products of extrusion. By canning the powder in a vacuum tight container, it can be degassed and extruded hot without contamination. Both densification and shaping are achieved in one operation. A schematic drawing of hot powder extrusion is shown in Figure 8.25. A small penetrator in the extrusion ram directly stresses the powder to avoid

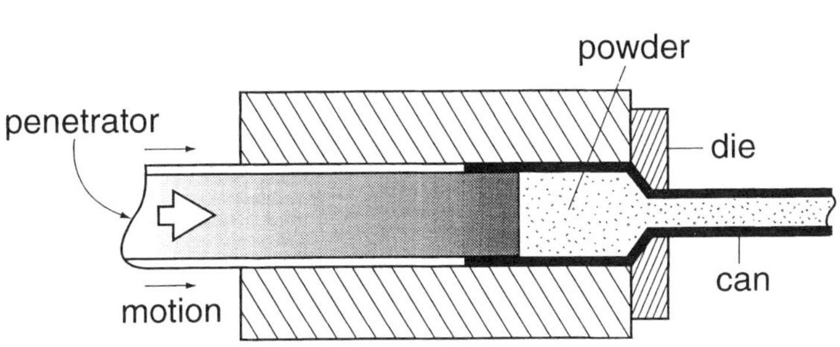

Figure 8.25. Hot powder extrusion involves forcing a can filled with powder though a die at the end of the extrusion barrel. Pressure is applied by a plunger or penetrator. With a large reduction in cross-sectional area, the powder is fully densified.

EQUATION BOX 8.3.

Extrusion flow depends on the extrusion ratio. The force F needed to continue densification and flow is proportional to the pressure P and starting cross-sectional area of the compact A_S

$$F = PA_S$$

in turn the pressure varies with the section size reduction and temperature, and the material. The reduction ratio is given by the ratio of the starting billet cross-sectional area and the final product cross-sectional area, A_S/A_F, *as*

$$P = \frac{\beta}{T} \ln\left(\frac{A_S}{A_F}\right)$$

where T is the temperature in centigrade and β depends on the alloy. For example, at 1000°C the value of β is near 240,000 MPa for low-carbon steel, but near 320,000 MPa for austenitic stainless steel.

buckling the can. The product might be a sheet, wire, tube, rod, or other simple shape.

In hot powder extrusion, the reduction in cross-sectional area usually has to be at least a factor of 10 to achieve full density. For some steel powders that resist densification, the ratio rises to 25. This means that if the input has a diameter of 100 mm, the output is 31.6 mm in diameter for a 10-fold area reduction ($100^2/31.6^2$), and 20 mm for a 25-fold area reduction ($100^2/20^2$). If the extrusion section thickness change is very large, the work of deformation will overheat the product. For a tool steel, such a condition occurs with an excessively high temperature (over 1125°C and 50-fold section thickness change).

Variants of the extrusion process include starting with loose powder, presintered powder, or powder densified by another process such as HIP. In one variant of powder extrusion, the powder is loaded into a cavity of a large billet, possibly with a complex shape, and the billet is extruded. Later the outer billet material is removed (possibly by chemical dissolution) to expose the inner core of extruded powder.

The design of the throat of the extrusion die is important to ensure balanced forces and defect-free products. Computer codes exist for the design of streamline dies, such as sketched in Figure 8.26. On the left is a drawing of a simple cylindrical reduction process, showing the projected powder flow and densification in passing through the die. The drawing on the right illustrates the cavity for an extrusion die intended to form a "T" section (upside down in this drawing). These complex contours keep the powder flow smooth, while ensuring uniform particle bonding and densification without cracking.

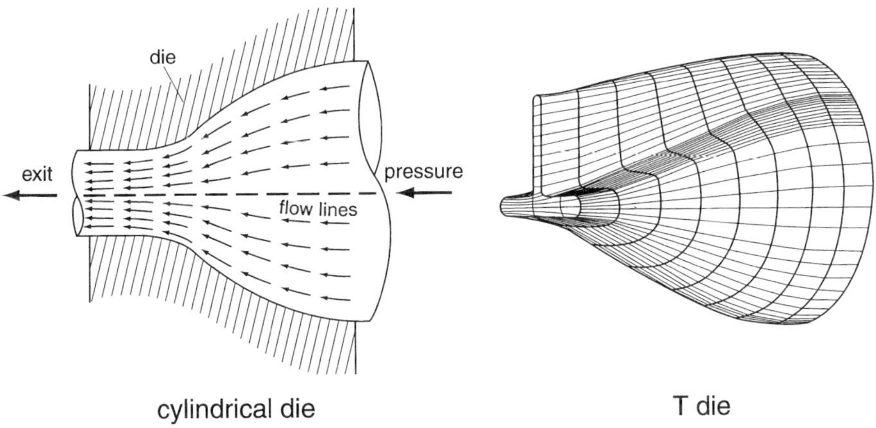

Figure 8.26. Streamlined die designs are used for smooth powder flow and densification. The contour changes in the die are calculated to ensure that the steel powder densifies into the final object shape without cracking. The left sketch shows an internal slice of a simple flow analysis, and the right sketch is an example die cavity for forming an extruded ''T'' section. (Based on work of H. Gegel, S. M. Malas, and S. M. Doraivelu.)

Although the steel composition affects the ease of extrusion, temperature is the most significant factor. An excessively high temperature damages the product and shortens the die life. Alternatively, an excessively low temperature makes extrusion difficult because of the higher pressures involved. Typically, steel powder extrusion is done at temperatures in the austenite phase range, over about 900°C (1650°F). The force to initiate extrusion flow is higher than the force to maintain flow, and increases with smaller particle sizes. Pressures in extrusion are typically in the 400–500-MPa (58–72 ksi) range, and temperatures range up to 1200°C (2200°F). One interesting application is in the production of seamless tubing for use in oil wells. Inert-gas atomized-steel powder is sealed in a steel container, and the container is hot-forged into a billet, which is subsequently hot-extruded.

8.5.4. Spray Processes

8.5.4.1. Spray Deposition. Spray forming techniques are another means to obtain full-density products. One idea of spray forming uses a gas atomizer and substrate closely positioned below the atomization nozzle. Many variants to this idea are in practice, but the most popular is known as the *Osprey process*. As illustrated in Figure 8.27, the spray of steel droplets is collected on the substrate, where they rapidly solidify after undergoing splat deformation. This drawing illustrates the production of continuous tubes. Some spray misses the substrate and becomes powder. Usually the spraying conditions are adjusted so that the droplets arrive in a semisolid state to fold on each other. The rapid heat extraction after impact gives microstructural homogeneity in the product.

Sometimes, the product is sufficiently dense for direct use, but usually it is subjected to hot forging, extrusion, or working. Spray forming is typically applied to deposition of highly alloyed steels that are difficult to fabricate by casting. Accordingly, the minimized segregation associated with the droplets is a clear advantage. Only simple shapes are possible using spray deposition. Some progress has been made in adding ceramic reinforcing phases to the spray deposit by cospraying ceramic particles into the atomization plume. Typical deposition rates are 0.5–2 kg/s.

8.5.4.2. Plasma Spray. In a manner similar to spray forming, the use of a plasma torch (or alternatively a flame torch), deposits dense coatings from semisolid droplets. Powder entrained in gas is fed into a plasma arc where the powder is rapidly heated to a molten state. A cross-sectional diagram of a plasma torch is shown in Figure 8.28. The plasma is generated by a voltage gradient between the electrodes where the temperature exceeds 5000°C (about 9000°F), with a theoretical limit of 30,000°C (54,000°F). A high gas flow between the electrodes extends the arc and

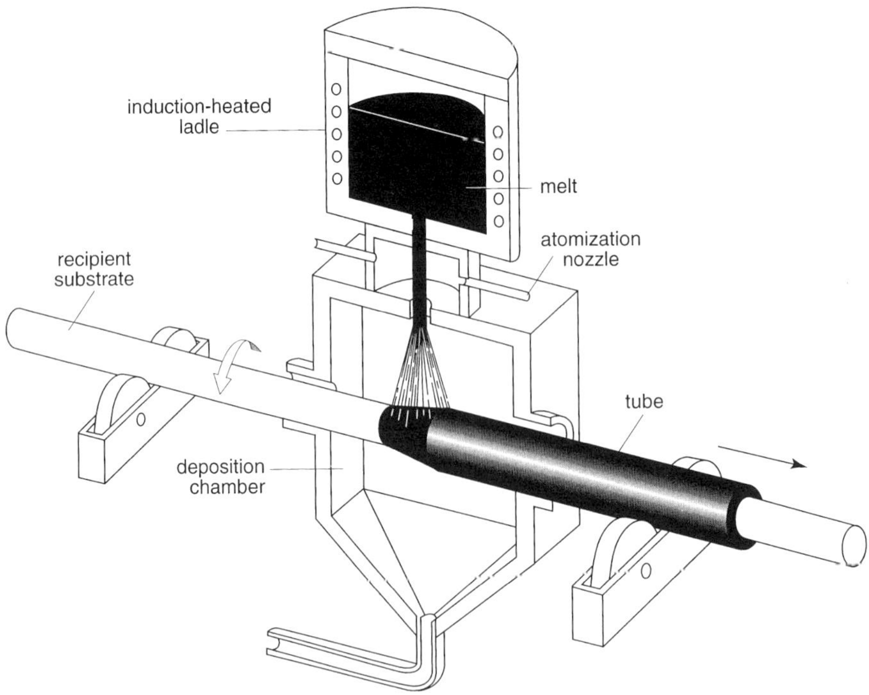

Figure 8.27. Spray forming involves placement of a solid substrate directly in the path of the droplets exiting from an atomization nozzle. The droplets splat and solidify onto the substrate to form a dense, rapidly quenched microstructure that is beneficial for simple shapes from high-alloy steels.

accelerates the powder through the arc. When the particles are injected into this high-temperature arc, they melt and accelerate toward the substrate as liquid droplets. On reaching the substrate, these droplets splat into a layered structure as sketched in Figure 8.29. Essentially, the process is like spray painting. If the droplets solidify during flight, they bounce off, while semisolid droplets adhere to the substrate. Spray deposition is especially useful for the fabrication of steel–ceramic composites such as illustrated in Figure 8.30.

Since there is a short dwell time in the hottest portion of the plasma arc, the particle size must be sufficiently small to allow melting before exiting the hot zone. However, very small particles evaporate in the arc. Generally, a particle size in the 40–80-μm range is optimal for steels. The porosity of the deposit can be as high as 15% if spraying is done under atmospheric conditions. A final densification step is often applied after spray deposition to eliminate residual pores. If spraying is performed in a vacuum, then deposit densities are near theoretical.

Plasma deposition has two general uses. The formation of dense protective coatings on wrought components is a main application. The deposit serves for corrosion,

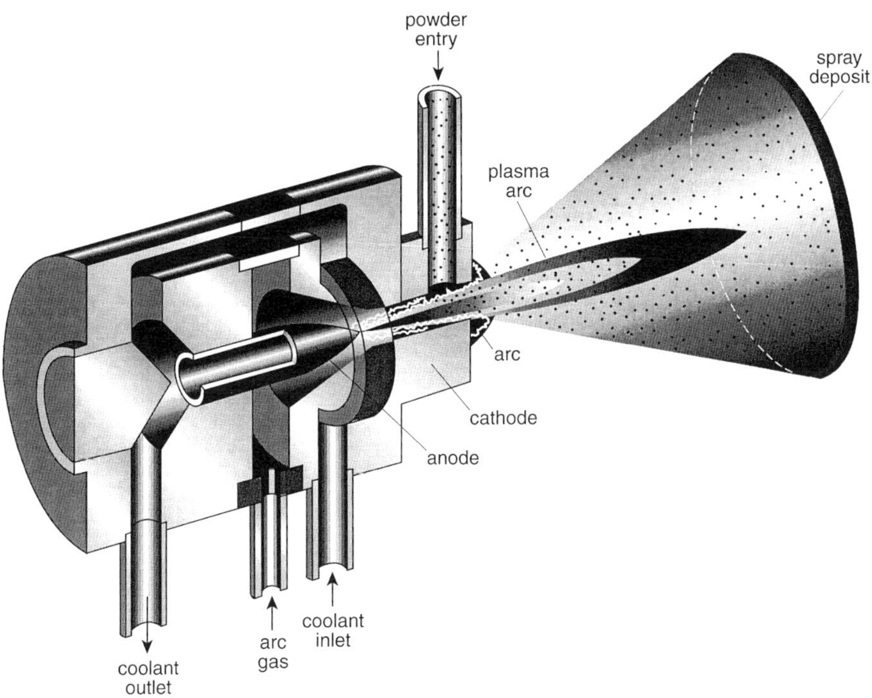

Figure 8.28. Plasma deposition of steel powder to full density relies on injecting the particles into a high-temperature plasma arc where they are accelerated. Before solidification, the molten droplets impact against a substrate to form dense layers. The process is best performed in a vacuum if full density is required. This sketch shows a cross section through the plasma torch. (Based on work of Herbert Herman.)

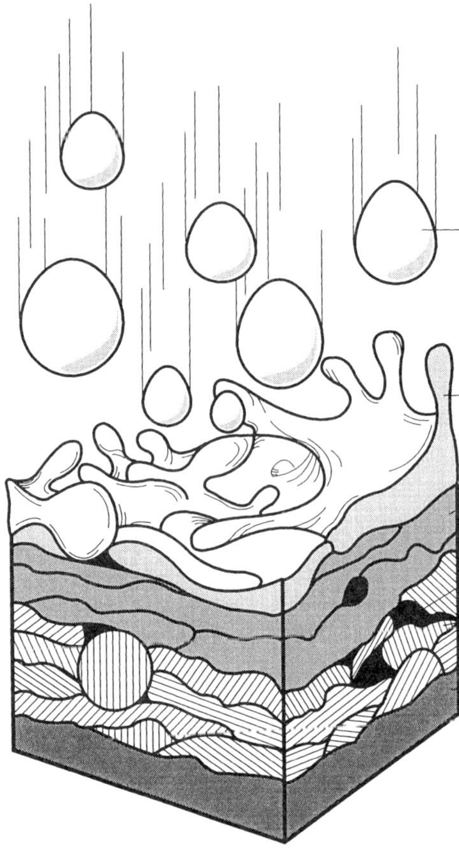

Figure 8.29. A schematic illustration of the events at the substrate in both plasma and spray deposition processes. Molten steel droplets impact on the already deposited layer to deform and quench onto the prior layers. (Based on work of Herbert Herman.)

oxidation, wear, or thermal protection of the substrate. The other use is in forming thin wall shapes, where the sprayed deposit is removed from the substrate after deposition.

8.6. RATIONAL SELECTION OF FULL-DENSITY PROCESSES

Techniques for eliminating porosity in full-density processing have been demonstrated for many steels. Usually the compact properties are improved by higher densities. The P/M processes are attractive for certain materials (typically high-alloy steels) and certain levels of complexity. Not only are complex shapes possible, but the productivity, material use, energy expenditure, precision, and microstructural

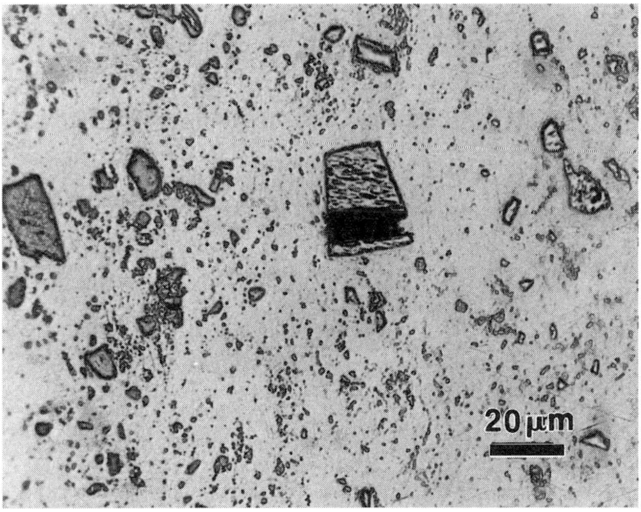

Figure 8.30. This micrograph shows an example of the full-density deposit possible by vacuum plasma spraying. This is a combination of stainless steel and titanium carbide useful for high-temperature wear applications. (Photograph courtesy of Ann Ritter.)

uniformity are improved by using P/M. When contamination is avoided, full-density P/M materials exhibit properties higher than attainable by other processes.

The emphasis has been on illustrating the common elements of full-density techniques that apply pressure while the powder is hot and weak. Today, the density and final compact shape can be predicted from computer models. In these models, we recognize that densification rates increase with higher applied pressures, and densification is further enhanced by higher temperatures, smaller powders, and longer hold times at the sintering temperature. Thus, changes that simultaneously alter one or more of these properties give higher densities. The actual mechanisms for applying pressure and temperature are highly variable, but the basic ideas are similar. Each process's engineering attributes are best suited for certain combinations of size, shape, tolerance, and complexity.

Full-density powder metallurgy is rapidly evolving, because of the benefits from high-performance steel structures. New techniques are invented frequently, but they rely on recognized combinations of pressure and temperature to either instantly (plastic flow) densify the powder, or slowly (creep) densify the powder. However, full density itself is not sufficient without use of properly prepared powders. The powders must be clean to avoid surface contamination. Beyond powder control, shear deformation during consolidation helps particle bonding. The success of full-density processing is not just in the final high density, but in the improved properties.

Applications for full-density processing are evident in several automotive components. Large annual production quantities of gears, sprockets, crankshaft components, and hydraulic components are produced to satisfy the requirements for endurance, wear resistance, and strength. In gear applications, full-density P/M steels exhibit twice the service life of wrought counterparts. One area needing attention for continued growth is improved dimensional control, since most of the techniques prove low in final precision. For example, powder forging is capable of ±0.2–0.4% tolerances (one standard deviation) on most features.

SUGGESTIONS FOR FURTHER READING

Arzt, E., M. F. Ashby and K. E. Easterling, "Practical Applications of Hot Isostatic Pressing Diagrams: Four Case Studies," *Metallurgical Transact.* **14A,** 211–221 (1983).

Apelian, D., M. Paliwal, R. W. Smith, and W. F. Schilling, "Melting and Solidification in Plasma Spray Deposition—Phenomenological Review," *Internat. Metals Rev.* **28,** 271–294 (1983).

Bose, A., "Overview of Several Non-Conventional Rapid Hot Consolidation Techniques," *Rev. Particulate Mater.* **3,** 133–170 (1995).

Eisen, W., "Mathematical Modelling of Hot Isostatic Pressing," *Rev. Particulate Mater.* **4,** 1–42 (1992).

Gegel, H., J. C. Malas, and S. M. Doraivelu, "Process Modeling of P/M Extrusion," in *Innovations in Materials Processing*, G. Bruggeman and V. Weiss (eds.), Plenum Press, New York, NY, 1985, pp. 137–159.

German, R. M., "High Density Powder Processing Using Pressure-Assisted Sintering," *Rev. Particulate Mater.* **2,** 117–172 (1994).

James, P. J. (ed.), *Isostatic Pressing Technology*, Applied Science, London, UK, 1983.

James, W. B., "Powder Forging," *Rev. Particulate Mater.* **2,** 173–213 (1994).

Koizumi, M. (ed.), *Hot Isostatic Pressing Theory and Applications*, Elsevier Applied Science, London, UK, 1992.

Kuhn, H. A., and B. L. Ferguson, *Powder Forging*, Metal Powder Industries Federation, Princeton, NJ, 1990.

Kulkarni, K. (ed.), *Powder Metallurgy for Full Density Products*, Metal Powder Industries Federation, Princeton, NJ, 1987.

Martins, S., and W. Misiolek, "Consolidation of Particulate Materials in Extrusion," *Rev. Particulate Mater.* **4,** 43–70 (1996).

Roberts, P. R., and B. L. Ferguson, "Extrusion of Metal Powders," *Internat. Mater. Rev.* **36,** 62–79 (1991).

Singer, A. R. E., "Recent Developments in the Spray Forming of Metals," *Internat. J. Powder Metallurgy* **21,** 219–234 (1985).

CHAPTER 9

Heat Treatment

9.1. COMMON HEAT TREATMENT STEPS

9.1.1. Overview

Approximately 60% of the current P/M steel production uses a heat treatment after sintering to attain the desired final strength and hardness. Sometimes, the heat treatment can be performed as part of cooling from the sintering temperature. More commonly, it is a separate operation that occurs after sintering. The heat treatment produces the desired phases of iron and carbon, giving the microstructure and properties required for successful service. Since the product from heat treatment is hard and strong, any reshaping or machining operations might best be done before heat treatment. For example, many P/M components are sized to adjust dimensions. If possible, then completing this task before final heat treatment would be best.

This chapter describes common heat treatment cycles used in powder metallurgy. Further, it considers the important role of residual porosity with respect to heat treatment. Table 9.1 provides a summary of definitions used in heat treating. Success in heat treating relies largely on understanding the changes in carbon solubility with temperature, recognizing that carbon solubility differs dramatically with the phase of iron.

9.1.2. Polymorphs of Iron

Iron is a *polymorph*, a metal that changes crystal structure with temperature and alloying. Heat treatments allow manipulation of the structure and properties for many common alloys by adjusting the type and quantity of each phase. There are many well-known transformations, including the notorious Dr. Jekyll and Mr. Hyde. For steel, the combination is more like ice and salt—the combination of two solids gives a totally new phase, a liquid we know as *saltwater*.

The best understanding of these heat treatment transformations is gained by examining the sintered microstructure. The types and amounts of each phase gives information on the expected properties. Of course, other considerations include the carbon content, porosity, alloying amount and type, and impurities. From these factors we can turn to roadmaps that show the phase changes and properties available with various heat treatments. Although complex, these transformation maps are valuable in predicting the impact of changes in the heat treatment cycle.

TABLE 9.1. Common Heat Treatments Applied to P/M Steels

Anneal—soften and remove residual strains by heating above the recovery and recrystallization temperatures, typically 25–50°C (45–90°F) over the eutectoid, followed by slow cooling

Carbonitride—a high-temperature modified carburization treatment in an ammonia-based atmosphere that promotes both carbon and nitrogen diffusion into the component surface for hardening, usually followed by quenching

Carburize—heat the component in contact with a carbon source, typically in a carbon-containing atmosphere, to provide additional carbon in the surface connected pores to harden the component

Case-harden—to use a packed powder or atmosphere gas to diffuse carbon into the exterior surface of the component to increase surface hardness

Nitrocarburize—a low-temperature [570°C (1058°F)] salt bath or gas treatment intended to nitride the component surface while increasing the carbon level

Normalize—similar to annealing in that the compact is heated to a temperature over the eutectoid temperature, but usually cooling is in air

Precipitation-harden—strengthening by growing precipitates via isothermal temperature hold where a second phase forms from a supersaturated solid solution; follows solutionization

Quench—rapid cooling from high temperature, typically using immersion of the heated steel in a fluid such as water, oil, or polymer

Sinter-harden—the generation of bainite or tempered martensite via controlled cooling from the sintering temperature, where the final microstructure is a mixture of martensite and bainite

Solutionize—heat to a single-phase temperature and allow alloying ingredients to homogenize, possibly followed by quenching to retain the structure; homogenization and austenitization

Steam treatment—a heat treatment typically above 500°C (932°F) in an atmosphere containing steam, used to grow a blue-black Fe_3O_4 tenacious surface oxide on steels

Stress relief—annealing to remove residual stresses from machining or working without altering the basic microstructure, followed by controlled cooling to avoid new stresses

Temper—postquenching heat treatment that transforms or modifies quenched martensitic via precipitation of cementite to improve toughness at the expense of hardness

Through-harden—solutionize and quench an alloy to form a hard phase throughout the component

9.1.3. Heat Treatment Steps

The simplest postsintering heat treatments are annealing, normalizing, and solutionizing. They involve heating the compact to a high temperature to remove residual stresses from machined or deformed components and to stabilize the microstructure. Annealing can also be used to homogeneously dissolve alloying ingredients. The cooling rate from the annealing temperature has a major role. A normalized steel is slow-cooled to form a soft, low-hardness microstructure good for subsequent deformation or working. On the other hand, strengthening is often the key concern in heat treatments. *Quenching* refers to quickly cooling the steel, usually resulting in a high hardness. It can be performed by plunging hot steel from temperatures near 1000°C (1832°F) into water or oil to form a phase termed *martensite*. The high hardness resulting from quenching usually gives a brittle steel, often too brittle for common applications. Consequently, *tempering*, which involves reheating to an intermediate temperature to relax the hard-quenched material, is used to soften and toughen the steel. Reheating allows partial transformation of martensite, produced by quenching, into phases that combine to produce a lower hardness and higher toughness. Tempering after quenching produces a desirable blend of strength and other mechanical properties.

Other heat treatment steps include precipitation hardening, where small second phase particles are grown in the steel. Initially, these precipitates add strength as they grow, although eventually the effect is lost with extended heat treatment times. Examples where precipitation gives increased strength are the iron–copper–carbon alloys (such as FC-0208) and precipitation hardened stainless steels (e.g., 17–4 PH). Most steels contain dispersions, either intentional, such as carbides; or unintentional, such as oxide inclusions. The dispersions increase strength if they remain small and uniformly dispersed. So, in a sense even carbides are precipitates, and their size and distribution impact on mechanical properties.

The important details in determining a heat treatment are as follows:

- Heating rate—this is seldom critical
- Soak temperature and time
- Cooling rate and atmosphere or quenching fluid used for cooling
- Intermediate hold temperatures and times during cooling
- Reheat temperatures and times
- Final cooling rate and medium

Figure 9.1 sketches two temperature–time plots associated with heat treatments of a steel. In cycle 1, cooling is halted at an intermediate temperature to ensure formation of a soft phase known as *pearlite*. In cycle 2, cooling is rapid to form martensite, followed by reheating to temper at an intermediate temperature. As a contrast of mechanical properties, cycle 1 delivers a lower hardness and strength with higher ductility.

A successful heat treatment cycle depends on identification of each of these six parameters. Effective decisions require knowledge of the alloy, carbon level, porosity, and desired properties. Usually handbooks can be consulted to select the best cycles. This chapter builds the background to understand the origin of the heat treatments and the changes caused by the common cycles.

9.2. PHASE CHANGES DURING HEAT TREATMENT

9.2.1. Crystal Changes

Major changes occur during heat treatment. These changes result from the differences in carbon solubility with changes in crystal structure for iron. The low-tempera-

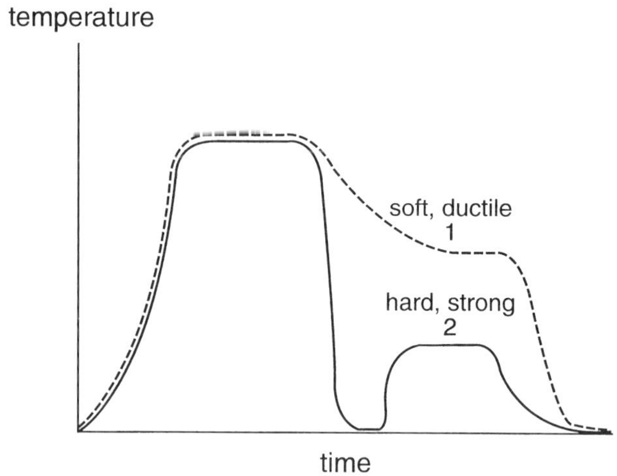

Figure 9.1. A sketch of two temperature–time plots associated with heat treatments and suggested differences in resulting mechanical properties.

ture form is alpha iron (ferrite), a body-centered cubic crystal structure. Ferrite has a low solubility for carbon. At elevated temperatures, the form is gamma iron (austenite), which has a face-centered cubic crystal structure. Austenite has a high solubility for carbon, up to about 2%.

So if carbon is nearly insoluble in alpha iron, yet many steels contain 0.4–0.8 wt% carbon, where does it go on cooling from austenite? The answer depends on the cooling rate. Slow cooling results in precipitation of iron carbide, a compound consisting of three iron atoms for every carbon atom—Fe_3C, which is called *cementite*. Although 25% of the atoms are carbon, the carbide only contains 6.67% carbon by weight. Alternatively, rapid cooling from austenite freezes the carbon in solution, resulting in a distorted body-centered cubic crystal structure that is now body-centered tetragonal, giving a new phase known as *martensite*. Thus, heat treatments revolve around the partition of carbon in the iron as it goes through a drastic solubility change on cooling. When a high-carbon-content alloy is cooled slowly, the carbon goes out of solution to form a lamellar mixture of ferrite and cementite known as *pearlite*. This gives a soft but ductile structure. In contrast, martensite is a hard, strong phase, but lacks ductility.

9.2.2. Alloying

On cooling, the transformation from austenite to ferrite is impeded by alloying additions. As the alloying level decreases, especially as carbon is removed from the alloy, the ability to manipulate properties decreases. Usually with ferrous alloys, a high hardness and high strength result from modest carbon levels in the range of 0.4–0.8%. Other alloying ingredients such as nickel, molybdenum, chromium, and manganese affect the heat treatment response, even at low concentrations.

An example of the variations possible with heat treatment is provided by Figure 9.2, which shows polished and etched microstructures of an injection-molded and sintered steel containing 2% nickel with two different heat treatments. The lower-hardness microstructure (80 HRB) consists of inhomogeneous pearlite that has a strength of 580 MPa (84 ksi). In contrast, the high-hardness microstructure (46 HRC) exhibits martensite and has a strength of 1260 MPa (183 ksi). The potent strengthening role with heat treatment is evident from this example.

Manipulation of the heat treatment response via alloying is well known in ferrous metallurgy. For each candidate addition, there are three key changes. The first relates to the relative effects on carbon dissolution in the various phases. Most metal additions stabilize either ferrite or austenite, but some form carbides. Second, the alloying addition changes the transformation temperatures. Thus, they can enhance or retard the formation of desirable phases in heat treatment. Finally, the rate of transformation is affected, and much interest exists in alloying additions that slow the change from austenite to martensite on cooling. The slower that transformation, the easier it is to form high-strength sintered components in heat treatment. Understanding the role of an alloying addition requires attention to how it interacts with iron and carbon, the two key ingredients in a steel.

9.2.3. Carbon Eutectoid

The phase diagram for mixtures of iron and small quantities of carbon is given in Figure 9.3. A *phase diagram* plots the phase ranges versus temperature and composition. Single-phase regions are shaded. The leftmost condition corresponds to pure iron with no carbon. Examination of the phase change versus temperature shows that on heating, iron transforms from ferrite to austenite at 910°C (1670°F). With continued heating, the austenite transforms into delta iron near 1392°C (2538°F). Finally, iron melts at 1537°C (2800°F).

As carbon is added to iron, the melting and transformation temperatures shift. The lines on this diagram show the transformation temperatures versus composition. A further complication is the formation of the carbide cementite, Fe_3C, at high carbon levels. Since carbon solubility in ferrite is small, a two-phase field is encountered where ferrite coexists with cementite. Likewise, other two phase regions are observed on the phase diagram (e.g., $L + \gamma$, $\gamma + C$, and $\alpha + \gamma$). Note the reduction in melting temperature for iron as carbon is added and the expansion of the temperature range over which austenite exists with carbon alloying.

With respect to heat treatments, a critical carbon content is at the eutectoid composition, which is 0.77% carbon for a simple steel. The corresponding critical temperature for this alloy, 727°C (1340°F), is termed the *eutectoid temperature*. At the eutectoid composition and temperature, austenite directly transforms into a lamellar mixture of ferrite and cementite called *pearlite*. There are then three types of steel based on the carbon level: hypoeutectoid with less than 0.77% carbon, eutectoid steels with about 0.77% carbon, and hypereutectoid steels with higher carbon levels. Only at the eutectoid composition and temperature is it possible to just form ferrite and carbide on cooling. During slow cooling a hypoeutectoid steel first forms ferrite

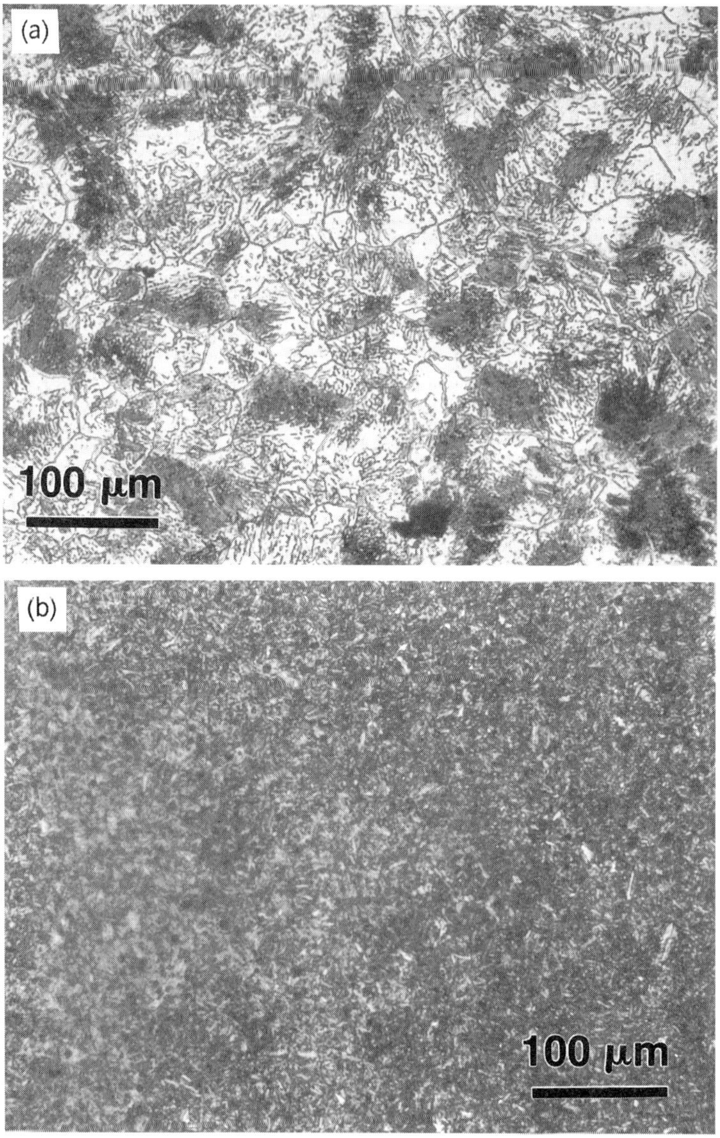

Figure 9.2. An example of the microstructure variation possible with a carburizing heat treatment for injection-molded and sintered 2% Ni steels with two different carbon levels: (a) lower-hardness inhomogeneous microstructure (145 Vickers hardness) with a yield strength of 580 MPa (84 ksi), (b) high-hardness microstructure (445 Vickers hardness) with full martensitic transformation with a yield strength of 1260 MPa (183 ksi). (Photographs courtesy of Haorong Zhang.)

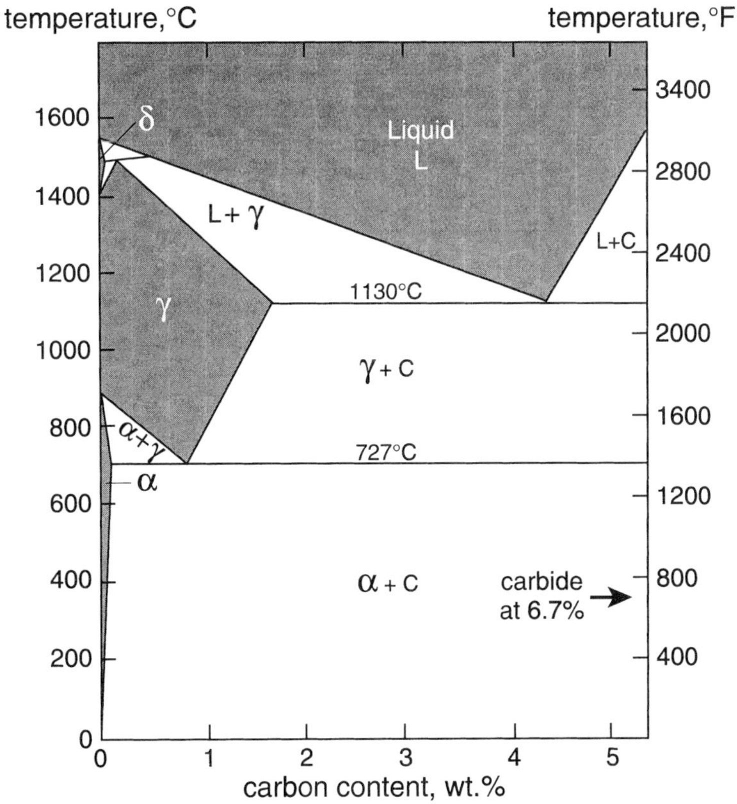

Figure 9.3. The binary phase diagram for iron and carbon. This is a map of phases that form for various carbon levels and temperatures. Iron is a polymorph, meaning that it changes crystal structure with both alloying and temperature. Consequently, it forms various phases with dramatically different solubilities. The two phase regions, which separate the single phase regions, are mixtures of the single phases.

on cooling, while a hypereutectoid steel first forms iron carbide. Any phase to form before the eutectoid transformation is designated as a *proeutectoid phase*.

In highly allowed steels, the carbon level corresponding to the eutectoid composition shifts to lower carbon levels for alloying ingredients such as nickel, manganese, chromium, silicon, tungsten, molybdenum, and titanium. Figure 9.4 plots the carbon content at the eutectoid composition and the eutectoid temperature versus alloying level for several additions.

9.2.4. Equilibrium Behavior

For illustration purposes, Figure 9.5 plots the slow-cooling behavior of three types of steel—hypoeutectoid, eutectoid, and hypereutectoid. Equilibrium indicates the most energetically favorable condition, usually attained after long times. Most steel heat treating is performed under rapid cooling conditions that prevent attaining equilibrium. Figure 9.5 shows the behavior for slow cooling from pure austenite, with nearly equilibrium conditions. The curves give the relative percentage of each phase versus temperature. When slow cooling takes place, the equilibrium phase diagram proves fairly accurate in predicting transformations.

For the hypoeutectoid steel (1 in Figure 9.5), the first event is for some austenite to transform to ferrite (α-Fe) and then for the balance of the austenite to transform into pearlite (ferrite and iron carbide) at the eutectoid temperature. The A_3 temperature is where the first ferrite forms, while the balance of the transformation occurs at the A_1 temperature. The phase that forms prior to the eutectoid transformation is the proeutectoid phase, which is ferrite here.

In contrast, the eutectoid steel (2) simply changes from austenite to a mixture of alpha iron and iron carbide, at the eutectoid temperature A_1. The lamellar structure

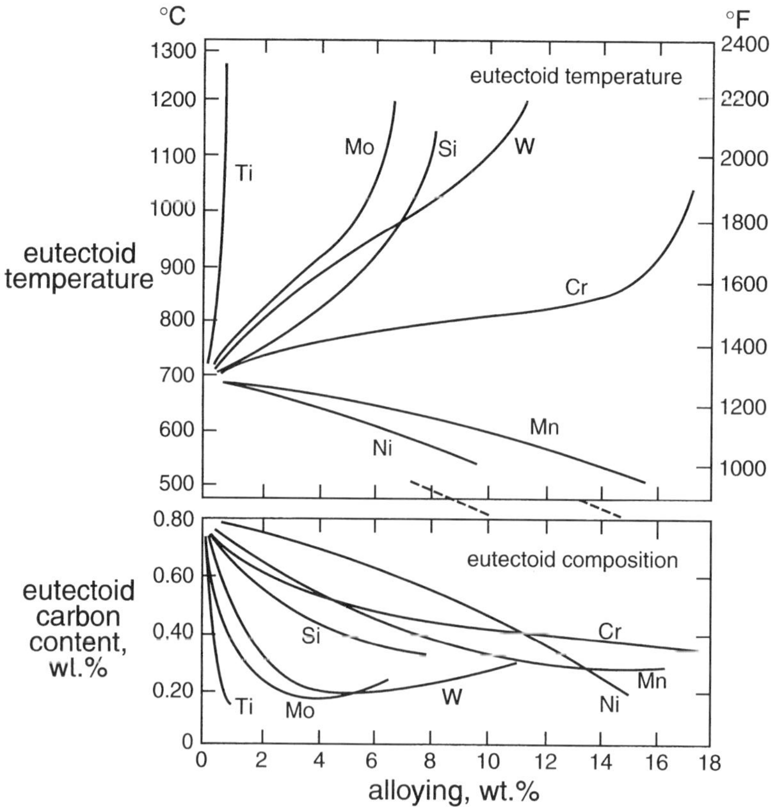

Figure 9.4. Plots of the carbon composition at the eutectoid and the eutectoid temperature versus alloying level for several additions. (Based on data of E. Bain and H. Paxton.)

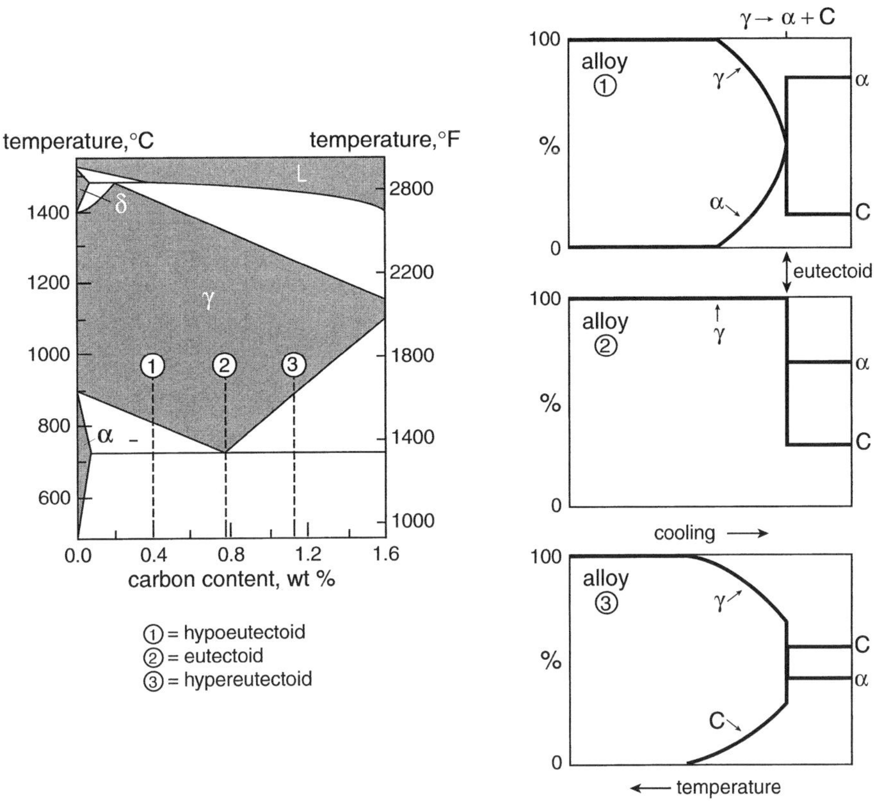

Figure 9.5. Plots of the phase changes for the slow cooling of three steels. The plots assume slow cooling from pure austenite γ, giving the relative percentage of each phase (ferrite α and cementite C) versus temperature.

of pearlite has a ''pearl-like'' luster. Figure 9.6 is a microstructure of a pearlitic steel showing the characteristic lamellar structure.

Finally, the hypereutectoid steel (3) first transforms from austenite to carbide during cooling, then at the eutectoid temperature the remaining austenite transforms into α-Fe and iron carbide—ferrite and cementite or pearlite. The proeutectoid carbide results in a brittle product, and is generally undesirable; consequently, high carbon levels are not commonly observed in P/M steels, except tool steels and friction materials.

Under factory processing conditions, some time is required for these phase changes. The relation between the time to transform and the temperature is of great consequence to heat treatment. For constant temperature situations, it is possible to plot the percentage of austenite transformed against time at various temperatures. Figure 9.7 captures a schematic of this behavior. We now need to understand the concept of undercooling. The equilibrium phase diagram shows what should happen

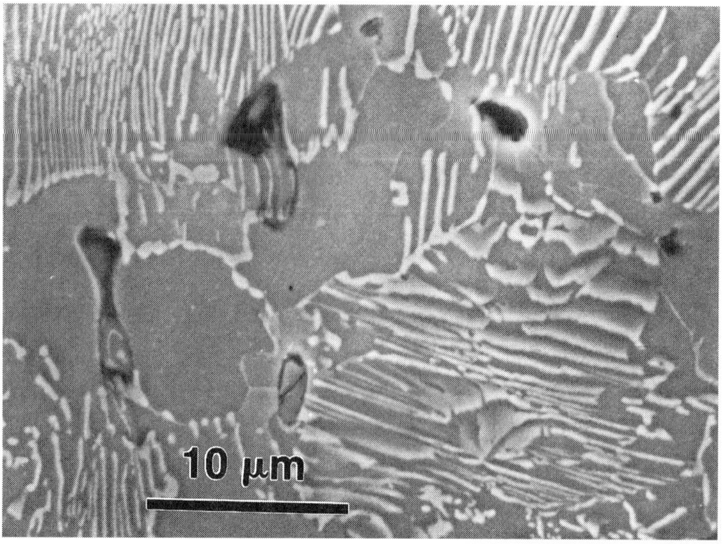

Figure 9.6. The SEM microstructure of a sintered steel showing the characteristic lamellar structure associated with pearlite. (Photograph courtesy of S. T. Paul Lin.)

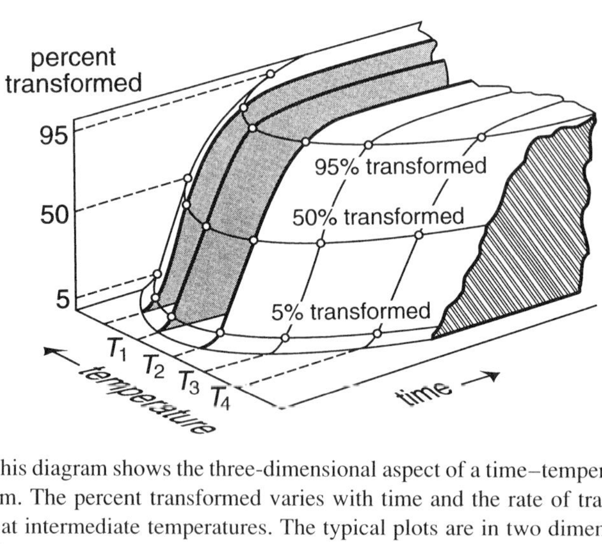

Figure 9.7. This diagram shows the three-dimensional aspect of a time–temperature transformation diagram. The percent transformed varies with time and the rate of transformation is usually faster at intermediate temperatures. The typical plots are in two dimensions with the phase range shown versus time and temperature. (Based on work of John Wulff.)

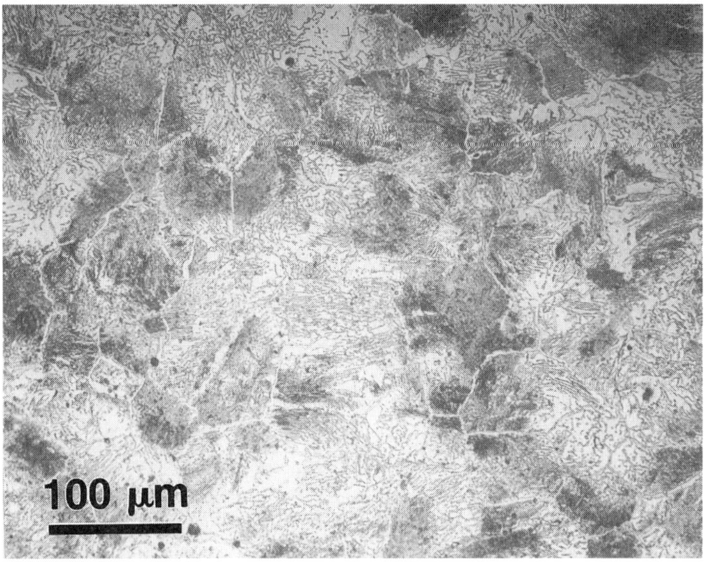

Figure 9.8. A photograph of proeutectoid ferrite on the grain boundaries in a sintered iron–nickel steel. As the cooling rate (but not quenching) increases, the smaller the scale of the pearlite and the stronger the product. (Photograph courtesy of Haorong Zhang.)

in terms of phases versus temperature for any carbon level if the steel is held for a long time at that temperature. When temperature changes take place, there is a time required for the atoms to rearrange into the equilibrium structure. Atomic motion is slower at lower temperatures, so longer times are required to attain equilibrium at lower temperatures. Consequently, on cooling, austenite exists for some time even though it should no longer exist according to the phase diagram.

Undercooling describes the condition where a phase exists when it should transform. With a small undercooling the transformation is slow because there is little departure from equilibrium. The greater the undercooling, the larger the driving force for transformation. The transformation requires less time at moderate undercoolings below the eutectoid temperature. However, with greater undercooling there is slower atomic motion, resulting in a delayed transformation. Such behavior is represented by a "C" curve that captures information on time, temperature, and amount transformed. A schematic of the three-dimensional plot is given in Figure 9.7. Note intermediate undercoolings give the shortest transformation time.

On slow cooling eutectoid steels form pearlite, hypereutectoid steels form proeutectoid cementite (iron carbide) and then pearlite, and hypoeutectoid steels form proeutectoid ferrite and then pearlite. Usually the proeutectoid phases form on grain boundaries. Figure 9.8 is a photograph of proeutectoid ferrite on the grain boundaries in a sintered iron–nickel steel. The more rapid the cooling (but not quenching) the smaller the scale of the pearlite and the stronger the product. Spacing between the carbide plates gives a measure of the cooling rate and strength; smaller spacings correspond to faster cooling and stronger steels. If pearlite is reheated to just below the eutectoid temperature, then the lamellar structure grows into spherical iron carbides in a ferrite matrix. This is called *spheroidization*, but it is not used in P/M

9.2.5. Cooling Rate Effects

Now consider some variations. Already we have seen that the carbon content affects microstructure and properties—slow-cooled hypereutectoid steels are brittle, rapidly cooled steels are stronger than slow cooled steels, and so on. An important treatment is to quickly cool the steel. Austenite is stable at temperatures near 1000°C (1832°F). If austenite is rapidly cooled to about 550°C (1022°F) and soaked at this temperature, a new microstructure will form. This is known as *bainite*. It is chemically the same as pearlite, but the structure is much different since the carbides are now very small. Thus, bainite is also a mixture of ferrite and iron carbide. However, the rapidly cooled austenite has now transformed via a cooling path that denies formation of the lamellar structure familiar in pearlite. Instead, the microstructure consists of feathery carbide regions in a ferrite matrix as illustrated in Figure 9.9.

In the extreme, water-quenching a steel starting in the austenite temperature range produces a new microstructure. The product from quenching has a high strength and is very hard, but ductility drops dramatically. This is because the quenched austenite cools so rapidly that carbon remains in solution while the iron is forced to transform to ferrite. The mixed signal result in ferrite with frozen-in carbon that

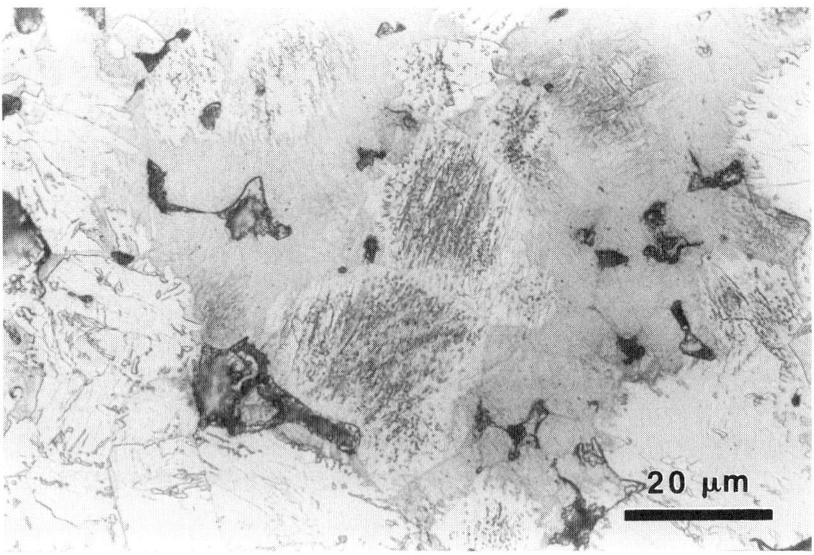

Figure 9.9. This micrograph of bainite shows feathery regions containing fine carbides in a ferrite microstructure. (Photograph courtesy of James Marsden.)

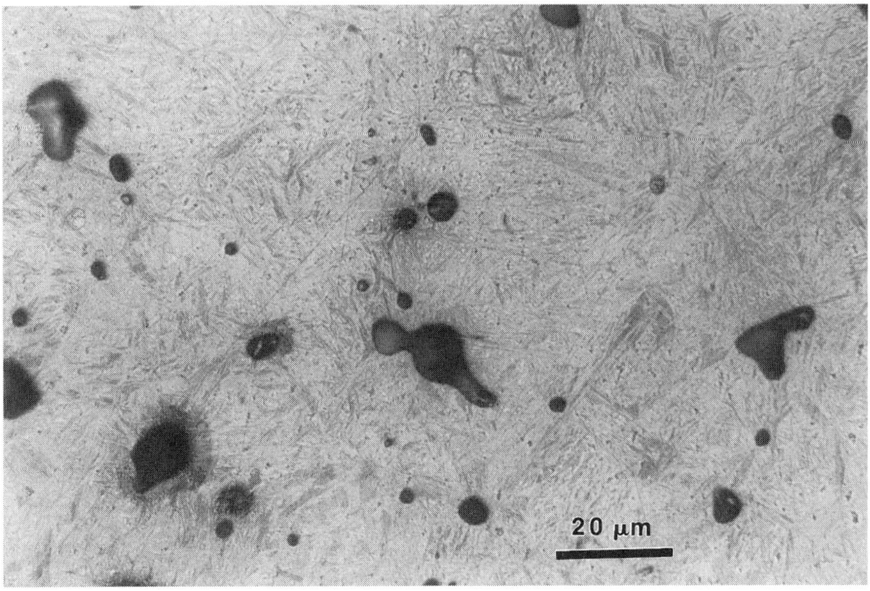

Figure 9.10. Martensite in a sintered steel, with the characteristic structure of strained ferrite due to the retention of carbon at low temperatures. (Photograph courtesy of Christoph Degoix.)

should not be there, forming a structure known as *martensite*. Normally, time is required to move atoms into the two phases of ferrite and carbide, but there is a delay in their formation during cooling. Because of this delay, martensite forms in quenching.

9.2.6. Martensite

Martensite is a hard structure with a distorted crystal structure from the excess carbon. With 0.8% carbon there is a 4% distortion to the ferrite crystals. This crystal distortion results from the frozen-in carbon and makes martensite hard and brittle as compared with ferrite. Figure 9.10 is an optical micrograph showing a sintered steel after quenching. The feathery structure is characteristic of martensite.

Thus, on slow cooling from austenite, carbon leaves solution to form a mixture of ferrite and carbide. In contrast, on rapid cooling carbon remains in solution and distorts the ferrite into martensite. Quenched iron without carbon will not form martensite. There are other "martensitic" alloys that do not contain carbon, and even some that do not contain iron. Hence, *martensite* is a general term applied to a phase that forms without diffusion. The new phase is not on the phase diagram.

For powder metallurgy steels, martensite is associated with rapid cooling. Its formation usually requires a quenching medium—oil, water, water–polymer, water–oil, or salt solution. Quench fluid selection depends on the alloying additions

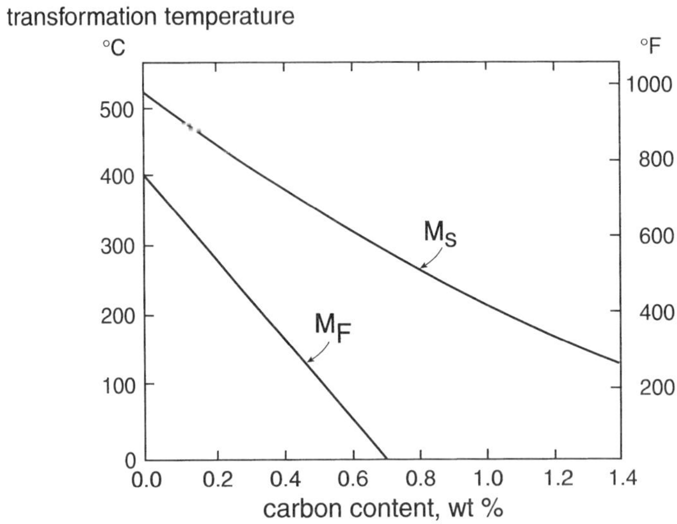

Figure 9.11. Plots of temperature for starting M_S and finishing M_F martensite formation on cooling austenite versus carbon content for plain carbon steels. (Based on data of E. Bain and H. Paxton.)

and carbon level. In some sintered steels it is possible to form martensite during cooling in the sintering furnace. This is of benefit since it eliminates postsintering heat treatments. The high hardness and strength of martensite provide major opportunities to strengthen a sintered steel. Unfortunately, most steels that are fully martensitic are low in toughness and can be as brittle as glass. Consequently, many quenched martensitic steels are reheated to improve toughness. This is the tempering process.

Martensite formation depends only on temperature during rapid cooling. Figure 9.11 plots the temperature of the start M_S and completion of martensite formation M_F versus carbon content for simple steels. Figure 9.12 further shows the amount of austenite and martensite versus temperature for an alloy of Fe–0.4C. The greater the carbon level, the lower the temperature needed to form martensite. If quenching is to a temperature above the martensite start temperature M_S, then ferrite and carbide will form. The sintered part must be cooled below the martensite finish temperature M_F to fully form martensite. Often this requires cryogenic cooling, since high alloying levels suppress the transformation temperatures.

The higher the carbon level, the lower the quench temperature needed to form martensite. In P/M steels this might be a problem since most alloys are formed from mixed-elemental powders. If the alloying ingredients are not homogeneous after sintering, then different compositions withing the sample will produce different phases, including retained austenite in the extreme. When properly formed, such mixtures result in exceptional properties, because of the dispersion of high-hardness regions in a high-toughness matrix.

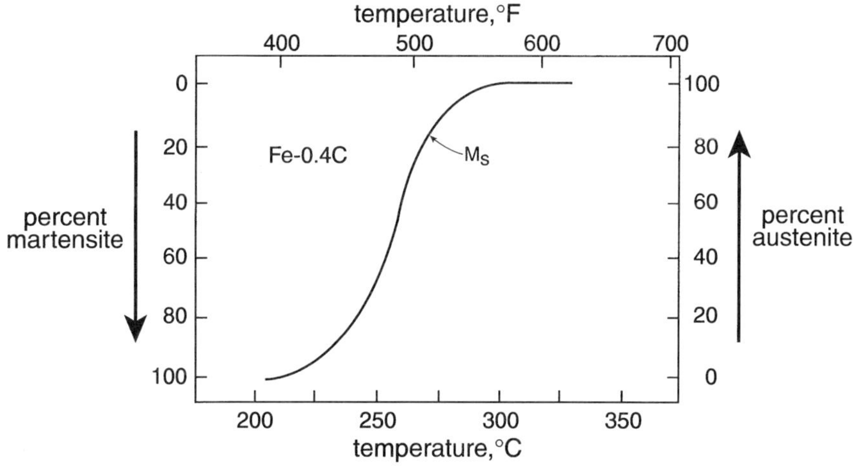

Figure 9.12. Temperature effects on the quenched austenite transformation into martensite for a hypoeutectoid steel (Fe–0.4C). The M_S temperature corresponds to the start of the transformation and is about 285°C (about 550°F) for this carbon content.

9.2.7. Transformation Diagrams

Transformation diagrams such as sketched in Figure 9.13 summarize several factors in one plot. This plot gives the percentage of each phase versus temperature and time. Such plots are widely accepted for graphically condensing transformation behavior.

Another version involves showing temperature and time, with curves representing the percent transformed. Two forms of this latter plot are in common use: time–temperature transformation (TTT) and continuous-cooling transformation (CCT) dia-

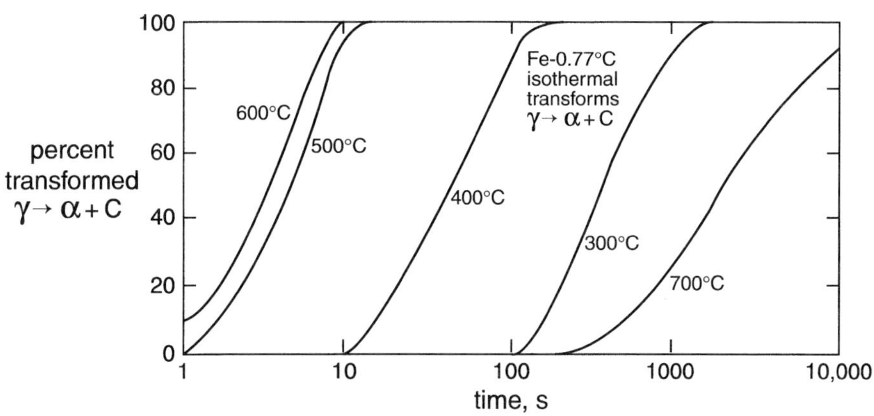

Figure 9.13. The time dependence of the eutectoid transformation of austenite into pearlite, a mixture of ferrite and cementite (α + C), for various temperatures.

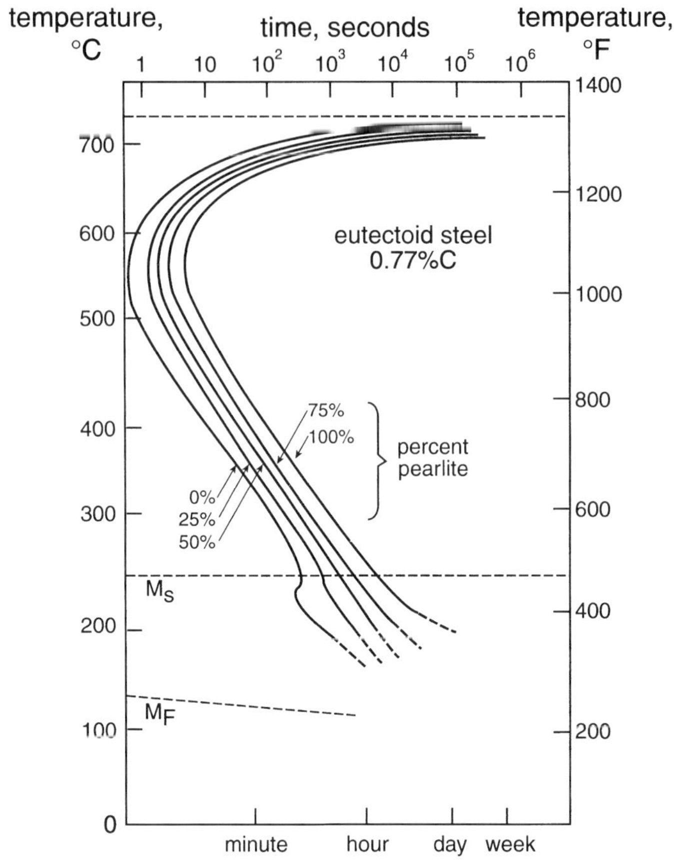

Figure 9.14. A common means for showing steel transformations is to plot the percent of each phase versus temperature and time on a TTT or CCT diagram (time–temperature transformation or continuous-cooling transformation). This is the TTT diagram for eutectoid steel (0.77% carbon).

grams. The plot given in Figure 9.14 is for eutectoid steel (0.77% carbon). Shown is a plot of temperature versus time on a logarithmic scale, mapping the conditions where various phases are observed. Above 910°C (1670°F) the structure is all-austenite. Assume that the steel is soaked in this temperature range sufficiently long to form homogeneous austenite. On cooling there are two potential paths. Slow cooling eventually forms pearlite at temperatures slightly below the eutectoid temperature. Between the start at 1% transformed and the finish at 99% transformed, the microstructure is changing but consists of a mixture of original austenite and newly formed pearlite, which is a mixture of ferrite and carbide.

The "nose" of the transformation diagram shows that with more rapid cooling cycles the onset of pearlite formation is at shorter times and lower temperatures.

Between the start and finish lines there is a proportionate increase in the amount of pearlite. Effectively, a steel captured between these two lines is continuously undergoing transformation and the start just represents the time to get organized while the finish shows the time needed to complete the transformation. This is like an auto racc. Bctween the green ''start'' flag and the checkered ''finish'' flag, all cars are at some point toward completion.

Now examine a rapid cooling path where the steel is instantly quenched. With a cooling rate over 500°C/s (900°F/s), cooling is sufficiently quick to miss the nose corresponding to the start of the pearlite formation curve. Now undercooling austenite temporarily exists at temperatures much below the eutectoid temperature of 727°C (1340°F). But it is unstable and eventually becomes either pearlite or bainite, or reaches the martensite start temperature M_S at about 220°C (about 430°F). Finally, on reaching room temperature, which is below the martensite finish temperature M_F at about 120°C (about 250°F), the remaining austenite transforms into martensite.

Figure 9.15 shows a CCT plot for a hypoeutectoid steel. There is a complication of proeutectoid ferrite formation. Given on the transformation diagram are the cooling rates associated with the five product regions. The hypereutectoid steels have a transformation nose that moves to longer times, making it possible to form martensite with slower cooling rates. From such plots it becomes evident that selected cooling rates can tailor a wide range of possible phases, microstructures, and properties.

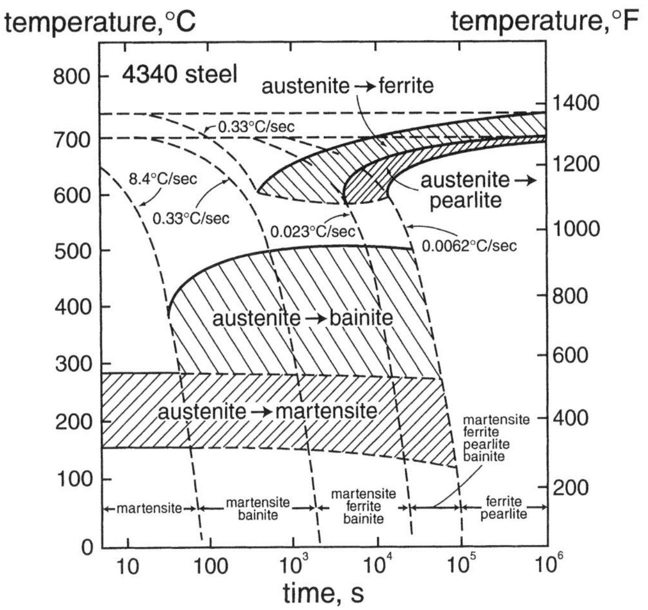

Figure 9.15. The constant cooling transformation CCT diagram for a hypoeutectoid 4340 steel. This plot shows the cooling rate effects on product microstructures observed at room temperature.

Alternative paths can be constructed to form bainite, or a mixture of phases such as partial formation of ferrite and carbide plus martensite. One example would be *up-quenching*, where a low temperature is first achieved to initiate the transformation into bainite, but to promote faster reactions, the temperature is progressively increased once bainite starts to form. Each path forms a different microstructure and gives different mechanical properties. This is the fun of heat treatment, but proves confusing on initial study.

9.2.8. Hardenability

Hardenability is an important idea in steels. Fundamentally, it is a measure of the ease in cooling a steel and still missing the nose of the ferrite–carbide transformation curve. On a practical level, it is usually measured by the depth in a quenched sample that still gives a high hardness. When quickly cooled at the surface, the steel far from the quenched surface undergoes slow cooling. Thus, it is more difficult to induce the martensite transformation at large distances from the quenched end.

The Jominy end-quench test is the most common means of assessing the hardenability. As sketched in Figure 9.16, a steel rod in the austenite condition is held in a water jet. Heat is rapidly extracted at the water-quenched end, but away from the water jet the steel cools slower. After the test, a sequence of hardness measurements are made on surfaces ground flat along the Jominy bar length. The change in hardness with distance provides a sense of the depth that martensite can form. At 700°C

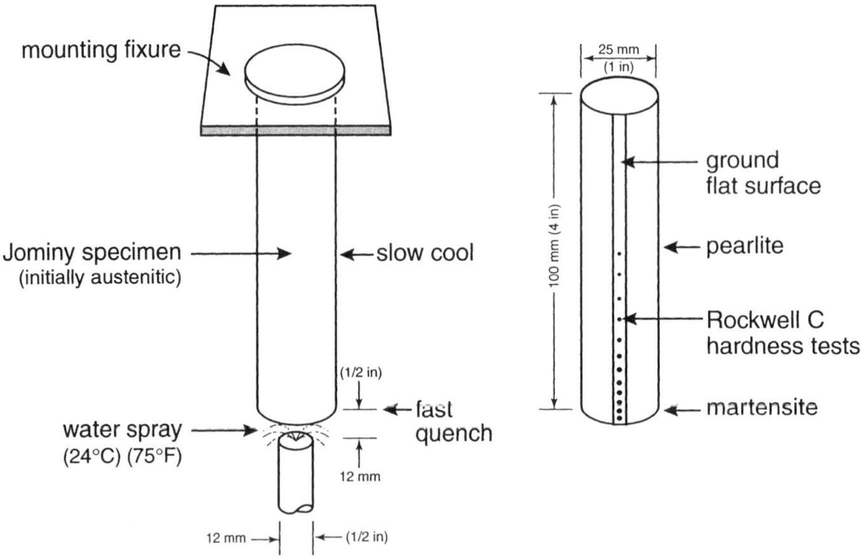

Figure 9.16. The Jominy end-quench test uses a 12-mm (0.5-in)-diameter rod in the austenite condition that is inserted into a water jet. Water cools only one end of the bar, so heat is rapidly extracted from the cold end, producing a range of cooling rates along the sample length for determination of hardenability.

EQUATION BOX 9.1

Several factors affect hardenability, including grain size and carbon content, but alloying effects are dominant. The most potent additives for steels are chromium, molybdenum, and manganese. Unfortunately these also require high sintering temperatures. Various indices are available to decide equivalent hardenability levels. For example, at an addition level of 0.5%, manganese is twice as effective as silicon, and chromium is twice as effective as nickel. A predictive technique uses multiplying factors to calculate the diameter of a bar that will produce 50% martensite at its center for a given alloy. The procedure requires that the alloying composition be known. Hardenability factor tables are consulted and a diameter of bar that will produce 50% martensite at the core is calculated as follows:

$$D_{50\%} = D_C F_{Ni} F_{Cr} F_{Mn} F_{Si} F_{Mo}$$

where D_C is the core diameter, which depends on the grain size and carbon level, and the F factors are determined from available tables, but depend largely on the alloying level. For example

$$F_{Mo} = 1 + 3.2\ X_{Mo}, \qquad F_{Cr} = 1 + 2.2\ X_{Cr}$$

where X indicates the concentration in weight percent. Such factors clearly show the potent effects from chromium, manganese, and molybdenum.

(1292°F), the cooling rate at the quenched end is probably about 500°C/s (900°F/s), while 20 mm (0.8 in) away the cooling rate is about 10°C/s (18°F/s), and at 40 mm (1.6 in) from the quenched end, the cooling rate is just 3°C/s (about 5°F/s).

High hardenability steels give a high hardness to a large depth. Thus, in spite of progressively slower cooling at large distances from the quenched end, the hardness is still high since the nose of the transformation to softer pearlite was missed. Alloying is a potent means to adjust hardenability. In P/M steels there is a difficulty because the alloying elements that provide a high hardenability also prove difficult to include in a steel sintered at 1120°C (2050°F). Consequently, sintering at temperatures between 1250 and 1300°C (2280 and 2370°F) is necessary to dissolve chromium or other desirable alloying ingredients.

9.2.9. Tempering

Since martensite is hard and brittle, it is directly useful only in applications where wear resistance is the main concern. Many P/M applications require densities over approximately 7.0 g/cm^3, with good impact, strength, and other attributes. For these reasons it is necessary to soften the martensite. This is achieved by tempering, a

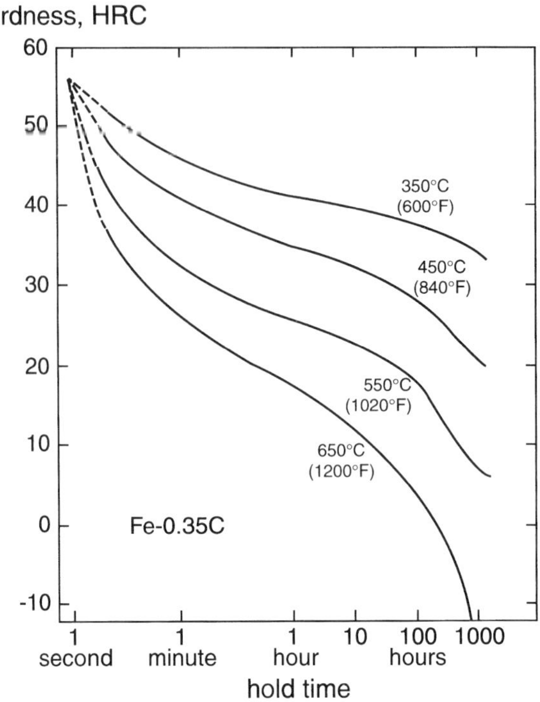

Figure 9.17. The change in hardness for a Fe–0.35C steel versus tempering hold time and temperature, after quenching to form martensite. The higher reheat temperature and longer hold time give a corresponding greater proportion of the softer ferrite and carbide mixture. (Based on data of E. Bain and H. Paxton.)

reheating that partially transforms martensite into ferrite and carbide. Depending on the desired properties, several different tempering treatments are employed after quenching, usually in the 200–500°C (about 400–900°F) range. Figure 9.17 plots the hardness on the Rockwell C scale against time for four tempering temperatures for a hypoeutectoid steel. Higher tempering temperatures and longer tempering times give a corresponding greater proportion of the softer phases. In turn, the strength and hardness decrease while the ductility improves.

One major difficulty with all heat treatments is dimensional change, since each phase has a different atomic volume. Any large-scale microstructure gradients lead to dimensional gradients and residual stresses. Consequently, it is necessary to sustain precise control of the heat treatment cycle to ensure minimized warpage or unpredicted dimensional change. Uniformity in heating and cooling during heat treatments are keys to fabrication of high-precision components.

9.3. EFFECTS OF RESIDUAL POROSITY

The heat treatment of P/M steels follows the patterns established for wrought materials, but is modified for the effects of residual pores. Pores reduce thermal conductivity, making it harder to extract heat fast enough to form martensite. Consequently, hardenability is reduced in proportion to porosity. Further, all mechanical properties are degraded by porosity, giving a lower hardness, strength, ductility, and toughness. Accordingly, there is greater difficulty in obtaining attractive properties at high porosity levels. Postsintering heat treatments are rarely applied to P/M steel components below a density of 6.8 g/cm^3.

Surface-connected pores trap quenching fluids that become contaminates in subsequent processing steps. A component with oil in the pores produces smoke on reheating and changes the carbon level during tempering; thus, thorough cleaning is required after quenching. An alternative is to use copper infiltration or repressing and resintering cycles to densify surface pores before heat treatment.

Hardenability is degraded by pores as illustrated in Figure 9.18. This figure plots the hardness on the Rockwell A scale against distance from the quenched end for a Fe–2Cu–0.8C alloy (FC-0208) Jominy test bar at three densities. The pores decrease the cooling rate and hinder martensite formation. Alloying is one option to offset this reduction in hardenablity. In a high-hardenability alloy, the hardness versus distance from the quenched end is flat, showing insensitivity to cooling rate. As illustrated in Figure 9.19, a high hardenability yields a high hardness to a large depth, especially as porosity is eliminated.

Certain alloying additions improve hardenability, making it easier to form martensite in a P/M steel. Common examples are manganese, chromium, molybdenum,

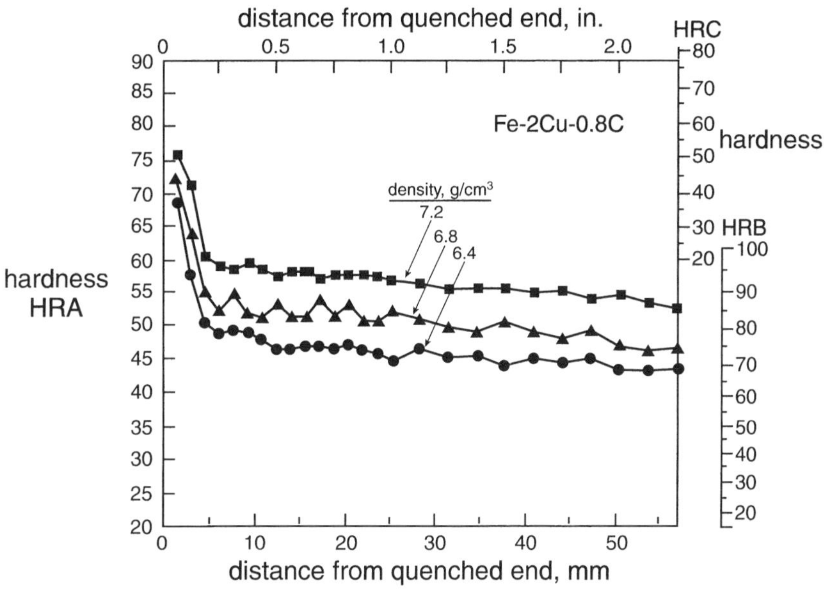

Figure 9.18. Jominy hardness traces versus depth for three Fe–2Cu–0.8C alloys (FC-0208) sintered at 1135°C (2075°F) and austenitized after sintering. Note the low hardenability of this alloy and the progressive decline in hardness at all depths at the lower densities.

and vanadium. Figure 9.20 plots the hardness on the Rockwell C scale against depth for common steels. A dramatic difference in hardenability is evident in this plot. For example, 1020, 1040, and 1060 steels respectively correspond to iron with 0.2, 0.4, and 0.6% carbon. On the other hand, the higher alloying levels associated with the 4140 and 4340 steels (both 0.4% carbon) give more uniform martensite formation.

Hardening the surface of a P/M steel improves wear resistance. Because the outer case is selectively hardened, the process is called case hardening. It is best done after all forming, sintering, and machining operations have been completed. Carbon is diffused into the surface from the processing atmosphere. Alternatively, case hardening of low-density P/M steel components commonly use carbonitriding to reduce carbon penetration depth via nitrogen additions. Figure 9.21 plots the Vickers hardness versus depth from the surface after one hour of vacuum carburization. Open pores increase the carbon penetration depth, since they allow the reactive atmosphere to penetrate deeper into the compact. As the porosity decreases, carbon penetration depends more on diffusion of carbon through the solid steel, a slower process. Thus, porosity has a mixed effect on the heat treatment of P/M steels; it increases case carburization depth, but decreases hardenability.

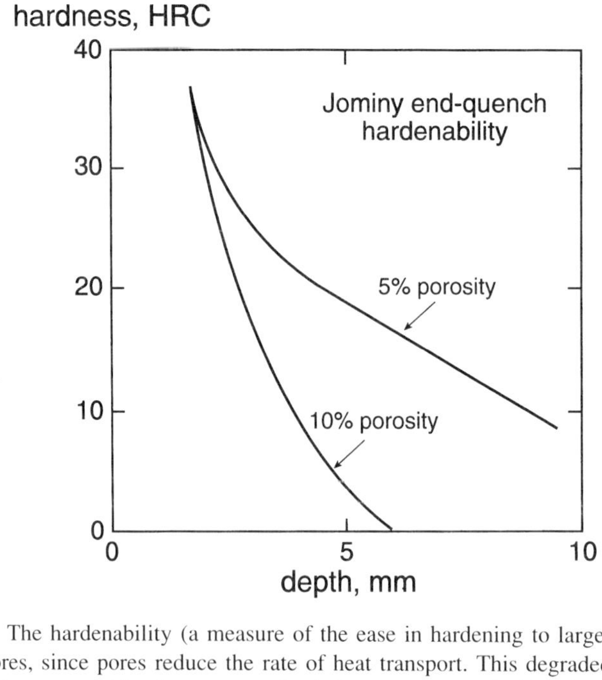

Figure 9.19. The hardenability (a measure of the ease in hardening to large depths) is degraded by pores, since pores reduce the rate of heat transport. This degraded hardenability is evident in the measured hardness versus depth for an end-quenched Jominy cylindrical test bar at two levels of porosity.

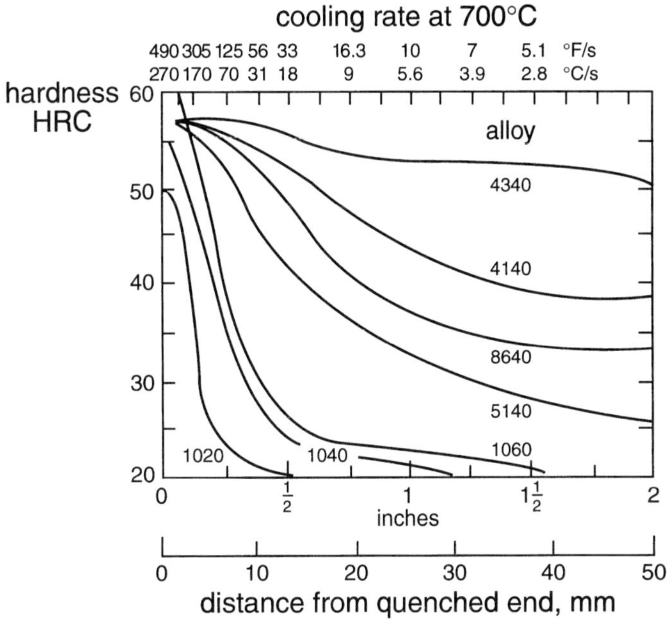

Figure 9.20. Jominy hardness traces for various common steels. These traces demonstrate the dramatic differences possible in hardenability in ferrous alloys. The 4340 steel consists of low levels of manganese, molybdenum, chromium, and nickel with 0.4% carbon, while the 1040 steel also has just 0.4% carbon.

9.4. COMPACT SIZE EFFECTS

Hardenability is a measure of the delays possible in cooling a steel while still forming martensite. A high hardenability implies easy martensite formation at slow cooling rates. This gives a high hardness deep into the component. Large structures take longer to cool, so a high hardenability is needed when heat treating large structures. Indeed, as the industrial revolution progressed, the ability to form high-strength steels in large section sizes was dependent on understanding hardenability.

One consequence of the component size effect is that large parts see different cooling cycles at the surface and core. The surface will be hard, but the core will be soft. Chromium is an effective hardening agent in ferrous P/M that offsets this problem. Figure 9.22 gives the hardness at both the surface and center of a 40-mm (1.6-in)-diameter rod containing 0.35% carbon and various chromium contents. Note the higher hardness associated with over 2% chromium, and the relatively close hardness at the surface and center. Typically such component size effects are not important to P/M, because the dimensions are small, ensuring uniform heating and cooling. However, when the section size exceeds a thickness of 25 mm (1 in), then problems occur.

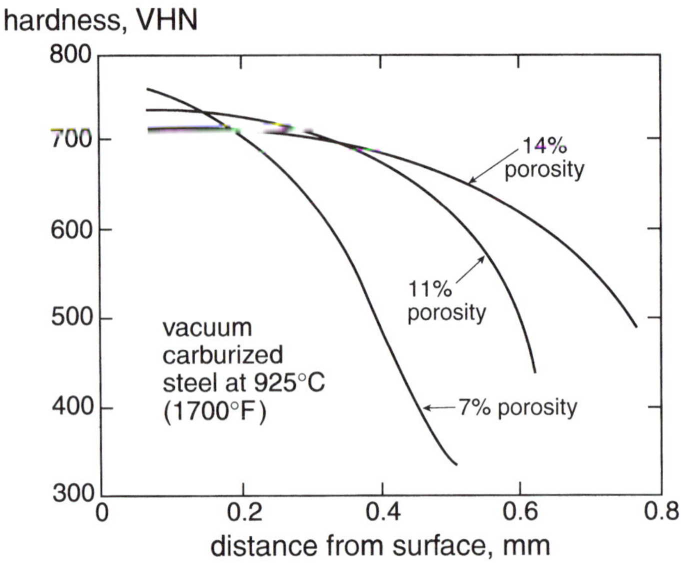

Figure 9.21. A high hardness surface is possible by vacuum carburization. The depth of this surface increases with porosity, since vapor penetration is easier with open pores, as illustrated by comparing the hardness traces for case-hardened steels processed at three porosity levels.

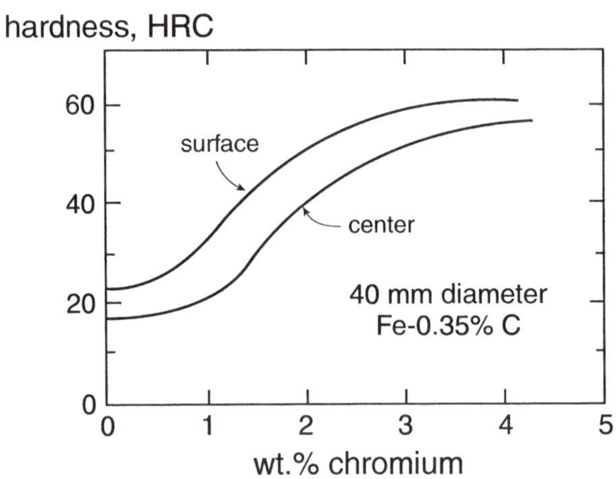

Figure 9.22. This plot shows the effective role of chromium in increasing the quenched surface and core hardness for a cylindrical steel sample 40 mm (1.6 in) in diameter containing 0.35% carbon. With just 2% chromium the alloy shows a high hardness at the component center.

9.5. ALLOYING EFFECTS ON PROPERTIES

9.5.1. Basic Effects

Alloying is employed in P/M steels to modify the mechanical properties, largely through the response to heat treatments. Both austenite and ferrite can be strengthened by alloying, but this is less important when contrasted with the shifts in heat treatment response. In addition, alloying can improve oxidation or corrosion resistance—such as with stainless steels. Some applications rely on alloying to secure special magnetic properties or high-temperature strength.

Alloying additions to a steel affect factors associated with heat treatment besides strength and hardness. For example, the eutectoid temperature and level of carbon at the eutectoid composition both depend on the alloying type and level. Most additions lower the carbon content at the eutectoid composition, but there is no systematic effect on the eutectoid temperature. As an example, nickel stabilizes austenite at room temperature when used at high concentrations. On the other hand, molybdenum increases the eutectoid temperature. For low-alloy steels, these composition effects are additive, so simple rules can determine the net changes from the combined alloying additions.

9.5.2. Phase Hardening

Alloying influences on hardness for ferritic steels are plotted in Figure 9.23. A range of responses is evident with phosphorous being very effective. From the standpoint

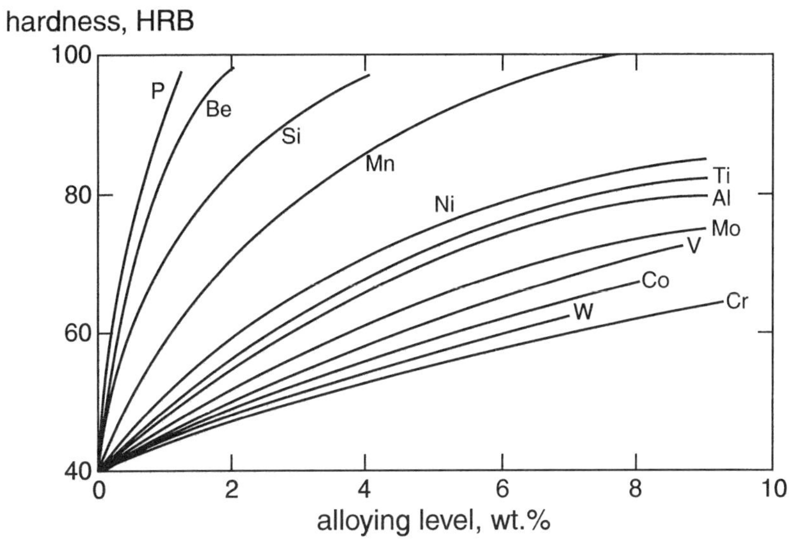

Figure 9.23. Hardness as a function of alloying level for several additions to a ferritic iron. The difference in effectiveness as a strengthening and hardening agent is traced to many factors, including atomic structure, atomic size, and chemical reactions in the steel. (Based on data of E. Bain and H. Paxton.)

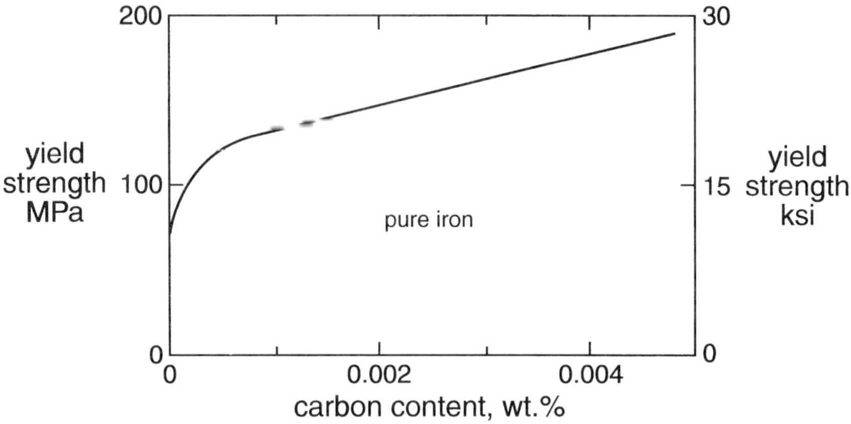

Figure 9.24. Tensile yield strength versus carbon content for ferrite (alpha iron), showing the initial strengthening due to low-alloying levels.

of effectiveness and cost, carbon is most useful. For example, Figure 9.24 plots the yield strength of ferrite versus the carbon content. Even with dilute additions there is substantial strengthening. As a further demonstration, Figure 9.25 gives the hardness of martensite as a function of the carbon level. A plateau occurs at slightly more than 65 HRC with about 0.6% carbon. For comparison, this figure also plots the hardness in pearlitic and spheroidized steels, showing less hardening of those phases.

Generally, additions that concentrate in ferrite, such as phosphorus, nickel, silicon, and aluminum, are applicable to strengthening P/M steels used without a heat treatment after sintering. Alloying additions that concentrate in the carbide, such as titanium, niobium, tungsten, and vanadium, are not as widely used in P/M, since the controlling factor is the transformation of austenite to martensite, rather than the hardness of the carbide.

9.5.3. Transformation Shifts

As alloying increases, the transformation diagrams become more complicated. As evident in Figures 9.26 and 9.27, a second transformation nose arises; one is associated with the pearlite reaction and the other, with the bainite reaction. For example, Figure 9.26 gives the transformation diagram for a 4340 steel, an alloy of 1.79Ni–0.80Cr–0.78Mn–0.33Mo–0.4C. Cooling at a rate of greater than 8.4°C/s (15°F/s) produces all martensite, while cooling slower than 0.0062°C/s (0.01°F/s) produces all pearlite. Intermediate cooling rates produce mixtures of martensite, bainite, and pearlite.

In contrast, Figure 9.27 shows the effect of a high carbon level where cementite is the first phase to form on cooling, producing a narrow region of austenite and cementite. Subsequently, on slow cooling the final structure is a mixture of proeutectoid cementite and pearlite. With rapid cooling the product is martensite. For each alloy, there is a critical rate that is the slowest constant cooling rate to produce 100% martensite. Figure 9.28 plots this critical cooling rate versus the carbon level as a function of carbon level, for full density (0% porosity) and for 10 and 20% porosity. The lower thermal conductivity of the porous material reduces hardenability. Because of this penalty from porosity, it is fruitless to heat-treat low-density P/M compacts.

To increase strength, the most useful change is to delay the transformation of austenite to pearlite, thereby assisting martensite formation with slow cooling. As an example, the effect of manganese is illustrated in Figure 9.29. This figure plots the time for 50% transformation of austenite to martensite at two temperatures versus the manganese level for a 0.55% carbon steel. The greater the manganese content, the longer the time before pearlite begins to form. However, the martensite starting temperature declines as the carbon level increases. As demonstrated in Figure 9.30, it can even move below room temperature with high carbon and manganese alloying

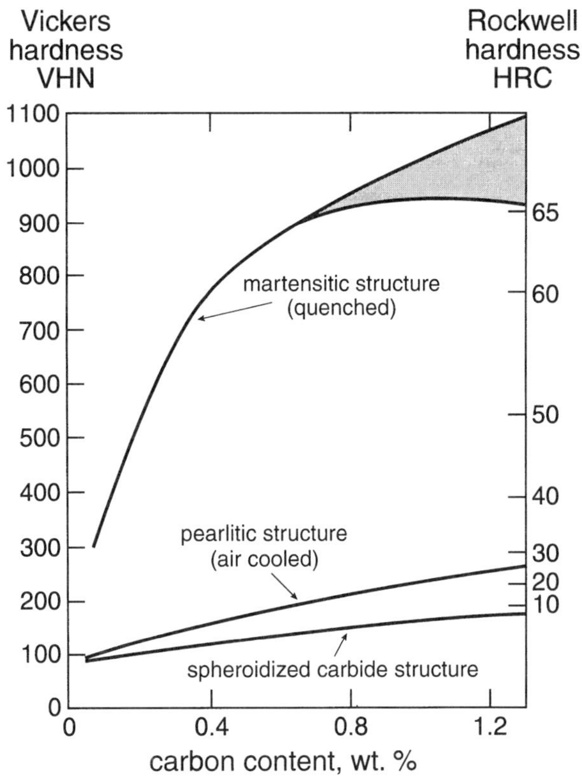

Figure 9.25. The hardness of martensitic, pearlitic, and spheroidized steels versus the carbon content. Martensite is much harder, but the hardness is unchanged by high carbon levels. (Based on data of E. Bain and H. Paxton.)

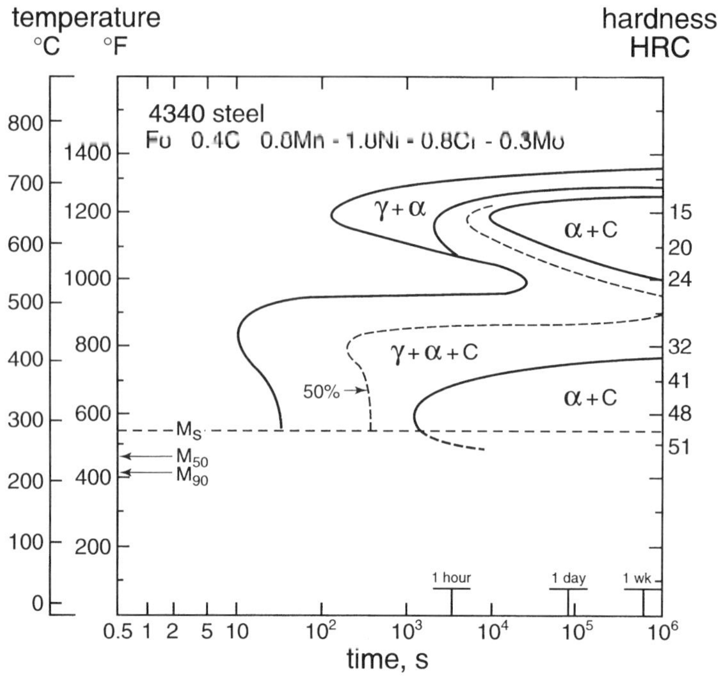

Figure 9.26. A transformation diagram for the cooling of a hypoeutectoid steel, in this case 4340 steel (Fe–1.8Ni–0.8Mn–0.8Cr–0.3Mo–0.4C). With very fast cooling, austenite directly transforms into martensite starting at about 300°C (about 550°F). With slow cooling, the first transformation product is ferrite as the proeutectoid phase and subsequent formation of eutectoid pearlite, a mixture of ferrite and cementite. At intermediate cooling rates the proeutectoid reaction is avoided, but the final product is still a mixture of ferrite and cementite. (Data from George Vander Voort.)

levels. Strength increases, and the ductility decreases as the manganese content increases.

9.6. SURFACE TREATMENTS

The heat treatment atmosphere must protect the steel against oxidation. Carbon or nitrogen can be added to the component during heat treatment by mixing methane, carbon monoxide, or ammonia into the atmosphere. These same atmospheres can be used to improve surface hardness or adjust the surface properties of sintered steels. Some special surface heat treatments are used for P/M steels, including oxidation in steam.

9.6.1. Steam

Rust formation on steels during service is avoided by growing an intentional film of black oxide during heat treating in a steam atmosphere. Steam treatment occurs by first heating in a dry atmosphere at temperatures below approximately 370°C (near 700°F). Then it is possible to form a stable and protective iron oxide called *hematite* (Fe_3O_4) at higher temperatures by introducing steam into the heat treatment furnace. As steam is introduced, the temperature is increased to approximately 520–570°C (near 1000°F), creating a tenacious surface oxide according to the reaction

$$3\ Fe\ (s) + 4\ H_2O\ (g) \rightarrow Fe_3O_4\ (s) + 4\ H_2\ (g)$$

Note the possible explosive result of the reaction due to the generation of hydrogen gas. If the temperature is too high, a nonprotective phase forms, so proper temperature selection is critical to the formation of hematite. Further, the process must be done in an atmosphere free of air and oil to avoid contamination. The hematite forms a strong surface coating and seals the surface pores as sketched in Figure 9.31. Such treatments are applied to P/M components used in rifles and shotguns, sporting

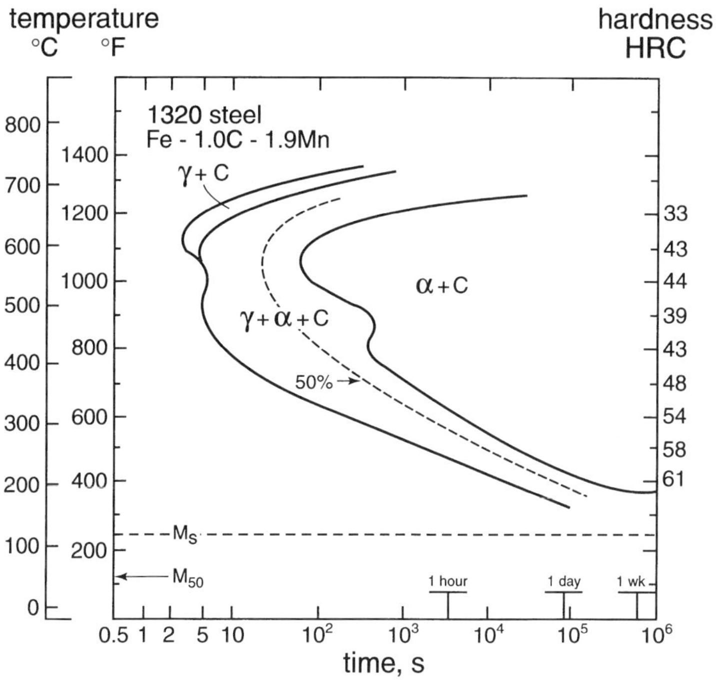

Figure 9.27. A transformation diagram for the cooling of austenite for a hypereutectoid steel, in this case Fe–1C–1.9Mn. Except very fast cooling rates, the first transformation product is cementite, the carbide of iron. With slow cooling, both cementite and ferrite form, giving a microstructure consisting of proeutectoid cementite and eutectoid pearlite. Only with rapid cooling can cementite be avoided to form martensite. (Data from George Vander Voort.)

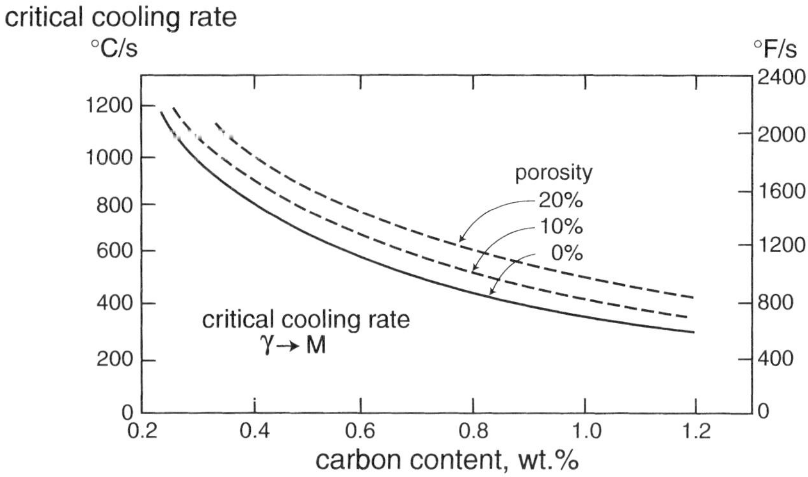

Figure 9.28. The critical cooling rate for complete transformation of austenite to martensite for a range of carbon contents. Slower cooling rates allow pearlite formation. Residual porosity mandates faster cooling due to lower thermal conductivity.

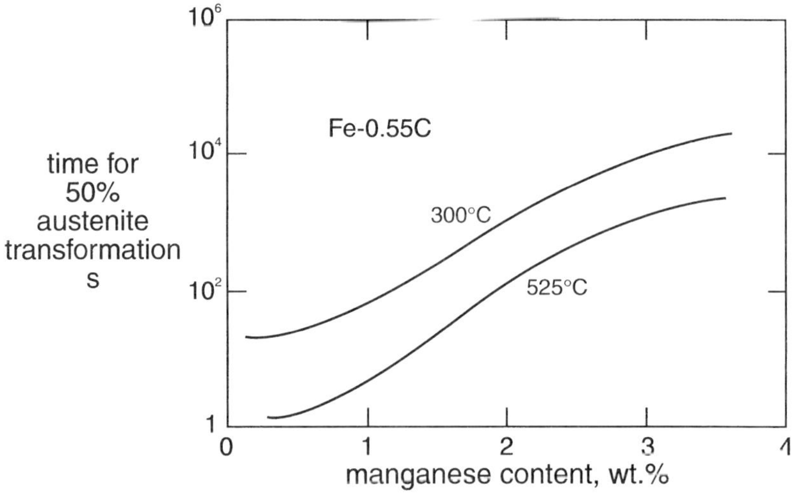

Figure 9.29. A desirable situation for high-strength steels is to delay the transformation of austenite to pearlite, thereby allowing formation of martensite with slower cooling rates. These plots give the effect of manganese alloying on the time for the transformation at two temperatures. The greater the manganese content, the longer the time before pearlite begins to form.

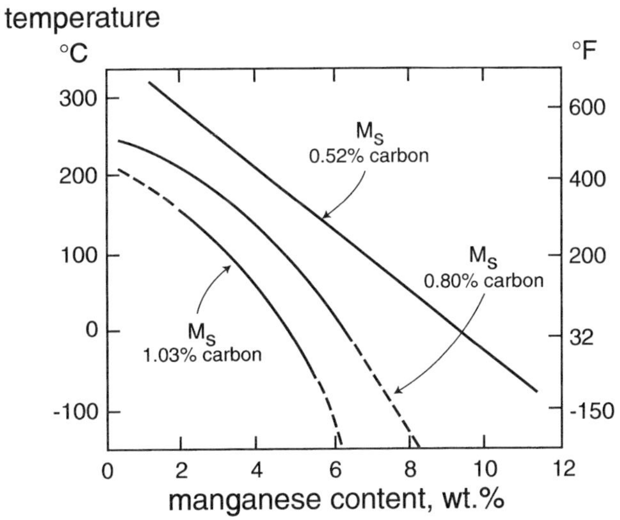

Figure 9.30. The martensite start temperature is suppressed by the addition of manganese, especially at higher carbon levels. (Based on data of E. Bain and H. Paxton.)

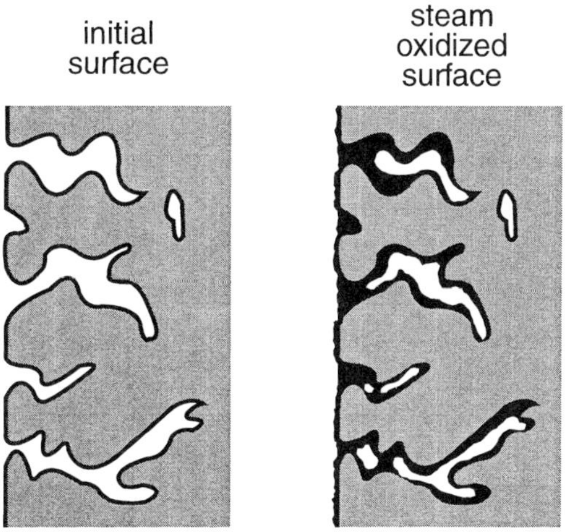

Figure 9.31. Steam sealing of the pores must be done in an atmosphere free of air and oil. At the optimal temperatures, hematite grows to a thickness of 0.5 mm (0.02 in) over the first 3 h to form a strong surface coating that seals the near-surface pores.

equipment, automobiles, appliances, office equipment, and consumer products. The adherent oxide prevents rusting and improves wear resistance.

Steam treatments can be performed in a batch mode or even in continuous-belt furnaces with loads of 100 mm (4 in) depth. For a furnace with a belt width of 300–400 mm (12–16 in), the output is generally 100–300 kg/h (220–660 lb/h) with a thick oxide coating. It is common to use forced gas cooling to ensure the preservation of the desired oxide after steam treatment. Color is probably the best measure of success—with both the deep black and blue-black colors are most desirable.

9.6.2. Carburizing

In another common surface heat treatment *case carburizing*, a high carbon content is incorporated in surface regions. Up to 1% C can be deposited from a reaction sustained by a carbon source in the heat treatment atmosphere. To obtain the desired carbon addition, the P/M components are heated for 60 min at 850–950°C (1560–1740°F) in atmospheres containing methane, butane, or propane, followed by an oil quench. The depth of added carbon depends on the porosity, as illustrated in Figure 9.32 for case-hardened steel.

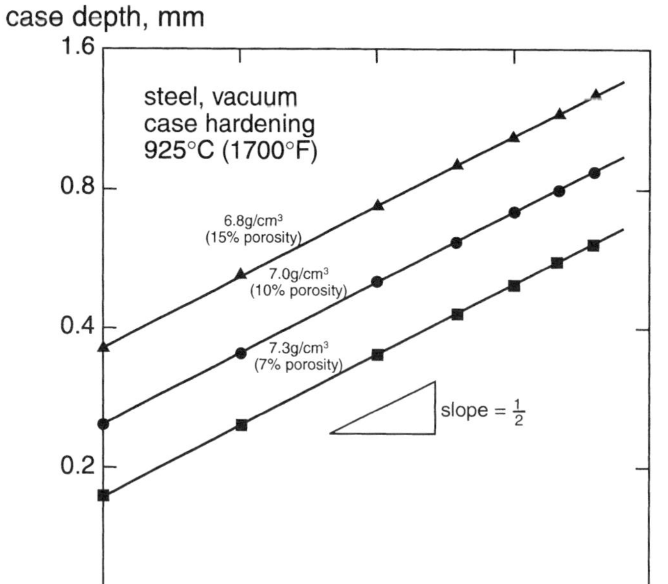

Figure 9.32. The depth of hard case increases with the square root of time during vacuum carburization. The depth also depends on the porosity, since vapor penetration is easier with open pores, as illustrated by comparing the results on case hardened steels processed at three porosity levels. (Based on data of Daniel Herring.)

EQUATION BOX 9.2

Case carburization occurs by the penetration of carbon into the surface from an atmosphere source. The depth of carburization X depends on the square-root of hold time t for a given situation:

$$X = K\sqrt{t}$$

The factor K is a rate constant and it depends on the temperature and density. Usually the case depth is measured by the distance into the compact that achieves a hardness of at least 50 HRC. As an example of the carburization rates in vacuum, a depth of 0.5 mm is formed for a 7.0 g/cm^3 compact heated at 840°C (1545°F) for one hour. Alternatively, the same compact density gives a depth of 0.6 mm after one hour at 925°C (1700°F).

Case carburization is not recommended when the porosity exceeds 15% (about 6.7 g/cm^3) or the base carbon level exceeds about 0.4%. With high carbon levels there is no benefit. The carburizing temperature must be high enough to stabilize austenite [typically in the 900–930°C (1650–1700°F) range]. During surface carburization, the depth of the carbon layer increases with the square root of the processing time. This means quadrupling the time for every doubling of the desired case depth. For example, if 40 min at temperature provides a 0.4-mm (0.016-in) case of high carbon, then time must be quadrupled to double the case depth; 160 min is required for a depth of 0.8 mm (0.032 in). The higher the temperature, the faster the case carburizes. One benefit is in improved fatigue strength due to the hard surface. Most carburization treatments are done in a protective atmosphere, but it is possible to use a vacuum furnace as well.

9.6.3. Carbonitriding

An effective surface treatment for P/M steels is *carbonitriding*, a treatment involving doping of the atmosphere with ammonia. Nitrides form on the compact surface as carbon and nitrogen are adsorbed into the steel to form martensite during cooling. The typical treatment temperature is in the 800–850°C (1470–1560°F) range for 30–60 min. The typical nitride depth is 0.5 mm (0.02 in).

A related technique is termed *nitrocarburization*, where nitrogen and carbon are reacted and deposited at the component surface. This technique is useful only for high-density components, typically more than 7.0 g/cm^3. Processing temperatures are between 570 and 600°C (near 1100°F) with an atmosphere of 50% ammonia and 50% endothermic gas (an air–methane mixture). The surface reaction slightly expands the component, but the greatly improved wear resistance is critical for sliding-wear applications. A typical surface hardness of 60 HRC is measured after nitrocarburization.

9.6.4. Induction Hardening

Induction hardening is useful for gears and objects where surface strength, hardness, and wear resistance are critical. The component core is usually a tough, ductile structure such as pearlite or bainite formed during cooling from the sintering temperature. As suggested in Figure 9.33, an induction coil is placed around the component. In this coil the oscillating magnetic field from a high-frequency electric current induces electric current in the sintered structure. Only the near-surface regions are heated; thus, the core remains largely unaffected while the surface heats to form austenite. When the component is quenched, the surface transforms into hard martensite. However, the core remains ductile and tough, since it never became very hot.

Because of the severe temperature changes, distortion or surface cracks form with water quenching. Therefore, oils and water-soluble polymers (such as acrylics) have been developed to mitigate the cooling rate. Induction hardening is especially useful

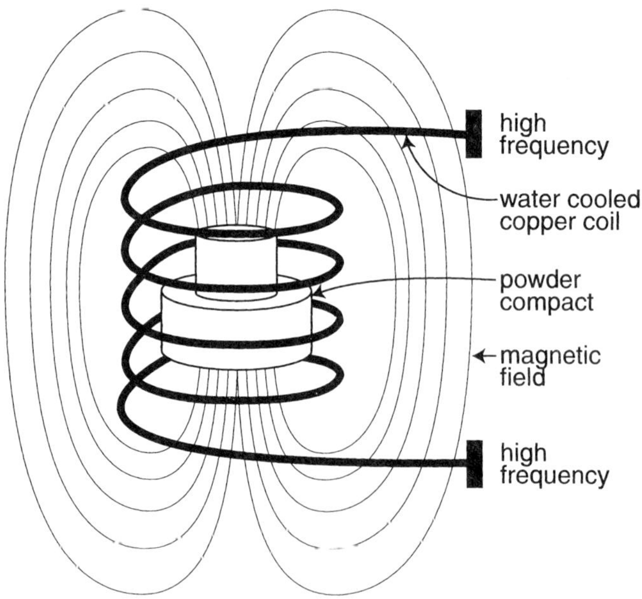

Figure 9.33. Induction heat treatment occurs by localized heating. The induction coil is placed around the component. In this coil, the oscillating magnetic field from high-frequency electric current leads to induced electric current flow in the part. Since only the near-surface regions are heated, the core remains unaffected while the surface heats and passes into austenite. When quenched, the surface transforms into hard martensite while the core remains ductile and tough.

TABLE 9.2. Time to Move a Part from the Heat-Treating Furnace into a Quenching Fluid

Density (g/cm^3)	Transfer Time (s)
6.4–6.8	8
6.8–7.2	12
≥7.2	25

for P/M components, because the pores reduce heat transport, making it easier to form a hard surface. For very large components, the unheated mass in the core is sufficient to quench the surface once the induction coil is turned off. Carbon can be added via the atmosphere surrounding the compact during induction heating.

9.7. PRACTICAL ASPECTS OF HEAT TREATMENT

9.7.1. Cooling Time and Rate

As the density of the P/M steel increases, the allowed time to move the part from the furnace to the quenching fluid increases. Table 9.2 provides a summary of these times between furnace exit to immersion in the quenching fluid. As a first recommendation, about 22–30 liters of quenching fluid is required for each kilogram of parts (about 3 gallons per pound). Because of porosity, it is necessary to remove any quenching oil before tempering. Similarly, if machining is performed, then oil in the pores must be removed prior to subsequent reheating. Residual oil will decompose, adding carbon to the process atmosphere, discoloring the material.

9.7.2. Mechanical Properties

Hardness is probably the most important measure of successful heat treatment. For a desired carbon level, a narrow range of mechanical properties will be associated with each hardness. As the cooling rate from the austenitization temperature increases and more martensite forms, the hardness climbs. Figure 9.34 plots the combinations of mechanical properties for a slow-cooled eutectoid steel (0.77% carbon) versus the final hardness. The strength increases, but the ductility declines. Indeed, plots of strength versus ductility usually reinforce this tradeoff between these two properties.

Pure iron is unaffected by quenching. However, as illustrated in Table 9.3, steels are greatly altered by cooling rates and heat treatments. Note the higher strength and lower ductility following quenching. If the quenched steel were tempered, the strength would decrease and the ductility would increase over time. As plotted in Figure 9.35, for 1 h tempering treatments at various temperatures, the hardness falls, but the overall steel is more ductile, tough, and fracture resistant.

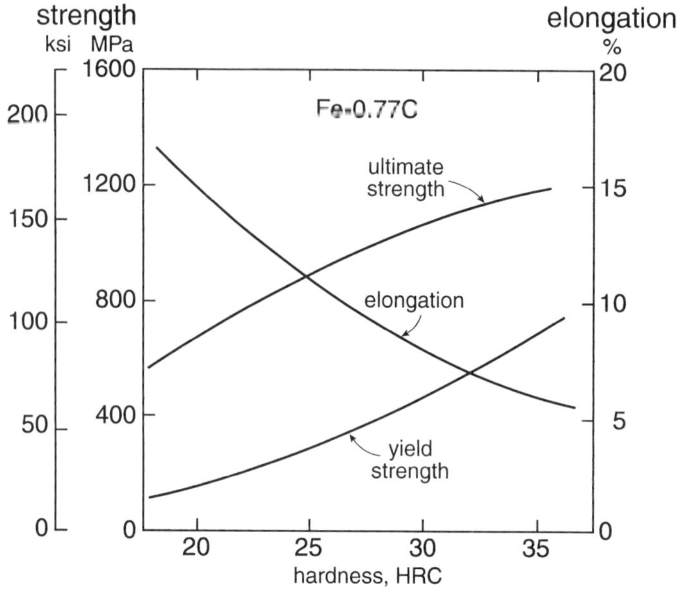

Figure 9.34. Strength, elongation, and hardness are related to one another for a typical eutectoid steel. This figure shows the loss of ductility associated with a heat treatment that produces a high hardness and strength.

9.7.3. Dimensional Control

Distortion is an unfortunate byproduct of heat treatment. Stresses from temperature gradients lead to distortion. Hot regions are softer and deform from thermal stresses induced by cold regions in the same part. Higher carbon levels or thicker sections result in poorer dimensional control. Rapid heating and rapid cooling increase the dimensional control problems. Slow heating helps avoid distortion, but often results in lower mechanical properties. Alternatively, heat treatments, especially quenching, induce stresses that can produce considerable distortion. Thus, one justification for alloying additions is to slow the necessary cooling rate to help reduce distortion. Although distortion might not be eliminated, the improved dimensional control and gains in the mechanical properties make a clear case for use of alloyed steels if postsintering heat treatment is required. Cost is the offsetting factor, since high hardenability compositions involve costly alloying additions.

TABLE 9.3. Cooling Rate Effects on Properties of an Eutectoid Steel (0.77% C)

Property	Slow-Cooled	Water-Quenched
Yield strength, MPa (ksi)	428 (62)	1380 (200)
Tensile strength, MPa (ksi)	773 (112)	1380 (200)
Elongation, %	10	1

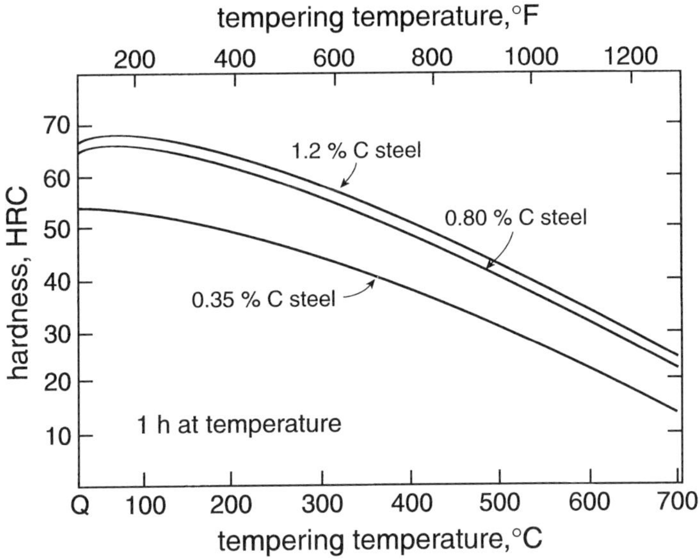

Figure 9.35. Hardness shown as a function of tempering temperature for three carbon levels for constant tempering times of 1 h. High carbon levels are ineffective in promoting higher harnesses.

9.7.4. Carbon Control

Carbon control is an important aspect of successful heat treatment. Repressing and resintering improve density and may be needed to close pores to reduce internal carburization during case hardening. If the carbon level is less than 0.2%, then case carburization is probably more effective in improving strength as compared with any quench and temper treatment.

For some stainless steels, it is desirable to eliminate carbon to ensure good corrosion resistance. In most structural steels the carbon level is adjusted in the initial powder mixture to hit a target value for the strength and ductility combination desired for an application. Proper control requires anticipation of decarburization from any oxygen in the powder or process atmosphere. In all heat treatments, it is important to minimize furnace overloading and to avoid overlapping of parts to ensure that uniform heat transfer occurs to give the same hardness to all regions.

9.7.5. Sinter-Hardening

Sinter-hardening is a concept where heat treatment is incorporated in the cooling cycle of the sintering furnace. The usual goal is to cool quickly to avoid the formation of pearlite and to generate a mixture of bainite and martensite, giving good strength and toughness. The critical cooling rate for sinter-hardening is usually between 1 and 8°C/s (2 and 14°F/s) over the temperature range from 850 to 200°C (1560 to

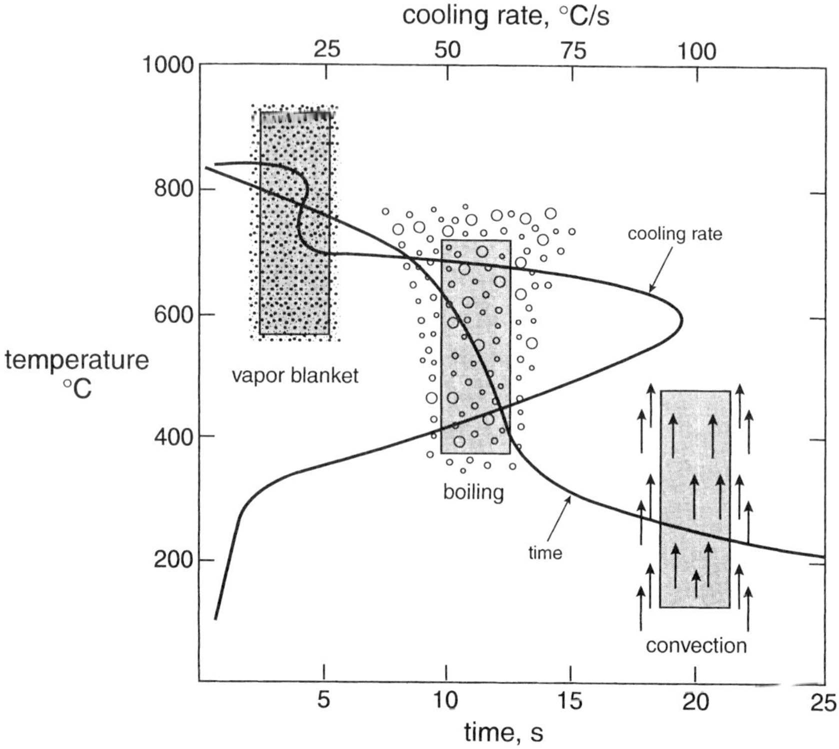

Figure 9.36. Three mechanisms contribute to cooling during quenching, with the highest cooling rate occurring at intermediate temperatures where boiling occurs at the component surface, typically near 600°C.

400°F). It is used mostly for high-density steels that exhibit good thermal conductivity. Compositions with chromium and manganese are favorites, but these alloys require sintering at temperatures in the range of 1200–1300°C (2200–2400°F).

9.7.6. Quenching Fluids

The cooling rate during quenching depends on the quenching medium. It is slow for air quenching, faster in salt baths, and progressively faster in oil and water baths. Usually, boiling occurs at the interface between the hot component and the cool fluid. As sketched in Figure 9.36, the highest cooling rate occurs near 600°C (near 1100°F), when the fluid boils from the compact surface. Polymer additions to the quench water control boiling and ensure uniform cooling. The polymers include most of the common water-soluble types, such as polyethylene glycol. Agitation of the quenching medium assists in cooling. Molten salt baths are problems with porous materials, since the salt can be trapped in open pores. The trapped salts will attack the steel over time, resulting in delayed corrosion.

TABLE 9.4. Examples of Heat Treatments for Common P/M Steels[a]

Alloy	Processing Route	Processing Cycle (°C/h/Gas)	Density (g/cm^3)	Heat Treatment	Hardness	Properties
17-4 PH stainless steel	Injection mold	1350/2/H_2	7.48	Soak 480°C, 1 h (soak 900°F, 1 h)	34 HRC	σ_Y = 923 MPa (134 ksi) σ_U = 969 MPa (140 ksi) ϵ_F = 7%
2Cu–0.5C	Die compaction	1150/0.5/N_2–H_2	6.8	Soak 1000°C, 1 h, WQ, temper (soak 1832°F, 1 h)	31 HRC	σ_U = 620 MPa (90 ksi) ϵ_F = 0.5%
2Cu–0.5C	Die compaction	1150/0.5/N_2–H_2	7.2	Soak 1000°C, 1 h, WQ, temper (soak 1832°F, 1 h)	36 HRC	σ_U = 700 MPa (100 ksi) ϵ_F = 0.5%
2Ni–0.8C	Die compaction	1150/0.5/N_2–H_2	6.7	Soak 1000°C, 1 h, WQ, temper (soak 1832°F, 1 h)	26 HRC	σ_U = 620 MPa (90 ksi) ϵ_F = 0.5%
2Ni–0.8C	Die compaction	1150/0.5/N_2–H_2	7.0	Soak 1000°C, 1 h, WQ, temper (soak 1832°F, 1 h)	35 HRC	σ_U = 1000 MPa (145 ksi) ϵ_F = 0.5%
2Ni–0.8C	Die compaction	1150/0.5/N_2–H_2	7.4	Soak 1000°C, 1 h, WQ, temper (soak 1832°F, 1 h)	42 HRC	σ_U = 1340 MPa (194 ksi) ϵ_F = 0.5%
2Ni–0.5C	Injection mold	1250/4/H_2	7.71	Soak 850°C (1562°F), 1 h, oil-quench, temper 205°C (401°F), 1 h	45 HRC	σ_Y = 1338 MPa (194 ksi) σ_U = 1722 MPa (250 ksi) ϵ_F = 9%
2Ni–0.5Mo–0.2Mn–0.4C	Injection mold	1300/1/N_2–H_2	7.5	Soak 900°C (1652°F), 0.5 h, oil-quench, temper 350°C (662°F), 1 h	45 HRC	σ_U = 1500 MPa (214 ksi) ϵ_F = 8%
2Ni–0.4C	Injection mold	1250/2/H_2	7.5	Soak 850°C (1562°F), 1 h, water-quench, temper 340°C (644°F)	31 HRC	σ_Y = 980 MPa (142 ksi) σ_U = 1125 MPa (163 ksi) ϵ_F = 6%
8Co–6W–5Mo–4Cr–3V–1.3C	Hot isostatic press	1150/1/200 MPa, Ar	8.0	Soak 1180°C (2156°F), quench, temper 560°C (1040°F), 1 h repeat 3×	67 HRC	—
18Cr–3V–1Mo–1Si–1.3C	Hot isostatic press	1150/3/200 MPa, Ar	7.8	Soak 1050°C (1922°F), 30 min, oil-quench, temper 400°C (752°F), 2 h repeat 2×	61 HRC	σ_Y = 2300 MPa (333 ksi) σ_U = 3000 MPa (434 ksi)

[a] *Abbreviations:* WQ = water-quench; σ_Y = yield strength; σ_U = ultimate tensile strength; ϵ_F = fracture elongation.

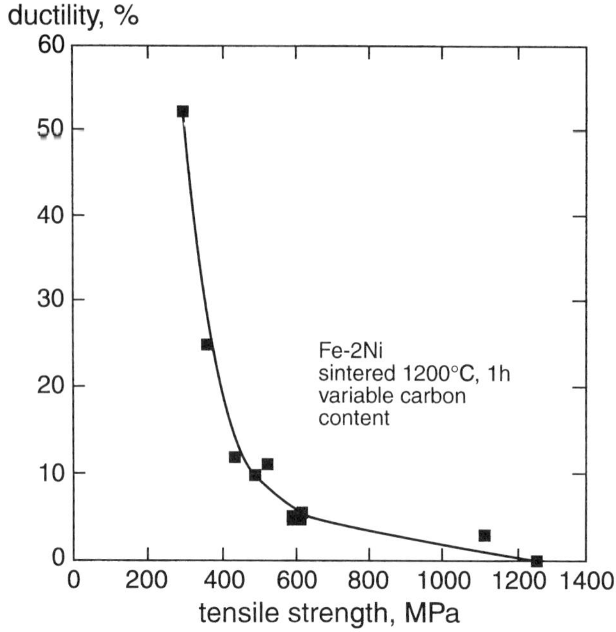

Figure 9.37. Plots of the effect of tempering temperature on strength and ductility following sintering at 1200°C (2192°F) for 1 h. Generally the strength has an inverse relation to ductility. (Data of Haorong Zhang and S. T. Paul Lin.)

The selection of the quenching fluid depends on the alloying level (which determines the TTT diagram), desired final mechanical properties, and component section size. The path on the transformation diagram determines the final microstructure and mechanical properties. As sketched in Figure 9.15, the range of cooling rates is broad, and the range of final phases depends on the cooling rate.

9.8. PROPERTIES ATTAINABLE THROUGH HEAT TREATMENT

The control parameters for final properties are the density, carbon content, microstructure, and heat treatment cycle. Table 9.4 summarizes cycles employed with P/M steels, giving the alloy, fabrication route, density, heat treatment cycle, resulting hardness, and other mechanical properties. The tabulation is only a first glimpse at the complexity of steels, cycles, and properties possible via heat treatment. These representative properties show that ductility suffers when a porous P/M steel is heat-treated to a high strength. Therefore, much current emphasis is given to high-density products via forging, hot isostatic pressing, high-temperature sintering, or injection-molding routes.

This chapter outlines the principles for heat-treating steels and tailored these principles to P/M compacts. Pores lower the thermal conductivity, retard hardening reactions, and can be problematic with respect to trapped quench fluids. If the pores are absent, then heat treatment of a P/M steel follows principles recognized for a bulk steel of the same composition. Consequently, an injection-molded steel behaves much like a wrought counterpart in terms of heat treatment response. For example, an iron–nickel steel (Fe–2Ni) illustrates the property interchange in tempering. Figure 9.37 shows the interplay between tempered strength and ductility following sintering at 1200°C (2192°F) for 1 h. This scatterplot emphasizes that heat treatments producing a high strength give a low ductility, and vice versa. Figure 9.38 contrasts the final microstructures, showing a case with no ductility (100% martensite) and one with a mixture of phases and a higher ductility, but lower strength.

Figure 9.39 provides a summary of the cooling paths, cooling fluids, final microstructures, and transformation events on cooling. This transformation diagram shifts with porosity, but the major features are evident.

As a final comment, exceptional mechanical properties are possible in heat-treated P/M products. High property levels have been attained via manipulation of the process atmosphere and final microstructure. As an example of such manipulation, Professor Miura of Kumamoto University in Japan has produced a heterogeneous steel via powder injection molding using mixed-element powders (iron, nickel, molybdenum, and manganese) with strength levels between 1500 and 1900 MPa (214 and 275 ksi), fracture elongations ranging from 1 to 8%, and fatigue endurance limits of 570 MPa (83 ksi). The alloy was tempered at just 350°C (662°F) for 1 h.

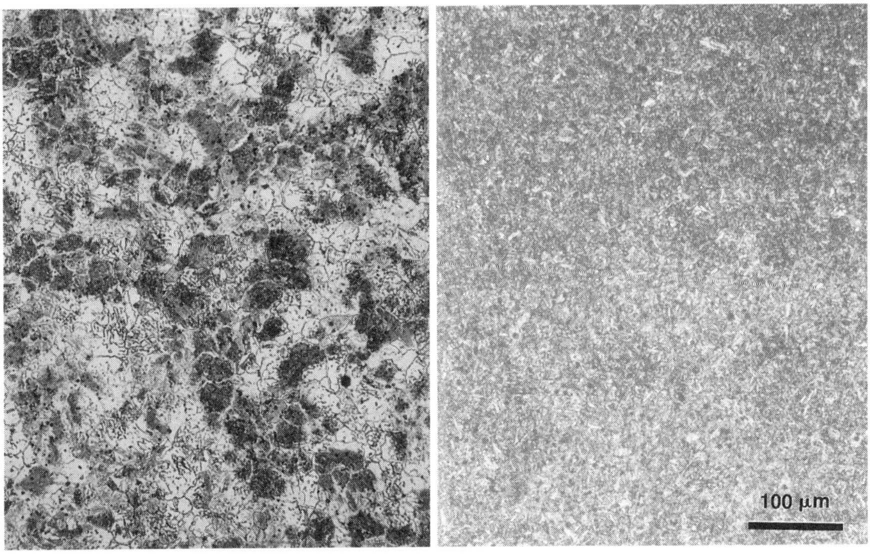

Figure 9.38. Optical micrographs providing a contrast in the final microstructures, showing a case with a mixture of phases that give a high ductility, but low strength, and a case with no ductility (100% martensite).

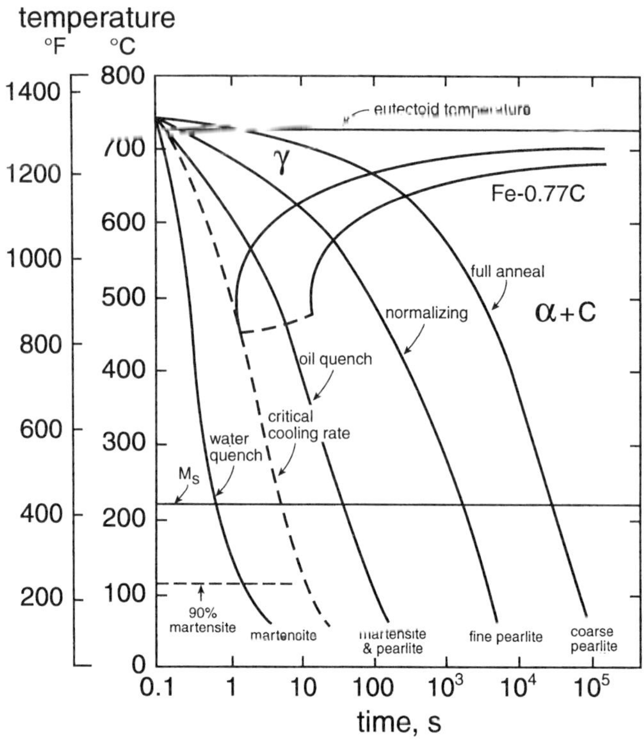

Figure 9.39. The transformation diagram for an eutectoid steel, mapping out the cooling medium effects on the final microstructure, showing how each path passes through the transformation diagram.

With controlled nitride and carbide formation, strengths of more than 3 GPa (430 ksi) have been attained. Such property levels enhance the opportunities for powder metallurgy in demanding applications; however, attaining these property levels requires an understanding of heat treatments after sintering.

SUGGESTIONS FOR FURTHER READING

Bain, E. C., and H. W. Paxton, *Alloying Elements in Steel*, American Society for Metals, Metals Park, OH, 1966.

Beiss, P., "Finishing Processes in Powder Metallurgy," *Powder Metallurgy* **32,** 277–284 (1989).

Furukimi, O., K. Yano, and S. Takajo, "Ultrahigh Strength Ferrous Sintered Components," *Internat. J. Powder Metallurgy* **27,** 331–337 (1991).

Heat Treating, Vol. 4, *Metals Handbook*, 9th ed., American Society for Metals, Metals Park, OH, 1982.

Herring, D., ''Vacuum Carburizing of P/M Steels,'' *Industrial Heating* **54**(9), 30–33 (1987).

Klar, E. (ed.), *Powder Metallurgy*, Vol. 7, *Metals Handbook*, 9th ed., American Society for Metals, Metals Park, OH, 1984.

Miura, H., T. Honda, K. Hens, and R. M. German, ''High Performance 4600 Steels by Injection Molding,'' in *Powder Injection Molding Symposium 1992*, P. H. Booker, J. Gaspervich and R. M. German (eds.), Metal Powder Industries Federation, Princeton, NJ, 1992, pp. 203–217.

Orzeske, P. J., ''Steam Treatment Improves P/M Part Quality,'' *Industrial Heating* **53**(5), 34–37 (1986).

Vander Voort, G. F. (ed.), *Atlas of Time-Temperature Diagrams for Irons and Steels*, ASM International, Materials Park, OH, 1991.

CHAPTER 10

Finishing Operations

10.1. GOALS

All P/M components are subjected to final finishing operations associated with adding various features or correcting existing features. The simple operations involve inspection, marking, and packaging the components. More exotic options include drilling holes, sealing pores, deburring, shot peening, electroplating, painting, and other derivatives from traditional manufacturing operations. Clearly, component production does not end with sintering and heat treatment, since additional steps are needed to meet the application requirements. Further, with the new structure of industrial operations—fewer vendors and more integrated production—successful execution of finishing operations is a required aspect of P/M.

Considerable value can be added to P/M products by tailoring the component to meet the application requirements. A chief benefit of P/M is the ability to form complicated, net-shapes in high-performance materials to high accuracy with fast production cycles. Accordingly, flexibility in tailoring the P/M product to the application is an important skill. Most of these operations are analogous to the manufacturing steps applied to other steels, but porosity alters the responses.

10.2. REPRESSING

Most P/M components are porous after sintering. Pores affect heat treatment, strength, fatigue life, and many properties. Consequently, much attention has been directed to the closure of surface pores or the selective strengthening of near-surface regions. For example, Figure 10.1 plots a hardness trace taken near the surface of a P/M steel subjected to a heat treatment geared to form a hard surface. Elimination of pores and adjustment of final dimensions is possible by sizing, coining, or repressing. Most repressing is used to reduce dimensional scatter for conformance to a specification while increasing density. The forces in repressing must exceed the steel's strength, thereby forcing flow and conformance to a die cavity. If repressing is at room temperature, the steel is strong and resists densification and deformation. By heating of the component, the strength decreases and the required forging stresses are lower. Both approaches are in use. If repressing occurs at room temperature, the compacts are usually sintered a second time.

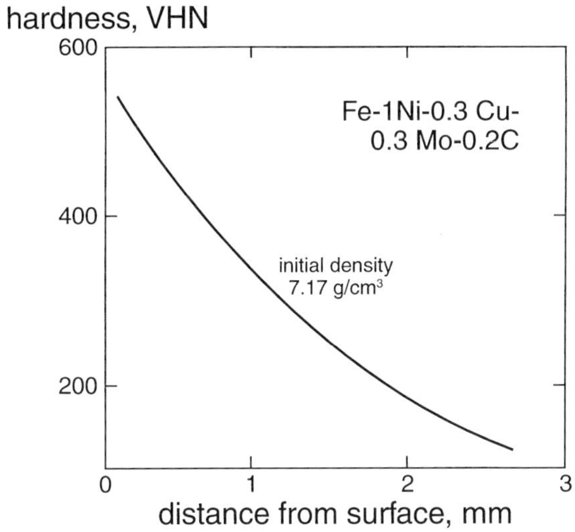

Figure 10.1. Near-surface hardness trace for a sintered and heat-treated alloy steel. (Data from Shigeaki Takajo.)

In repressing, the shape change is usually small, such as in flattening or straightening, but some cases achieve large deformations. Depending on the degree of deformation, these postsintering mechanical deformation treatments are known as *repressing, sizing,* and *coining*. The phrase *double press–double sinter* (DPDS) refers to the cases where most of the concern is with densification while sustaining dimensional precision.

Repressing involves less deformation than encountered in forging or extrusion. It adjusts final dimensions without large shape changes. Indeed, upset forging gives more pore collapse and densification when compared with repressing, as illustrated by the hot-forming data in Figure 10.2. The sintered compact is placed in a closed die and subjected to a hammer impact by the upper ram (high-strain-rate pressing). For repressing, the compact and tool sizes are the same and the flow is in one direction. Often the first sintering temperature is low to intentionally keep the compact weak [possibly as low as 750°C (1400°F) before carbon goes into solution.], but the repressing pressure exceeds the original compaction pressure.

In repressing, pores near the surface are sealed, but internal pores may be unaffected. As shown in Figure 10.3, hot forging improves the near-surface hardness, but has less effect on bulk properties. This plot gives data taken from a corner after upset forging at two temperatures. The higher temperature is more effective since the steel is softer.

With a repressing treatment it is most common to see the strength increase, but ductility and toughness decrease. More important, fatigue strength increases. Figure 10.4 plots the repressing pressure effect on a sintered steel at room temperature, showing the density increase with both the initial density and the repressing pressure.

Because of the restrike, the compact is densified, strengthened, and adjusted for dimensions and surface finish. A high initial density gives a lower relative density gain with repressing.

Often, the repressed component is sintered a second time to obtain a higher density and strength. The gain in density obtainable through repressing is typically 0.2 g/cm^3 (e.g., from 7.0 to 7.2 g/cm^3), while powder forging usually obtains full density. A common term is *double press–double sinter*. This cycle implies first pressing at possibly 500–800 MPa (35–60 tsi), sintering at 750°C (1400°F), repressing at 400–700 MPa (30–50 tsi), followed by final sintering at 1120°C (2050°F) or higher. Such an approach gives a density near 7.6 g/cm^3. For comparison, repressing probably costs only half that of hot forging, but the final properties are below those obtained by forging.

Sizing is targeted at improved dimensional precision instead of densification. A key use is to remove distortion associated with sintering or heat treatment. For example, it is used to give a precise bore diameter to a bushing. Although the handling associated with sizing adds to the manufacturing steps, it is an effective means for attaining high-accuracy final dimensions and smooth surfaces. Component roundness or flatness is improved by sizing, so it is applied to components such as in bearings or clutch plates.

Coining is a technique for adding surface patterns to a sintered blank. The term comes from the historic use of the technique for fabrication of coins and medallions. Figure 10.5 illustrates a surface impression obtained by coining; it is commonly used for impressing identification marks into the surface of P/M products. The sint-

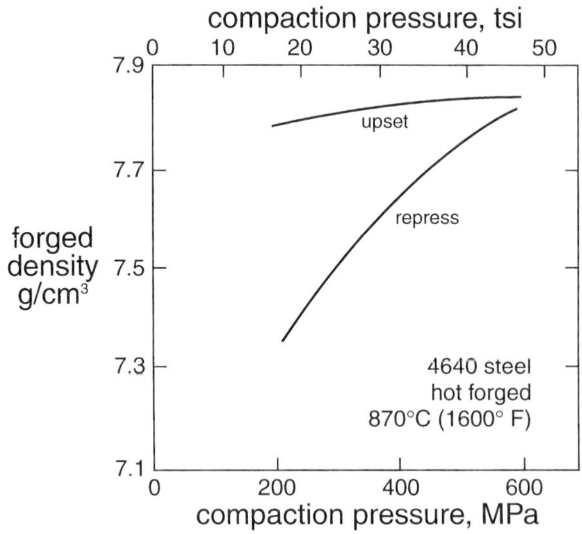

Figure 10.2. A comparison of the density versus applied pressure for hot repressing and upset forging. The constrained condition from the die wall in repressing limits the densification when compared with upset forging. (Data from W. Brian James.)

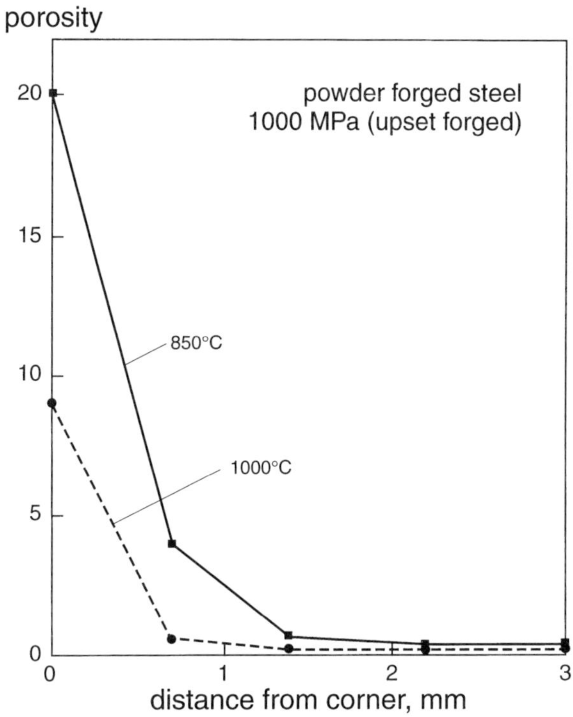

Figure 10.3. Porosity variation with distance from a compact corner after upset forging at 1000 MPa (145 ksi) and 850 or 1000°C (1560 or 1830°F). The higher temperature allows more effective elimination of porosity at the corner of the compact.

ered blank must be aligned precisely to avoid tool damage, and the surfaces must be lubricated to reduce tool wear.

Restrike operations are useful for closing surface pores, improving surface finish, increasing surface hardness, increasing precision, and adding detail to the component. However, they are expensive largely because of the handling and tool cost. Presses for repressing are similar to those used for die compaction. Some high-performance operations rely on screw presses. These employ a treaded shaft to drive a punch into the work. Very high pressures are possible with excellent dimensional control.

10.3. MACHINING

Machining involves the removal of mass from the P/M product to add holes, threads, undercuts, grooves, and special features that cannot be incorporated into the original tooling. It is estimated that 30% of all P/M parts are machined after sintering. Those adjustments are made using high-speed, low-feed cutting techniques. It contrasts

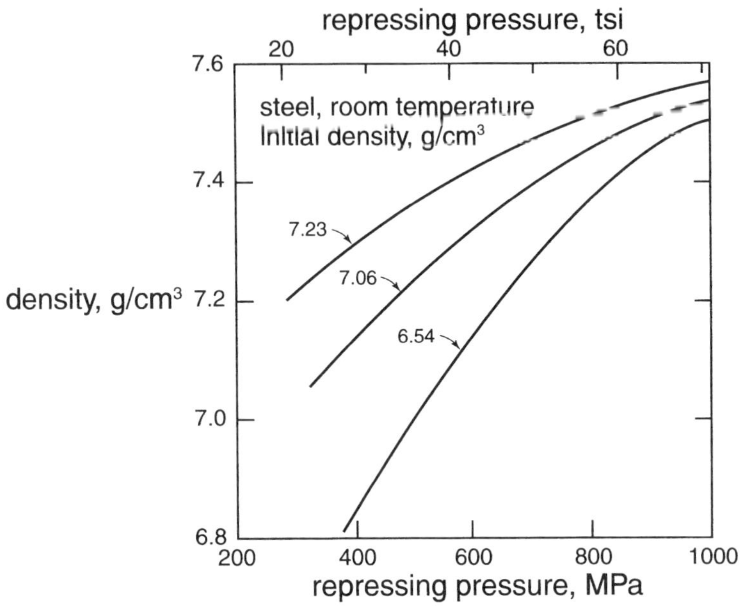

Figure 10.4. The density versus repressing pressure for steel components of three initial density levels. The density increases with initial density and repressing pressure.

Figure 10.5. An photograph of a coin that demonstrates localized deformation from surface coining an impression into a sintered blank to form a medallion. (Photograph courtesy of Julian Thomas.)

TABLE 10.1. Example Machining Operations

Boring—single-point rotating tool, stationary work, used to cut large holes or large-diameter circular contours
Broaching—bar with multiple-shaped tool tips of increasing size, statinary work, reciprocating push–pull motion of tool over work cuts grooves, holes, or other features
Burnishing—reciprocating round hard tool, stationary work, surface smoothing by friction that causes shear displacement to harden, smooth, or finish surface
Drilling—rotating fluted tool (usually with multiple tips), stationary work, cutting rotation removes mass to form hole
Electrical discharge machining—electrical spark machining across a small gap filled with a dielectric fluid, progressive sparks evaporate and expel material to form tool profile in work
Grinding—moving abrasive wheel, stationary or rotating work, used to generate a smooth surface (similar operations are honing and lapping)
Hobbing—rotating cutting tool with teeth arranged along helical thread, used for cutting gear teeth by synchronized rotation of both the hob tool and work
Honing—an abrasive finishing operation that uses fine-grit rotating abrasive wheels to produce accurate dimensions and excellent finishes
Lapping—fine abrasive grit loaded into a rotating flat disk, work moves over rotating disk to attain flat surface and smooth finish
Milling—rotating tool with many tips, stationary or reciprocating work, generates flat surfaces
Planing—single-point stationary tool, reciprocating work motion, used to form flat surfaces
Reaming—rotating tool with cutting edges on fluted outer diameter, stationary work, used to make small cuts to size, contour, or improve hole accuracy
Shaping—similar to planing except the work is stationary and the tool moves in a reciprocating motion to form a flat surface
Tapping—rotating conical cutting tool with threaded periphery, used to cut screw threads in a hole
Turning—stationary tool, revolving work, used to remove surface to form circular profiles

with repressing, which changes the component shape without changing its mass. Several examples of machining processes are identified in Table 10.1, and a few common situations are sketched in Figure 10.6. Those are the same machining steps applied to wrought materials. The intent is to extract mass by turning, milling, shaping, broaching, or grinding.

For P/M steels, both hardness and residual porosity affect tool life, with longest life when the porosity is either very high or very low. At intermediate densities, the life of the cutting tool is short since the P/M component has a low thermal conductivity, leading to heat erosion of the cutting tool. Also, in machining a porous structure, the stress at the tip of the cutting tool is cyclic since the cutting path is interrupted by pores. The tool is stressed by solid, but this stress is relieved on contact with a pore. Consequently, cyclic loading makes for shorter tool life than experienced with constant cutting forces. Accordingly, the cutting tool undergoes fatigue failure in

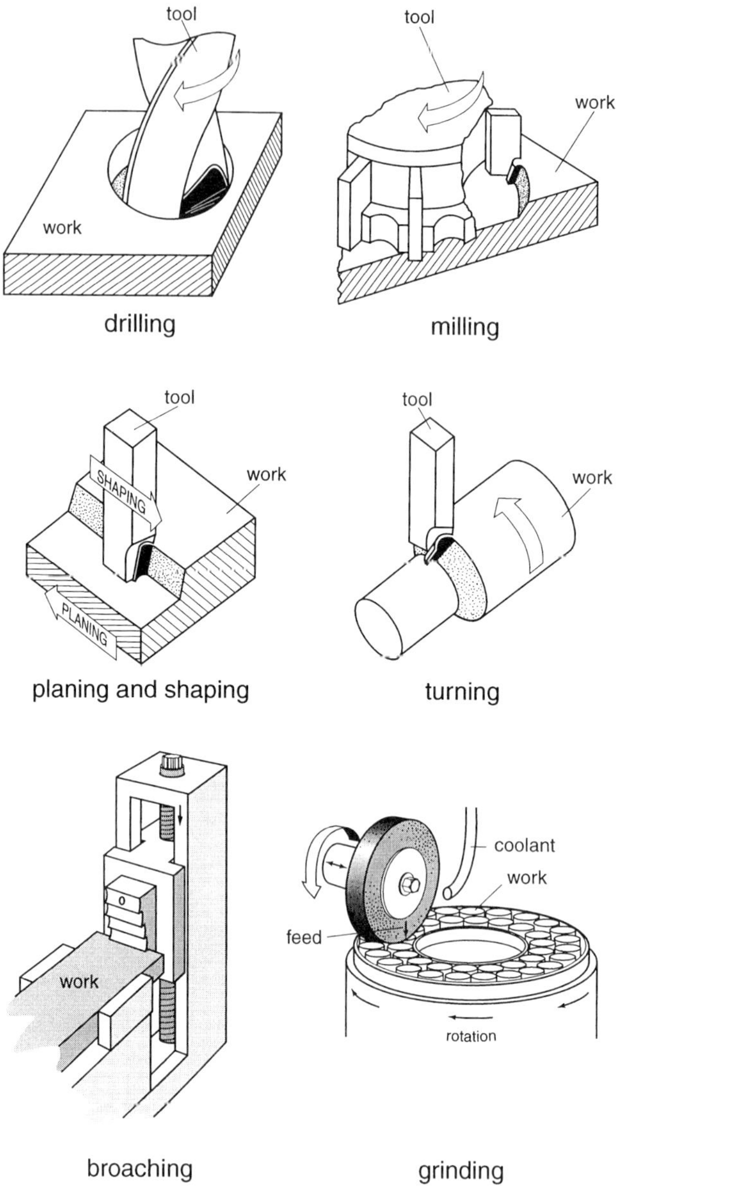

Figure 10.6. Examples of common machining practices where metal is removed by drilling, milling, planing, shaping, turning, broaching, or grinding the P/M component.

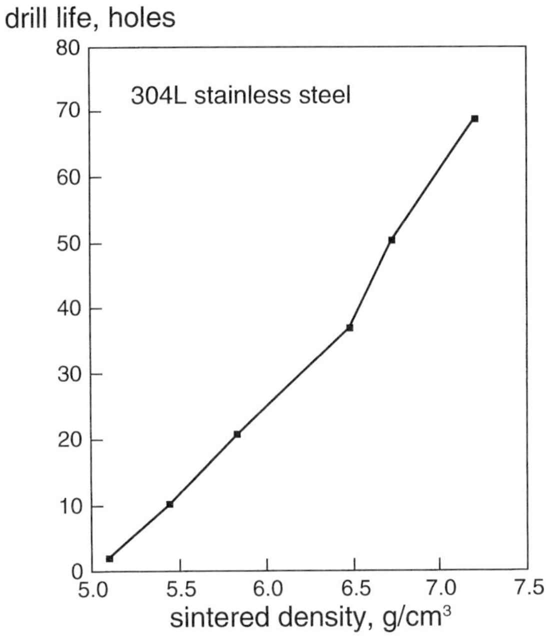

Figure 10.7. Drill life (measured by the number of holes drilled prior to failure) for sintered 304L stainless steel. This plot shows the weaker material is more difficult to machine. (Data from John Agapiou.)

machining a porous steel, especially when the pores are open and connected (density typically below 7.2 g/cm^3).

Figure 10.7 plots the drill life (number of holes drilled before failure) versus density for sintered 304L stainless steel. Note that drill life improves with a higher density. Stainless steels are more difficult to machine than most other steels. For comparison, data on drilling sintered Fe–2Cu–0.8C (FC-0208) shows that 10 times more holes are possible at a 6.7 g/cm^3 density. At lower densities, the low tool life can be offset by infiltrating the pores with a low-melting-temperature metal before machining, for example, by copper infiltration. At 20% porosity (about 6.7 g/cm^3), tool life increases by a factor of 100 if the remaining pores are infiltrated before machining.

Cutting fluids are used in most machining operations, but care must be exercised to reduce contamination by any fluids trapped in the pores. That requires cleaning after machining, especially if the component is to be heat-treated.

Additives to the sintered steel are possible for improved machinability. These can be included in the initial powder mixture, as is common for components requiring postsintering machining. A favorite for improved machining is MnS (manganese sulfide). Figure 10.8 compares the life of drills in machining a sintered Fe–2Ni–0.5C (FN-0205) steel with three MnS contents and two sintering atmospheres. During cutting, the manganese sulfide coats the cutting tool to provide lubrication. There is a slight decrease in mechanical properties from the MnS addition, typically less than 20 MPa (3 ksi) strength loss at a 0.5% concentration. Other additives for improved machinability include tin, tellurium, lead, and molybdenum compounds. Further improvements in machining generally are observed by using high speeds and low feeds. The tools last longer and the machined surfaces are smoother.

In P/M steels, both alloying and heat treatment influence the ease of machining. Very soft and very hard materials are the most difficult to machine; thus, there is a sensitivity to carbon level. Sometimes it is possible to rough-machine the component in the green or presintered conditions to reduce final material removal. However, best dimensional precision is attained by grinding the component after heat treatment. Machining deforms pores connected to the surface. This might be beneficial in reducing corrosion, for example, but in applications where the pores are useful, such as in filters, they should be opened by electroetching after machining.

Dimensional scatter is an inherent aspect of all fabrication processes. In some P/M operations, machining is used to ensure close dimensional control on critical surfaces. The final dimensions on most P/M components have a natural scatter, near $\pm$ 0.05% (one standard deviation) about the mean dimension. Machined surfaces reach surface roughnesses near 2 μm (0.0001 in) and even better (by a factor of 10)

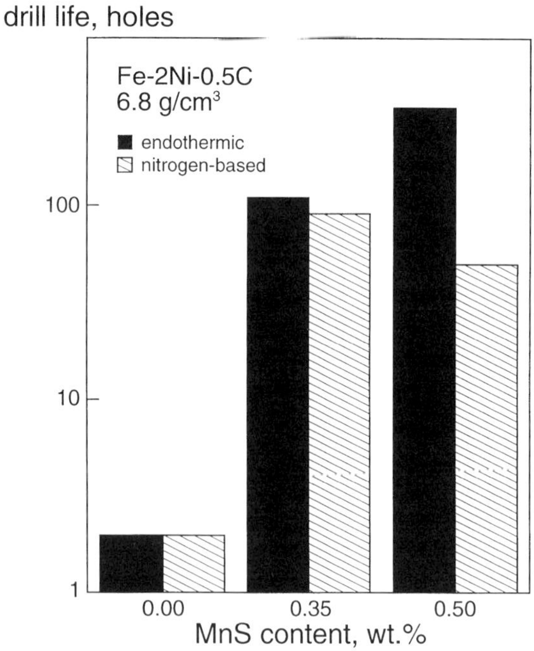

Figure 10.8. A demonstration of manganese sulfide effect on drill life for a low-alloy steel. (Data from Robert Causton.)

for polished surfaces. Pores connected to the surface create a rough-appearing surface, although the solid portion of the surface might be very smooth.

10.4. JOINING

Assembly of P/M components is a necessary aspect of most applications. Many techniques are possible, ranging from interference fits to welding. Permanent bonding is usually the goal, except for components joined using fasteners, such as bolts or screws. There are several variants, including joining porous materials to solid steels, porous materials to porous materials, and full-density P/M steels to wrought steels. The latter is the most straightforward, since there are no pores to interfere with bonding and the material properties are similar to handbook values. However, pores degrade bond strength and create difficulties in using liquid metal bonds, such as brazes and welds, since the pores wick the molten metal away from the joint.

10.4.1. Sinter-Bonding

A simple form of assembly is via an interference fit. A slightly oversized component is inserted into another, usually by heating the smaller component or by slight deformation in assembly. The lower elastic modulus of porous P/M steels makes it easier to press-fit as compared with full-density structures. However, this porosity lowers the strength and can induce cracking with large differences in sizes. A size difference of about 0.2% is typically most useful for a press fit, and this can be accommodated by heating the smaller component to 200°C (about 400°F). Once a good interference fit is in place, strong bonds are formed by resintering the component. Further, brazing compounds can be added to enhance bonding.

Another variant of sinter-bonding uses green structures and is a process unique to P/M steels. Green compacts are mechanically joined before sintering. During sintering the mating surfaces diffusion bond and form a strong assembly. The related process involves joining solid structures to cast or forged components. A good mechanical fit is required to ensure bonding at the interface. Sinter-bonding is unique since it can be applied to joining components during the sintering cycle. Bond strengths between components in the range from 375 to 420 MPa (55 to 60 ksi) are typical for sinter bonds.

Most P/M components fabricated in this manner have specifications for minimum bond failure loads. Note that shrink fits are not as strong as sinter bonds. Thus, resintering is an effective means to create strong joints.

10.4.2. Joining Processes

After fabrication, several components may be joined to form an assembly. The assembly can use different materials that could not be combined by sinter bonding.

TABLE 10.2. Joining Processes for P/M Components and Materials

Adhesive bonding—bonding between clean surfaces by the action of an acrylic or epoxy giving a low-strength chemical bond

Arc welding—consumable (MIG) or nonconsumable (GTA) arc is formed to melt material at the contact, the metal flows and bonds the objects prior to solidification

Brazing—a low-melting-point liquid is formed by heating the component, the liquid wicks into a gap of approximately 0.5 mm width and near-surface pores to create a chemical and mechanical bond

Diffusion bonding—contacting parts are heated to a high temperature under low loads such that atomic diffusion forms a metallurgical bond; also known as *sinter bonding*

Electron-beam welding—local melting of contacting materials by use of a high-energy electron beam

Friction welding—inertial energy from a moving or spinning object is converted into heat at a frictional contact to cause bonding

Laser weld—local melting to form a bond between two contacting parts by use of high-energy lasers

Press fit—a mechanical bond formed by the localized plastic defomation at an interference zone between two close-tolerance components with approximately 0.2% misfit

Resistance welding—electric current passes through the contact between simple shapes to cause heating and subsequent bonding

As identified in Table 10.2, all of the usual joining processes are applicable to P/M materials. The examples range from mechanical interference fits to bonding using high-energy lasers or electron beams. Generally, stair-step-shaped lap joints are the most successful.

Bonding porous P/M steels by welding or brazing requires special care. Pores reduce thermal conductivity and can wick molten metal away from the area being joined, so they must be cleaned. Braze, diffusion, and adhesive bonds are all weaker than other joining techniques, while electron-beam and laser welding can cause distortion in porous components. Adhesive bonding is one means to join P/M components, but this is used in rare situations. The pores provide cavities that trap the adhesive, but useful strength is low, near 30 MPa (4 ksi). The typical small P/M component size gives a small bonding area that lowers the loads that an adhesive bond can withstand. Rivets and other simple fasteners (screws and bolts) can bond components together, assuming that the stress concentration is not deleterious to function or performance.

At the other extreme, laser bonding is used to form high-quality bonds for thin sections. Figure 10.9 is an example of the high-quality bonds possible via lasers. This is a cross section through a sintered stainless steel formed by injection molding and sintering to full density before laser welding, to show the quality of the resulting joint. On the other hand, brazing works best on porous structures. Warpage, porosity, and alloy composition factors influence the selection of the joining process. At low

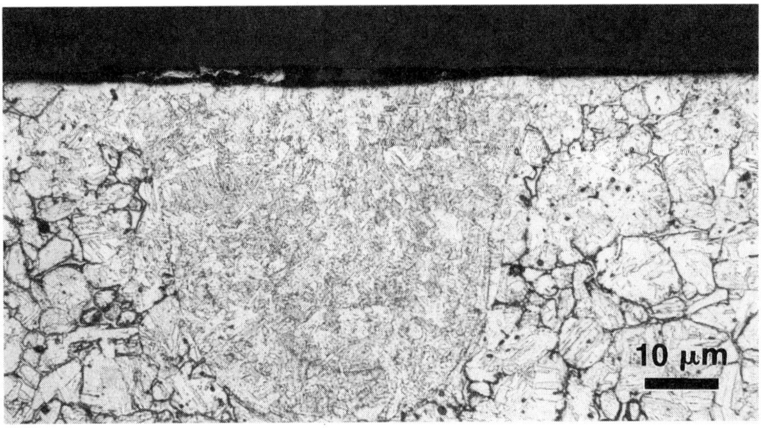

Figure 10.9. A cross section through laser-welded 316L stainless steel formed by injection molding. The region melted by the laser is fine-grained and well bonded to the sintered steel. (Photograph courtesy of Rene Michael Cooper.)

porosity levels, below approximately 8%, no major difference exists between joining procedures for P/M and wrought materials.

For porosities between 10 and 20%, techniques that form molten metal at the weld work well, but must be carefully controlled to reduce the melt quantity. With gas tungsten arc welding, a filler metal is required to fill the pores, while the feed wire in the metal–inert-gas technique helps fill pores during welding. At higher porosity levels, molten metal joining techniques should not be used because of distortion and residual stress generated by the contracting weld. It is important that the porous steels be free of oils, cooling fluids, or other contaminants before welding.

All the techniques involving melting produce a heat-affected zone near the weld with a reduced strength. The thermal stresses associated with the bond can cause distortion, cracking, or accelerated corrosion in this zone. Stress-induced cracking is a common problem in joining P/M steels. The best practice is to use as high a sintered-density component as possible to ensure a strong weld. Lower-carbon contents reduce cracking near the weld.

As a final note, many metal powders are used for joining, especially in forming welding rods, brazing compounds, and laser weld fillers. In the latter area, a prealloyed spherical powder fills a gap and precise laser melting is performed to bond two components with minimized melting of the parent structures.

10.5. SURFACE TREATMENTS

Several surface treatments can be applied to improve function or aesthetics. Examples of postsintering surface treatments include deburring, coating, spraying, painting, polishing, cleaning, anodizing, plating, sealing, and laser glazing. Many surface treatments are encountered as part of heat treatment cycles, including case carburization and steam blackening.

10.5.1. Deformation and Mechanical

Deburring removes sharp edges, flash, burrs, or other surface irregularities. It is performed by tumbling or vibrating the P/M component with abrasive media, or applying surface abrasion (such as sandblasting). Simple deburring techniques are wire brushing, tumbling, or vibration of the P/M components with abrasive media and oil. They are useful in removing surface flash without damaging the component. Further, a corrosion inhibitor may be coated onto the component during deburring. Zinc powder can be mechanically vibrated onto the surface to form a coating resistant to corrosion. After tumbling, surface contaminants should be removed by washing in a solvent.

An important means of improving the fatigue life of a P/M component is through shot peening. Compressive forces are generated on the surface by bombarding the component with small-diameter shot. As sketched in Figure 10.10, impacts leave small indentations that create a compressive stress on the surface. Repeated impacts flatten the surface and provide uniform compression. This greatly retards fatigue cracking, since most porous P/M steels exhibit surface nucleated fatigue fracture. For example, a sintered P/M steel with a 7.1 g/cm^3 density and 1.5% carbon has a

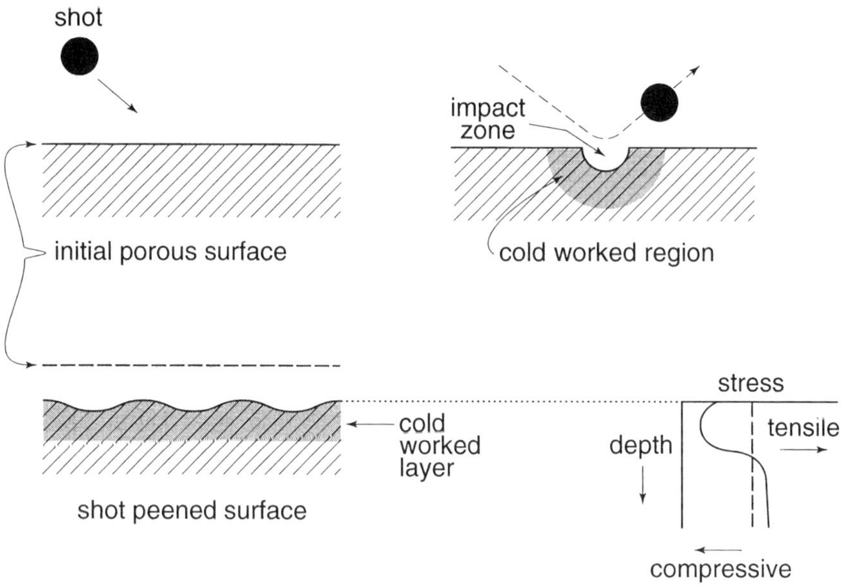

Figure 10.10. Shot peening involves repeated impacts by high-velocity shot onto the surface of the P/M component. The impacts create cold-worked regions with compressive residual stresses for extended component life in fatigue.

fatigue endurance limit of 130 MPa (19 ksi). After shot peening, the same material has a fatigue endurance limit of 190 MPa (28 ksi). Although shot peening improves fatigue life, it decreases dimensional precision because it creates a rough surface.

Burnishing involves the application of a shear rolling force that exceeds the yield strength to provide surface deformation. It is used to correct dimensions, especially on bores. It is also used to improve the rolling contact fatigue life of gears, allowing a 40% load increase. Burnishing is most effective when applied with several small cycles rather than one large cycle.

10.5.2. Coatings

Coatings are used to improve wear life and provide corrosion resistance. For P/M components the coatings range from paints to hardfacing layers, but the most typical is surface carburization or plating. Surface carburization diffuses carbon into the component during heat treatment in a carbon-rich atmosphere. Mechanical plating is achieved by tumbling the component in a mixture of zinc powder and glass beads, causing cold welding of the zinc onto the component surface. Similar coatings are possible using polymers, especially inert polymers.

Electroplating is applicable to components of over approximately 7.3 g/cm^3 density. For lower densities, it is appropriate to fill the pores with a resin to avoid plating chemicals from becoming trapped in the pores. Burnishing or other surface deformation techniques such as shot peening or buffing are successful in sealing pores as well. Typical coatings for corrosion resistance are based on chromium, nickel, copper, zinc, or cadmium. Subsequent corrosion problems may occur if the electroplating solution is trapped in the surface-connected pores. Thus, thorough cleaning after coating is crucial. Plating with copper is usually followed by nickel electroplating or coating with a protective lacquer, since copper will naturally corrode. Nickel plating is used for wear resistance or corrosion resistance, and might provide a needed base for chromium electroplating. Chromium is selected for corrosion and wear resistance. Finally, zinc or cadmium are sometimes used for corrosion protection. Although cadmium is toxic and much more expensive, it is a favorite on electronic components.

10.5.3. Sealing

Impregnation is widely employed for sealing pores connected to the outer surface. Variants fill the pores with oil or Teflon for low-friction, epoxylike fluids for improved machining, or polymers for corrosion resistance. Usually impregnation is performed on heated components under a vacuum to aid in fluid penetration. Oil impregnation is done by forcing lubricating oil into evacuated pores, usually at elevated temperatures, so the oil viscosity is low. The impregnating fluid spreads to eventually fill all the open pores. Large pores have less resistance to filling, so they fill most rapidly. Once a resin is in the pores, it is polymerized or cross-linked, but a volume change must be avoided to prevent damage to the porous structure. At 90°C (194°F) a typical resin cures in about 10 min. One use for impregnation is to eliminate open pores so that the component can be used in an application where leaks are not tolerable.

A technique being explored for improving corrosion behavior of full-density, high-performance P/M materials is laser glazing. A laser beam is rastered over the component surface to locally melt the alloy. Because the bulk material below the molten surface is not heated, the melt is rapidly quenched into a surface glaze. Sometimes, the quenching process is sufficiently rapid to produce an amorphous coating that exhibits excellent corrosion resistance. In a similar manner, various plasma and ion treatments are under development to alter the surface state of a component without degrading the bulk properties. Generally, those treatments affect only the first 100 μm of the surface.

10.5.4. Infiltration

Infiltration relies on heating the compact to a temperature where copper or a copper alloy melts. At temperatures over 1083°C (1980°F), molten copper flows into the iron or steel structure through the surface pores, leading to a dense, pore-free compact. Figure 10.11 is a micrograph that captures the copper infiltration process, where the white phase is copper. This micrograph was obtained by quenching the compact during infiltration.

There is preferential dissolution of the steel at the surface where copper first melts, so some precision is lost on this surface during infiltration. Also, depending on the carbon level, the compact might swell during infiltration. Typical swelling is about 0.25%. If an excess of copper is infiltrated, then the compact distorts. Most useful is a slight deficiency in copper, where there is not enough to fill all pores. This condition gives superior dimensional control. Further, a higher strength occurs by filling the pores. For example, at 7.0 g/cm^3 a steel with 0.6% carbon has a strength of 1080 MPa (157 ksi), but infiltration with copper to give a density of 7.5 g/cm^3 results in a strength of 1240 MPa (187 ksi). In other examples, infiltrated steels have delivered fourfold impact strength gains.

10.6. INSPECTION

Component inspection is an important activity following fabrication. Typical fluctuations in production might be as much as a 0.5% variation in important features such as density. The inspection techniques must be geared to detect changes that influence performance. As a testimonial of the quality possible in P/M, the reported failure rate for powder-forged automotive connecting rods is in the 0.03-ppm range (one out of 33 million). Even the exceptional failures are not necessarily due to P/M processing, since use conditions often play a dominant role.

10.6.1. Statistical Process Control

In recent years, inspection has been integrated into each production step. The idea is to capture errors as they occur rather than after fabrication. An important technique

EQUATION BOX 10.1.

Assuming capillary tubes for pores and ignoring the action of gravity, the depth of fluid impregnation h depends on the pore diameter d, fluid viscosity η, and filling time t

$$h^2 = \frac{\gamma t d}{4\eta}$$

where γ is the surface energy for the liquid–vapor interface. Higher-viscosity fluids and smaller pores require more time for impregnation. If the fluid is pressurized (which is the usual case), then the rate of impregnation increases

$$h^2 = \frac{d^2 t}{16\eta}\left(\frac{4\gamma}{d} + P_f\right)$$

where the pressure on the impregnation fluid is P_f. Thus, an applied pressure ensures homogeneous, rapid filling of all pores.

Since P/M components contain a distribution of pores, varying in both size and shape, the preceding infiltration equations are only approximations. The largest pores fill first while small pore filling is delayed; thus, because of the pore size distribution we deal with the average depth h of pore filling, which follows an empirical parabolic equation

$$h^2 = At + B$$

where the constants A and B depend on the pore size and fluid density, and vary inversely with the fluid viscosity. For example, water impregnates a porous compact rapidly. The viscosity of water is approximately 10^{-3} Pa · s at room temperature and the surface energy is 0.074 J/m^2. For a pore of 10 μm diameter, the time to penetrate 10 mm depth is calculated as follows:

$$t = \frac{4\eta h^2}{\gamma d}$$

which gives a time of 0.54 s. The addition of one atmosphere external pressure (assuming evacuated pores) reduces the time to 0.12 s. Note the time scales directly with the fluid viscosity. Hence, a change to a polymer with a viscosity of 1 Pa · s will increase the penetration time to 120 s, if the surface energy remains constant.

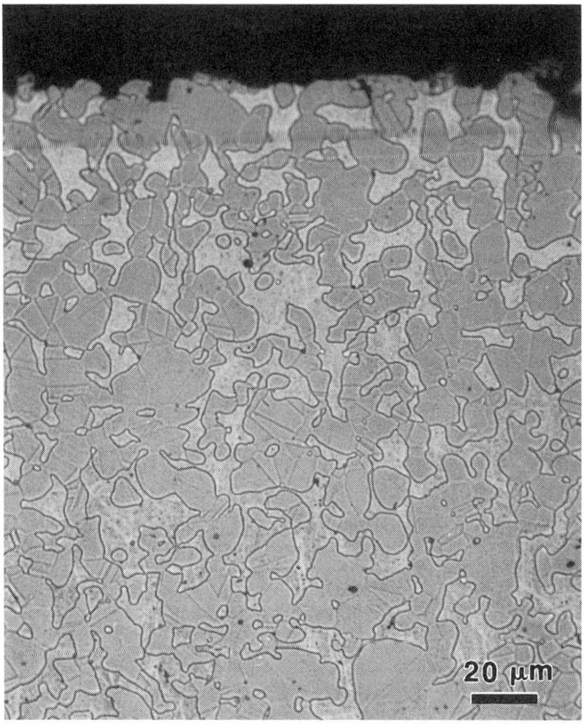

Figure 10.11. Copper infiltration as captured in this metallographic cross section using a sample quenched during infiltration. (Photograph courtesy of Jean Murray Stewart.)

is statistical process control (SPC). It is used at intermediate manufacturing steps to assess drifts from standard conditions, such as in component weight or size. For example, Figure 10.12 illustrates the apparent density scatter associated with an incoming powder. The SPC charts track changes over time, with preset upper and lower control limits. Here, lot 10 would have been rejected before pressing parts. The goal is to avoid fabrication of components that deviate from the accepted control range. Some variability occurs in any P/M operation, but SPC continually senses drifts that may take a process out of the acceptable range.

Formerly, many P/M operations performed inspection after the entire process. This is open-loop control, which has fallen out of favor, because much waste is possible. Open-loop control has been replaced by continuous process monitoring and adaptive control concepts. Statistical process control represents a form of adaptive control, where changes are made immediately after a rejected part is formed or as a process drifts out of control. That greatly reduces the number of rejected parts.

Currently, the idea of closed-loop feedback control is emerging, where each fabrication step is instrumented and controlled to make immediate adjustments as a component is formed. That is essentially in situ control and greatly improves quality

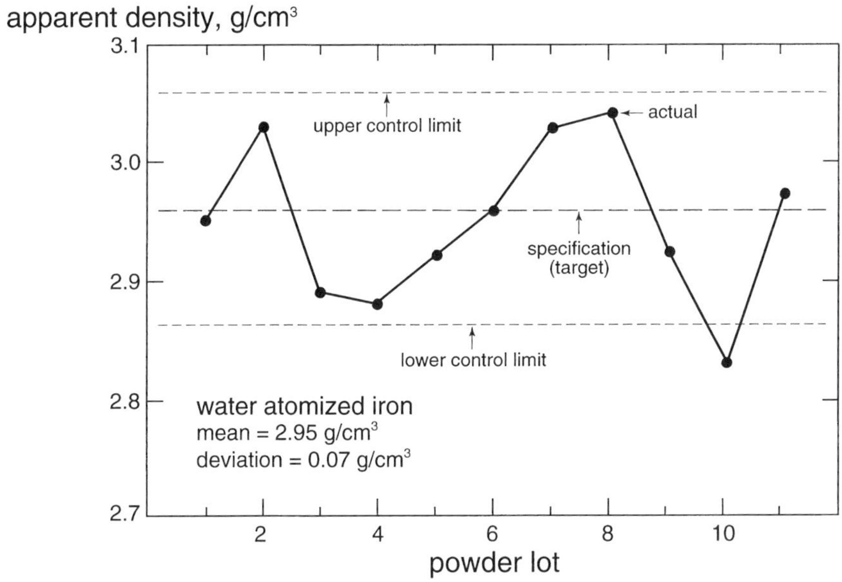

Figure 10.12. An example statistical process control (SPC) chart for incoming powder inspection. This situation relies on the apparent density to detect trends. (Data from Jack Bonsky.)

by ensuring that each part is correct. In the future, feed-forward control would use computers to analyze and correct delinquent process trends. Such an approach can anticipate machine or tool repairs before loss of production. Various computer-based artificial-intelligence schemes are evolving to assess production data for this purpose.

In any of the inspection procedures, four possible situations must be recognized. Accepting the defect-free components and rejecting the defective components is the goal. The worst case is to accept a defective component; the rejection of a defect-free component is also undesirable. Thus, careful attention must be directed to the criteria for both the accept and reject decisions.

10.6.2. Testing Techniques

Although visual inspection is important, much effort is being made to automate and quantify inspection tasks. The focus can be on any of the available nondestructive tests, including density, weight, dimensions, or measurement of properties. Defects are sought using dye penetrant, radiography, ultrasonic, laser, or eddy-current techniques. X-ray radiography provides an image that allows detection of defects. It can be combined with image analysis computers for automated identification of defects. Likewise, sonic waves propagating through a material are scattered by cracks; hence, ultrasonic, laser vibration, and resonant vibration inspection techniques are applicable to high-density P/M steels.

EQUATION BOX 10.2.

The classic distribution in a measured parameter is described by the "normal curve" or "bell curve," where there is a spread in readings grouped around a center value. The average of the distribution is the mean and is calculated by adding all the observations and dividing by the number of observations. In school this was also called the "average" and is calculated mathematically as

$$U_A = \frac{1}{N} \sum_{i=1}^{N} U_i$$

which says to add the individual readings U_i from the first to last and divide the sum by the number of readings N to obtain the average U_A. For example, the average of 7, 8, 11, 21, and 23 is 70/5 or 14.

The standard deviation σ is calculated as the variation about the mean U_A and is also described by a mathematical formula:

$$\sigma = \sqrt{\frac{\sum_{i=1}^{N} (U_i - U_A)^2}{(N - 1)}}$$

It depends on the departure of each reading from the average. Today, many automotive P/M components are specified by ± 3 standard deviations. This implies that 99.7% of the components must be within a prescribed range of values.

The process capability C_p is a measure of the ability of a P/M operation to stay within a specified limit. It measures the ratio of the process spread in terms of three standard deviations and compares this process spread with the tolerance allowed spread

$$C_p = \frac{U_M - U_A}{3\sigma}$$

where U_M is the maximum (or minimum) control limit, U_A is the average, and σ is the standard deviation.

If the mean value is centered between the upper and lower bounds, then the process can have a wider allowed variation. However, if the process skews toward one end or the other, then there is less allowed variation. This distribution is usually adjusted to reflect the actual process distribution about the mean value. The distance from the mean of the measured values to the nearest specification limit provides a modification parameter, leading to a process control capability designated as C_{pk}. The value of C_{pk} can never exceed C_p. High values of the process control capability C_{pk} are desirable, and values of 1.33 are generally most impressive.

Eddy currents as induced by an oscillating magnetic field do not pass through cracks. Thus, they provide another signal that can be used for nondestructive inspection. New techniques are being developed for inspection of P/M materials, including computerized axial tomography (CAT) scan, gamma-ray attenuation, electrical potential drop, and scanning acoustic microscopy. Further, destructive tests may be applied on a random basis to measure strength, hardness, toughness, microstructure, chemistry, or other properties.

Ultrasonic inspection is widely applied to low-porosity components where the residual pores give low scattering of the acoustic signal. It is an important qualification step in the production of hot isostatically compacted aerospace components. Test frequencies of up to 100 MHz are used to detect cracks, voids, and inclusions. Higher frequencies are required when the compact is porous. Strength (normally measured with a destructive test) can be assessed using nondestructive ultrasonic velocity measurements—the ultrasonic velocity scales directly with strength and density. Generally the porosity will be homogeneous ($\pm$ 0.5%) in a P/M component, so the ultrasonic signal can be used to qualify the component quality. However, in large components the porosity variation may be as much as $\pm$ 5%. Thus, the ultrasonic assessment must include allowance for porosity variations. Inclusions over 20 μm and voids more than 50 μm can be detected by ultrasonic techniques. As the section thickness increases, the ability to detect smaller defects decreases.

SUGGESTIONS FOR FURTHER READING

Agapiou, J. S., and M. F. DeVries, ''Machinability of Powder Metallurgy Materials,'' *Internat. J. Powder Metallurgy* **24,** 47–57 (1988).

Beiss, P., ''Finishing Processes in Powder Metallurgy,'' *Powder Metallurgy* **32,** 277–284 (1989).

Bocchini, G. F., ''The Role of Quality Control in Powder Metallurgy,'' *Metal Powder Report* **39,** 129–135 (1984).

Boyer, H. E., ''Secondary Operations Performed on P/M Parts and Products,'' in *Powder Metallurgy*, Vol. 7, of *Metals Handbook*, 9th ed., E. Klar (ed.), American Society for Metals, Metals Park, OH, 1984, pp. 451–462.

Charlton, B., and J. M. Newton, ''Application of Gamma-Ray Attenuation to the Determination of Density Distributions within Compacted Powders,'' *Powder Technol.* **41,** 123–134 (1985).

Hamill, J. A., ''P/M Joining Processes, Materials and Techniques,'' *Internat. J. Powder Metallurgy* **27,** 363–372 (1991).

James, P. J., ''The Machinability of Sintered Steels,'' *Powder Metallurgy Internat.* **22**(6), 19–23 (1990).

Madan, D. S., ''An Update on the Use of Manganese Sulfide (MnS) Powder in Powder Metallurgy Applications,'' in *Advances in Powder Metallurgy—1991*, Vol. 3, Metal Powder Industries Federation, Princeton, NJ, 1991, pp. 101–115.

Patterson, B. R., K. L. Miluus, and W. V. Knopp, ''Ultrasonic Evaluation of the Strength of PM Materials,'' *Metal Powder Report* **39,** 145–147 (1984).

Salak, A., *Ferrous Powder Metallurgy*, Cambridge International Science, Cambridge, UK, 1996.

Tonshoff, H. K., H. G. Wobker, and A. Fritsch, ''Machining of Powder Metallurgy Materials,'' in *Advances in Powder Metallurgy and Particulate Materials—1995*, Metal Powder Industries Federation, Princeton, NJ, 1995, pp. 8.105–8.128.

Whittaker, D., ''Choosing a Surface Treatment for PM Parts,'' *Metal Powder Report* **51,** 20–26 (May 1996).

CHAPTER 11

Properties

11.1. OVERVIEW

The properties of sintered steels can approach those observed via other metalworking techniques. One difference is with the property decrements due to porosity. Thus, porosity and density are important parameters. Beyond porosity, the behavior of a sintered steel also depends on the microstructure and alloying. High-density compacts are most desirable in stressed components.

For simple geometries, porosity is measured by determining the weight and dimensions and comparing the density with the theoretical value. For more complex shapes or inhomogeneous structures, applying microscopy techniques may be necessary. A common approach is via point counting as performed on a polished metallographic cross section. Here a grid is laid over the microstructure and the percent of grid points falling on pores is taken as the porosity. The pore structure must be preserved during specimen grinding, polishing, and etching. Distortion by excess working smears material over the pores, giving the appearance of low porosity. Etching, on the other hand, enlarges pores, leading to an overestimation of the porosity. Common etchants for steels include nital (1–5% nitric acid in methyl or ethyl alcohol), picral (4 g picric acid in 100 ml methyl alcohol with possible supplemental additions of hydrochloric or nitric acid), and other reagents based on acidic solutions.

The Archimedes technique of water immersion is useful for measuring density, if the pores can be sealed to prevent water penetration during the analysis. Various sealants, including silicone oil, are available for that. A series of weight measurements determines the density. One weight is the dry compact, while a second is after filling the pores with oil, and the third is while the compact is immersed under water. The weight change with immersion gives the volume, since water has a density of 1 g/cm^3, so each gram of weight reduction (from the oil-impregnated state) corresponds to 1 cm^3 of volume. The first weight divided by the volume gives the density.

Pore shape is another factor that can affect sintered properties. At a given porosity there is considerable possible variation in the pore shape, ranging from spherical to highly angular. The latter are needed for filters, but prove detrimental to strength in highly stressed components. Open pores allow fluid flow through the sintered structure and are most useful for self-lubricating bearings, flow restrictors, and filters.

EQUATION BOX 11.1

The Archimedes technique for obtaining density for a porous material requires preventing water penetration of open pores on the compact surface. If the compact is very dense, then the pores are closed and there is no difficulty. Nevertheless, if the density is below about 7.5 g/cm^3, the pores must be sealed before water immersion. For the latter case, the component is weighed dry (W_1), after oil impregnation (W_2), and immersed in water (W_3). Usually a wire is used to suspend the sample in the water and its weight W_w must be measured in water, too. Then the density ρ in grams per cubic centimeter is calculated from the weight determinations as follows:

$$\rho = \frac{W_1}{W_2 - (W_3 - W_w)}$$

assuming a density of 1 g/cm^3 for water.

Thus, as a first point in understanding properties of P/M materials there is a requirement that porosity and pore structure differences are measurable. Creating a favorable pore shape depends on processing conditions. Generally, those actions that improve sintering are favorable to the mechanical properties. At low densities, the strength is limited by the interparticle bond size. For the higher densities, the strength has a complex dependence on the pore structure and the particle boundaries.

11.2. POROSITY EFFECTS ON MECHANICAL PROPERTIES

Structural applications are dominant in ferrous P/M, so most of this chapter is devoted to the mechanical properties such as hardness, strength, ductility, and impact energy. Accordingly, strength is a first concern when dealing with structural components. Pores degrade the mechanical properties, partly because of the reduced area supporting the load. Another factor is the stress concentration at each pore. Ductility is also sensitive to porosity and pore shape, as are most other mechanical properties; impact energy, fracture toughness, and fatigue strength have the greatest sensitivities. Even in those materials possessing full density, inferior properties can occur as a result of microstructural defects. In evaluating a P/M steel, it is logical to first inspect hardness and strength. If these appear satisfactory, then attention should be directed to ductility. Finally, if strength and ductility are satisfactory, then focus on the dynamic properties.

11.2.1. Quasistatic Properties

Pores reduce the effective cross-sectional area and negatively affect hardness and tensile properties. Quasistatic properties are those measured under low loading rates.

Major determinations from quasistatic testing are the transverse rupture strength, yield strength, ultimate tensile strength, elongation to fracture, reduction in area, elastic modulus, and Poisson's ratio. Likewise, hardness tests are usually done on undeformed portions of tensile samples. Several attempts have been made to find predictive relations between these various measures of mechanical properties. However, such efforts prove invalid. For example, the ratio of transverse rupture strength to tensile strength varies from 1.5 for martensitic steels to 2.3 for austenitic steels.

All the quasistatic mechanical properties vary with the density and microstructure. Furthermore, the presence of defects lowers strength. A scanning electron micrograph of the fracture surface from an incompletely densified stainless steel compact is shown in Figure 11.1. Particle bonds have ruptured, and most of the fracture surface shows little evidence of particle bonding. Since pores lack strength and concentrate stress to cause failure at low stresses, the elimination of pores is an obvious goal for high properties.

Just as porosity degrades strength, it also degrades ductility, but there is an additional sensitivity to pore shape and placement. Smooth pores at large spacings are less detrimental than small, closely spaced pores. The negative effect of pores on ductility is due primarily to the initiation of cracks at the pores. Thus, beyond a porosity effect on ductility, there is a further sensitivity to the pore shape.

Interestingly, cast and wrought materials often possess pores. However, a small level of porosity is not detrimental, unlike the situation in P/M steels. This is due

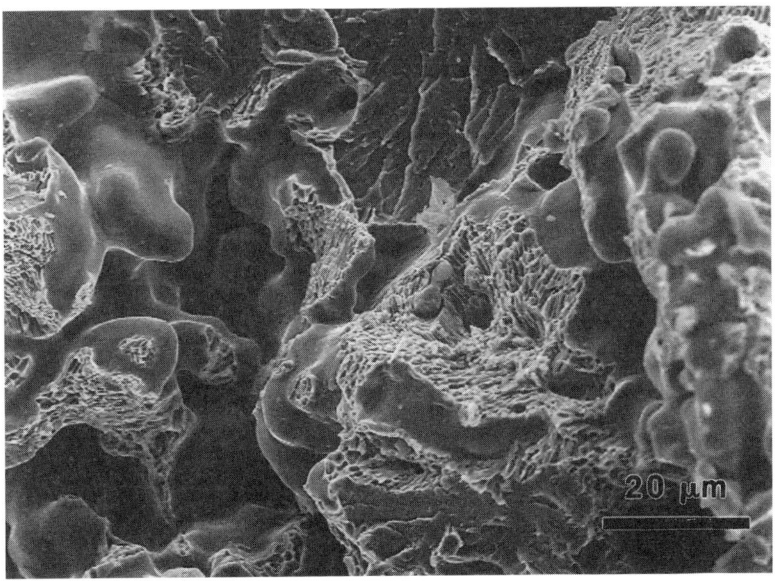

Figure 11.1. Scanning electron micrograph of the fracture surface of a 304L stainless steel fractured during sintering, showing the development of sinter bonds. (Photograph courtesy of Tracy Potter.)

EQUATION BOX 11.2

Lacking microstructural problems, strength σ varies with density as follows:

$$\sigma = \sigma_0 K \left(\frac{\rho}{\rho_T} \right)^m$$

where σ_0 is the wrought strength of the alloy, K is a geometric and processing constant similar to a stress concentration factor, m gives the exponential dependence on density, ρ is the density, and ρ_T is the theoretical density. The ratio ρ/ρ_T is the fractional density. Elastic modulus follows a similar relation, with m typically near 0.35. For brittle steels, the normal strength measure is via three-point bending in the transverse rupture test. Typically, transverse rupture strength is 1.6 times the ultimate strength, and theoretically it should be 1.4 times.

to the close spacing of pores in sintered materials. Thus, the shape, spacing, size, and placement of pores contribute to a high level of variability in ductility in P/M materials. With greater than 15% porosity, ductility is often negligible.

In a study involving hundreds of sintered Fe–2Cu–0.8C samples, the standard deviation in measured hardness was 0.8 HRB, tensile strength was 32 MPa (4.5 ksi), while the ductility variation was 1.7%. This scatter in tensile tests suggests the expected accuracy in mechanical property testing.

11.2.2. Dynamic Properties

Although quasistatic mechanical properties are degraded by porosity, a more severe detriment occurs in the dynamic properties. These are the properties of impact strength, fracture toughness, and fatigue strength. They are also sensitive to microstructure. At low sintered densities, it is unwise to subject sintered steel to cyclic stresses or impact loads. At high densities, the behavior depends on the microstructure, and is affected by the pore size and pore spacing.

For a common P/M alloy of Fe–2Cu–0.8C (FC-0208), the fatigue endurance limit at 10^7 reverse cycles is 265 MPa (38 ksi) for a density of 7.15 g/cm^3, which is approximately 10% porosity. That is 35% of the tensile strength, and many porous P/M materials exhibit similar low ratios of fatigue strength to tensile strength. When the sintered density is increased to 7.5 g/cm^3 (5% higher density), the fatigue endurance limit increases by 50% to 400 MPa (57 ksi). The role of porosity is well known in low-alloy (Ni–Mo–C) steels; for example, the fatigue endurance strength for hot forging and 0% porosity is 650 MPa (95 ksi), injection molding with 3% porosity is 575 MPa (83 ksi), and die compaction with 12% porosity is 390 MPa (57 ksi). Because the fatigue properties are low for porous P/M steels, many tests are done on polished samples to give higher values. Note that the as-sintered surface gives a

15% lower fatigue strength. Further, notches reduce the fatigue strength dramatically. Generally the fatigue strength is 20–25% of the tensile strength for notched porous P/M materials. For example, the Fe–2Cu–0.8C (FC-0208) fatigue strength drops by 40% when a notch of 0.5 mm radius is present.

One growing application for ferrous powder metallurgy is in gears. During the operation of a gear, the loading on each tooth produces a cyclic bending stress. The allowed stresses are low if a gear is to survive for millions of revolutions. Little data are published on this subject, but much effort has been directed toward estimating the maximum allowed stress. A real concern is that the fatigue data usually only report 50% survival, which is unsatisfactory since half fail. Accordingly, conservative design rules estimate that 99% survival occurs at a stress 15% lower than the normally reported fatigue strengths. Further, the maximum stress in a gear can be only about 32% of that stress. The crucial point, allowed stresses in gears are far below the ultimate tensile strength. Further, if the gear is subjected to sudden loading or high vibration, then the allowed stresses are now as low as 10% of the ultimate tensile strength. Unfortunately, P/M gears can tolerate less stress than can wrought and cast steels, where the base strength is higher and the allowed fatigue stress is 50%, not 35%, of the ultimate tensile strength. The trend toward long-term warranties creates much concern over possible delayed fatigue failure.

Pores play a role in both the formation of fatigue cracks and their propagation through the structure. Under dynamic loading, the crack grows ever so slightly on each stress cycle. Thus, with millions of stress cycles the crack eventually extends to cause failure. Many efforts are used to improve this weakness in P/M steels, including shot peening after sintering, copper infiltration, high-temperature sintering, and surface carburization. However, pores are the major difficulty.

The impact toughness and fracture toughness are also sensitive to pores. The impact toughness measures the energy needed to fracture a sample, while the fracture toughness measures the energy needed to grow a crack in an already defective structure. Porous P/M steels are poor in these tests. Most impact toughness data are taken on unnotched samples, since the standard Charpy V-notch gives almost no toughness. The toughness can be improved by attaining full density.

11.3. PROCESSING FACTORS

11.3.1. Density

Density is a dominant factor with respect to the properties of P/M steels. Figure 11.2 illustrates this characteristic using the tensile properties for an alloy steel fabricated by two techniques. Note that the strength increases by nearly 25% with a density change from 7.0–7.6 g/cm^3, while impact toughness increases by a factor of 6. Although not shown, ductility increases by a factor of 10. Table 11.1 shows similar effects for a 316L stainless steel. The table and figure make a compelling case for a density role on mechanical properties.

Studies with different iron powders find some differences. For example, with six

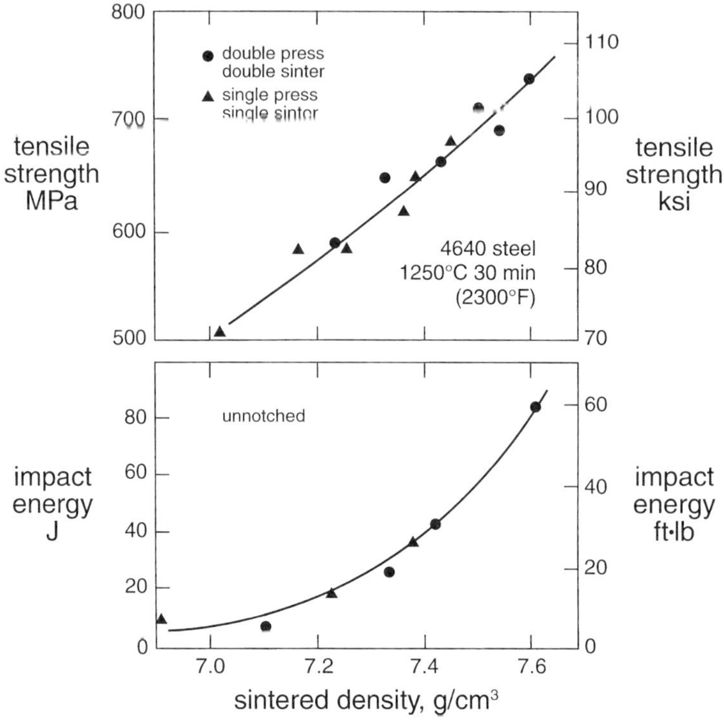

Figure 11.2. Tensile and impact property changes with sintered density using several different processing technologies, demonstrating the dominant effect of density. The alloy is a 4600 steel with 0.4% carbon, sintered at 1260°C (2300°F) for 30 min in a nitrogen–hydrogen atmosphere. (Data from Tina Cimino.)

different iron powders, the fracture elongation for a 7.0 g/cm^3 density varied by a factor of 2. However, for all powders, a density change from 6.0 to 7.0 g/cm^3 gave over a fivefold increase in the ductility. Similar findings show there is a role of powder, sintering, and other factors, but essentially nothing can make up for density. Only in applications where porosity is useful, such as in bearings, is there a perfor-

TABLE 11.1. Density Effects of Tensile Properties of a Sintered 316L Stainless Steel

Density (g/cm^3)	Porosity (%)	Elongation (%)	Yield Strength (MPa)	Tensile Strength (MPa)	Elastic Modulus (GPa)
6.31	20.6	11.1	176	308	103
6.66	16.2	14.2	271	417	117
6.82	14.2	15.9	280	444	128
6.94	12.7	18.5	289	468	144

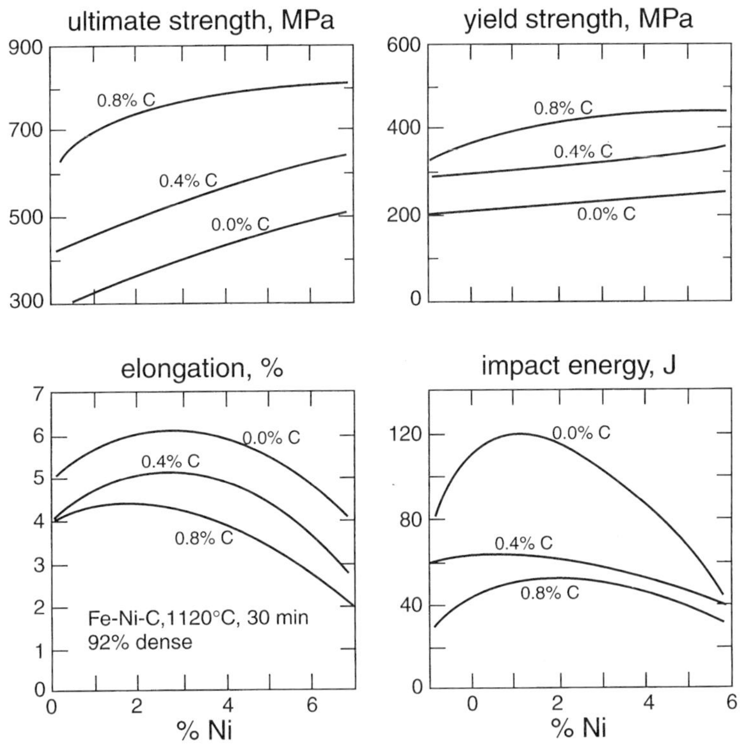

Figure 11.3. Composition affects on tensile and impact properties of iron–nickel–carbon alloys sintered to a density of 92% densities using a temperature of 1120°C (2050°F) for 30 min.

mance loss with higher densities. Because most P/M production is targeted to high performance levels, the goal is higher sintered densities.

11.3.2. Composition

Evidence for a density role in the mechanical properties is apparent throughout P/M. Usually the alloying effect is less clear. For example, Figure 11.3 plots the combined nickel and carbon effects on mechanical properties at a constant density of 92% of theoretical (with the varying alloying the theoretical density changes). Strength increases with alloying, but ductility and toughness peak at 2% nickel and decrease with high carbon contents. Another example using the iron–copper–carbon composition is presented in Figure 11.4 as a topographical map of transverse rupture strength. Peak strength occurs at 5% copper and 1% graphite. Thus, unlike density, alloying has a less systematic impact on mechanical properties. In part this traces to microstructure changes with various alloying elements and concentrations.

As a partial view of composition effects, Figure 11.5 shows the relative ranking

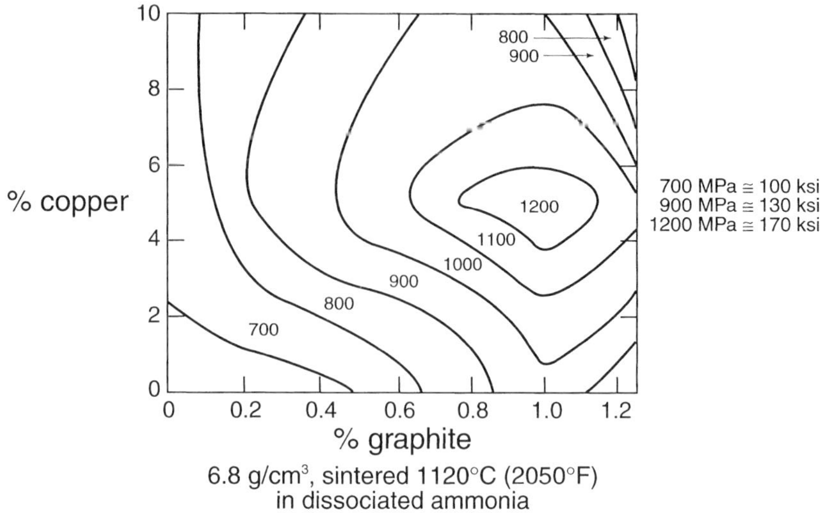

Figure 11.4. A contour map of equal transverse rupture strength for iron–copper–carbon alloys pressed to 6.8 g/cm^3 and sintered in dissociated ammonia at 1120°C (2050°F). (Data from C. Durdaller.)

of tensile strength for many steels. To normalize the density effect, each alloy is presented at a density of 7.0 g/cm^3. Although ignoring many other factors, this plot helps rationalize the role of various alloying additions. An offsetting concern is in the increased processing difficulty for each alloy. Likewise, consideration might be required for corrosion or magnetic response, cost, or heat treatment response. Some alloys prove temperamental and can add to dimensional control problems in production, making final alloy selection a subjective decision. Many ferrous alloys are heat-treated, so final properties can be effectively doubled. For example, the Fe–2Cu–0.8C alloy heat-treats to a tensile strength of 925 MPa (132 ksi) versus the 500 MPa (73 ksi) observed after sintering.

New alloys are emerging for sinter-hardening response, meaning that they are spontaneously heat-treated during cooling from the sintering temperature. Such alloys usually contain chromium or manganese for improved properties after heat treatment. For example, an alloy with 0.85Mn–1.25Mo–1.1Ni–0.8C with a small amount of chromium delivers a tensile strength of 830 MPa (120 ksi) in the as sintered condition. Additionally, higher densities and higher alloying levels deliver heat-treated strengths of 3000 MPa (425 ksi).

11.3.3. Microstructure

An understanding of processing effects on properties is often available via the microstructure. Even for a constant density there is considerable variation in pore size, grain size, and second-phase content. Pores can be rounded or angular, but the former

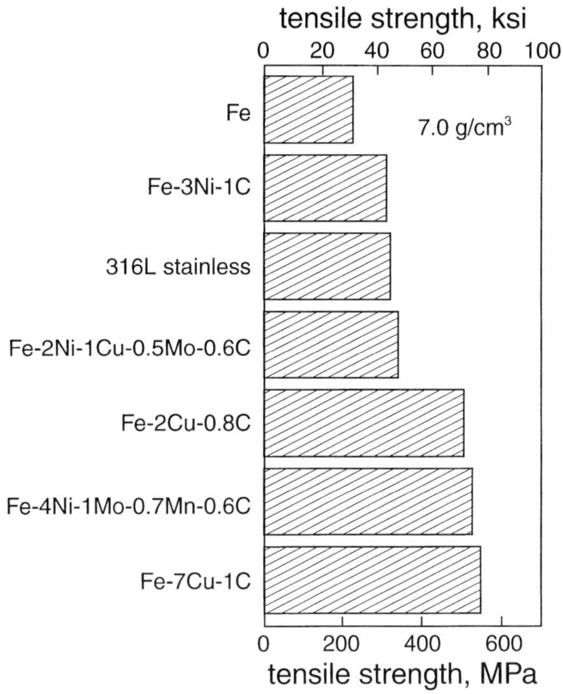

Figure 11.5. The relative strength of various alloys available by P/M. The actual ranking depends on many factors, including sintered density, sintering temperature, alloying uniformity, postsintering heat treatment, microstructure, and impurity content.

are most desirable and often require high sintering temperatures. Figure 11.6 is an optical micrograph from a cross-sectioned sintered steel with large and angular pores. This material exhibited low strength and poor fatigue resistance.

High-temperature sintering increases the mechanical properties, especially dynamic properties, largely because of changes in microstructure homogeneity and pore shape. Inhomogeneous alloying is observed in steels formed from mixed powders and sintered for too short a time. Even the cooling rate from the sintering temperature affects properties, with fast cooling giving twice the tensile strength of slow cooling for alloyed steels. For instance, a prealloyed 4600 steel (2Ni–0.9Mo) with 0.5% carbon and 2% copper sintered to 7.0 g/cm^3 in vacuum [1250°C (2300°F)] has a strength of nearly 600 MPa (85 ksi) when slow-cooled, but 1190 MPa (170 ksi) when rapidly cooled.

Other detrimental features include inclusions, such as oxides, or contamination in the microstructure. For example, Figure 11.7 shows how small concentrations of oxide inclusions lower the impact toughness of hot-forged 4340 P/M steel. A dramatic property loss is associated with a low contamination level. Consequently, the purity specifications are higher on powders, lubricants, and graphite additions used in hot forging as compared with die compaction.

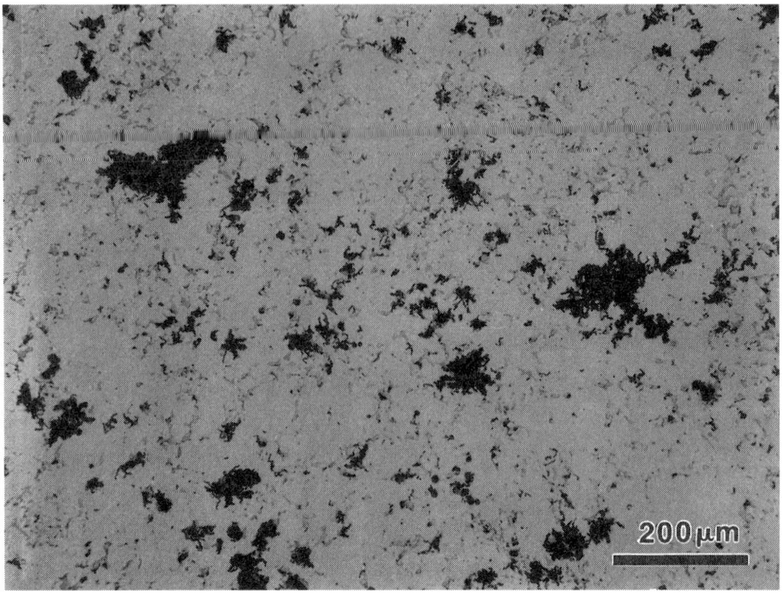

Figure 11.6. Angular pores in a sintered iron–copper–carbon steel. This micrograph was prepared by polishing the cross-sectioned material with epoxy infiltrated in the pores. (Photograph courtesy of Kevin Christian.)

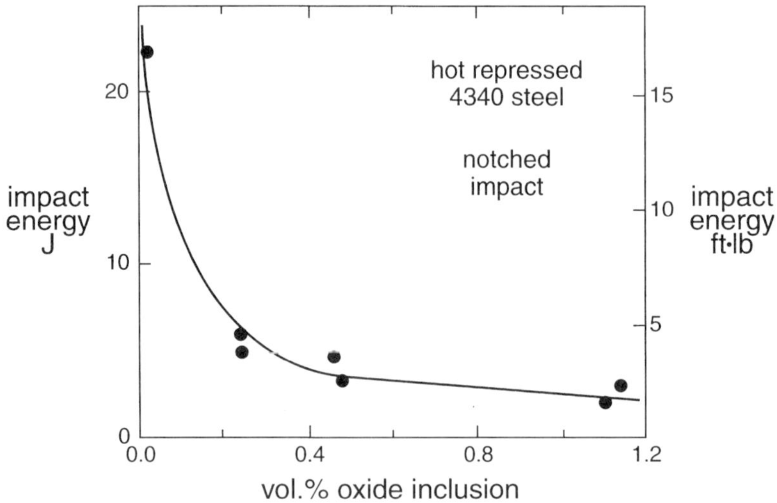

Figure 11.7. Impact toughness as a function of the oxide inclusion content in a hot-repressed 4340 steel, showing the negative effect from a small level of contamination. (Data from G. Bockstiegel and C. A. Blande.)

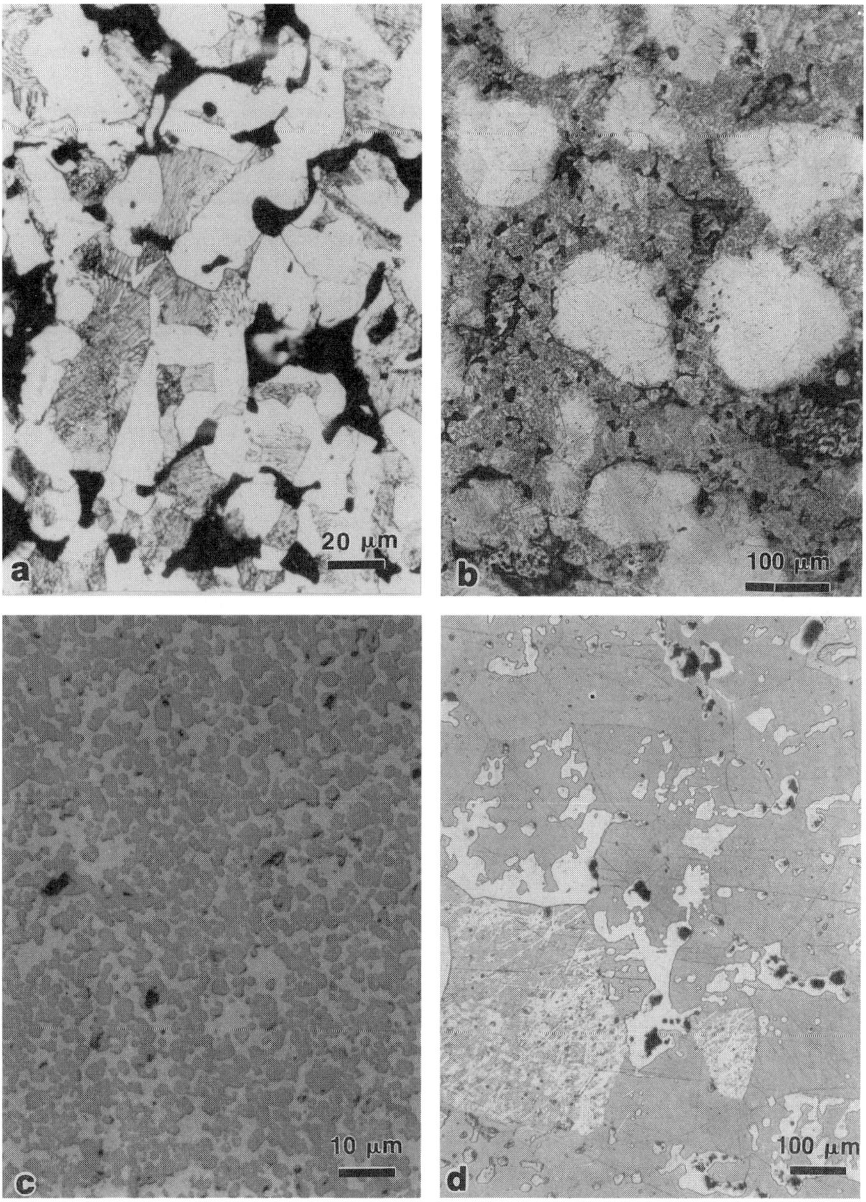

Figure 11.8. Optical micrographs illustrating the diversity of structures possible in ferrous P/M: (a) a low-alloy steel with approximately 0.4% C and a mixed pearlite–ferrite microstructure (pores are black) at a density of 6.8 g/cm^3; (b) mixed iron–manganese powders sintered to produce a two-phase microstructure; (c) mixed tool steel–titanium carbide composite formed by liquid-phase sintering; (d) iron–silicon sintered to near full density. (Photographs courtesy of James Marsden, James Dunlap, and Barbara Lograsso.)

Microstructural analysis starts by preparation of a sample cross section, which is polished, etched using acids, and finally examined in a reflected light microscope (known as a *metallograph*). The image observed in the microscope can be digitized and quantified to measure many parameters that influence mechanical properties—including pore size, pore shape, pore spacing, and grain size. For iron and steels, the common enchants include ethanol and nitric acid mixtures (nital) or other acids such as dilute hydrochloric acid or dilute copper sulfate or ferric chloride. The diversity of observable microstructures is large, as illustrated by four examples in Figure 11.8. These show a moderate carbon steel, a heterogeneous iron–manganese structure, a composite of tool steel and titanium carbide, and a liquid phase sintered iron–silicon alloy. Interpretation of the microstructure in light of the properties is important for improved P/M products.

An alternative is to rely on electron microscopy, especially scanning electron microscopy. This, too, can be coupled to an image-analyzing computer. Fracture surface analysis is best done using the scanning electron microscope. Here differences in fracture mode help explain property differences. Figure 11.9 is a scanning electron micrograph of the fracture surface from a brittle sintered steel. Flat fracture

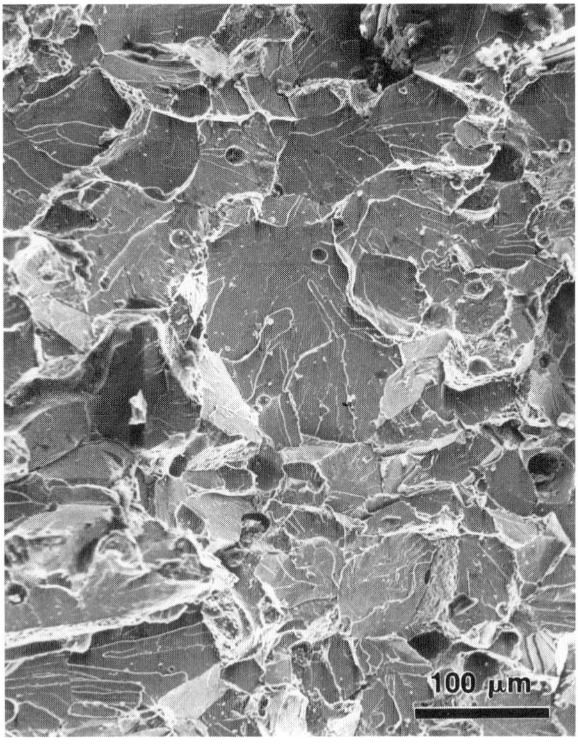

Figure 11.9. A scanning electron micrograph of the fracture surface of a brittle sintered steel, showing cleavage of the grains. (Photograph courtesy of James Dunlap.)

surfaces are evident at the boundaries between iron particles, showing that the grain boundaries were embrittled in sintering. The properties presented in the balance of this chapter represent nominal processing. Variations in density, alloying, or microstructure have important effects on these properties.

11.4. MECHANICAL PROPERTIES

In this section the mechanical properties of various P/M steels are summarized. Much data exist on this subject, especially as affected by processing changes. Without detailing these effects, the following tabulations are geared to providing an overview of the main alloys, properties, and behavior trends. References listed at the end of this chapter provide details. Further, most manufacturers provide tables of typical properties based on their particular processing cycles. However, not all such data are collected using the same procedures. Many companies use machined and polished tensile bars to obtain artificially high properties. It is recommended that only as-sintered surfaces be used for valid testing. For this treatment on mechanical properties, we will subdivide the subject by processing differences.

11.4.1. Die-Compacted

The classic die compaction–sintering cycle is a cost-effective production route for P/M. If the mechanical properties and component attributes (shape, precision, features) satisfy the application requirements, this is usually the process of choice. Assuming good manufacturing practices, we can examine some typical alloys and mechanical properties.

Table 11.2 summarizes the main alloy classes encountered in ferrous press and

TABLE 11.2. Common Ferrous P/M Alloy Classes

Alloy Class	Designation	Alloying	Ranges	Common Composition
Pure iron and steel	F-00xx	Fe with x.x% C	Max 1% C	0.8% C
Copper steel	FC-xxyy	Fe, xx% Cu, y.y% C	1–22% Cu, max 1% C	2% Cu, 0.8% C
Iron–nickel	FN-xxyy	Fe, xx% Ni, y.y% C	1 to 3% Ni, max 2.5% Cu, max 0.3% C	2% Ni, 0.6% C
Nickel steel			1–8% Ni, max 2.5% Cu, max 1% C	2% Ni, 2% Cu, 0.6% C
Low-alloy steel	FL-abcx FLN-abcx FLC-abcx	Fe; abc designates alloy class, 0.x% C N = Ni, C = Cu	Max 4% Ni, max 1% Mo, max 0.8 C	2% Ni, 0.8% Mo, 0.6% C
Infiltrated steel	FX-xxyy	Fe, xx% Cu, y.y% C	Max 25% Cu, max 1% C	15% Cu, 0.6% C
Stainless steel	Various	300 series contain Ni, Cr 400 series contain Cr	Max 22% Cr, max 13% Ni, max 3% Mo	18% Cr, 8% Ni
Tool steel	Various	M = molbdenum, T = tungsten	Many combinations of W, Mo, V, Cr, Ta, Nb, and max 1.5% C	6% W, 5% Mo, 4% Cr, 2% V, 0.8% C

TABLE 11.3. Representative Mechanical Properties of Die Compacted and Sintered Steels

Alloy	Condition[a]	Density (g/cm^3)	Yield (MPa)	Tensile Strength (MPa)	Elongation (%)	Impact (J)[b]	Hardness
0.3C (1030)	AS	7.4	180	275	15	34	30 HRB
0.8C (1080)	AS	7.0	260	390	3	7	70 HRB
0.8C (1080)	HT	7.1	630	660	Nil	7	35 HRC
1Cr–0.7Mn–0.3Mo–0.7C	HT	7.1	—	1200	1	30	42 HRC
2Cu–0.8C (FC-0208)	AS	7.0	410	610	2	8	77 HRB
2Cu–0.8C (FC-0208)	HT	7.1	490	720	Nil	7	43 HRC
3.5Mo	HT	7.6	230	310	25	—	105 HRB
2Ni–0.5C (FN-0205)	AS	6.9	210	350	3	16	59 HRB
2Ni–0.5C (FN-0205)	HT	6.9	650	830	Nil	6	29 HRC
2Ni–0.9Mo–0.5C (4650)	AS	7.3	460	750	4	8	93 HRB
2Ni–0.9Mo–0.5C (4650)	HT	7.2	—	1070	Nil	9	39 HRC
2Ni–1.5Cu–0.5Mo–0.6C	AS	7.2	530	700	3	—	95 HRB
0.45P (45P)	AS	7.4	245	370	14	45	60 HRB
12Cr (410L)	AS	6.9	190	340	20	80	50 HRB
16Cr (430L)	AS	7.2	240	380	22	145	62 HRB
17Cr–1Mo (434L)	AS	7.1	250	375	18	100	57 HRB
18Cr–8Ni (304L)	AS	6.5	280	390	10	34	62 HRB
19Cr–11Ni (304L)	AS	6.8	315	480	15	66	67 HRB
17Cr–12Ni–2Mo (316L)	AS	6.6	140	280	19	47	20 HRB
17Cr–13Ni–2Mo–0.3N (316LN)	AS	7.2	240	380	22	145	62 HRB
17Cr–12Ni–2Mo–2Sn–1Cu (316LSC)	AS	6.9	170	345	22	95	39 HRB

[a] AS = as-sintered; HT = heat-treated.

[b] Unnotched impact energy per square centimeter.

sinter powder metallurgy. Already in this book, mechanical properties for several of these alloys have been used as examples of processing and density effects. Table 11.3 lists several alloys and typical mechanical properties. When the alloy exhibits less than 1% elongation to failure, the yield and tensile strengths are essentially the same. This is a common feature of porous P/M alloys. Accordingly, the yield strength is not reported. In such cases, it is appropriate to rely on the transverse rupture strength, since this is applicable to brittle alloys. Figure 11.10 plots the correlation between the measured tensile strength and the transverse rupture strength for many ferrous alloys. As evident, the transverse rupture strength is probably 1.6 times the ultimate tensile strength, but highly scattered.

The last entries in Table 11.3 are various stainless steels, 430L, 304L, 316LSC, and so on. The 400 series are largely iron–chromium, while the 300 series contain nickel (austenitic stainless steels). The designation of L indicates low carbon (below 0.03%), the N designates nitrogen-alloyed, while the SC indicates copper and tin alloying (C for Cu and S for Sn). Although not complete, this tabulation provides some overview on the relative property levels obtainable.

The elastic modulus scales with density and is not significantly different between these alloys. At a density of 6.2 g/cm^3 the elastic modulus is about 110 GPa (16 Mpsi) and Poisson's ratio is 0.25 with a transverse rupture strength about 1.5 times the ultimate tensile strength. When the density reaches 7.2 g/cm^3, the elastic modulus is 155 GPa (22 Mpsi) and Poisson's ratio is 0.28 with a transverse rupture strength

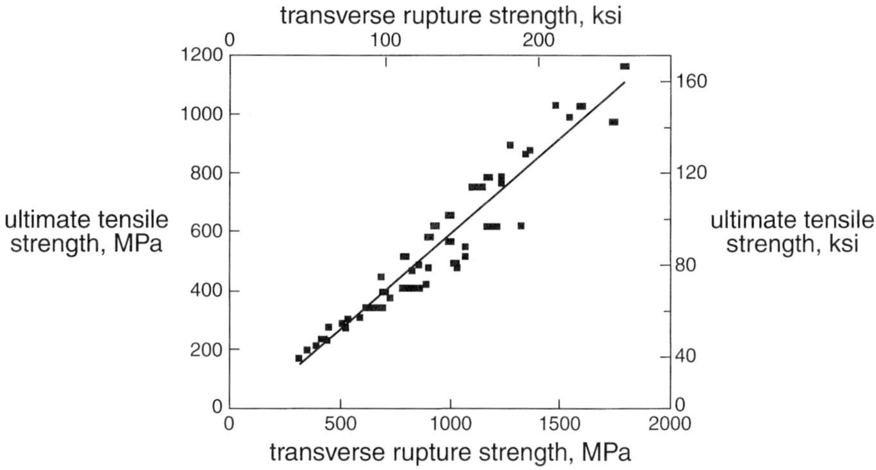

Figure 11.10. A correlation between the measured transverse rupture strength and tensile strength for many sintered steels. At high strengths the correlation is poor, but at low strengths there is a good relation. Generally, the transverse rupture strength is 1.6 times the tensile strength, with a 93% confidence level.

about 2.0 times the ultimate tensile strength. For comparison, full dense steel has an elastic modulus near 200 GPa and Poisson's ratio is near 0.3. Mosca provides the most comprehensive tabulation of mechanical properties for die-compacted steels, and he suggests a monotonic relation between Poisson's ratio and density and elastic modulus and density, independent of the alloy.

Little is known about the mechanical properties of porous P/M steels at elevated temperatures. One study measured the tensile properties after one hour in air at 675°C (1250°F) using two 400 series stainless steels. At densities in the 7.1 g/cm^3, the tensile strength was in the 70–80 MPa (about 10 ksi) range, which represented a fivefold reduction from the room-temperature tensile strength. It is suspected that longer exposure times will greatly degrade strength.

The properties of die-compacted steels are highly variable, with many processing factors affecting behavior. Thus, the tabulated data show only general trends. Even the lubricant can change sintered strength by 20%, the graphite source can change strength by 8%, and the iron powder source will change strength by 30%. Thus, depending on such factors, a twofold shift in properties occurs between manufacturing operations.

11.4.2. High-Temperature-Sintered

Sintering temperature is important and is the subject of much study. For example, a common stainless steel (316L) exhibits a twofold increase in yield strength when the sintering temperature goes from 1120°C (2050°F) to 1200°C (2200°F). Consequently, much attention is being devoted to the higher sintering temperatures. Table

TABLE 11.4. Mechanical Properties of High Temperature Sintered Alloys

Composition	Yield Strength (MPa)	Tensile Strength (MPa)	Elongation (%)	Comments
3.5Ti	—	375	19	Corrosion-resistant
4Al	—	300	10	Corrosion-resistant
2Cu–0.8C	—	830	3	High strength
2Cu–2Ni–0.9Mo–0.8C	700	830	4	High toughness and hardness
3Mo–2.4Ni–0.2B	—	650	5	High impact toughness
12Cr–1Mo–1Ni—0.5Mn–0.2C—0.02B	670	800	16	Elevated temperature strength
22Cr–18Ni–3Mo–1Si–0.8B	—	670	23	Corrosion-resistant

11.4 illustrates some ferrous alloys and their strengths when sintered at high temperatures.

Fatigue strength is highly variable in the die-compacted ferrous alloys. Tables 11.5 and 11.6 provide values from two fatigue studies. The first shows the effect of density for the Fe–2Cu–0.8C (FC-0208) alloy, and the second includes heat treatment effects for a higher-copper-content Fe–10Cu–0.3C (FC-1003). Rotating-beam tests are the most frequently applied to die-compacted steels. Unfortunately, these results poorly reflect service behavior, since the rotating-beam test uses a machined and polished surface, but service usually relies on the as-sintered or heat-treated surface that might contain notches. Accordingly, useful fatigue strengths are probably 20% lower than those typically reported in industry brochures.

As an additional concern, the fatigue strengths are for 50% survival, meaning that half the samples failed at that strength. Normally the error in this value is about 5%. Because of these factors, much conservatism is required when using fatigue data for design purposes. In light of these cautions, Table 11.7 summarizes several fatigue evaluations based on the rotating-beam test. Most fatigue strengths range near 35% of the tensile strength. However, stress concentration notches in the component and surface pores greatly lower these values. Thus, many P/M materials in

TABLE 11.5. Example Density Effects on the Mechanical Properties of Fe–2Cu–0.8C

Density, g/cm^3	6.65	6.85	7.15
Yield strength, MPa	360	400	415
Tensile strength, MPa	425	495	620
Elongation, %	1.3	1.8	2.5
Rupture strength, MPa	890	1025	1325
Fatigue strength, MPa	168	198	266

Source: Data from Kevin Christian.

TABLE 11.6. Density Effects on Properties of a P/M Structural Alloy (Fe–10Cu–0.3C)

Density (g/cm^3)	6.4	6.4	7.1	7.1
Thermal condition	Sintered	Heat-treated	Sintered	Heat-treated
Hardness	50 HRB	25 HRC	80 HRB	40 HRC
Yield strength, MPa	280	—	395	655
Tensile strength, MPa	310	380	550	690
Elongation, %	0.5	0.5	1.5	0.5
Fatigue strength, MPa	115	145	210	260
Impact energy, J	4	Nil	11	Nil
Elastic modulus, GPa	90	90	130	130

TABLE 11.7. Fatigue Strengths (50% Survival at 10^7 Fully Reversed Stress Cycles)

Composition	Processing	Density (g/cm^3)	Fatigue Strength (MPa)
0.6C (1060)	1120°C, 0.5 h	6.9	210
0.8C (1080)	1120°C, 0.5 h HT[a]	6.9	250
1Cr–0.7Mn–0.5Mo–0.7C	1250°C, 1 h	7.0	375
2Cu–0.5C (FC-0205)	1120°C, 0.5 h	7.1	125
2Cu–0.8C (FC-0208)	1120°C, 0.5 h	6.7	165
2Cu–0.8C (FC-0208)	1120°C, 0.5 h	7.0	235
2Cu–0.8C (FC-0208)	1330°C, 1 h	7.1	270
2Cu–2Ni–0.8C	1120°C, 0.5 h, HT	7.0	240
3.5Mo	1120°C, 0.5 h, HT	7.6	170
2Ni–0.8C (FN-0208)	1120°C, 0.5, HT	7.1	160
2Ni–0.8C (FN-0208)	1120°C, 0.5	7.3	240
2Ni–0.8C (FN-0208)	1120°C, 0.5, HT	7.4	450
2Ni–0.5Mo–0.5C (4650)	1120°C, 0.5 h, HT	7.0	350
4Ni–1.5Cu–0.5Mo–0.6C	1250°C, 0.5 h	7.1	150
4Ni–1.5Cu–0.5Mo–0.6C	1250°C, 2 h, HT	7.1	390
4Ni–1.5Cu–0.5Mo–0.6C	1250°C, 2 h	7.1	195
4Ni–1.5Cu–0.5Mo–0.8C	1250°C, 2 h, HT	7.1	420
7Ni–2Cu–1Mo–0.6C	1250°C, 0.5 h	7.0	390
12Cr (410L)	1230°C, 0.5 h	6.9	130
16Cr (430L)	1230°C, 0.5 h	7.2	195
17Cr–1Mo (434L)	1230°C, 0.5 h	7.0	180
19Cr–11Ni (304L)	1290°C, 0.5 h	6.9	160
17Cr–13Ni–2Mo–0.3N (316LN)	1280°C, 0.5 h	6.6	120

[a] Heat-treated.

fatigue-sensitive applications are subjected to high-temperature sintering (pore rounding) and shot peening (pore closure) to improve fatigue life. The highest fatigue strengths are obtained by selected combinations of alloying, densification, high-temperature sintering, heat treatment, and surface finish.

11.4.3. Warm-Compacted

Closely related to die compaction is the option of heating the powder and die to about 150°C (300°F) during compaction. Usually lubricants are dispersed on the powder surface to assist in bonding the particles in the die, giving higher green densities as compared with traditional compaction. Various polymers are employed in warm compaction, with the quantity of polymer being critical to success. Good properties can be obtained using just lithium stearate, but more sophisticated systems exist. This is a newer compaction option, but has similar compositions, densities, and properties as compared with die compaction.

Consider an alloy consisting of Fe–4Ni–1.5Cu–0.5Mo–0.8C, sintered at 1120°C (2050°F). At a sintered density near 7.2 g/cm^3, the alloy properties are given in Table 11.8. When compared on an equivalent-density basis, the two techniques give similar mechanical properties. Indeed, with the lower carbon level the conventional process gives higher mechanical properties. Heating melts the polymer to make compaction easier, although the equipment- and polymer-treated powders are more expensive. At very high densities, there are some property advantages with warm compacted powders, but tool life is reduced and high-temperature sintering is needed to attain the high properties.

11.4.4. Injection Molded

Most injection-molded steel is processed to near full density. Thus, the mechanical properties converge to handbook values. The elastic modulus and Poisson's ratio are close to wrought values, and properly heat-treated steels behave like wrought

TABLE 11.8. Comparison of Warm and Conventional Compacted Steels[a]

Alloy and Type	Density (g/cm^3)	Yield (MPa)	Tensile Strength (MPa)	Elongation (%)
0.3% C conventional	7.23	450	705	3
0.3% C warm	7.22	375	630	3
0.8% C conventional	7.18	345	705	3
0.8% C warm	7.21	365	720	3

[a] Fe–4Ni–1.5Cu–0.5Mo, two carbon levels, sintered 1120°C, 0.5 h, 25% nitrogen–75% hydrogen.

Figure 11.11. An optical micrograph of a vacuum-sintered eutectoid steel, showing almost no porosity and a very homogeneous sintered microstructure. (Photograph courtesy of S. T. Paul Lin.)

materials. For example, Figure 11.11 is an optical micrograph taken from a vacuum sintered eutectoid steel. There is almost no porosity, and the microstructure is fully pearlitic. Further, the fracture surface evidences much ductile tearing before failure, as shown in Figure 11.12. This is a scanning electron micrograph of the fracture surface of a Fe–2Ni alloy sintered at 1200°C (2200°F) for 1 h, with a subsequent quench and temper heat treatment. The final strength was more than 1120 MPa (160 ksi) with a hardness of 35 HRC. The favorable characteristic of ductile dimples is evident in this fractograph.

A summary of mechanical properties of injection-molded steels is provided in Table 11.9. Note that the densities are high in comparison with the die-compacted steels. With the high densities are high ductilities. Because of the high sintered densities, most injection-molded steels are subjected to postsintering heat treatments to maximize the final properties. However, sensitivity to binders and sintering options give variable final properties. Several reports give the properties of injection-molded 316L stainless steel. Sintered densities range from 7.4 to 8.0 g/cm^3, and the average yield strength is near 220 MPa (32 ksi), but the reported values range from

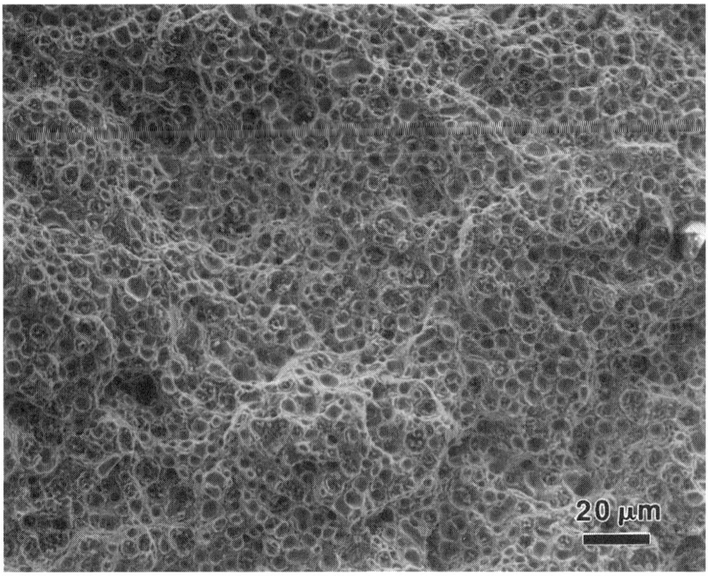

Figure 11.12. Scanning electron micrograph of the fracture surface from an injection-molded and sintered iron–nickel steel with a tensile strength of 1120 MPa (160 ksi) and 3% ductility. Note the uniform structure of dimples and much evidence of material flow in deformation. (Photograph courtesy of Haorong Zhang.)

170 to 345 MPa (about 25 to 50 ksi). Likewise, the elongation to failure ranges from 18 to 81%, with a mean of 45% and a standard deviation of 17%. Beyond a density effect, such property scatter reflects the lack of standardized processes, with concomitant variations in impurities and microstructure.

The dynamic properties, such as fatigue and impact toughness, depend on the pore structure. The limited porosity improves the resistance to crack propagation. Unfortunately, dynamic properties are not well tested. Early reports give fatigue endurance strengths of 219 MPa (32 ksi) for Fe–7Ni, 238 MPa (34 ksi) for Fe–2Ni–0.5C and Fe–7Ni–0.5C, 575 MPa (83 ksi) for 4640 steel, 230–275 MPa (33–40 ksi) for 316L stainless steel (vacuum versus hydrogen sinter), and 517 MPa (75 ksi) for 17–4 PH stainless steel. Hot tensile tests have been performed on injection-molded 316L stainless steel, showing a steady decline in strength as temperature increases, with yield strengths of 258, 177, 121, 71, and 62 MPa at temperatures of 180, 300, 500, 700, and 900°C. The ductility remained high, with more than 25% elongation up to 900°C.

11.4.5. Infiltrated

Infiltration of the pores in a sintered steel with copper is used to form intermediate-performance alloys with reduced corrosion rates and better machining. The copper

TABLE 11.9. Selected Mechanical Properties of Injection Molded and Sintered Steels

Composition (wt%)	Material	Density (g/cm^3)	Yield (MPa)	Ultimate (MPa)	Elongation (%)	Hardness
Fe–0.2C	1020 steel	7.5	185	380	23	—
Fe–1Cr–0.4C	4140 (HT[a])	7.4	1240	1380	2	40 HRC
Fe–2Cr–1Ni–1Mn–0.4C	4340 steel (HT)	7.6	480	620	6	20 HRC
Fe–2Ni–1Mo–0.4C	4640 steel (HT)	7.7	1400	2000	3	30 HRC
Fe–36Ni	Invar	7.8	240	425	40	65 HRB
Fe	Iron	7.5	105	220	35	50 HRF
Fe–2Cu–0.8C	Iron–copper steel	7.5	—	700	10	92 HRB
Fe–5Mo	Iron–molybdenum	7.9	210	410	34	66 HRB
Fe–8Ni	Iron–nickel	7.6	255	440	24	75 HRB
Fe–50Ni	Iron–nickel	8.0	170	420	20	50 HRB
Fe–2Ni–0.5C	Iron–nickel steel (HT)	7.4	1230	1230	1	45 HRC
Fe–2Ni–0.9C	Iron–nickel steel	7.6	450	650	9	90 HRB
Fe–7Ni–0.5C	Iron–nickel steel (HT)	7.7	1420	1460	1	46 HRC
Fe–0.6P	Iron–phosphorus	7.8	260	280	2	80 HRB
Fe–3Si	Iron–silicon	7.8	345	520	25	85 HRB
Fe–29Ni–17Co	Kovar or F15	8.3	350	520	42	60 HRB
Fe–16Cr–4Ni–4Cu	Stainless steel 17-4 PH (HT)	7.7	965	1140	12	35 HRC
Fe–18Cr–8Ni	Stainless steel 304L	7.7	240	480	35	85 HRB
Fe–17Cr–12Ni–2Mo–2Mn	Stainless steel 316L	7.7	220	510	45	75 HRB
Fe–21Cr–9Ni–3Mo–2Mn	Stainless 316L duplex	7.6	230	540	43	80 HRB
Fe–11Cr–0.5C	Stainless steel 410L (HT)	7.6	410	650	5	20 HRC
Fe–13Cr–1Mn–1Si	Stainless steel 420 (HT)	7.4	690	1440	6	47 HRC
Fe–17Cr–1Ni–1C	Stainless steel 440C (HT)	7.7	410	620	2	43 HRC
Fe–6W–5Mo–4Cr–2V–1C	Tool steel	8.1	—	2000	0	66 HRC

[a] Heat-treated.

level is usually in the 8–25% range. Table 11.10 lists the properties of common infiltrated steels. After infiltration the density is high, nearly equal to theoretical, but the mechanical properties are intermediate. This is because copper is a weak metal. For example, a steel alloy at 7.07 g/cm^3 with a tensile strength of 490 MPa (70 ksi) and 5% ductility, exhibits a tensile strength of 750 MPa (107 ksi) and 8% ductility after infiltration. However, most ferrous alloys do not exhibit this large a strength increase, although the impact toughness can increase dramatically. One case with a boron-doped copper infiltrant has delivered an impact toughness greater than 300 J.

TABLE 11.10. Example Mechanical Properties of Copper-Infiltrated Steels

Property	Fe–11Cu–0.5C	Fe–20Cu–0.5C
Yield strength, MPa	340	410
Ultimate strength, MPa	530	520
Elongation, %	4	2
Elastic modulus, GPa	160	145
Poisson's ratio	0.28	0.24
Impact energy, J	18	11
Transverse rupture strength, MPa	1090	1020

TABLE 11.11. Mechanical Properties of Hot-Forged or Repressed P/M Steels (All Values in Heat-Treated Condition)

Process and Alloy Type	Composition	Density (g/cm^3)	Yield (MPa)	Ultimate (MPa)	Elongation (%)	Hardness (HRC)
DPDS[a] modified 4600	4Ni–1.5Cu–0.5Mo–0.8C	7.4	475	1030	3	27
DPDS 4600 base	2Ni–0.5Mo–0.2Mn–0.3C	7.3	700	910	—	—
DPDS nickel–copper steel	3Ni–2Cu–0.6C	7.4	305	405	14	—
DPDS 4600 base	2Ni–0.5Mo–0.5C	7.5	590	800	19	—
Hot-forged 1040	0.4C	7.8	—	760	4	12
Hot-forged FC-0208	2Cu–0.8C	7.9	630	1025	10	33
Hot-forged FN-0204	2Ni–0.4C	7.7	—	1200	4	30
Hot-forged H-02	0.5Ni–0.6Mo–0.3Mn–0.3C	7.8	—	1035	5	35
Hot-forged 4640	1.8Ni–0.5Mo–0.2Mn–0.4C	7.9	1340	1620	7	50
Hot-forged 4660	1.8Ni–0.5Mo–0.2Mn–0.6C	7.8	1150	1630	3	55
Hot-forged 4680	1.8Ni–0.5Mo–0.2Mn–0.8C	7.8	1590	1970	3	50

[a] DPDS = cold repressing using a double-press and double-sinter cycle.

11.4.6. Repressed and Hot-Forged

Major growth in the use of P/M steels is occurring in high-stress, cyclic loading applications in the automobile. Connecting rods are a prime example. Accordingly, these are processed to near-full-density conditions by double pressing (cold repress forging) and double sintering or hot forging. Most of these components are heat-treated to further maximize mechanical properties. Table 11.11 summarizes common alloys and tensile properties from these routes. Note the high final densities and corresponding high mechanical properties.

One of the most popular alloys by powder forging is FC-0208, containing 2% copper and 0.8% carbon, which is also a favorite alloy for traditional die compaction. A strength gain of 30–40% occurs with hot forging, but the ductility is dramatically improved. Further, the fatigue and impact properties are greatly increased by the elimination of residual porosity. For example, the fatigue strength is near 500 MPa (72 ksi) in hot-forged FC-0208, giving a fatigue strength : tensile strength ratio near 0.45 as compared with a ratio of 0.33 for the porous counterparts. Most of the powder-forged steels peak at endurance strengths of less than 600 MPa (85 ksi), but a few have reached nearly 800 MPa (114 ksi). On the other hand, double-pressed and double-sintered alloys with lower densities peak at fatigue strengths near 450 MPa (64 ksi). Always these products are given a postsintering heat treatment to promote the higher properties.

Although impact toughness is not outstanding, powder-forged steels are much improved over traditional P/M alloys. Impact energies in the range 3–60 J are reported for the various alloys with a notch. Typical values for powder-forged 4600 steels are near 30 J (22 ft · lb). However, this alloy and process can deliver impact toughness values ranging from 5 to 80 J (3 to 60 ft · lb), depending on heat treatment, oxygen contamination, and carbon level. Finally, the crack propagation resistance, a property known as the *fracture toughness*, shows hot-forged P/M steels are competitive with wrought alloys.

TABLE 11.12. Tensile Properties of Full-Density Hot Isostatically Pressed Steels

Alloy	Composition	Heat Treatment	Yield Strength (MPa)	Tensile Strength (MPa)	Elongation (%)
1080	0.8C	None	—	385	18
1080	0.8C	Quenched	—	730	12
4600 steel	1.8Ni–0.5Mo–0.5C	None	590	800	19
A286	30Ni–14Cr–2Ti–1.3Mo–0.4V	None	380	700	38
A286	30Ni–14Cr–2Ti–1.3Mo–0.4V	Aged	740	1115	25
Maraging	19Ni–8Co–5Mo–0.4Ti	Aged	1515	1600	14
Tool steel	12Cr–1Mo–1Ni–0.6Mn–0.3V–0.24C	—	655	845	20
Tool steel	5Cr–1.4Mo–1V–1Si–0.4C	Tempered	1310	1430	21

11.4.7. Full-Density

Powder-forged compacts have 0 to 5% porosity, so they are close to the performance levels associated with wrought steels. Most of the data available for full-density products are for hot isostatically compacted steels. Usually the properties converge on those seen in handbooks for the same compositions as delivered through other metalworking routes.

Table 11.12 gives the mechanical properties of a few ferrous alloys processed to full density by HIP. For example, an alloy of Fe–2Ni–1.5Cu–0.5Mo–0.5C was consolidated to full density by HIP at 1160°C (2120°F), 105 MPa (15 ksi) with a 3-h hold. After heat treatment the resulting fatigue strength was 480 MPa (70 ksi), which is 3 times that obtained with processing based on die compaction. Thus, although the tensile strength is not much improved, dynamic properties are greatly enhanced.

Especially unique to P/M are the full-density, high-alloy tool steels and various particulate-reinforced composites. For example, excellent wear resistance is possible with a composite of titanium carbide TiC dispersed in a tool steel matrix. At 45 vol% carbide, the composite is near impossible to consolidate, so it is first liquid-phase-sintered and then hot isostatically pressed to full density. Properties of such composites are as follows: density of 6.5 g/cm^3, elastic modulus of 280 GPa (41 Mpsi), transverse rupture strength of 1280 MPa (185 ksi), tensile strength of 1400 MPa (200 ksi), compressive strength of 3300 MPa (480 ksi), and unnotched impact strength of 9 J (6 ft · lb). Since its main application is in forming tools, it is seldom subjected to tensile loading.

New composites rely on HIP borides in a tool steel matrix. These materials exhibit wear resistance superior to cemented carbides. One composition is a 50 vol% iron–molybdenum–boride in a tool steel matrix (high molybdenum grade) formed by HIP, giving a hardness of 81 HRA, transverse rupture strength of 1800 MPa (260 ksi), and elastic modulus of 350 GPa (50 Mpsi).

Use of the HIP tool steels is based on cutting and wear properties as opposed to tensile properties, so there is less emphasis on strength, fatigue, or toughness. They are consolidated and heat-treated to hardness levels in the 50–70 HRC range. Depending on the alloy composition, strengths range from 1200 to 5000 MPa (175 to 725 ksi), but with low ductilities. Impact toughness is usually less than 11 J (15 ft · lb). Fatigue strengths for the HIP tool steels can approach 1000 MPa (140 ksi). Very successful HIP tool steels use high vanadium contents to form a hard vanadium carbide phase that promotes wear resistance.

Many stainless steels processed to near full density are used in corrosion and high-temperature processing. Alloys such as iron-based superalloys are employed for jet engines, and various HIP composites have been developed for plastic processing machinery. In each case, the mechanical properties are of less concern than are the performance characteristics in an aggressive environment.

11.5. MAGNETIC PROPERTIES

Great interest exists in the soft magnetic properties of ferrous powder compacts. At low impurity levels and high densities, the magnetic properties are very attractive, especially for complicated shapes. Characterization of the magnetic response is determined by applying a magnetic field. The test usually involves initial magnetization and then reversal magnetization. In practice, this means increasing the current and then reversing the current direction in the applied magnetic field, which causes the north and south poles of the magnet to switch. An industry-standard test is based on generating a hysteresis loop (current applied in both directions in a cyclic manner). The measurement is taken from a toroidal test geometry as sketched in Figure 11.13. It is 50 mm (1.969 in) in outer diameter, 40 mm (1.575 in) in inner diameter, and 4 mm (0.157 in) in thickness. Primary and secondary coils are wound around the toroid, and the applied and induced magnetic fields are measured. A schematic response is shown on the right of the test setup. After the first magnetization, the hysteresis curve becomes apparent, with the area between the outer curves representing energy loss on field reversal.

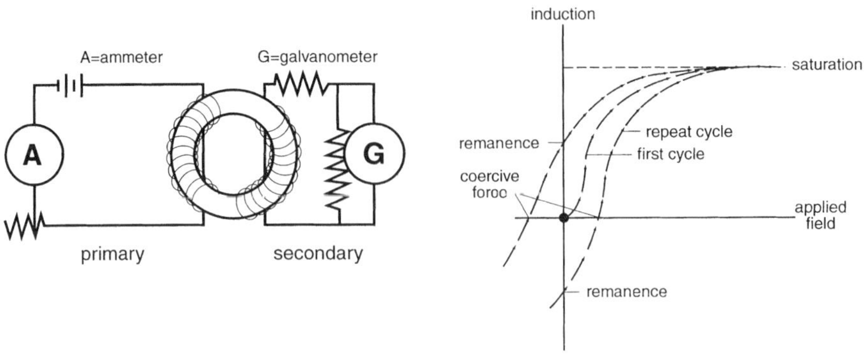

Figure 11.13. A schematic layout for the standard toroid test for magnetic performance and a typical cycle showing the definition of saturation magnetization, remanence magnetization, and coercive force.

Various magnetic properties are determined, including the initial rate of magnetization, maximum magnetization, and characteristics of the residual magnetization after the applied field is removed. Pores and impurities degrade magnetization. Most P/M materials are designed for soft magnetic properties, meaning that the material has no residual magnetization once the applied field is removed. This is useful in oscillating fields, since a high saturation makes a powerful magnetic when turned on, but gives no stray fields when turned off.

On the other hand, the typical magnet used to attach notes to the refrigerator door is a permanent or hard magnet. For direct-current motors and other application of that sort, we want a magnet that retains as much magnetization as possible when the field is removed. Only a few P/M alloys are useful in this regard, with considerable current interest in magnets derived from rapidly solidified iron–neodymium–boron powders.

The soft magnetic characteristics usually emphasize pure iron and a few simple alloys, including Fe–2Ni, Fe–3Si, Fe–0.45P, Fe–0.6P, and Fe–50Ni. Of these, the Fe–50Ni alloy has the most attractive combination of high magnetization and low coercive force. Table 11.13 compares the magnetic properties of P/M compositions. Unfortunately, experience shows considerable variability in these properties, especially the magnetic force needed to demagnetize an alloy. Part of this variation is due to differences in sintered density, but impurity control is also a major factor. Like strength, the magnetic characteristics are sensitive to contaminants.

One new use of iron powder is in a bonded structure for use in hand tools and automotive electrical systems. A polymer coating on the iron powder prevents electrical conduction, yet the powder responds magnetically to an applied field. These

TABLE 11.13. Typical Magnetic Properties of P/M Alloys[a]

Alloy	Resistivity ($\mu\Omega \cdot$ cm)	Maximum Magnetization (T)	Remanence Magnetization (T)	Coercive Force (Oe)	Maximum Permeability
410L	90	1.2	4.1	3.5	960
430L	93	1.1	4.8	2.7	630
434L	104	1.1	5.7	2.3	1300
Fe	20	1.6	1.2	1.4	4000
Fe–49Co–2V	40	2.0	5.6	3.0	3000
Fe–50Co	60	1.7	0.56	2.0	2000
Fe–5Mo	—	1.6	1.3	1.8	2900
Fe–2Ni	15	1.5	1.2	1.0	—
Fe–50Ni	40	1.3	0.9	0.3	25000
Fe–0.45P	21	1.4	1.2	1.1	4000
Fe–0.8P	23	1.8	1.3	0.4	6100
Fe–3Si	45	1.4	1.3	0.9	4500
Fe–6.5Si	81	1.3	1.2	0.3	4000

[a] Except for 400 series stainless steels at approximately 6.9 g/cm^3, all others are high-density products over 7.4 g/cm^3.

powders are compacted at 700 MPa (50 tsi), but not sintered. The polymer cures on heating to 175°C (315°F), giving a density near 7.2 g/cm^3, strength of 140 MPa (20 ksi), electrical resistivity of 200 $\mu\Omega \cdot$ m, and magnetic properties that mimic steel laminations used for alternators and other automotive applications.

11.6. CORROSION PROPERTIES

Corrosion involves chemical attack of the steel. We see corrosion as rust. Many tests and environments are used to test P/M materials, with most of the attention on sintered stainless steels. Tests range from immersion in household chlorine bleach to extended salt spray tests. The results depend on the test environment, so often a variety of chemicals, times, temperatures, and immersion-drying cycles might prove necessary to fully characterize a material. Some more popular tests involve saltwater or seawater immersion, saltwater sprays, nitric acid immersion, sulfide or sulfuric acid exposure, or ferric chloride immersion, with times of up to a few weeks. Corrosion resistance can be quantified by the time to first rusting, amount of surface rusted after a time, weight loss per area per time, or loss of strength after prolonged immersion. Stress corrosion is the most aggressive test, where the material is subjected to stress while immersed in the corrosive fluid, possibly at an elevated temperature (to simulate uses in petrochemical plants). Typically, only the stainless steels consolidated by HIP are subjected to this test.

When fully dense, P/M steels exhibit excellent corrosion resistance. Figure 11.14

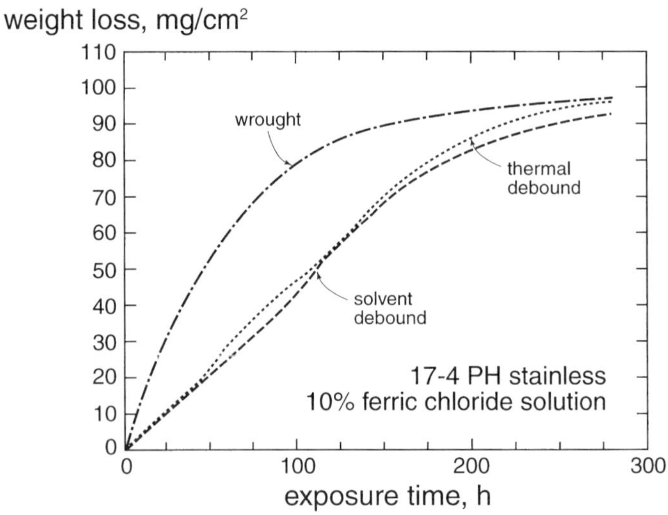

Figure 11.14. Weight loss during corrosion testing at room temperature of wrought and injection-molded 17-4 PH stainless steels in a 10% ferric chloride solution. (Data courtesy of Donald Kubish.)

illustrates such a situation by comparing the corrosion weight loss [milligrams per square centimeter (mg/cm^2) of sample area] versus time for 17–4 PH stainless steel in a 10% ferric chloride solution. The injection-molded samples were about 7.5 g/cm^3 after sintering, and there was no significant difference between debinding routes.

More typically, the corrosion resistance of powder metallurgy products is inferior to equivalent wrought materials. The higher rate of attack results from the pores that give more surface over which attack occurs. Also, the sintering cycles induce chemical reactions that are detrimental. A good example of the problem is encountered in P/M stainless steel. In a nitrogen-, oxygen-, or carbon-containing atmosphere, the chromium in a stainless steel reacts to form a compound such as a chromium oxide, nitride, or carbide. Consequently, the chromium is not available for protection against corrosion, resulting in accelerated attack, as apparent in Figure 11.15. Also noted is the strengthening effect with nitrogen additions of up 0.3% adsorbed nitrogen. These weight-loss results in nitric acid are from sintered 316L stainless steel with varying adsorbed nitrogen levels. The preservation of corrosion resistance in a sintered stainless steel requires uniform chromium distribution without the formation of compounds.

Table 11.14 gives the corrosion resistance of a few stainless steels produced by P/M. These are all die-compacted and high-temperature-sintered compositions. Alloys with the L after the number indicate low carbon, a necessary criterion to ensure corrosion resistance, meaning that total carbon is below 300 ppm or 0.03%. The LN designation indicates low carbon, but intentional nitrogen alloying from the sintering atmosphere for strengthening. Alloys with the LSC designation are low in

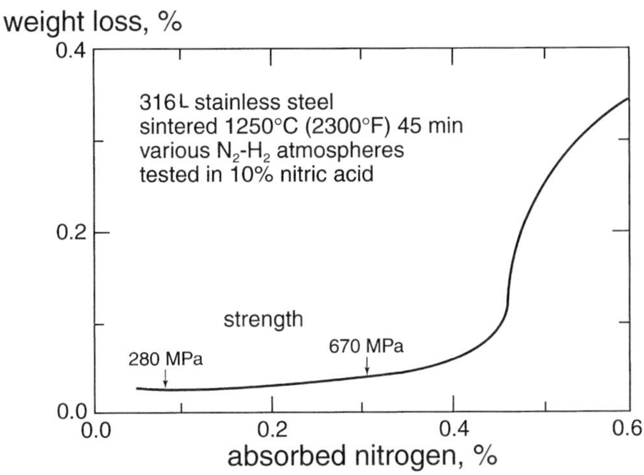

Figure 11.15. The absorbed nitrogen effect on corrosion in 10% nitric acid for sintered 316L stainless steel, processed at 1250°C (nearly 2300°F) for 45 min in various atmosphere mixtures of hydrogen and nitrogen. Attack accelerates at nitrogen contents over about 0.4%, but at lower nitrogen levels there is opportunity for a large strength increase. (Data of G. Lei.)

TABLE 11.14. Corrosion Ratings for Die-Compacted and High-Temperature Sintered Stainless Steels

Alloy	Density (g/cm^3)	Time (h) for 25% Rusting in 5% NaCl Solution	Iron Ion Release (ppm) in 5 NaCl Solution after 72 h	5% Salt Spray Time (h) to Rust 25% of Surface	2% Sulfuric Acid Immersion Weight Loss in 24 h in g/m^2
304L	6.5	984	4.19	>16	13
304L	6.9	528	1.57	>4	12
304LSC	7.0	1212	0.12	>48	−17
316L (impregnated)	6.5	696	—	—	2
316LN	6.6	48	4.05	>48	50
316LN	6.8	60	3.57	>36	13
316LSC	6.6	1632	0.07	4	—
316LSC	6.9	708	0.08	4	—
430L	6.9	18	—	—	4510
430L	7.2	48	—	—	3390

Source: Data from Howard Sanderow.

carbon and doped with tin (Sn) and copper (Cu) to improve corrosion resistance. Usually doping levels are 2% Sn and 1% Cu. The tabulation includes the time for 25% of the surface to be coated with rust during immersion in a 5% saltwater solution. Seawater is a more complicated environment, but would probably give similar results. Nitrogen alloying (316LN) is detrimental to corrosion resistance. The 316LSC probably is the most corrosion-resistant alloy of those tabulated, but it showed 25% rusting in just 4 h of salt spray exposure. On the other hand, the 400 series stainless steels are almost not stainless, especially when compared to the 300 series.

Enhanced corrosion resistance requires attention to three factors: density, sintering atmosphere, and composition. A high sintered density reduces the pore area over which corrosive attack occurs. The solubility for contaminants like nitrogen is reduced at higher sintering temperatures. Both factors point to better corrosion resistance with higher sintering temperatures. A sintering atmosphere consisting of pure hydrogen gives maximum corrosion resistance; however, the strength is lower since nitrogen strengthens stainless steels. Fast cooling from the sintering temperature also preserves a homogeneous chromium distribution for better corrosion resistance. Alternatively, pores impregnated with a polymer after sintering reduce the surface area in contact with a corrosive fluid and prolong component life. When properly processed without contamination, the high-density P/M products are corrosion-resistant, and are often superior to wrought materials for thin sections.

11.7. WEAR PROPERTIES

Like corrosion, wear is a parameter that depends on the interaction of the P/M material with its environment. Several forms of interaction are possible, but abrasive materials are usually involved. A standardized test for wear involves rotating a rubber wheel against the test sample with sand trapped between the rubber wheel and sample. The sand abrades and removes the test sample, with a higher rate of mass loss as the wear resistance decreases. After a predetermined number of revolutions by the rubber wheel (1000 or 10,000), the test is stopped and the mass loss measured and reported. Other tests might involve sliding, scratching, or rubbing two materials—one is a standard and the other is the test P/M steel. The test environment affects behavior.

A simple P/M steel shows rapid wear. If surface-hardened, for example, by heat treatment or case carburization, then the wear rate is greatly reduced. As an illustration, sintered iron with 0.5% carbon at a density of 7.1 g/cm^3 exhibits a 14,000-fold decrease in wear rate when heat treated in special cycle. This involves putting the sintered compact in coarse chromium granules and heating for 8 h at 1000°C (1832°F) to harden the compact surface. Generally, higher hardnesses give more wear resistance. Hence, many P/M steels are subjected to postsintering case carburization or other surface-hardening heat treatments to improve wear resistance. Tool steels exhibit best wear resistance when heat-treated to hardnesses near 60 HRC.

11.8. POROSITY CHARACTERISTICS

Only the open pores contribute to the flow-related properties such as filtration and permeability. Many porous stainless steels are used for high-temperature or corrosion-resistant filters, such as in filtering hot reactive gases. The flow of a gas or liquid through a porous material depends on the pore structure. To characterize the material, flow rate is measured versus the applied pressure. Because of the complex pore geometry of sintered steels, the flow is tortuous. As the fluid passes through the compact, the curves and twists in the pores prove effective in filtering particles from the gas or liquid.

Permeability is a measure of the flow resistance as determined by fluid flow as a function of the applied pressure. The permeability coefficient increases with pore size and porosity. Thus, it gives indirect information on the pore structure. Another test, termed the *bubble-point test*, is based on the pressure needed to push a liquid out of the pores. Both tests are used extensively in the specification of porous stainless steels. Table 11.15 lists the properties of various sintered stainless-steel filters.

TABLE 11.15. Filtration Characteristics of Porous P/M Stainless Steels

Filtration Rating, μm	0.5	2	5	10	40	100
Density, g/cm^3	6.0	5.3	4.9	4.6	3.9	2.7
Porosity, %	25	34	39	43	51	66
Average pore size, μm	8	18	20	33	79	136
Largest pore size, μm	10	21	26	42	114	347
Permeability, m^2	$8 \cdot 10^{-14}$	$8 \cdot 10^{-13}$	$2 \cdot 10^{-12}$	$6 \cdot 10^{-12}$	$3 \cdot 10^{-11}$	$1 \cdot 10^{-10}$
Strength, MPa	164	145	113	79	34	12
Elongation, %	3	2	2	2	1	2
Elastic modulus, GPa	61	54	35	34	17	12

EQUATION BOX 11.3

The relation between flow rate and pressure drop in a porous material leads to the property known as the *permeability coefficient* α determined from Darcy's law

$$Q = \frac{\alpha A}{\eta t}\left(\frac{P_1^2 - P_2^2}{2P_2}\right)$$

where Q is the gas flow rate (m^3/s), A is the sample cross-sectional area of the material, t is the sample thickness, η is the gas viscosity, and P_1 and P_2 are the upstream and downstream pressures, respectively. This equation is valid for low flow rates at moderate pressures, for example, when using pressures near atmospheric pressure.

The pore size from the bubble test is determined by soaking the filter in a wetting fluid, such as an alcohol. Then gas pressure is slowly applied to the pores (e.g., by pressurizing the inside with gas) and observation is made on the pressure required to first form a bubble on the surface. The corresponding maximum pore size d_P is determined by the pressure P where the first bubble forms

$$d_P = \frac{4\ \gamma \cos(\Theta)}{P}$$

where Θ is the wetting angle (usually zero degrees) and γ is the surface energy of the liquid.

Most stainless-steel P/M filters are formed by die or cold isostatic compaction to intermediate densities. An irregular pore shape results from use of water-atomized powders. The compacts are sintered to relatively low densities and are not very strong; thus filters are typically not subjected to high stresses. Fatigue resistance is quite poor, and fatigue is a main cause of failure in service.

11.9. THERMAL PROPERTIES

The thermal and electrical conductivities of P/M steels are lower than wrought counterparts, due to pores or nonconducting inclusions. Likewise, the coefficient of thermal expansion for a porous steel is lower than that for the equivalent wrought composition. Relatively few measures of thermal properties are known, and even these data are low in accuracy.

For the typical ferrous alloy with a density near 7.0 g/cm^3, the thermal expansion

EQUATION BOX 11.4

Conductivity κ depends on porosity as expressed by the following correlation:

$$\kappa = \kappa_0 \frac{1 - \epsilon}{1 - \chi\epsilon^2}$$

where κ_0 is the conductivity of the fully densified material and ϵ is the fractional porosity. The coefficient χ expresses the sensitivity to the presence of pores. This equation lacks internal structure-dependent parameters, but analysis of several stainless-steel powder compacts, representing a variety of pore sizes and shapes, gives a best-fit value of 11 for χ. In the low-porosity region, the relative conductivity follows a linear behavior with fractional porosity ϵ; thus

$$\kappa = \kappa_0(1 - \omega\epsilon)$$

where ω is between 1 and 2. This second model is most appropriate at porosities less than 30% for either electrical or thermal conductivity.

The thermal expansion coefficient C_T is also sensitive to pores and varies with density ρ as follows:

$$C_T = C_0\left(\frac{\rho}{\rho_T}\right)^{1/3}$$

where C_0 is the bulk thermal expansion coefficient corresponding to a theoretical density of ρ_T.

coefficient is between 11 and 13 $\cdot$ 10^{-6} °C^{-1} (6 and 7 $\cdot$ 10^{-6} °F^{-1}), thermal diffusivity is near 6 mm^2/s, thermal conductivity is near 50 W/(m $\cdot$ °C) for steels and 15 W/(m $\cdot$ °C) for stainless steels, and the specific heat is near 0.58 kJ/(kg $\cdot$ °C). Injection-molded iron–nickel alloys are used when a low-thermal-expansion coefficient is needed—these invariant-length alloys are known as invars. Injection-molded and sintered iron with 42% nickel has a sintered density of 8 g/cm^3, strength of 300 MPa (44 ksi), elastic modulus of 205 GPa (30 Mpsi), thermal conductivity of 20 W/(m $\cdot$ °C), and thermal expansion coefficient of 5 $\cdot$ 10^{-6} °C^{-1} (3 $\cdot$ 10^{-6} °F^{-1}). At full density, the thermal properties of P/M steels converge to equal those of wrought compositions.

SUGGESTIONS FOR FURTHER READING

Esper, F. J., and C. M. Sonsino, *Fatigue Design for PM Components*, European Powder Metallurgy Association, Shrewsbury, UK, 1994.

German, R. M., and R. A. Queeney, "Fatigue and Fracture Control for Powder Metallurgy Components," in *Fatigue and Fracture*, ASM International, Materials Park, OH, 1997, pp. 337–344.

Jenkins, I., and J. V. Wood (eds.), *Powder Metallurgy—an Overview*, The Institute of Metals, London, UK, 1991.

Jenkins, I., and J. V. Wood (eds.), *Selected Case Studies in Powder Metallurgy*, The Institute of Metals, London, UK, 1991.

Klar, E. (ed.), *Powder Metallurgy*, Vol. 7 of *Metals Handbook*, American Society for Metals, Metals Park, OH, 1984.

Kulkarni, K. M. (ed.), *Powder Metallurgy for Full Density Products*, Metal Powder Industries Federation, Princeton, NJ, 1987.

Lall, C., *Soft Magnetism: Fundamentals for Powder Metallurgy and Metal Injection Molding*, Metal Powder Industries Federation, Princeton, NJ, 1992.

Luk, S. H., H. G. Rutz, and M. A. Lutz, "Properties of High Density Ferrous P/M Materials After Various Processing Methods," *Industrial Heating* 53–59 (Oct. 1995).

Materials Standards for P/M Structural Parts, Metal Powder Industries Federation, Princeton, NJ, 1997.

Mosca, E., *Powder Metallurgy Criteria for Design and Inspection*, Associazione Industriali Metallurgici Meccanici Affini, Turin, Italy, 1984.

Salak, A., *Ferrous Powder Metallurgy*, Cambridge International Science, Cambridge, UK, 1996.

Schaefer, R. J., and M. Linzer (eds.), *Hot Isostatic Pressing: Theory and Applications*, ASM International, Materials Park, OH, 1991.

CHAPTER 12

Design Considerations

12.1. OVERVIEW

A component designed for powder metallurgy inherently has a cost and performance advantage over any component converted from an alternative production scheme. In general terms, Table 12.1 captures the design advantages for steel P/M. Much information is available on the design of components for each of the powder metallurgy processes. Likewise, many alloy offerings exist. This chapter builds a view of P/M in terms of performance attributes, geometric design, material selection, process rationalization, and fabrication economics. Such an approach establishes a foundation for attaining the desired behavior from a component, with no surprises. Thus, as illustrated in Figure 12.1, design relies on three interlinked information groups. A critical consideration is the component and its function, desired properties, required features, costs, and failure conditions. Embedded in the component definition is knowledge of the overall system design, reliability, and consequences of failures. For example, failure of a washing-machine transmission gear is less critical than failure of an aircraft brake.

Material selection examines databases, failure conditions, and similar designs to identify suitable compositions and property combinations. Finally, with these packets of information, a process is defined, where the fabrication route is selected on the basis of production quantities, costs, tooling, process advantages, and compatibility with the required properties and geometry. Underlying much of this activity is prior knowledge, generated by production experience with similar components or by ongoing research and development activities. A contingency analysis allows identification of weaknesses in the overall plan and assures that critical testing is performed before full production.

The last chapter provided information on the properties available by various P/M production routes. This chapter focuses on the intersection of design with the fabrication capabilities of P/M. The next chapter completes the story by showing many P/M components. Such a combination of information on properties, design, and applications provides a comprehensive base for P/M designs.

As an introductory overview, the design rules for P/M are as follows:

1. Keep the part as simple as possible.
2. Design for the process, including long tool life.

TABLE 12.1. Design Advantages of P/M Steels

Nest-shape forming—reduced machining steps
Competitive properties
Low material loss
Close dimensional control on several features
Some flexibility in composition, alloys, and microstructures
Good surface finish
Multiple phase microstructures—wear resistance, toughness
Controlled porosity—lubricating oil storage, filters, flow restrictors
Shape complexity
Moderate to very high production quantities
Established performance data in diverse applications

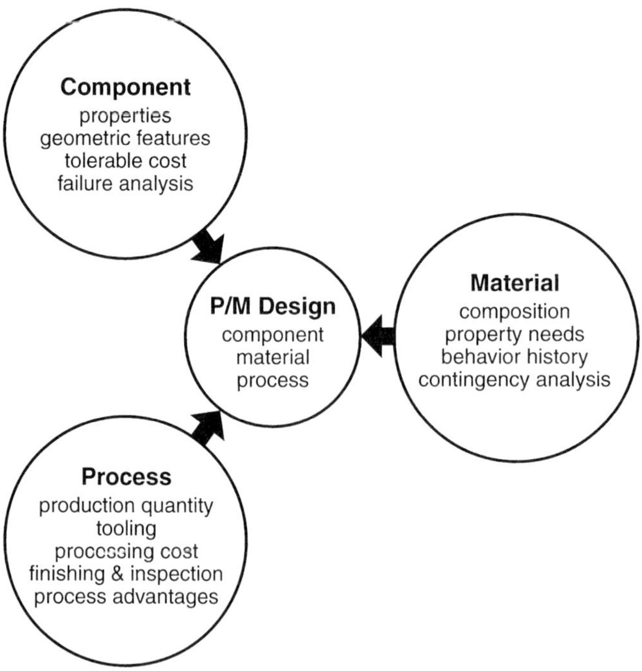

Figure 12.1. An overview of the P/M design process, which includes component function and definition, materials selection, and process specification activities.

3. Ensure that the component ejects from the tooling without difficulty.
4. Keep thickness as uniform as practical.
5. The powder needs to be selected for both the component and the production process.
6. Rationalize component shapes and features to the process's attributes.
7. Consider the materials, porosity, and design attributes required for proper function.
8. Understand the attributes of competitive processes.
9. Always attend to issues of quality, environment, waste, and reliability.

12.2. GEOMETRIC ASPECTS

12.2.1. Overview

The first consideration is the component size. Most P/M components range from about 1 mm (0.04 in) to 300 mm (12 in) in the maximum dimension. If we examined the range on a mass basis, this translates into components ranging from 1 g to 2 kg (0.002 to 2.2 lb). Exceptions exist, but probably 95% of P/M production is in the range of 10–500 g (about 0.02–1 lb), and only occasional components reach 15 kg (33 lb) or more. The largest reported parts are about 1 ton, but these are specialized products such as large filters, military hardware, and oil-well drilling equipment. Microcomponents are a growing part of P/M for microelectronic, military, and biomedical applications. Some ferrous structures with masses ranging within 5–240 mg (10^{-5}–$5 \cdot 10^{-4}$ lb) are in production, and a few microcomponents formed by injection molding are so small that 100,000 can be produced from a handful of powder.

One basis for analyzing the component production techniques is the complexity. This is defined in terms of the information needed to specify the part. The information content embedded in a P/M product is moderate, with less than 100 geometric features on the engineering drawing. Although this represents a high level of shape complexity, only certain combinations of features prove economical. Several other technologies compete in this product size, mass, and complexity range. These include forging, machining, casting, stamping, and plastic injection molding. It is a competitive arena, yet powder metallurgy excels on the basis of satisfaction of engineering properties (strength, hardness, magnetic response) in a low-cost fabrication route. The low cost typically comes from the elimination of machining while attaining high precision or good surface finish. Although powder cost may be high when compared to cast iron, the higher precision and reduced wastage in P/M can result in a lower production cost.

Two classes of components ideal for P/M emerge from this definition: porous materials, where porosity and pores are intentional features; and high-performance components, where pores are a detriment. This latter class, focused on mechanical, magnetic, or wear components, is the largest portion of ferrous P/M. Here, P/M is best applied when it reduces the number of parts by combining previously machined pieces into an assembly. The powder metallurgy process is attractive if the component is not too simple, since precision stamping or fine blanking proves more cost-effective and precise. Consequently, holes, steps, tapers, and functional features such as studs, slots, groves, and chamfers contribute to P/M being competitive. However, too many features, especially undercuts, perpendicular holes, close-tolerance threads, and parallel walls, detract from selecting P/M as the production route.

12.2.2. Die Compaction

12.2.2.1. Features. Component production by die compaction is cost-effective, assuming that the part is compatible with the process. Accordingly, as Table 12.2 shows, various features are best suited to this approach, especially stepped shapes and through-holes. Quantitative design attributes available in ferrous die compaction are given in Table 12.3 in terms of minimum, maximum, and typical properties. Further, Figure 12.2 sketches some component features that are difficult to die-compact with suggested redesigns that are better suited to the technology. Each is described in Table 12.4 in terms of the problem and recommended solution.

TABLE 12.2. Component Design Guidelines for Die Compaction

Restrictions
No inside closed cavities
No undercuts perpendicular to pressing direction
Corner radius >0.075 mm
Smallest hole diameter 0.1 mm
Minimum thickness 0.2 mm
Weight range 0.02 g–15 kg
Desirable features
Squat structure, low height
Slight taper for ejection
Largest dimension <200 mm
Weight <100 g
Wall thickness <30 mm
Assemblies in one piece
Flat, parallel surfaces
Allowed design features
Controlled porosity
Added second phases for machining or wear
Multiple through-holes in pressing direction
Hexagonal, square, round holes
Indents, marking, lettering
Multiple levels in pressing thickness
Helical outer profiles
Tapered side walls

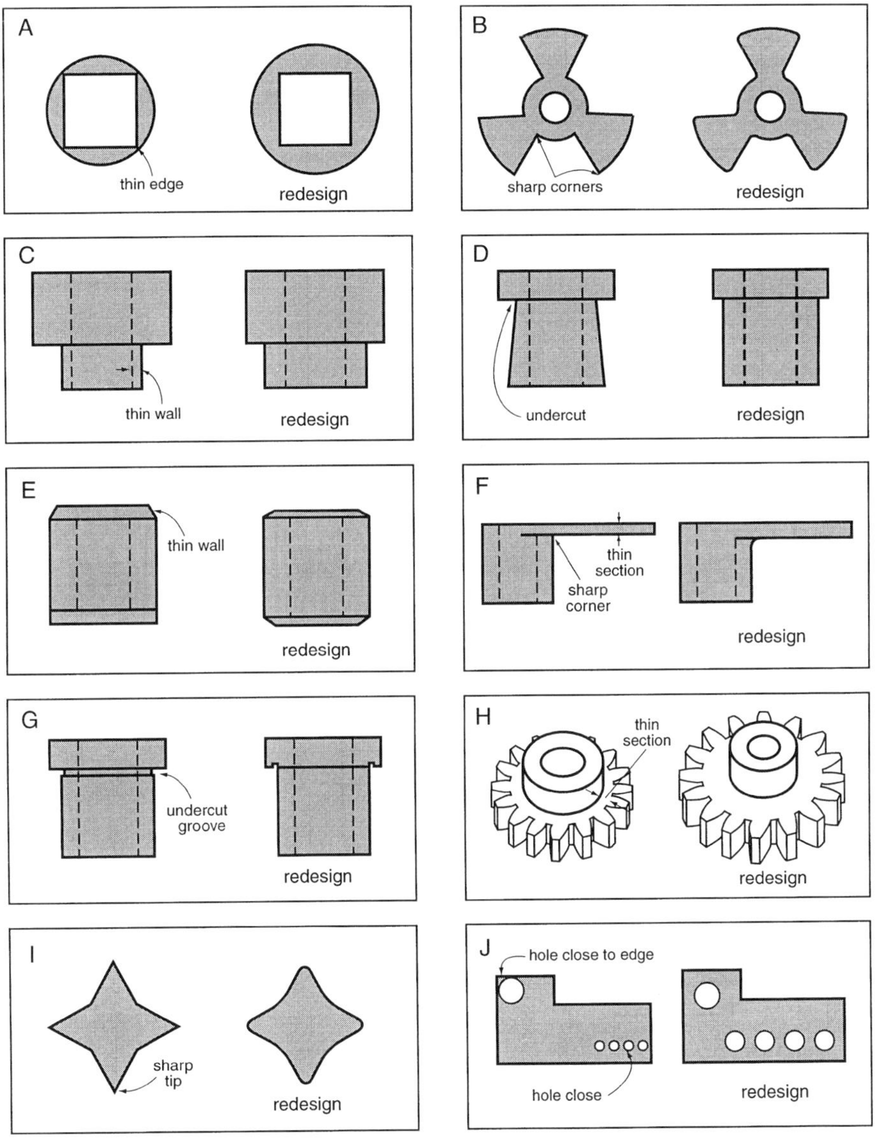

Figure 12.2. Features that prove difficult or costly to include in die compacted components and suggested changes to improve suitability for P/M. Table 12.4 discusses each redesign.

TABLE 12.3. General Geometric Design Features Available with Ferrous Die Compaction Technologies

Attribute	Minimum	Maximum	Typical
Thickness parallel to pressing direction, mm	0.6	100	15
Thickness perpendicular to pressing direction, mm	0.2	200	30
Pressing area, cm^2	0.02	150	15
Thickness variation	None	10×	2×
Size or dimension, mm	0.2	200	25
Height: diameter ratio	0.1	6	0.5
Tolerances, % (1 deviation)	0.03	2.0	0.3
Dimensional specifications	20	100	40
Mass, g	0.02	15,000	40
Production quantity per year	200	20,000,000	150,000

The first requirement is that the shape be ejected from the compression tooling. Although multiple levels are common, each level adds to the tool and press cost. Holes are incorporated by core rods in the punches, and steps are incorporated via multiple punches. The hole diameter is usually more than 20% of the component depth with a diameter of more than 1.5 mm (0.06 in). Such features are limited by

TABLE 12.4. Explanation of Design Changes for Die Compaction in Figure 12.2

Drawing	Problem	Recommended Redesign
A	Thin walls near edges, low strength, possible punch damage	Make walls thicker, especially near sharp corners
B	Avoid sharp corners to prolong tool life	Design in radii where possible
C	Large change in wall thickness, thin wall, possible component cracking	Expand feature thickness, enhance powder fill
D	Undercuts and reentrant angles cannot be ejected from tooling	Use parallel walls
E	Radii on corners lead to tool damage	Make chamfer with >60° angle use flat to eliminate thin punch
F	Thin portion of punch at junction of sections	Enlarge section to avoid thin piece
G	Undercuts do not allow ejection	Move groove to allow ejection
H	Hub and gear tooth root are too close	Shrink hub diameter to allow easier compaction
I	Sharp tips that prove difficult to compact to high densities and are dimensional problems	Put radius on tips to eliminate sharp tools
J	Holes close to edge and each other	Move holes, wider spacing, enlarge holes

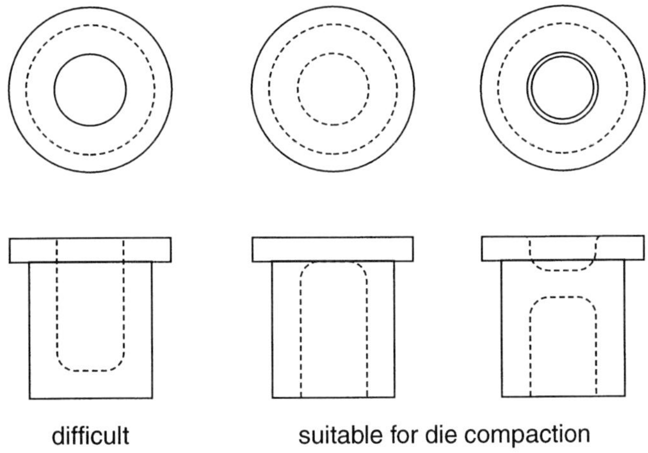

Figure 12.3. Preferred designs for blind holes in die-compacted components. The two redesigns reduce chances of tool damage.

the tool and press, so not all P/M operations have the same capabilities. Round holes are least expensive, but other shapes such as keyways, half-circles, and squares are possible. Blind and tapered holes can be incorporated in certain combinations. Figure 12.3 illustrates two preferred designs for blind holes. Side holes, perpendicular to the pressing direction, must be drilled after sintering. Exceptions occur when the side hole is exactly at the center of the compact, but even then side core rods are easily damaged in pressing.

Flat sections are the easiest to compact, but most compacts incorporate several profile features, such as sketched in Figure 12.4. This set of cross-sectional sketches illustrate the increasing difficulty in die compaction with various combinations of holes, recesses, and levels. The more difficult compacts have tapers and nonperpendicular features. The wall thickness is usually more than 1.5 mm (0.06 in), but must be larger on thicker parts. Thin tool components are subject to easy failure and should be avoided. Because of tool damage at thin edges, it is typical to combine a belly band with a spherical cap. For example, Figure 12.5 illustrates the two possible upper punch designs for compaction of a spherical segment. The preferred component design has a flat tip on the punch that is at least 0.25 mm (0.01 in) wide.

Tapers are possible, but must be aligned with the ejection direction. Tapers require a short section parallel to the pressing direction to prevent the upper punch from pressing into the tapered portion of the die. Spherical depressions are easily incorporated into P/M tools. Chamfers are preferred on component edges (as contrasted with a corner radius) to avoid trapping powder between the tool pieces. A fillet radius is possible, and larger radii are most desirable. Other allowed features include countersunk holes, but these cannot lead to thin punches, and hubs, but these must not be thin. The minimum allowed section thicknesses are usually 1.5 mm (0.06 in), but thinner sections are possible under special conditions. Long and thin components

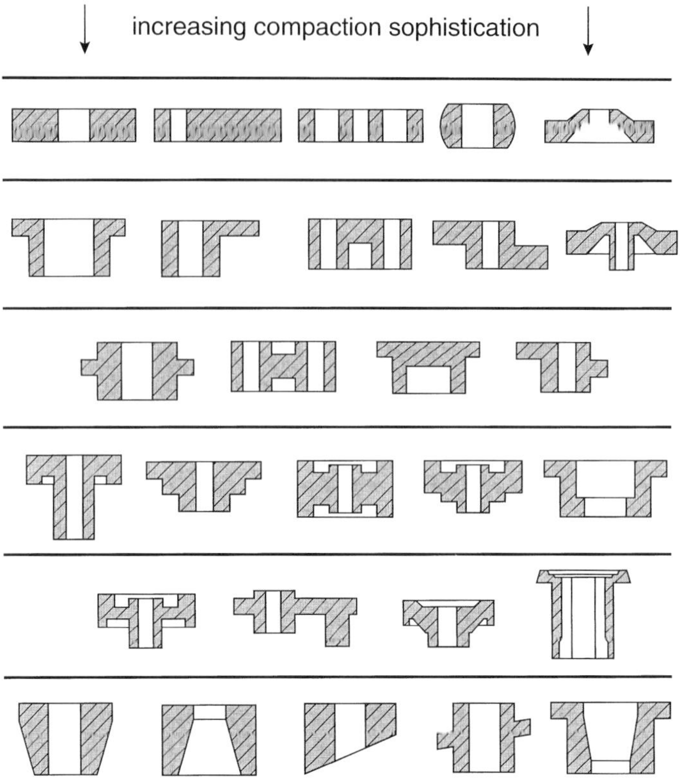

Figure 12.4. Axial sections through die-compacted profiles to illustrate the variety of possible shapes. These cases move to greater complexity and pressing sophistication, as evident by the larger number of levels, cores, and holes.

generate density gradients and lead to lower repeatability in sintered dimensions. Figure 12.6 provides an idealized die-compacted component that incorporates the several desirable features for extended tool life.

12.2.2.2. Tolerances. Nominal tolerances for die-compacted and repressed compacts are given in Table 12.5. This tabulates the typical industry practice and that possible with special care. It also shows how the as-sintered and sized features compare. Repressing, or double compaction and double sintering cycles are more precise when compared to single-pass cycles. On the other hand, heat treatment and surface-steam-treated steels are less precise. For higher-alloy steels, the scatter is larger. Critical dimensions can be sized, coined, or repressed after sintering.

Many processing factors impact on the observed dimensional precision. For example, dimensions perpendicular to the pressing direction vary less in processing than do dimensions aligned with the pressing direction. Additionally, the size, location, and tooling conditions influence tolerances. Surface finish is a difficulty if there are

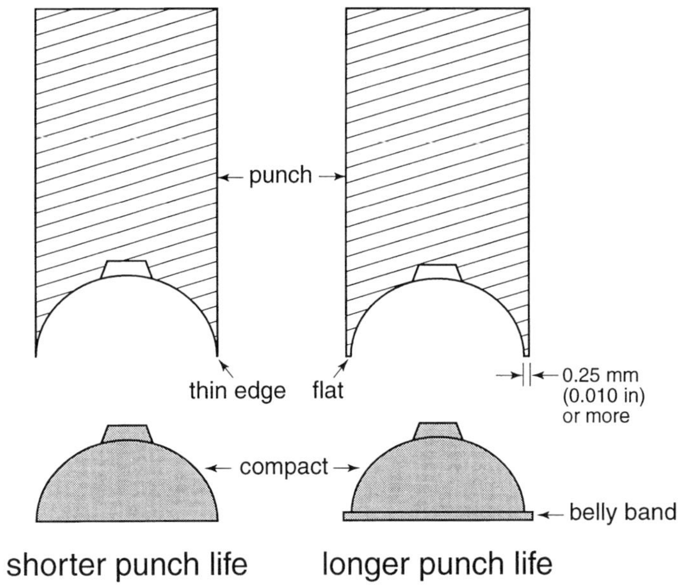

Figure 12.5. An upper punch design suggestion for pressing a spherical segment. Tool life is expanded by the addition of a flat on the tooling that results in a belly band on the component.

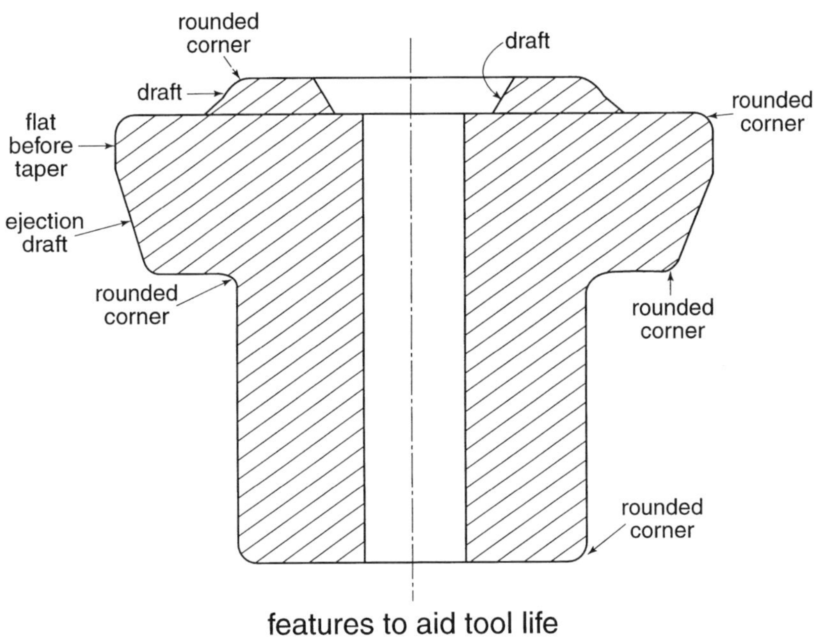

Figure 12.6. Features in component design that will help to prolong tool life, including tapers, drafts, and rounded corners.

TABLE 12.5. Dimensional Tolerances Practical and Possible in Die-Compacted Ferrous Alloys after Sintering and after Sizing (Postsintering)

Feature	Practical As-Sintered	Possible As-Sintered	Practical Sized	Possible Sized
Radial size, mm	±0.04	±0.02	±0.03	±0.01
Length, mm	±0.13	±0.07	±0.10	±0.07
Flatness, mm	±0.05	±0.03	±0.03	±0.03
Parallelism, mm	±0.04	±0.03	±0.03	±0.03
Diameter, mm	±0.15	±0.03	±0.03	±0.01
Concentricity, mm	±0.07	±0.05	±0.05	±0.03
Roughness, μm	0.1	0.06	0.06	0.04

pores connected to the surface. Between the pores the finish is very good, typically with a roughness of just 0.1 μm (32 μin), but surface pores produce a roughness independent of processing. Special attention can be given to troublesome dimensions. The growth in P/M gears is pushing the process capability. Studies on process yield show an increase in rejection rate due simply to an increase in the number of closely toleranced dimensions. Consequently, only tolerances needed should be requested, since production costs are very dependent on the requested component precision. The barrier in widespread application of P/M to gears arises because there are many simultaneous design specifications, including pitch, outside diameter, root diameter,

EQUATION BOX 12.1

Precision in P/M components is related to the dimensional variation as a ratio to the tolerated range of dimensions. Statistical distributions are used to decide the average and standard deviation for a production process. Usually the assumed distribution is a normal or Gaussian distribution—the bell curve. Within six standard deviations (three above the mean and three below the mean) exist 99.7% of the samples. Precision is described by the process capability as a ratio to the total tolerance range

$$P = \frac{X_U - X_L}{\sigma}$$

where P is an index of precision, X_U is the upper allowed size or property, X_L is the lower allowed size or property, and σ is the standard deviation measured for the process. Processes with a ratio greater than 8 for P are designated as high precision, while below 6 is low precision. High-precision processes are unlikely to produce products that are out of the specification range.

TABLE 12.6. Capabilities of Steel Injection-Molding Technologies

Attribute	Minimum	Maximum	Typical
Thickness, mm	0.2	25	10
Thickness variation	None	100×	2×
Longest dimension, mm	2	1000	100
Tolerances, % (1 deviation)	0.03	2.0	0.3
Dimensional specifications on engineering drawing	20	1000	100
Mass, g	0.02	1000	40
Alloy	Iron	Composites	Alloys
Properties	Unimportant	Highest attainable	Handbook
Cost per part, $	0.20	400.00	2.00
Production quantity per year	2000	20,000,000	150,000

involute contour, and fillet radius at the tooth root. Although most all features of a P/M compact can be adjusted in finishing operations, these add to the overall expense.

12.2.3. Injection Molding

12.2.3.1. Shapes. In a simple sense, if a shape can be formed from plastic, then steel injection molding is viable. Tables 12.6 and 12.7 provide quantitative and qualitative component design guidelines. These include criteria relating to wall thicknesses, overall component size, component weight, and internal cavity shape. Particle size dictates the minimum section thickness. No dimension can be smaller than about 10 times the particle size. Particle size also limits the sharpness of all corners, since the minimum particle radius is actually the edge radius, but more typically several particles are needed to fill out the dimension. A practical limit on an edge radius is probably 0.05 mm (0.002 in) and on a corner radius, 0.1 mm (0.004 in).

Table 12.7 lists options in the design of injection molding compacts. These include square holes, flat-bottomed holes, knurled surfaces, twists, curls, helical geometries, and external or internal threads. Although these features are not needed to make a compact viable, they are possible, and add to the design flexibility. Certain attributes, such as component size and mass, will increase as experience and manufacturing optimization occur. Components of up to 10 kg (22 lb) and 1000 mm (39 in) maximum length have been demonstrated by injection molding. These require special efforts and are not routine products.

The wall thickness is usually small, and might be less than 10% of the largest dimension, with the typical value near 10 mm (0.4 in). Larger thicknesses are in use, but present problems with slow debinding. Further, the variation in thickness over the part is usually small. Scissor blades are an example of where the length is much greater than the thickness, and the thickness is not highly variable; it is thickness that determines debinding time and is the main limitation of injection molding. With respect to changes in the section thickness, a ratio of 2 is typical; however, compacts have been successfully fabricated with thickness changes of 100 : 1.

TABLE 12.7. Nominal Injection Molding Component Design Guidelines

Restrictions	
	No inside closed cavities
	No undercuts on internal bores
	Corner radius >0.075 mm
	2° draft on long parts for ejection
	Smallest hole diameter 0.4 mm
	Minimum thickness 0.2 mm
	Weight range 0.02 g–1 kg
	Allowance for parting line, gate, and ejectors
Desirable features	
	Gradual section thickness changes
	Largest dimension <100 mm
	Weight <100 g
	Wall thickness <10 mm
	Combine assemblies in one piece
	Flat surfaces for support
	Small-aspect-ratio geometries
Allowed design features	
	Holes at angles to one another
	Hexagonal, square, blind, and flat bottom holes
	Stiffening ribs
	Knurled and waffle surfaces
	Protrusions and studs
	Helical geometries allowed
	External or internal threads
	"D"-shaped and keyed holes
	Part number or identification in die

Another gauge of injection-molding design candidates is through the number of callouts or dimensions on the engineering drawing. A simple shape, like a flat washer for a bolt, has only a few features, while a microcomputer circuit has millions of features; both would be poor applications for injection molding. Common successes involve several dimensions—a wristwatch case is one example that matches well with the technology.

12.2.3.2. Tolerances. Tolerances are a common concern. Table 12.8 summarizes the minimum and typical tolerances possible in injection molding without secondary operations. Generally, the ability to hold tolerances increases with production experience. Rapid advances in control devices make newer installations more precise. Shot-to-shot molding variations and warpage in sintering are the largest contributors to dimensional variation. Because thermal debinding allows more distortion, most contemporary operations have moved to solvent or catalytic debinding for improved precision. With little effort, dimensional tolerances can be held within 0.5% (one

TABLE 12.8. Typical Standard Deviation of Injection-Molded Steel Components

Feature	Best Possible	Typical
Angle	0.1°	2°
Density	0.2%	1%
Weight	0.1%	0.4%
Dimension	0.05%	0.3%
Absolute dimension	0.04 mm	0.1 mm
Hole diameter	0.04%	0.1%
Hole location	0.1%	0.3%
Flatness	0.1%	0.2%
Parallelism	0.2%	0.3%
Roundness	0.3%	0.3%
Perpendicularity	0.1% or 0.1°	0.2% or 0.3°
Average roughness	0.4 μm	10 μm

standard deviation). A controlled process with solvent, wicking, or catalytic debinding allows dimensions to be held with a deviation of 0.3% or less, and critical dimensions can be held within 0.1%. Newer binder systems and integrated process controls improve this to 0.05% or better. Large dimensions exhibit more scatter. Fortunately, component density can be held within ±1% and weight variation to ±0.1%.

12.2.3.3. Design. Besides dimensional variability, a successful injection-molding component design must allow for noncritical placement of the parting line, ejector pin marks, and gate. These features are usually moved to areas where they will not detract from the operation or aesthetics. It is also desirable to include a slight draft or taper of 0.5–2° to help in ejection from the mold.

Typical injection-molded components are small enough to be held in a hand, but can be as small as the tip of a pencil. This implies that very large objects are difficult to produce or are expensive by injection molding as contrasted with other forming technologies. A major penalty is the raw powder cost as compared with other processing routes. For many components in the 5-g range, raw materials are a trivial aspect of the production cost, but for larger components the powder price becomes a factor. Over 100 g (0.2 lb), the powder cost becomes a disadvantage. Also, processing rates are slow with thick sections, adding to the economic disadvantages.

Component design greatly influences the success of injection molding. Early identification of injection molding as a production process allows design modifications to ease manufacturing. Possible design shifts that give essentially the same functional features are sketched in Figure 12.7. Two alternatives are shown in each case. Both alternatives achieve the same objectives, but one is better aligned with the specifics of the approach. In cases *a* and *b*, coring maintains a constant section thickness, minimizing distortion, and speeding debinding. For case *c*, the smooth transition from a thick to a thin section is preferred to eliminate molding stresses.

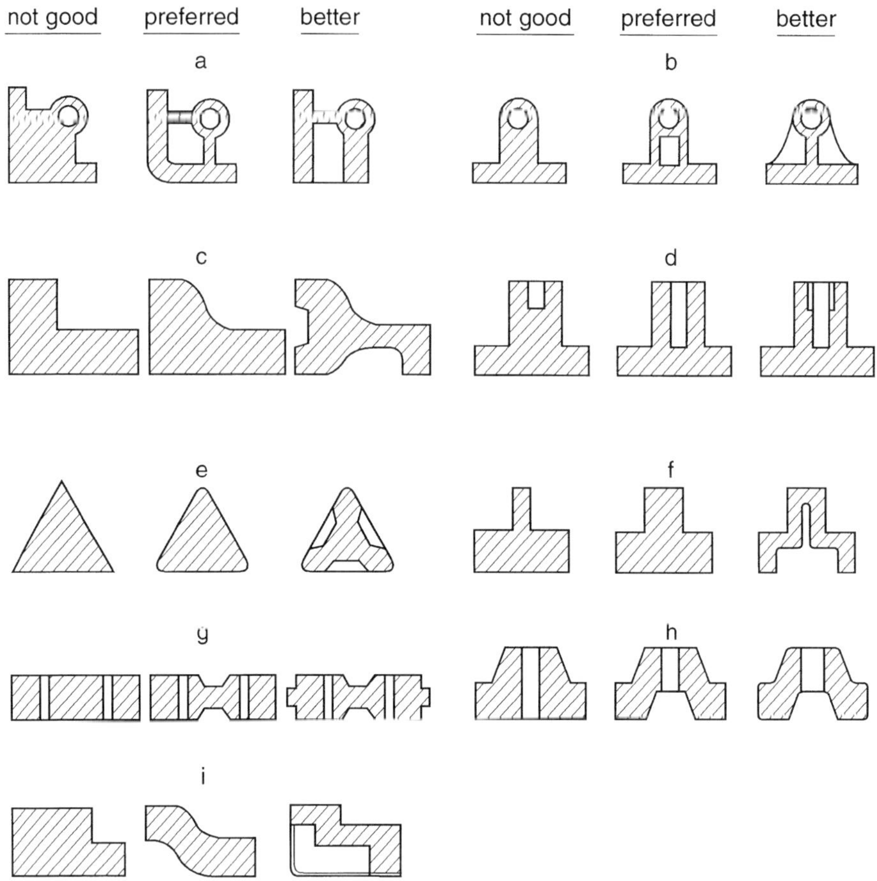

Figure 12.7. Component design suggestions that will ease the cost and processing difficulty in the production of injection-molded steels. Three cases are illustrated for each feature. The first case is a typical design, while the second phase is a partial improvement, but the final case shows a design that aids molding and debinding.

In case *d*, the blind hole is extended to avoid a thick section thickness under the hole. Case *e* shows the preferred replacement of sharp corners by rounded corners. Case *f* illustrates expansion in the size of a thin post to help maintain uniform section thicknesses. The through-holes in cases *g* and *h* create thick–thin regions that can be avoided by redesign. Finally, case *i* is an example of where a curved section change avoids a thick–thin junction, giving a desirable uniform wall thickness.

To appreciate the shape complexity possible with use of injection molding, Figure 12.8 sketches several viable geometries and combinations of features. This collection illustrates possible combinations of holes, pins, fins, slots, grooves, arms, and so on.

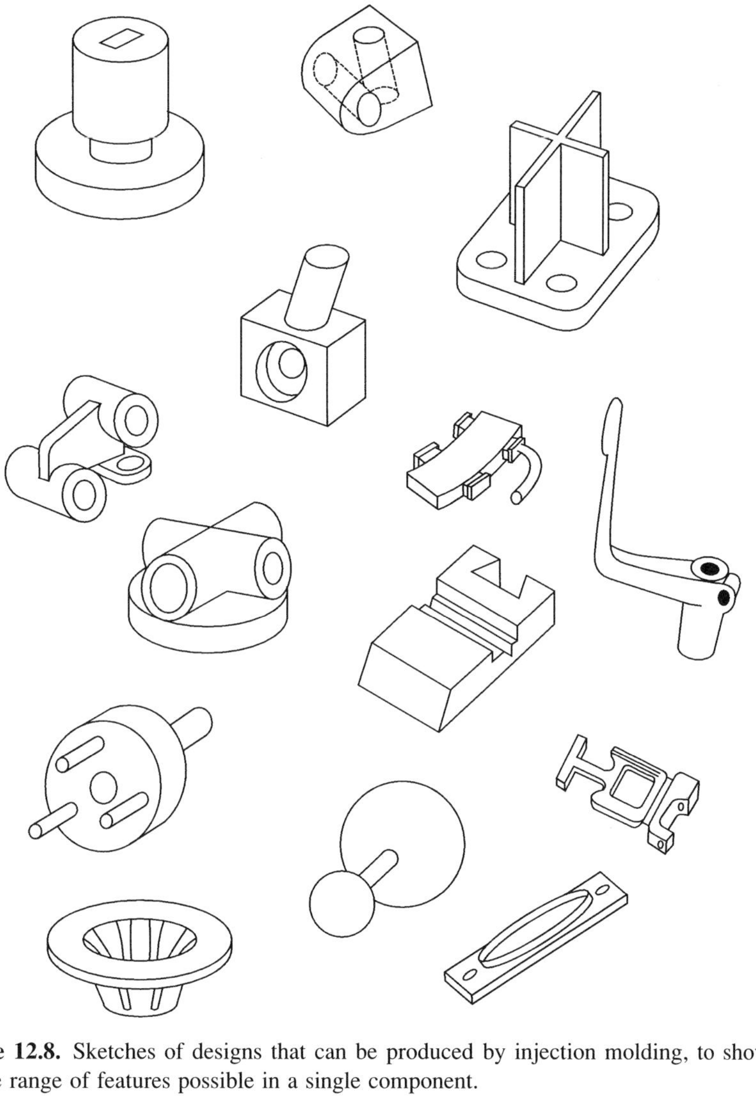

Figure 12.8. Sketches of designs that can be produced by injection molding, to show the unique range of features possible in a single component.

12.2.4. Powder Forging

Powder forging produces a nearly full dense structure. It usually relies on a high compaction pressure and a heated P/M preform. The preform for forging affects the final component dimensions as a function of mass flow in the forging die. Design of the preform for flow without cracking is an important technology. Powder forging is practiced by only a few, large operations. Thus, preform design data are not widely available. The high-temperature/high-pressure combination needed for densification results in tool deflection during forging that makes the process less accurate than

TABLE 12.9. Typical Tolerances on Powder Forged Components

Feature	Best Possible	Typical
Density	0.2%	1%
Weight	0.1%	0.4%
Length, mm	0.1	0.4
Hole diameter, mm	0.2	0.3
Outside diameter, mm	0.2	0.4
Parallelism, mm	0.1	0.2
Concentricity, mm	0.1	0.2
Roundness, mm	0.1	0.2
Surface roughness, μm	0.4	0.8

injection molding or die compaction. Consequently, often the forged component is machined on critical surfaces. The process is best suited to squat, rounded components. Radii on corners are large to reduce tool wear. On the outside corners, a radius of at least 1 mm (0.04 in) is generally required.

Table 12.9 shows typical tolerances for powder-forged components. One problem with holding tight tolerances is the rapid tool wear associated with the high temperatures and large forces involved in forging. Accordingly, the cited tolerances are larger than associated with most other powder metallurgy processes, except isostatic compaction.

12.2.5. Isostatic Compaction

Simple porous components do not require tooling, but can be formed by pouring powder into a ceramic cavity and sintering to attain strength. Loose-powder sintering approaches are capable of holding dimensions in the sintered object to ±2%. This sets in place a baseline against which isostatically pressed compacts are judged.

Cold isostatic compaction produces stainless steel and tool steel components with considerable dimensional variation. Dimensional scatter results from the soft tooling used in the process. Diameter tolerances for CIP are often ± 5%, but only ±2% on wall thickness. It is common in ceramics and cemented carbides to machine the green CIP compact to attain dimensional accuracy. However, this is not common in CIP of P/M steels. Improved dimensional control is possible using rigid supports on the flexible molds and by evacuation to removed air. Cores and hard inserts are used to improve certain dimensions.

Likewise, HIP is imprecise unless special precautions are taken to ensure uniform powder filling and rigid container construction. The general protocol is to machine the compact after consolidation, except where containerless HIP is applied to previously sintered compacts. Reported variation in densification shrinkage is as much as 6% between locations, giving a typical dimensional scatter of 1%. There is less scatter in a single dimension. This can be improved, and a dimensional scatter of ± 0.2% is possible. Much current effort is addressing computer simulation techniques to predict final dimensions. When HIP is applied to already sintered components in the containerless mode, dimensional scatter is much reduced and only about twice that for die-compacted alloys. Because of limited production experience, little data are available on the other full-density techniques.

12.3. PERFORMANCE ATTRIBUTES

Only in a few instances are mechanical or physical properties unimportant, such as in wristwatch cases. More typically the goal is to attain high properties with the selected material. Likewise, only rarely are properties pushed to the limits for any material. In these cases, hot isostatic pressing or other treatments might be employed to ensure best performance. Wrought materials tend to displace P/M materials if high properties are required. From a cost consideration, typical P/M prices are moderate, but that assumes a high production volume to justify the cost of tool construction. Always, dimensional tolerances must be compatible with the inherent process variability.

More than any other factor, the performance of a P/M component is dependent on final density. This is the unique aspect of the process. With casting, forging, machining, and other metalworking technologies, density is invariant for a given material. However, powder consolidation determines the final density and final properties. Consequently, many components are designed with porosity as a desirable feature. These include flow restrictors, filters, wicks, fluid storage reservoirs, air bearings, self-lubricating bearings, fluid bed substrates, and impact absorption materials. Intentionally porous metals range within 10–75% of theoretical density. Very low-density structures are fabricated from fibers or by mixing pore formers (plastic particles) with the powder, while the higher-density structures come from compacted powders. Bearings usually have densities in the range from 70–85% of theoretical. Structural components overlap with this range (most are at least 85% dense) and go to full density.

The properties of alloys fabricated to densities of 95% or higher are often competitive with wrought steels. However, when dealing with low-ductility compositions, the mechanical properties decrease rapidly with a density below about 7.5 g/cm^3. The last chapter tabulated and discussed many property changes with processing and density, so those remarks need not be repeated here. Recognize that heat treatment, carbon level, and microstructure interact with the density level to affect properties. For example, injection-molded and sintered Fe–2Ni steels can exhibit wide carbon shifts traced to the debinding cycle. When the carbon level is almost zero, the sintered yield strength is low at 190 MPa (28 ksi), but the fracture elongation is approximately 30%. When the carbon level is increased to 0.5%, heat treatment produces a yield strength of 1230 MPa (175 ksi) with 1% fracture elongation. The dramatic shifts in strength and ductility are due to the carbon and phase changes.

Most engineering properties decrease with porosity; however, a few are not affected. These are the basic material thermodynamic properties such as melting temperature, specific heat, transformation enthalpy, and transformation temperatures or

EQUATION BOX 12.2

Porous bearings are one of the high-use structures formed by P/M. The strength of a bearing is usually measured by a radial crush test. A cylindrical bearing of length L, outer diameter d_O and wall thickness t, is placed on its side. Force is applied until the bearing crushes, giving a peak load as P. The radial crush test leads to a K value that is the radial crush strength as follows:

$$K = \frac{3\,P\,(d_O - t)}{\pi\,L\,t^2}$$

Values for K, the radial crush strength, do not agree with other strength measures, but they are useful in testing and comparing various bearings. For common P/M bearings, such as pure iron and iron–copper–carbon, the K values range from 140 to 280 MPa (20 to 40 ksi) and densities range from 5.6 to 6.8 g/cm^3.

solubilities. On the other hand, most engineering properties are sensitive to the component density, including strength, ductility, elastic modulus, thermal conductivity, thermal expansion, electrical conductivity, magnetic permeability, toughness, fatigue life, corrosion resistance, wear resistance, and hardness. Suitable performance depends on understanding the allowable density range to obtain the needed properties.

The properties attainable with high-temperature sintering, powder forging, injection molding, and hot isostatic pressing are generally equivalent to those possible via other metallurgical techniques, such as casting. Summaries of attainable properties were provided in the previous chapter. In most cases, the strength will vary (one standard deviation) by 20 MPa (3 ksi) between production lots, and the elongation will have a standard deviation of approximately 1%.

12.4. MATERIALS SELECTION

As evident from several tables included in this book, a range of steels is available via P/M. The common examples include pure iron, plain carbon steels, low-carbon steels, stainless steels, and alloys with varying levels of copper, nickel, carbon, molybdenum, and special alloys such as invar. Carbon steels by P/M are inexpensive, but have limited strengths and heat treatment responses. They are good for strength levels below about 400 MPa (55 ksi). Copper steels, especially those containing 2% copper, are widely used in P/M. At low sintered densities they serve as bearings, become structural components at intermediate densities, and are high-performance alloys when forged. As a hybrid, copper infiltration is used to fill and improve impact toughness, but this does not significantly affect tensile strength. Iron–nickel steels are frequently selected for small components that are heat-treated. Peak tensile strengths are probably near 700 MPa (100 ksi) for the alloys with nickel and carbon, but higher levels are attainable with molybdenum, chromium, and manganese alloying. Nickel-based alloy steels allow heat treatment to higher strength levels, reaching peaks of 1200 MPa (170 ksi) with carbon levels near 0.8% and postsintering heat treatment. For magnetic applications, nickel contents of up to 50% are used with pure iron, but no carbon is included. Other magnetic ferrous alloys might contain phosphorous, silicon, or aluminum.

Stainless steels used in P/M come in three varieties: austenitic, martensitic, and precipitation-hardenable. These all contain at least 12% chromium and are largely selected for corrosion resistance instead of strength. Depending on the composition, microstructure, and carbon content, stainless steels can be processed to high hardness levels for use in cutting tools. Other high-alloy steels include microelectronic alloys, tool steels, and high-temperature alloys, including FeCrAlY compositions and iron-based superalloys. These latter alloys are not used very often.

Properties are influenced by the production process. Alloying is a potent means to improve properties, but usually increases cost. Iron is one of the lowest-cost metallic materials, so any alloying addition impacts on the economics. Property gains are possible if higher costs are acceptable. Figure 12.9 is a plot of yield strength and unnotched impact toughness, showing the alloying level associated with various property combinations. This plot is for a constant density of 7.0 g/cm^3. Mutual

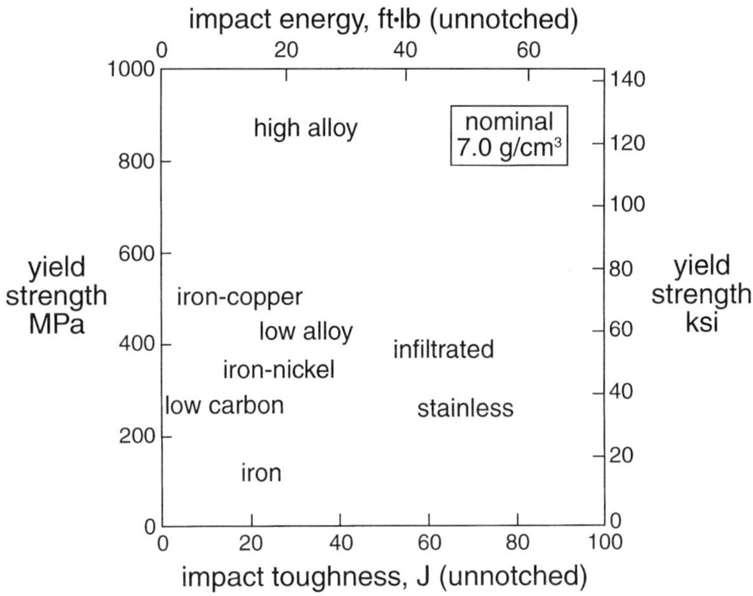

Figure 12.9. Yield strength and impact toughness (unnotched) for die-compacted steels, showing for a constant density of 7.0 g/cm^3 the need for higher alloying levels if both properties are to be improved.

TABLE 12.10. Cross-Comparison of Production Processes and Various Engineering Attributes

Shaping Technique	A	B	C	D	E	F	G	H
Binder assisted extrusion	3	3	4	2	3	2	4	2
Cold isostatic compaction	1	3	4	3	4	2	4	4
Die compaction	1	2	4	4	2	4	2	1
Double press	1	2	2	4	2	3	2	1
Gas forging	1	4	1	4	3	2	2	4
Granular forging	1	4	1	2	3	2	3	4
High-temperature sinter	2	3	3	4	2	4	2	4
Hot extrusion	3	3	1	1	4	2	3	1
Hot forge	1	3	1	2	2	2	2	1
Hot isostatic pressing	1	4	1	2	4	1	4	2
Hot pressing	1	4	2	3	2	2	2	3
Injection molding	1	3	3	4	1	4	2	1
Loose powder sinter	2	1	4	4	1	4	4	4
Roll compaction	4	3	2	1	4	3	4	3
Slip casting	1	3	4	3	3	2	3	4
Spray forming	3	4	1	1	4	1	4	4
Tape casting	4	3	4	2	3	4	4	4

[a] *Key:* A—constant process; B—materials range; C—controlled porosity; D—smallest mass; E—largest mass; F—smallest size; G—largest size; H—tool cost; I—tool life; J—equipment cost; K—production rate; L—powder cost; M—production cost; N—product quantity; O—density level; P—final property; Q—typical tolerance; R—surface finish; S—shape complexity; T—axial ratio.

increases in both properties come only with alloying. However, besides the direct cost of the alloying ingredients, an increased difficulty in processing the alloys further adds to the overall cost.

Other performance attributes are not as easy to quantify. If corrosion is a concern, then stainless steels are the most common alloys. If soft magnetic properties are the prime concern, then high iron contents are the common choice. Subtle property combinations contribute to longer service life, better aesthetics, or easier production. In this regard, machining behavior, surface finish, corrosion or tarnish resistance, or ease of welding prove important. It is common to find that two vendors have essentially the same material (according to the specifications) yet exhibit 100-fold differences in machining rates or service life.

Density provides another dimension to the material's specification and can be selected for beneficial attributes in lubrication, sound attenuation, filtration, or flow restriction. Thus, it should not be ignored as part of the materials selection process. For porous materials, porosity, pore size, filtration rating, or permeability become the dominant features. Steel bearings are favored over bronze bearings when strength is a major dictate. In comparison with bronze bearings, the coefficient of friction is not as low, but the strength is higher. Most iron-bearing alloys are either iron–carbon or iron–copper–carbon compositions. Most metallic filters are formed from stainless steels.

I	J	K	L	M	N	O	P	Q	R	S	T
4	3	3	2	3	2	1	2	2	3	1	3
2	2	2	4	3	2	2	2	2	2	2	4
3	3	4	4	4	4	2	2	3	3	3	2
3	2	3	4	3	3	3	4	4	4	3	2
2	1	2	3	2	1	3	4	2	2	3	1
2	1	2	3	2	1	4	4	1	1	2	1
3	2	3	3	3	3	3	3	3	3	3	2
2	1	2	2	1	2	4	4	2	3	1	3
3	1	3	4	2	3	4	4	3	2	2	1
1	1	1	1	1	2	4	4	2	2	2	2
2	2	2	3	2	2	4	4	3	2	1	2
4	2	3	2	3	4	3	3	3	3	4	3
4	4	4	3	4	3	2	1	3	1	2	1
4	1	2	3	2	3	4	3	4	4	1	4
2	4	1	2	4	2	3	3	2	3	4	3
1	1	3	2	1	1	3	3	1	1	1	3
4	3	4	2	4	3	3	2	3	4	1	4

Composites are a new area for ferrous P/M. Many new compositions are possible by mixing ferrous powders with various ceramic phases. Examples are tool steels with mixed titanium carbide, tungsten carbide, iron–molybdenum boride, chromium boride, or alumina particles. These ceramic reinforced composites possess high strength, hardness, wear resistance, high-temperature strength, or exceptional performance in reactive environments. New polymer–powder composites are being used for magnetic applications. These are termed *bonded structures*, since the polymer glues the particles together. High-performance magnets are the most exciting example, but the products also include practice bullets, toys, and low-stress consumer items. Many of these are formed using compression molding or transfer molding, hybrids between die compaction and injection molding.

12.5. PROCESS RATIONALIZATION

Each P/M technique has specific attributes that help rationalize product design with the production route. Several questions go into rationalization of the P/M processes. A first concern is with residual porosity. Some products are designed for controlled porosity, so they are not viable when formed using a full-density technique. Usually porous metals are generated by loose-gravity sintering, die compaction, tape casting, slip casting, or cold isostatic compaction. Some sheet material is formed by tape casting, but this would be an inappropriate route to the fabrication of discrete objects. Further, large components are seldom fabricated by injection molding, forging, or die compaction. On the other hand, cold isostatic pressing, slip casting, and hot isostatic pressing are more applicable to large bodies.

Tooling in P/M is a high expense. Accordingly, most production is geared to at least 100,000 components per year for applications in automobiles, lawn and garden equipment, appliances, business machines, sporting goods, musical instruments, consumer products, or firearms. Powder metallurgy is costly for lower production volumes, largely because of tooling amortization. Curiously, ferrous P/M rarely seems to attain the very large production quantities. Once a product grows to large consumption, it seems to shift to plastics. Production rates of 40 million compacts per day are seen in the P/M production of tantalum capacitors and molybdenum heat sinks, but these are not encountered in ferrous P/M.

To sort out the technologies and identify their strengths, Table 12.10 brings together much of the information organized in this book. It forms a matrix of the production processes and their attributes. For each attribute there is a numerical value assigned to show the relative strength of the production route. As indicated in Table 12.11, the scaling is not linear, to help discriminate between techniques. The organization of these tables is subjective, and for every entry there is an exception. Obviously, such simplicity hides many underlying details. So some caution is required in analyzing the rankings. Simple shapes can be produced by most techniques, but they will not be cost-effective. For example, because of the higher powder cost

TABLE 12.11. Approximate Parameter Ranges Used to Rank Process Attributes

Attribute	Level 1	Level 2	Level 3	Level 4
Constant process	No, discrete	Possible	Some	Yes
Material range	Iron	Low alloy	High alloy	Composite
Controlled porosity	Not used	Rare	Some	Yes, useful
Smallest mass	>1 kg	>100 g	>10 g	<1 g
Largest mass	<1 kg	<2 kg	<10 kg	>10 kg
Smallest size	>10 mm	>5 mm	>1 mm	<1 mm
Largest size	<10 mm	<30 mm	<70 mm	>70 mm
Tool cost	>$10,000	<$10,000	<$5,000	<$1,000
Tool life	Single use	Limited, 100	100,000 parts	1,000,000 parts
Equipment cost	>$1 million	About $400,000	About $100,000	<$100,000
Production rate	One per hour	Few per hour	Few per minute	Several per minute
Powder cost	Special order	High	Intermediate	Low
Production cost	Very high	High	Intermediate	Low
Production quantity	Small; 100s	Limited; 10,000	High; 100,000	>1,000,000
Density level	80%	90%	95%	100%
Final properties	Low	Intermediate	Cast	Forged
Typical tolerances	Poor	Low	Good	Excellent
Surface finish	Very rough	Sandblast finish	Smooth	Polished
Shape complexity	Very simple	Two levels	Multiple levels	Undercuts, cores
Axial ratio	Squat	Equiaxed	Long, thick	Long, thin

and higher number or production steps, it would be unwise to injection-mold a component that can be die-pressed. Considerations in making a production decision are wide ranging, and include the surface finish, tool life, component size and shape, product complexity, properties, and density.

12.6. FABRICATION ECONOMICS

12.6.1. Competitive Techniques

Competing technologies influence the viability of P/M. Table 12.12 organizes a brief comparison of the advantages and limitations of various shaping technologies against which P/M routinely competes. Each is best suited to a particular set of materials,

TABLE 12.12. Advantages and Limitations of Various Manufacturing Practices

Process	Advantages	Limitations
Cold forging	High mechanical properties, good surface finish, high productivity	High tool wear, shape limited by material flow
Die casting	High production rate, shape complexity	Requires draft angles, low mechanical properties, many defects
Fine blanking	Smooth surface and edge, high repeatability, high productivity	Applied only to single-thickness components
Hot forging	Excellent mechanical properties, high productivity	Limited forms, shapes, details; low precision; large parts; high tool wear; postforging machining required
Investment casting	Good shape complexity, wide size range possible, high-performance materials	Slow production cycles, waste, poor surface quality, possible ceramic contamination
Machining	Excellent tolerances, inexpensive tooling, widely employed, applied to all materials, wide dimensional range	Costly for large quantities, waste, low production rate
Permanent mold casting	Wide shape range, widely employed, internal cores possible	Poor surface quality, porosity and defects
Plastic injection molding	Greatest shape range; large size range; high precision, close tolerances; excellent surface finish	Low mechanical properties, expensive tooling, low-temperature materials only
Powder metallurgy	Materials flexibility, materials utilization, self-lubrication, high automation, net-shaping	Powder costs; porosity degrades properties; size, shape, precision limits; tooling cost
Sand casting	Wide shape range, low materials and tool cost, applied to large shapes	Low mechanical properties; defects, pores, inclusions; poor surface finish; waste

tolerances, sizes, shapes, and properties. For example, casting techniques excel in shape complexity, but lag in surface finish. At high production quantities, machining techniques suffer with material wastage and production costs. It is not the goal to go into detail here; however, note that all technologies are evolving, making the technological competition a dynamic situation. For every component that P/M takes away form machining or casting, plastic injection molding probably takes away a previously formed P/M component, because of lower cost and lower density.

Cost is the critical parameter in determining the feasibility of P/M production. It depends on many factors, especially initial tooling cost. Tools for P/M are customized to the final component. This is a barrier to small or trial production runs. Many other factors enter into the analysis, including surface roughness and required tolerances. The greater the number of production steps, the higher the fabrication cost. For example, in converting a business machine ratchet from machining to P/M, the number of steps was reduced by a factor of 2 and manufacturing cost was reduced by a factor of 6.

Attractive applications for P/M involve large production quantities of complicated shapes fabricated from materials that are difficult to machine or cast. To succeed against competitive technologies, P/M must satisfy the engineering requirements at lowest cost. It has strengths in the variety of available materials, component shape complexity, surface finish, and ability to replicate details. On the other hand, P/M is weak with respect to tool costs and production costs.

12.6.2. Cost Guides

Some simple guides help rationalize production costs for a typical component. As a gross generalization, powder constitutes about one-fifth to one-third of the manufacturing costs, processing steps account for about the same proportion, and finishing or heat treatment might have a similar impact. The production cost advantage emerges at large quantities, typically as production approaches one million items per year. Beyond this level, less advantage is apparent, largely because tooling has to be replaced and equipment reaches capacity.

Figure 12.10 is a schematic of the interplay between component complexity and production volume for several forming approaches. Two levels are illustrated, one for porous structures and the other for high-performance structures. On each level are shown the shape complexity and production quantity to suggest the more successful technologies. The partitions between fabrication approaches are for illustrative purposes, since many factors impact these decisions, including the material and design details. Die compaction is most productive with moderate mass components produced in quantities of more than 100,000 per year. Injection molding should not be applied to high production quantities of low complexity shapes where die compaction is operating. However, as the component complexity increases, injection molding becomes viable, especially when secondary finishing operations can be avoided. Although some products are generated in quantities as small as 5000 per year, most efficiency occurs with production quantities over 50,000 per year.

To help appreciate the cost features, Figure 12.11 shows a scatter diagram of

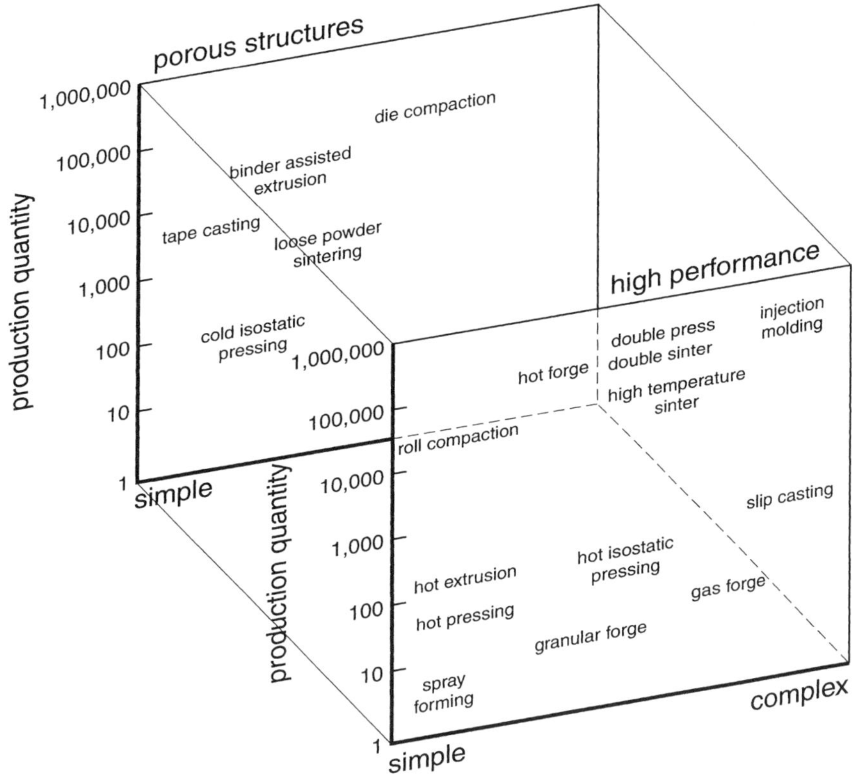

Figure 12.10. A rationalization of production volume, fabrication process, and shape complexity in ferrous P/M. The two levels are for high-performance structural components and low-density, controlled-porosity components. Each production technique is placed to show typical areas of application based on shape complexity and production quantity.

relative cost versus component density. Again this is a schematic drawing, showing that actual cost is highly dependent on the component, alloy, and other details. Still, for illustrative purposes, this scatter diagram provides an early introduction to process selection and how it impacts on component costs.

Powder cost is not a significant factor for masses below 10 g. However, in contrasting P/M with alternative metalworking approaches such as investment casting, powder cost is higher than the casting feedstock. Thus, the elimination of machining steps is one of the best ways to justify P/M when compared with investment casting. Another advantage is the high materials utilization in P/M. It is estimated that 97% of the powder is delivered in final components.

Many successes in converting components to P/M show that at least a 30% lower production cost is required in comparison with alternative techniques. In the P/M production of military components, several case studies show a 40% cost reduction

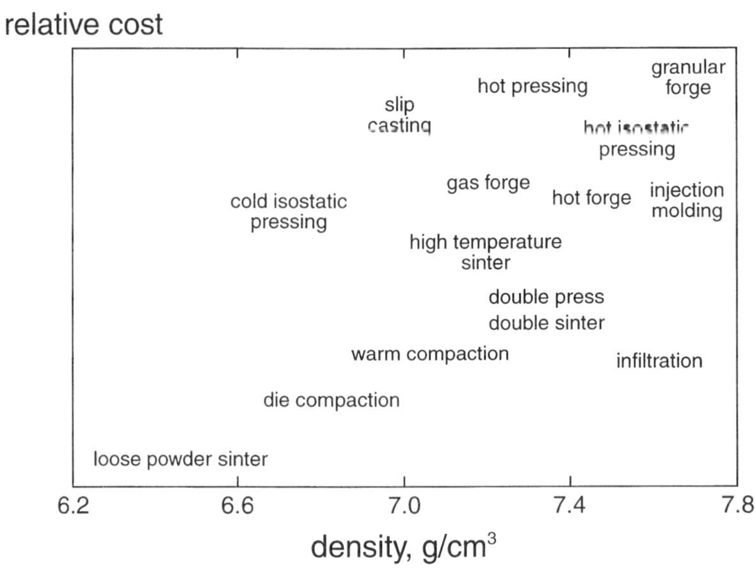

Figure 12.11. Relative cost versus steel density for various production techniques, giving a basis for evaluating the production techniques introduced in this book.

in manufacturing via P/M as compared with machining. In cases of difficult to fabricate materials, injection molding has shown at least a 75% cost savings as compared with machining. Case studies in automotive, computer hardware, business machines, medical devices, and industrial hardware show similar trends. As an example, a machined 40-tooth ratchet used in office machines had an 84% cost reduction in shifting from machining to P/M. The device successfully performed 5,000,000 cycles without wear. In some situations, the production costs are widely different for P/M versus the alternative approach, but the tooling cost becomes decisive. Occasionally, cost reductions over 20-fold are reported.

12.7. DESIGN CONCEPT

Understanding the relation between materials properties, geometry, economics, and process choice is a part of design. For some geometry and property combinations, the selection is simple, while for others there is a need to synthesize an evaluation criterion. This might be difficult for components that fall between two choices of materials or processes. However, when final decisions are made, often a few concerns are dominant. Examples are strength and corrosion resistance, high-temperature strength and cost, shape complexity and biocompatibility, and pore size and corrosion resistance. Consequently, one useful analysis technique is to create a cluster of properties. Essentially these are design parameters that combine the dominant factors, and evaluation is based on this cluster of design-specific factors. Such an

EQUATION BOX 12.3

Material and process selection relies on combinations of parameters that are dominant in a design. For example, consider the selection of a material for precision devices that can hold shape in spite of ambient temperature fluctuations. A high stiffness is desirable to resist stress, but a high thermal conductivity minimizes heating and cooling cycle effects. A low thermal expansion coefficient is useful in minimizing temperature-induced changes. Thus, the most suitable material will possess a high thermal conductivity κ, low thermal expansion coefficient α, and high stiffness as measured by the elastic modulus E:

$$I = \frac{E\kappa}{\alpha}$$

For this case, the index of merit I leads to the conclusion that diamond would be the ideal material, ignoring price. A second choice would be silicon, and for the metallic materials invar (Fe–36Ni) is the best material.

approach is well established in aerospace design, where specific strength (strength divided by weight) and specific stiffness (elastic modulus divided by density) are dominant concerns.

Consider the combination of strength, cost, and corrosion resistance. The typical goal is a high corrosion resistance, high strength, and low cost. When quantitative data are available, these factors might be combined into a cluster of the desirable features in the numerator and the undesirable features in the denominator. Ranking of candidate materials and processes (assuming that each is viable for the component or design) allows isolation of the best two or three choices. Now further evaluations are possible to find missing details and qualify vendors.

Essentially, each design leads to a cluster of key concerns and uses numerical values to rank the candidates. This approach is of great value in eliminating candidates that match only one or two attributes. For example, dispersion-strengthened platinum would be very good for corrosion resistance and high-temperature strength, but not cost; so it would be immediately eliminated. However, if we require operation at 1200°C (2200°F) in air, then platinum becomes the only option—a fact well recognized in glass processing.

Design requires understanding of the various processes and their limits and strengths, as well as the combination of design features important to functional success. Several recent articles and books on design and process selection have been helpful in describing the procedure—see, for example, the book by Ashby or the article by Dewhurst and Reynolds listed below.

With evolving computer databases, this procedure is not a burden, assuming that most common processes and materials are in the databases. Unfortunately, P/M is poorly documented as compared with other forming techniques, especially plastics. Sources of raw data include handbooks, computerized services, subscription services, journal articles, magazines, books, company data sheets, and laboratory tests. Much of these data are computerized, including those available on several Internet sites. A few that focus on powder metallurgy are the European Powder Metallurgy Association, Japan Society of Powder and Powder Metallurgy and Japan Powder Metallurgy Association, Metal Powder Industries Federation, National Center of Excellence in Metalworking Technology, and *Metal Powder Report.*

SUGGESTIONS FOR FURTHER READING

Anonymous, *Guide to the Design of Sintered Parts*, Assinter, Torino, Italy, 1996.

Anonymous, *Powder Metallurgy Design Manual*, 3rd ed., Metal Powder Industries Federation, Princeton, NJ, 1997.

Anonymous, *Powder Metallurgy Parts Design and Implementation for Economic Reliability*, publication SP-535, Society of Automotive Engineers, Warrendale, PA, 1983.

Ashby, M. F., *Materials Selection in Mechanical Design*, Pergamon Press, Oxford, UK, 1992.

Boyer, H. E., and T. L. Gall (eds.), *Metals Handbook Desk Edition*, ASM International, Materials Park, OH, 1985.

Dewhurst, P., and C. C. Reynolds, ''A Novel Procedure for Selection of Materials in Concept Design,'' *J. Mater. Eng. Perform.* **6,** 359–364 (1997).

Edwards, L., and M. Endeau, *Manufacturing with Materials*, Butterworth-Heinemann, Oxford, UK, 1995.

German, R. M., and A. Bose, *Injection Molding of Metals and Ceramics*, Metal Powder Industries Federation, Princeton, NJ, 1997.

German, R. M., *Powder Metallurgy Science*, 2nd ed., Metal Powder Industries Federation, Princeton, NJ, 1994.

Meadows, J. D., *Geometric Dimensioning and Tolerancing*, Marcel Dekker, New York, NY, 1995.

Mosca, E., *Powder Metallurgy Criteria for Design and Inspection*, Associazione Industriali Metallurgici Meccanici Affini, Turin, Italy, 1984.

Salak, A., *Ferrous Powder Metallurgy*, Cambridge International Science, Cambridge, UK, 1996.

CHAPTER 13

Products

13.1. OVERVIEW

The design of a component seldom starts with powder metallurgy as the target fabrication technology. Instead, the designer seeks to satisfy a function, and the design is subsequently identified as a candidate for P/M. Interest in using P/M evolves around the three classic attributes of delivery, quality, and price. Powder metallurgy is selected because it delivers lower price components as compared with alternative technologies, within the design specification (dimensions, properties, and performance), in a productive and timely manner. The cost issue reflects the net-shape fabrication with fewer steps, little scrap, lower labor content, and less energy input as compared with other metalworking routes.

This body of advantages provides a base for the fabrication of many useful objects. This chapter focuses on overviewing the applications by selecting examples from many areas. By no means is it possible to include a comprehensive listing, but the spirit of P/M and the diverse applications become evident in these examples. This chapter ties together the many principles developed in the previous chapters. Although ferrous P/M is dominated by automotive applications, there are many uses in aerospace and aircraft structures, agricultural systems (especially processing equipment), business machines and computers, laboratory instruments, hardware and hand tools, consumer products, hydraulics systems, musical instruments, and filtration equipment.

While P/M is continually seeking to displace other metalworking technologies, the competition is constantly seeking new successes at the expense of P/M. The competition includes injection-molded plastics, die casting, investment casting, automated machining, fine blanking, stamping, closed-die forging, extrusion, and cold drawing. Figures 13.1–13.4 provide a basis for analysis of P/M and the competitive technologies. Usually the key advantage is the lower component cost.

As illustrated in Figure 13.1, tolerances by many competitive technologies are similar or better. Many competitive technologies can produce components with the same complexity and size as P/M. Thus, the decision in favor of P/M versus alternative technologies evolves from other attributes. One is the surface finish. Figure 13.2 compares surface roughness, defined as the profile area between peaks and valleys along a line length (units of area over length, giving roughness in units of length). Statistics show that a finish better than 3.2 μm becomes very expensive.

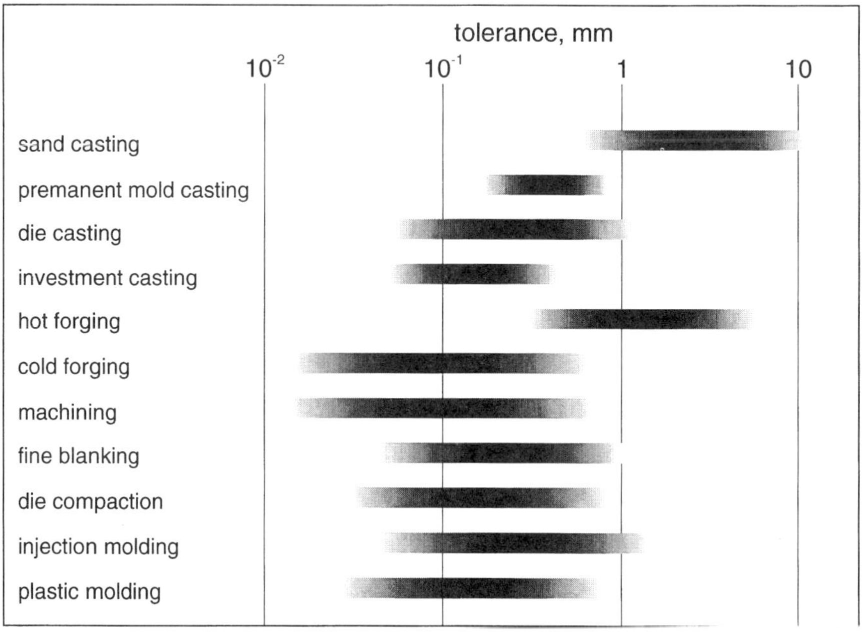

Figure 13.1. Typical tolerance ranges possible via the main production technologies for discrete components.

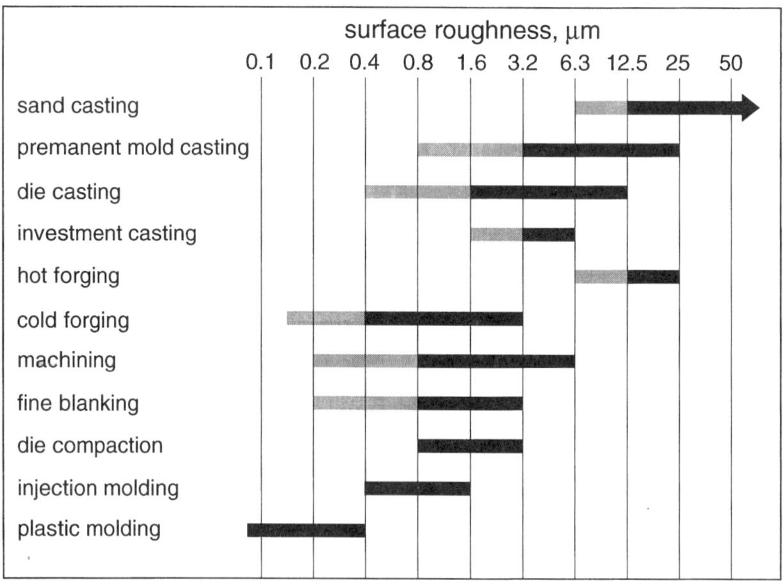

Figure 13.2. Surface roughness comparison for the main production technologies, showing the relative merits of P/M.

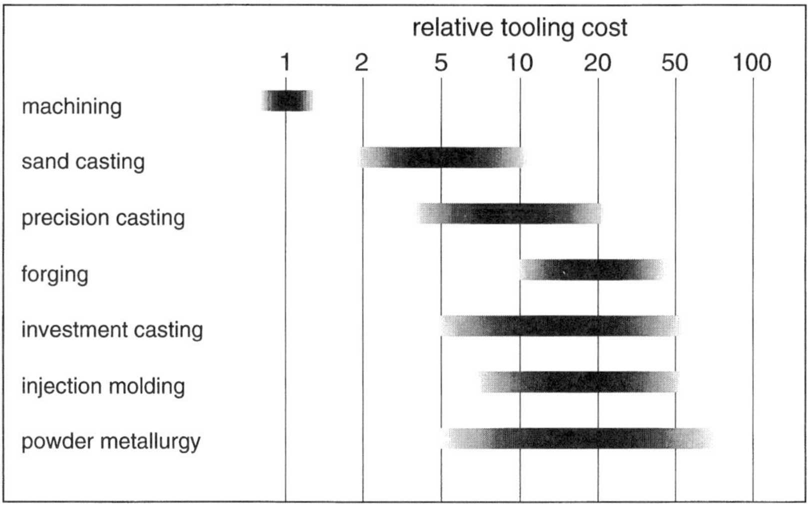

Figure 13.3. Relative tooling costs for various production technologies, showing the high cost position of P/M technologies versus competitive routes, making P/M more useful for high production quantities.

The cost difference in production to obtain a 0.4-μm surface finish is probably 4 times that needed to obtain a 3.2-μm finish. Thus, P/M technologies are not attractive for highly polished surfaces.

Tooling costs in P/M are relatively high, as schematically demonstrated in Figure 13.3. Machining is very attractive in this respect, because there is no tool cost. However, the compensating factor is a high unit production cost by machining. Alternatively, high-production-volume techniques require investments in tooling.

The comparison in Figure 13.4 helps understand the various production technologies as to attainable strength. The competition includes alternative materials and processes. The P/M materials are higher in strength as compared with zinc, aluminum, and copper, but do not reach the levels attainable in forged full-density steels.

Thus, P/M is usually justified for higher-volume production, where tool costs can be distributed over many components. The strength and complexity levels are attractive, but not by themselves outstanding. Tolerances and surface finish provide additional evaluation criteria for deciding the viability of each process. Because of the technological competition, documenting the successes for P/M is important. The examples cited here provide a sense of the diverse uses.

13.2. MARKETS FOR P/M

Humans inherently value metallic components. Accordingly, in spite of economic successes by plastics, powder metallurgy has been growing as a base for the fabrica-

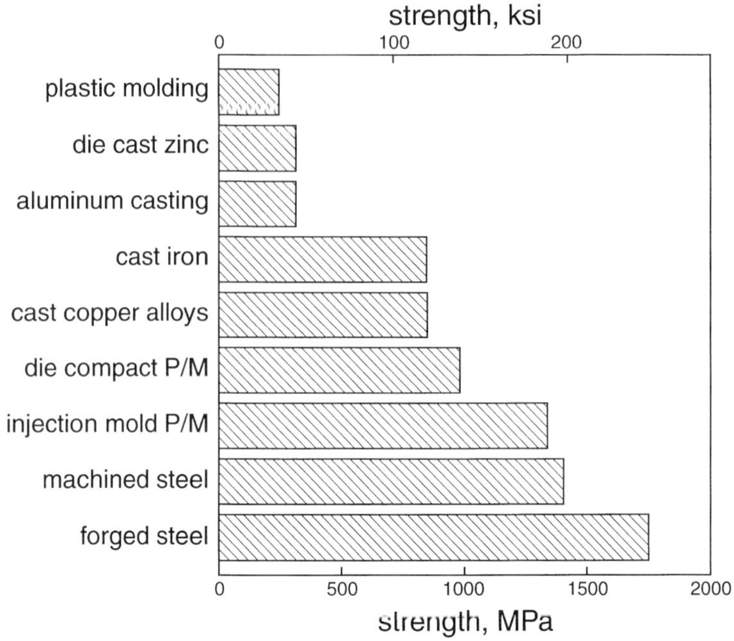

Figure 13.4. A comparison of strength levels attainable from the common discrete component production technologies. The die-compacted and injection-molded P/M options rank high, but do not attain the levels produced in wrought steels.

tion of discrete metallic objects with complicated shapes. As part of the growth, much attention has been given to setting material standards and design guides. This proves necessary, since powder metallurgy is a minor business as compared with competitive technologies such as casting. Considering the worldwide production of raw steel at approximately 750 million tons, P/M is relatively small.

Traditional P/M relies on die compaction and is restricted to moderately complex components. About six companies supply most of the iron powders, but in North America more than 400 companies use this powder for the fabrication of components. Of these, only 10 are beyond the small business status. These 400 companies constitute a sales volume of press and sinter components near $1.5 billion per year in North America. A related group of companies participates in advanced powder metallurgy technologies, and these constitute about $500 million in annual sales in North America. The situation is similar in Japan, where three companies make the iron powder and three companies produce 60% of the P/M components. A few other companies constitute the next 30% market share, and the balance (10%) is divided between almost 100 parts manufacturers. European production has a similar character, with a few dominant parts fabricators, but only three major powder producers.

The sales per employee in traditional powder metallurgy averages $104,000 (standard deviation of $20 thousand), which is low compared with competitive technologies. For example, traditional metals production industries exceed sales of $250,000 per employee in 1997.

On the basis of tonnage, ferrous powders constitute nearly 80% of the P/M activity. Likewise, automotive applications account for about 80% of ferrous powder metallurgy sales. One problem directly traces to the reliance on automotive sales. Historically, the typical profit for an automotive company has been near 4% of sales, yet sales per employee is 3 times that for the P/M companies. The vendors to the automotive industry are under constant price pressures. Thus, the P/M companies are constricted on profits, making it difficult for the industry to grow.

P/M has sustained an overall annual growth rate of 8–10% in recent years. However, the traditional ferrous die compaction has grown at a sustained 4–6% per annum level (5.8% projected up to the year 2000). Thus, advanced materials and fabrication systems carry a large share of the industry growth. Here the forming technologies tend to emphasize full density via hot isostatic pressing, powder rolling, extrusion, forging, injection molding, and hot pressing. Consequently, there is little communication between traditional ferrous powder metallurgy devoted to automotive applications and advanced powder metallurgy devoted to high-performance applications.

13.3. EXAMPLE APPLICATIONS

13.3.1. Aerospace

Applications involving critical flight components usually avoid ferrous P/M, except hot isostatically pressed or hot-forged, high-alloy steel components. One reason is the need for excellent fatigue and stress rupture properties, which require fully dense alloys and careful testing. Many structures produced via powder metallurgy have residual porosity, unless sintered with a liquid phase or subjected to postsintering densification. Porosity is unacceptable for stressed aerospace components. Unfortunately, any added postsintering steps, such as hot isostatic compaction, are detrimental to the production economics. Another difficulty with aerospace applications is the need to withstand elevated temperatures in aggressive environments. This requires nickel, cobalt, or titanium alloys, although a few stainless-steel components are in use. However, in unmanned structures and low-stress mechanical components, P/M has been successful.

13.3.2. Appliances

In many regards, P/M applications in household appliances reflect similar uses in automotive systems. Many pulleys, levers, sprockets, bearings, and gears are fabricated for power transmission in washing machines and other common household

Figure 13.5. Transmission gear from a washing machine, pressed from an iron–copper–carbon powder mixture. (Photograph by Julian Thomas.)

devices. Most are fabricated from low-alloy steels. Figure 13.5 shows a gear in a transmission system for a home washing machine. The success of this product is based on fabrication to close tolerances in high production volumes. A critical requirement is for the gear to have precision on the tooth profiles to ensure smooth, quiet operation.

Another award-winning appliance application was an ice-cube ejector for the automatic icemakers in home refrigerators. Components near food, or in contact with humans, are fabricated to a closed-pore condition using stainless steels. Components involved in load transmission are fabricated from common P/M alloys such as iron–copper–carbon. Another unique application was a hot plate fabricated with a heating element embedded in iron powder. To form the compact, the coiled heating element was first inserted into a bed of iron powder in the compaction die for pressing. After sintering, the heater is infiltrated with copper. It provides uniform heat distribution over a wide area.

13.3.3. Automotive

The range of automotive applications includes motorcycle, off-road, light truck, and even heavy-duty truck and industrial tools (forklifts and other mobile machines). Throughout this book automotive alloys and components have been used to illustrate P/M technologies. Although opportunities abound, only a few components sustain long-term production. Several successes are found in timing-belt pulleys, seat-belt locks, antilock brake systems, locking mechanisms, shock absorbers, door levers, brake pads, clutches, steering mechanisms, camshaft lobes, oil pumps, engine bearings, transmission synchronizers, seat gears and racks, connecting rods, airbag sensors, mirror mounts, sprockets, signal-control instruments, and air conditioners. As new, higher-performance components are developed, older, lower-performance applications disappear. However, every year there are more types, sizes, and complexities to the automotive applications. Typically, the newer components are expected to last longer, and are more complex. All the products are moving to higher densities to improve performance.

One recent success has been the speed sensor ring associated with the antilocking brake system. A picture of one design is given in Figure 13.6. These are toothed rings attached to each wheel to provide magnetic pulses every time a tooth passes a stationary sensor. A microprocessor calculates the relative rotational velocity of each wheel to decide the appropriate braking pressures. Tight tolerances exist on the outer diameter and the tooth structure, while the inner spline locks the ring onto the axle via a press fit. Since there is little load beyond that involved in installation, the major requirements are in dimensional precision and magnetic performance. The typical ring is uniaxially die-compacted from an iron–phosphorous powder mixture (iron and ferrophosphorous or Fe_3P) with final phosphorous levels from 0.45 to 0.80%. Compaction pressures ranges to more than 500 MPa (35 tsi) using 0.6 wt%

Figure 13.6. Sensor ring for the antiskid brake sensor for an automotive control system. (Component courtesy of Thomas Haberberger and photograph by Julian Thomas.)

of stearic acid as lubricant. The green compact is typically 7.0 g/cm^3, and sintering is done at 1120°C (2050°F) for 25–30 min in a nitrogen–hydrogen atmosphere. After sintering, the compact is tumbled to remove burrs, impregnated with sealant polymer, and electroplated with zinc for corrosion resistance. Tensile strength is about 300 MPa (40 ksi). Tests for these components include corrosion, usually in a salt spray, strength, and hardness, and no visible cracking can occur during installation. Compared with machining, the P/M sensor ring is produced at a 40% lower cost.

Other P/M automotive components are pictured in Figures 13.7–13.9. The first is a die-compacted copper-steel flange pulley, fabricated to a density of 6.6 g/cm^3. It consists of the outer pulley with an embedded sensor function on the face. It has a sintered yield strength of 270 MPa (39 ksi), tensile strength of 390 MPa (56 ksi), and 55 HRB hardness.

The next example, shown in Figure 13.8, is a die-compacted 6.9-g/cm^3 complex nickel–steel component that serves as two halves of an inner axle differential case for a heavy-duty truck. The function is to distribute input power between the front and rear axles in a tandem truck. The component weighs 0.9 kg (2 lb), and two halves are arc-welded to form the case, as illustrated in this figure. Mechanical properties include a hardness of 85 HRB, yield strength of 410 MPa (60 ksi), tensile strength of 710 MPa (103 ksi), and fatigue endurance strength of 270 MPa (39 ksi). Additionally, the transverse rupture strength was measured at 1340 MPa (195 ksi) and the unnoteched impact energy was 27 J (20 ft · lb).

Figure 13.7. A copper–steel automotive flange pulley that includes a sensor on the face. (Photograph courtesy of Metal Powder Industries Federation.)

Figure 13.8. Multilevel nickel–steel components are welded to form an assembly as a case on the axel of a tandem truck. (Photograph courtesy of Metal Powder Industries Federation.)

Figure 13.9 shows a group of automotive transmission sprockets fabricated using high-temperature sintering. The teeth are densified after sintering by a cold repressing operation followed by surface carburization in a vacuum. This selective densification of the teeth leads to a core density of 7.0 g/cm^3 , while the teeth are 7.8 g/cm^3. Mechanical properties are not uniform in the component, but hardness of the teeth reach 60 HRC with a tensile strength of 860 MPa (125 ksi). Most importantly, the fatigue endurance strength exceeds that attainable via alternative fabrication processes.

Future growth in automotive uses of P/M will come from several sources. First is simply an expansion of current applications, as more manufacturers accept P/M components. This will occur rapidly in Japan and Europe, where the use of powder metallurgy is behind that of North America. A second development is improved precision, allowing P/M to take orders on more complicated gears. The use of stainless steel in automobiles is opening opportunities for corrosion-resistant P/M stainless steels. Further, as fatigue life improves via higher final densities, there will be

Figure 13.9. A sequence of automobile transmission sprockets formed by repressing the teeth to near full density, followed by vacuum carburization to increase the fatigue strength. (Photograph courtesy of Metal Powder Industries Federation.)

new opportunities that are sensitive to fatigue. The few examples shown in this section only hint at the extensive opportunity for ferrous P/M in automotive systems.

13.3.4. Business Machines

Business-machine applications are a mature area for P/M. Most of the sintered components for copiers, postal meters, and printers were selected for their low production cost. Early examples were found in high-speed mechanical printers. These were formed by die compaction with the characters impressed on the print faces. The low density proved a problem with fatigue life. Similar ideas were employed for the production of gears, drive components, and various motion components. Figure 13.10 is a photograph of a tool steel component used in a postage machine. The teeth are used to tear off paper tapes after imprinting the postage. A high hardness and corrosion resistance is required for long service life.

Electronic systems are replacing electromechanical equipment to form faster, more reliable, and lower-cost business equipment. Accordingly, traditional P/M applications are on the decline in favor of components with greater complexity, such as through injection molding. One early success occurred in an injection-molded steel used in electric typewriters. It was formed from a mixture of iron and nickel powders using a wax–polymer binder. Other injection-molding applications include cutters from tool steels, various gears and levers, and many printer components. As P/M expands to include wear-resistant materials, the applications will grow for high-density, fatigue-resistant printing devices. However, the overall field is not growing.

13.3.5. Computers

Because of the declining use of P/M for business machines, one growth area has been in computer equipment. These are small components. A major success has

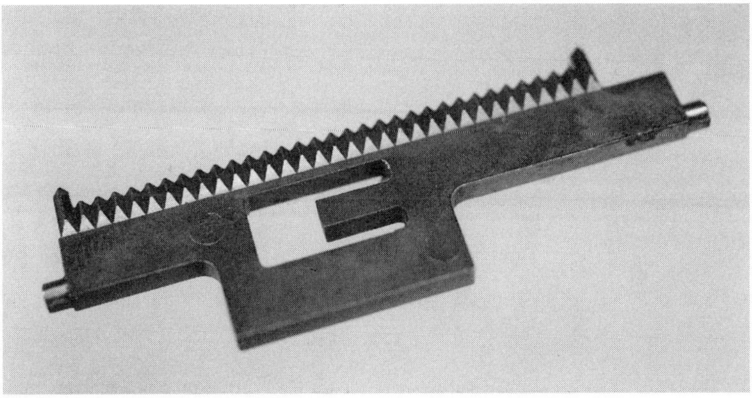

Figure 13.10. A portion of the paper tape dispenser on a postage meter. This P/M part is formed from tool steel to have high hardness, wear resistance, and corrosion resistance. (Photograph courtesy of Julian Thomas.)

been in magnets for disk drives. These applications rely on closer tolerances as compared with investment casting. One driving force is product miniaturization. In the early 1980s a disk drive was nearly 80 mm high. Reductions in disk size resulted in smaller heights and weights, yet more memory and lower cost.

Disk drives use injection molding for various magnets, arms, and supports. By the early 1990s, disk-drive component production was a major activity for injection molding. Today the production quantity is high and the shapes match well with the attributes of the technology. Figure 13.11 shows a disk-drive magnet formed from iron. It was molded using a 58 vol% solids loading, subjected to thermal debinding, and vacuum-sintered to a density near 7.7 g/cm^3. One difficulty is the rapid progression in designs, making it difficult to keep a shape in production for an extended period. However, the savings versus investment casting is significant; machining is 22 times more expensive, and investment casting is 6 times more expensive. Thus, cost and property goals drive the expanded use of injection-molded steels, with most compositions being formed from carbonyl iron powder. Other parts are produced for computers and computer peripherals include mechanical components, sockets, junctions, heat sinks, and connectors.

13.3.6. Consumer Products

Powder metallurgy has slowly penetrated the consumer market through the fabrication of items ranging from cooking utensils to musical instrument components. However, the rough surface finish has been a difficulty. Powder metallurgy has found applications in eyeglass hinges, sunglass frames, watch bands and watch cases, cosmetic cases, small gears for electric toothbrushes, can openers, electric knives, kitchen tools, and kitchen blenders. These are largely formed from steels or stainless steels. An application with considerable appeal is in cutting instruments—scissors,

Figure 13.11. Magnetic component for a computer hard-disk drive unit, fabricated by injection-molding pure iron. (Component courtesy of Mary Ann Wright and photograph by Julian Thomas.)

knives, hair trimmers, and other cutting tools. The emerging use of injection molding in sporting equipment is also exciting—for example, in golf clubs, and running and golf-shoe cleats.

These applications represent a large potential use of P/M, but shifts in consumer taste create rapid changes in production rates over short times. For example, a small gear used in electric toothbrushes is shown in Figure 13.12. These were injection-molded from 17-4 PH stainless steel to a density of 7.6 g/cm^3 in production rates reaching 60,000 per day. Molding was performed in a 16-cavity tool set. High-strength stainless steel was required for abrasion and corrosion resistance associated with toothpaste and water. The item was replaced by a plastic component and rapidly went out of production. Current high production volumes are associated with watch bands and watch cases from stainless steel.

13.3.7. Defense

One early success for ferrous powder metallurgy was a rotating band on artillery shells. This became a critical technology during the 1940s. Although the defense industry still pushes the performance level of P/M materials, most of the growth is occurring away from defense applications. Critical weapon components are always in production, but the quantities are usually small compared with other components.

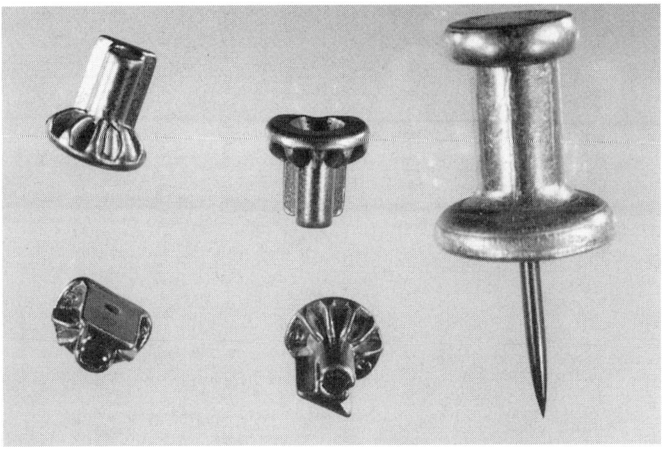

Figure 13.12. Injection-molded gear for an electric toothbrush formed from 17-4 PH stainless steel. (Components courtesy of Matt Bulger and photograph by Julian Thomas.)

In some instances, short-term production volumes are large. For example, some artillery shells have arming fuses and large production occurs while stockpiles are accumulated, but otherwise the activity is small.

Figure 13.13 is a guided-missile baseplate support formed by die compaction of 316L stainless-steel powder. It is a three-level part that requires secondary machining

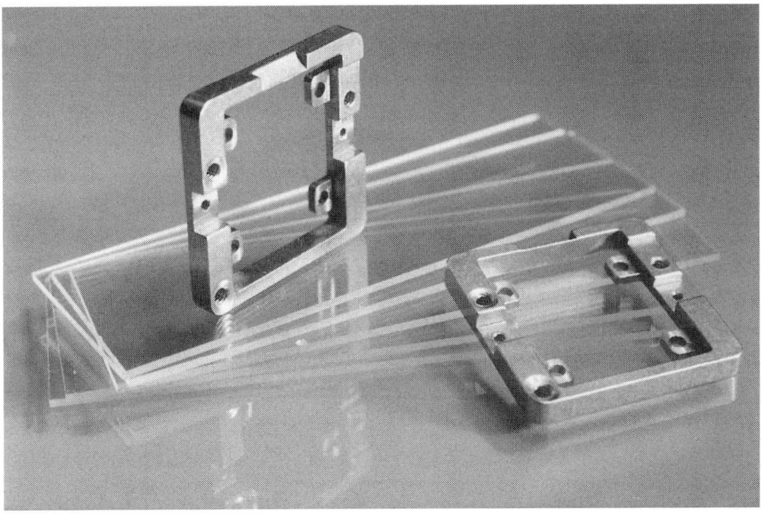

Figure 13.13. Support base for a guided missile formed from 316L stainless steel by die compaction, sintering, with final machining. (Components courtesy of Lynn Youngberg and photograph by Julian Thomas.)

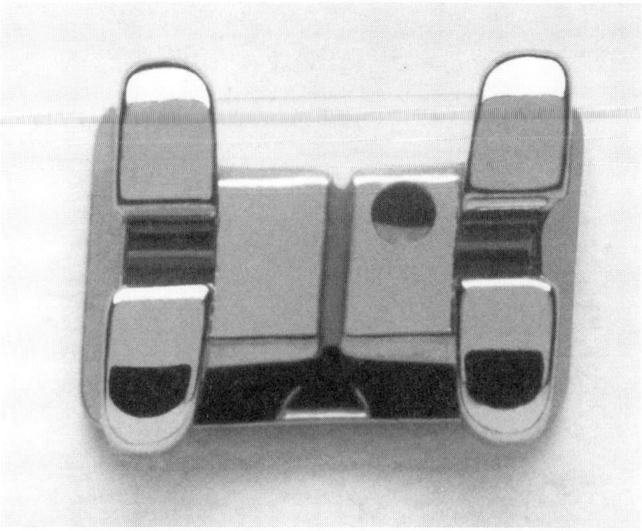

Figure 13.14. Orthodontic bracket formed from stainless steel by injection molding. (Photograph from Animesh Bose.)

for flatness and threading of the holes. Water-atomized stainless steel powder was pressed to a density near 6.5 g/cm^3 and sintered in hydrogen at 1200°C (2200°F). After sintering, the component was coined, plastic-impregnated, and lapped on one face for flatness. Threads were added by machining. These components are inspected for several features, including dimensions, flatness, surface burs, and corrosion resistance. Other applications are found in stabilizer fins, guidance components, and fragmentation devices.

13.3.8. Dental Components

Corrosion-resistant materials are fabricated using P/M techniques, usually from near-full-density stainless steels. The orthodontic bracket is an example of an application dependent on corrosion resistance. An example bracket is shown in Figure 13.14. The bracket is affixed to the tooth surface using a biocompatible adhesive to align teeth by applying constant stress. It is fabricated using gas-atomized powder (nominally 15 μm) mixed with a thermoplastic binder and injection-molded into the oversized shape. Because of the small size, debinding and sintering can be combined into a single operation, with a final sintering temperature near 1325°C for 2 h in hydrogen. The final density is more than 98% of theoretical, and the component exhibits good corrosion resistance and excellent mechanical properties (tensile strength of 500 MPa and 25% elongation). Higher-strength brackets are formed using 17-4 PH stainless steel.

The application of PIM to orthodontic brackets is due to miniaturization and

EQUATION BOX 13.1

For sintered steels, the ductility is sensitive to residual porosity. This variation follows a model described by Haynes and others as follows:

$$Z = \frac{(1 - \epsilon)^{3/2}}{(1 + c\epsilon^2)^{1/2}}$$

where Z is the relative ductility, c is an empirical constant, and ϵ is the porosity. The relative ductility is the ratio of the porous material to the equivalent wrought material. The coefficient c represents the sensitivity of the ductility to the presence of pores. High-strength steels are very sensitive to residual pores, and a value of the coefficient c is as high as 10^5 for powder forged steels. On the other hand, stainless steels are tougher and less sensitive to porosity, so a value of the coefficient c near 100 is more appropriate.

improved aesthetics. Until the early 1970s, stainless-steel orthodontic brackets were made from welded bands. Those early appliances were large and ugly. Smaller brackets and new adhesives allow direct attachment of the bracket to each tooth. Several companies currently produce these devices, and most are fabricated by the dental supply companies for direct sale to the dentist. The shape is intricate and the production volumes in the United States are more than 100,000 brackets per day. This activity will double as global dental health care improves, and such growth is expected to involve continued use of injection molding; however, much effort is going into nickel-free stainless steels and titanium as replacement alloys.

13.3.9. Electrical–Mechanical Systems

Permanent magnets and soft magnets are two major classes of magnetic P/M materials. They differ in response to an applied field. In an alternating current transformer, most desirable is to have minimal energy loss with each cycle of induced magnetization. Here a soft magnet like pure iron is used. Alternatively, in a direct-current motor, such as the starter motor in an automobile, a permanent magnet with a large remanent magnetization is required.

The major drawback with P/M-produced magnetic materials is porosity, since that disrupts the magnetic field. Other factors such as pore size and material purity (especially those impurities adsorbed from the sintering atmosphere) can negatively affect the magnetic properties. The principal properties of concern are the coercive force (the magnetic force necessary to demagnetize the material), the maximum permeability (the largest magnetic response), the remanent magnetization (the degree of magnetization remaining after applying an aligning field), and the energy loss (energy consumption per alternating cycle of applied magnetization). Impurities

Figure 13.15. A magnetic pole piece consisting of two layers of iron–phosphorous with a separating layer of stainless steel. (Component courtesy of Terry Cadle and photograph by Julian Thomas.)

like oxygen or nitrogen are detrimental and are best eliminated by high sintering temperatures in a reducing atmosphere. It is important to control the impurity level to 100 ppm or less and to obtain a high sintered density. Sintering can be at temperatures of up to 1400°C (2550°F) to give a final density of 7.1–7.5 g/cm^3. Attaining shape precision and a high density requires a double press–double sinter route. Applications range from compact-disk players to industrial motors.

A soft magnetic component is shown in Figure 13.15. This is a two-material magnetic-pole component created from stainless-steel and iron–phosphorous layers. It is part of an automotive variable assist steering system, formed by simultaneous compaction of the two powders to give annular segments. The inner and outer poles are formed from the soft magnetic alloy, while the thin separation layer is formed from 304L stainless steel. Compaction of this geometry required three lower punches and two upper punches and two powder fills. The overall density of the magnetic portion is 7.0 g/cm^3, but the stainless-steel layer is just 6.2 g/cm^3.

Many alloys are available for these applications, including iron containing nickel, cobalt, phosphorous, silicon, and other combinations. High-permeability alloys consisting of nearly equal proportions of iron and nickel are used in housings for microwave systems over a range of frequencies from 2 to 12 GHz. Iron–silicon alloys are used for printer pole pieces in dot-matrix printers. These alloys are selected for alternating current applications where low core losses are important. Core losses (hysteresis and eddy-current losses) are high for iron because of its low electrical resistivity. Silicon is an effective addition for increasing the electrical resistivity of iron. It also reduces the hysteresis loop, which is desirable. Sendust alloys based on Fe–Al–Si are the basis for magnetic-card readers, and current P/M efforts use injection molding. Another hard magnetic materials is based on the compound $Fe_{14}Nb_2B$. Rapid growth is occurring in bonded forms of the powder, where a polymer is used to glue the particles in place, with no sintering required.

13.3.10. Filtration Systems

A natural application for P/M is the fabrication of controlled pore structures for bearings, filters, flow restrictors, air-distribution surfaces, sound absorbers, heat pipes, and biomedical implants. Pore size control is achieved by classifying the powder into a narrow size fraction. Three features control the success, porosity, pore size, and permeability. The latter measures the ease of passing a fluid through the pores. Materials selection depends on the use environment and the desired properties. For filtration applications, corrosion-resistant stainless steels are preferred.

A common application for porous metals is in barrier filtration, where debris in a fluid must be retained. A small pore size is needed in a material with a high flow permeability. To aid filtration and reduce the pressure loss over the filter wall, high-surface-area designs are preferred. Various stainless-steel filters are pictured in Figure 13.16. These are used for flow control, polymer filtration, and flow distribution. Most structures are fabricated by compacting water-atomized stainless-steel

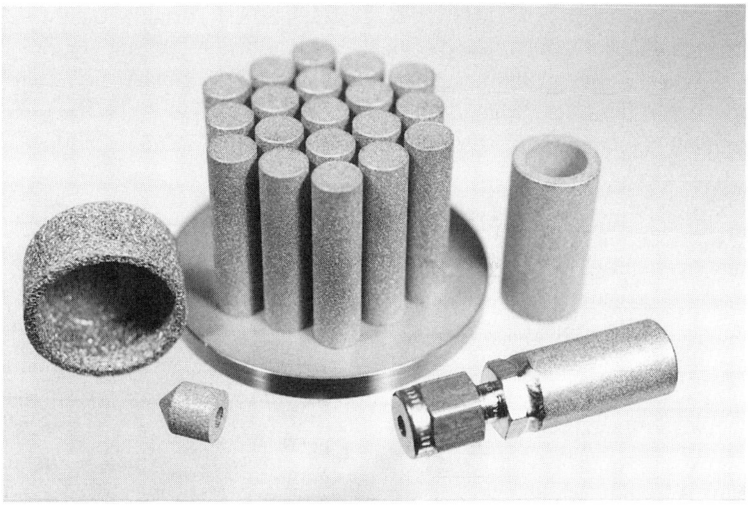

Figure 13.16. Various filtration designs formed from stainless steel using pressing and sintering technologies. For high surface area, multiple filter elements are combined into a single structure to give more filtration. (Samples courtesy of Mark Eisenmann and photograph by Julian Thomas.)

EQUATION BOX 13.2

On the basis of porous stainless steel fabrication studies by Deepak Kapoor, an empirical correlation between the filter rating and pore characteristics is given as follows:

$$\alpha = B(\epsilon F)^N$$

where F is the filtration rating and is usually given in terms of the nominal particle that will not pass through the filter, ϵ is the porosity, and α is the permeability coefficient. Both B and N are empirical constants that depend on the processing cycle. Irregular pores prove most useful in industrial filters.

powder followed by sintering near 1200°C (2200°F) using pure hydrogen. The strength of the sintered stainless steel allows periodic cleaning to reduce the pressure drop at the filter.

One key aspect of forming porous structures is to anticipate the porosity and pore size in selection of the powder and compaction conditions. The smaller the desired pores size, the smaller the particles. Ratings on the filters are based on the sizes of debris that cannot pass through the filter. For example, a filter rated at 2 μm captures debris over 2 μm in size, but has pores as large as 21 μm. This filter is formed from −270/ + 200-mesh water-atomized powder (64 μm mean size). The powder is cold isostatically compacted [150–200 MPa (20–30 ksi)] or tape-cast into a sheet, depending on the component shape, and sintered to give a final density of 5.3 g/cm^3. Sintering is near 1200°C (2200°F) for at least 30 min in hydrogen. This sintered material has a tensile strength of just 150 MPa and only 2% elongation, so most applications avoid high stresses.

13.3.11. Firearm Components

Handguns, shotguns, and rifles consist of many components fabricated by P/M. Examples include triggers, trigger guards, magazine boxes, sights and sight mounts, firing pins, and various stressed components. For this industry, steel parts are most desirable. Most of the growth in this area is occurring in the fabrication of handguns for police and private use. Firearm applications continue to expand as designers find more components suitable for P/M.

Figure 13.17 is a picture of a trigger used on a sporting shotgun. It was formed using a 75-μm water-atomized iron powder to a green density of 7.0 g/cm^3 and sintered at 1120°C (2050°F) for 1 h in dissociated ammonia. The serrations are added in a postsintering repressing operation. Other finishing operations include final machining of a hole and slot, black oxide formation via steam treatment, and pore sealing via polymer impregnation. The key justification for P/M rests with cost.

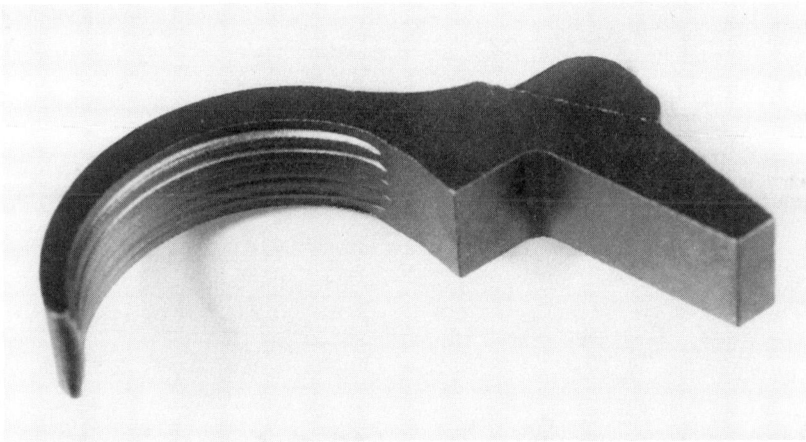

Figure 13.17. Die-compacted trigger for a sporting firearm, fabricated by die compaction, sintering, and machining, with several secondary operations. (Picture courtesy of Louis Baum.)

13.3.12. Friction Materials

Brake pads and clutches are compacted and sintered from mixtures of iron and other powders. These are required for higher performance (heavy duty) applications such as in aircraft brakes, heavy-machinery clutches, and brake linings on racing cars. Figure 13.18 is a picture of an aircraft brake pad. Similar designs exist for many automotive and truck components. These structures convert mechanical energy into heat. Clutches perform the same function in accelerating components up to speed. Friction materials are produced by a combination of compaction and sintering, or warm pressing. A typical composition consists of mixed iron, graphite, copper, and tin powders. For high-performance applications, such as race cars, a nominal composition might consist of iron and copper, with additions of various hard and soft phases, such as coke and molybdenum disulfide.

The powders are mixed in a double-cone blender with an oil binder. That mixture is compacted at pressures between 150 and 300 MPa (11 and 21 tsi). The compacted mixture must be sintered to attain sufficient strength for handling and use. Because chemical homogeneity is not desired during sintering, it is sufficient to heat to temperatures ranging from 700–1000°C (1290–1830°F) to melt the tin. The sintered pad is often grooved and brazed onto a backing plate for assembly into the brake system. When compared with the typical organic brakes, the sintered metal brakes have a higher coefficient of friction with a lower wear rate and higher thermal stability, and prove insensitive to humidity changes.

13.3.13. Industrial Equipment

Ferrous P/M alloys are widely used in industrial components. This diverse field includes everything from pump components to valve handles. As an example of the

Figure 13.18. Aircraft brake pad sintered from a mixture of powder, including sponge iron, using die compaction and low-temperature sintering. (Component courtesy of Patricia Cuzzo-Taub and photograph by Timothy Weaver.)

applications, the screws for plastic injection-molding machines are fabricated by hot isostatic pressing steel powders. These screws are tapered to compress and transport plastic in the barrel of an injection molder. When polymers filled with minerals or glasses are molded, screw life can be shortened by the combination of chemical reaction (especially with chlorine- and fluorine-containing polymers such as polyvinyl chloride, polyvinyl fluoride, or polytetrafluroethylene), wear (from the glass filler), and heat. Hot isostatic compaction is used to form clad screws by filling powder into an oversized container around the core. The powder is consolidated at 950°C (1740°F) with pressures in the 100–200-MPa (15–30-ksi) range over times of 1–3 h. The core is a tough steel, while the outer powder layer is densified and bonded to the solid core. Because of the high alloy content, the HIP layer contributes to longer screw life. The high-performance alloys contain high vanadium levels (10–20%), with 22% chromium and 5% carbon. After HIP, they have a hardness between 60 and 72 HRC, depending on the heat treatment. Consequently, these screws provide up to 100 times longer life when compared with stainless-steel or cobalt alloy screws. Boride–tool steel composites are liquid-phase-sintered as the newest form of hard coating. A microstructure of the interface between the steel core and boride composite is shown in Figure 13.19.

Similar ideas in forming wear- or corrosion-resistant alloy layers are emerging in many industrial applications. These include tools for drawing, rolling, or extrusion,

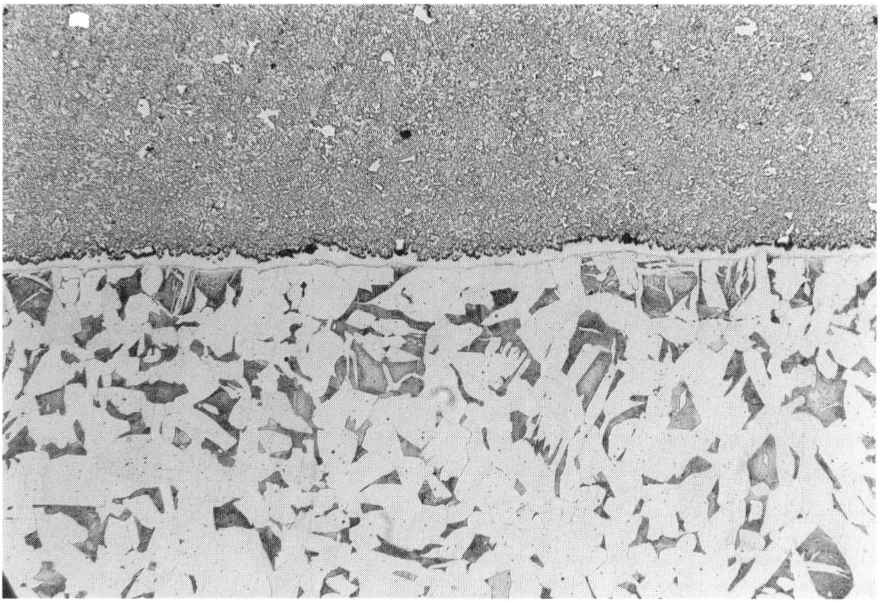

Figure 13.19. Micrograph of the interface between a low-carbon steel and a liquid-phase bonded steel–boride composite for wear resistance. (Photograph courtesy of Karthik Sivaraman.)

as well as components and bearing surfaces for materials in contact with high stresses, temperatures, or aggressive environments. Good examples are found in nuclear power equipment, submarines, agriculture devices, oil-well equipment, high-pressure valves, and polymer mixers.

13.3.14. Lawn and Garden

Powder metallurgy applications in lawn and garden equipment are similar to automotive and appliance applications. Most of the successes are associated with power transmission, so P/M lawn and garden components are often gears, cams, clutches, pulleys, or sprockets. Figure 13.20 pictures garden tractor components that are part of the power transmission system in a riding lawnmower. They were formed by warm compaction using an alloy steel (4405). Compaction was to a density near 7.4 g/cm^3. The warm compaction process was followed by sintering at 1275°C (2326°F) and heat treatment to bring out high final properties. The final tensile strength exceeds 1250 MPa (180 ksi), with a hardness ranging from 65 to 78 HRA, and an impact toughness (unnotched) of 25 J (18 ft · lb). A key performance criterion is a high impact strength to avoid failure in situations where sudden loading occurs.

13.3.15. Medical

Medical instruments represent a very large opportunity for P/M. This application area is driven by desires to eliminate reusable items that might be contaminated

Figure 13.20. Nickel–steel bevel gear for use in the transmission of a riding lawn tractor. (Component courtesy of Kevin Couchman and photograph by Julian Thomas.)

with blood. Early indicators promised rapid growth, but cost pressures have forced adoption of plastics for many components. Current projections better realize the possibility for slow, but progressive growth in the medical tool and instrument areas. An early surge in interest occurred with new surgical stapling components produced by injection molding. Unfortunately, growth has been slow for these devices, but an example is given in Figure 13.21. This is a surgical stapler fabricated by injection-molding prealloyed powders, with solvent debinding and subsequent sintering to near full density.

Applications for P/M components in the human body will be very slow in emerging and will typically move to titanium. The fatigue requirements represent a barrier, and open porosity is not tolerated. However, injection-molding stainless-steel surgical tools can be used in low-stress applications, because of their cost, corrosion resistance, and closed porosity. Die compaction is not a viable fabrication route for medical components because of the open pores.

13.3.16. Microelectronics

Potential applications for low-thermal-expansion and high-thermal-conductivity materials in microelectronics are extensive. Future telecommunications and avionics systems will depend on integrated packaging with a low-thermal-expansion coefficient and high thermal conductivity to extricate heat.

Although the market for P/M microelectronic packaging is only developing, the use of some iron-based low-thermal-expansion alloys such as Fe–Ni and Fe–Ni–Co are already known. A recognized low-thermal-expansion alloy is Fe–36Ni. This

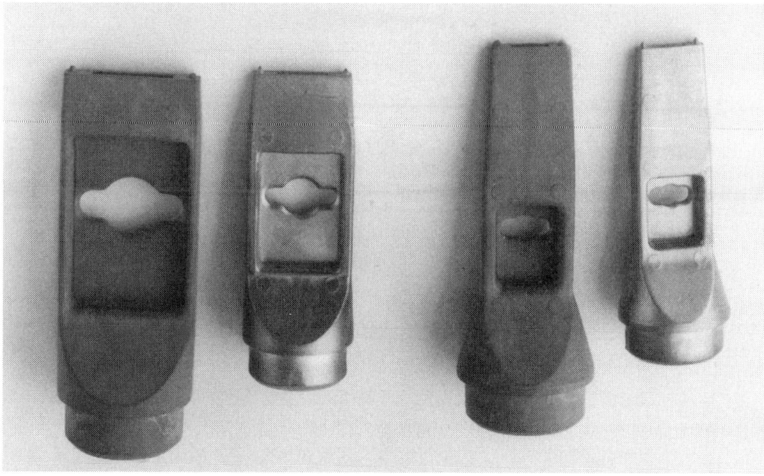

Figure 13.21. Surgical staplers (molded and sintered) formed for medical use by injection molding stainless steel. (Photograph courtesy of Animesh Bose.)

alloy exhibits anomalous thermal expansion behavior near room temperature with an expansion of just $1.7 \cdot 10^{-6}\ °C^{-1}$. Alloying with 5% cobalt further lowers the expansion coefficient. Among the Fe–Ni–Co-based alloys, the Fe–29Ni–17Co composition is most popular. It is employed in applications ranging from thermostats to gyroscopes. One primary use is in matching the thermal expansion of various semiconductor or glass sealing materials.

Components for fiberoptic communication and computer networking are important new applications. Figure 13.22 shows one such device used for computer connections. The housings are often formed from low-thermal-expansion steels, because precise dimensions are required during use. Packaging structures for high-performance microprocessors is another possible P/M application.

13.3.17. Sporting Equipment

Much attention has been directed to high-performance sporting equipment. Often weight limitations eliminate ferrous P/M from consideration. However, there are a few examples of success in sporting equipment. One of the most widely discussed options is in cold isostatically pressed or injection-molded golf-club heads. These use stainless-steel powders to fabricate wedges or hollow drivers. Other applications are in sporting cleats, shotguns, and outdoor equipment. The possibility of forming replacement shot for hunting, out of steel powders, is being considered, since lead shot contaminates the water and is eaten by ducks, causing poisoning.

13.3.18. Tooling

Dies and cutting tools are fabricated from tool steels by a variety of techniques. Much of the high-performance tooling is derived from rapidly quenched alloy parti-

Figure 13.22. An injection-molded iron–nickel computer connection. (Photograph courtesy of Julian Thomas.)

cles produced by nitrogen gas atomization. These powders are loaded into large containers and consolidated by hot isostatic pressing or extrusion to full density. The densified products are hot-rolled similar to traditional tool steels. The P/M varieties allow higher vanadium alloying levels, up to 15%, and give homogeneous microstructures and properties. Other tool steel production routes include cold isostatic pressing and sintering of water-atomized powders, spray deposition, and injection molding and sintering of gas-atomized powders. The latter variant is attractive, because it produces high mechanical properties in net-shaped yet complex components.

Figure 13.23 is a simple cylindrical component formed by injection molding and sintered to full density using supersolidus liquid-phase sintering. Most tool steels via P/M attain mechanical properties that are close to those in standard handbooks. Another application is to sinter forming dies. The fabrication process begins with mixing the powder with 0.5% stearic acid, with subsequent compaction at 600 MPa (43 tsi) to a green density of 75% of theoretical or about 6.2 g/cm^3 from a −325-mesh water-atomized, prealloyed powder. The green strength is about 15 MPa (2 ksi). During sintering the compact shrinks by 8–9% to give a final density near 99% of theoretical. Sintering is done in a batch vacuum furnace at a temperature near 1245°C (2275°F). The final carbon content is 0.92%. When properly sintered, the microstructure consists of many small, dispersed carbides. The final wear properties are obtained by a combination of austenitization [at 900°C (1650°F)] and tempering [below 500°C (932°F) for 3 h]. Tempering might be repeated to ensure uniform

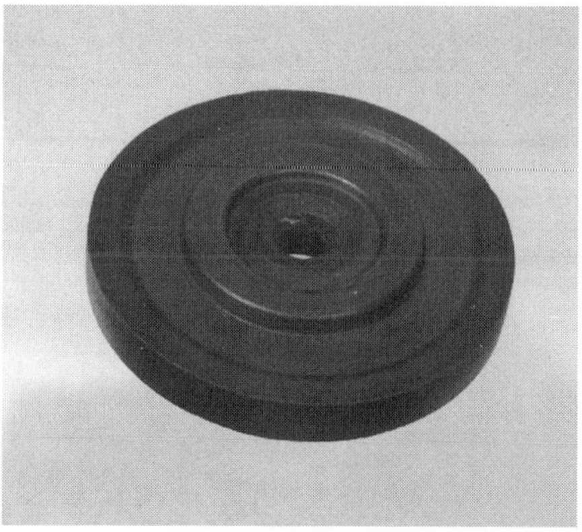

Figure 13.23. Tool steel spindle formed by injection molding and sintered to full density using supersolidus liquid phase. (Component courtesy of Neal Myers and photograph by Timothy Weaver.)

carbide distribution. Carbon control in the final product requires measurement of the carbon and oxygen levels in the powder with appropriate graphite additions. Success is measured by the production of close tolerances with controlled carbon levels with high harnesses.

13.3.19. Wear

Many custom compositions are used in wear applications. These include the common alloys and several exotic cermets (combinations of steels and ceramics). Of all ferrous alloys designed for wear applications, tool steels account for the largest use. The success of a tool steel depends on the microstructure. A fine, homogeneous distribution of carbides with a small grain size is most successful. Such a material exhibits high strength, toughness, and wear resistance.

Powder metallurgy tool steel ingots are formed by atomization of a vacuum-melted alloy, often giving a mean particle size near 100 μm and a tap density of 45% of theoretical. The powder is hydrogen-reduced and cold isostatically pressed to form a green billet that is consolidated via hot isostatic pressing. An alternative is to use hot extrusion at a high reduction ratio. Because the carbides form low melting phases, care is necessary during processing to avoid carbide coarsening. Some tool steels are processed using supersolidus liquid-phase sintering; atomized powder is sintered at a temperature where liquid nucleates within the particles. Precise temperature control is needed to obtain densification while avoiding carbide coarsening.

The properties of tool steels produced by powder metallurgy are very sensitive to oxygen. An oxygen level below 800 ppm is needed in the consolidated material. Usually, the P/M-produced tool steels have desirable carbide microstructures that contribute to uniform properties, high strengths, and good wear resistance. Conse-

Figure 13.24. Wear components formed from tool steel reinforced with titanium carbide and machined to form a screw segment, pelletizing knife, and bearing. (Components courtesy of Paula Slagle and photograph by Julian Thomas.)

quently, the P/M tool steels provide at least twice the service life of comparable cast tool steels. One application for the press and sintered compositions is in valve seats for automobile engines, where production volumes of more than one million per day are envisioned.

Cermet composites are important wear-resistant materials. Widely used are the composite alloys consisting of TiC (titanium carbide) or WC (tungsten carbide) dispersed in a steel matrix. These are hot-pressed, hot isostatically pressed, or liquid-phase-sintered to full density and subsequently machined to final dimensions. Example components formed from such cermets are shown in Figure 13.24. The items shown include a screw segment, underwater pelletizer knife, and bearing. The composite is machined to final tolerances.

13.4. OPPORTUNITIES

Besides continued expansion of known applications, several new opportunities are emerging. Many depend on improved densification technology, reduced production costs, and the availability of higher-quality powders. Growth in constantly offset by advances in competing technologies. Early identification of a P/M component in the design stage ensures better cost and production success. Many demonstration components have not moved forward into production. Early examples were coins and tokens that reached low levels of production, but the application never matured.

New technologies will help move P/M into a wider range of applications. Injection

Figure 13.25. Automotive airbag mechanical components formed by injection molding. (Photograph by Julian Thomas.)

molding and powder forging have contributed significantly to the technology growth in the late 1990s. Early successes for injection molding came from military components, but now automotive production is growing. Figure 13.25 shows a few mechanical components from the airbag assembly. These are in large volume production and will continue to grow. Likewise, powder forging has grown slowly since the 1960s, but now is an accepted fabrication technique for selected shapes and high performance levels.

Coupled with the newer production processes, there are new powders offering higher compressibilities and better sintering. Prealloyed powders with high compressibilities enhance the final properties by ensuring homogeneous heat treatment response. Molybdenum is a common alloying addition that improves heat treatment response, yet does not substantially affect compressibility. Alloys with about 1% molybdenum are now available for die compaction. When mixed with copper, graphite, and nickel, these powders sinter to high strength and toughness levels.

Another idea under investigation is deformation of the sintered component during cooling. While the steel is still in the austenitic phase, the structure is soft and easily worked. Thus, alloy steels can be warm-worked into full-density structures. Some anticipated successes are in the fabrication of automotive transmission gears. Likewise, much growth in P/M will occur when dimensional precision is improved in heat treatment. This requires attention to the whole process, including more uniform powder loading into the die cavity, and more uniform heating and cooling in sintering. With sinter-hardening concepts, the sintered structure will be in the final heat treatment condition, allowing for fewer production steps and delivery of the desired high-strength microstructures.

All P/M production processes are being influenced by increased computerization, including automated inspection, closed-loop feedback control, and constant process supervision. One novel notion is condition-based maintenance, where the press or furnace is constantly monitored to predict needed maintenance before producing parts out of specification.

Rapid prototype ideas have been applied to iron powders. Here a compact is created directly from a computer image via laser, inkjet, or other pinpoint fabrication techniques. No tooling is required, so first parts can be produced within days of conceptualization. Although the early demonstrations are crude and rough, given continued research, the rapid prototypes will improve and become competitive with machining for low production volumes. One exciting variant is the rapid production of tools. First-demonstration tool sets have been fabricated in times of just a few days, but current hardness, strength, stiffness, and wear resistance is lower than those associated with traditional tooling.

Future applications will build on P/M's strengths as more product acceptance takes place. Improved tolerances, a wider range of compositions, and lower production costs with fewer defects are major areas for gains. The applications for ferrous P/M are everywhere. Experience shows that shifts from existing manufacturing routes requires at least a 30% price reduction. This savings can be delivered by integrating design and development activities, using automated production, fast tool production, and large production volumes.

What are the markets for P/M? With the diverse range of existing applications, it appears that P/M could potentially grow to occupy 5% of the total production of metal. It is applied to selected shapes, sizes, and performance levels, so it will not displace many larger and well established production processes for steels. However, once a component has been identified as a candidate for P/M, much progress is possible. Thus, the market for P/M is probably 3 times its current size and is limited only by the imagination of the designers. Some interesting opportunities and current products are obtainable on the World Wide Web, since many companies, laboratories, universities, and professional societies maintain updated home pages.

SUGGESTIONS FOR FURTHER READING

Froes, F. H., (ed.), *1991 P/M in Aerospace and Defense Technologies*, Metal Powder Industries Federation, Princeton, NJ, 1991.

Haynes, R., "A Study of the Effect of Porosity Content on the Ductility of Sintered Metals," *Powder Metallurgy* **20,** 17–20 (1977).

Jenkins, I., and J. V. Wood (eds.), *Selected Case Studies in Powder Metallurgy*, The Institute of Metals, London, UK, 1991.

Klar, E. (ed.), *Powder Metallurgy Applications, Advantages and Limitations,* American Society for Metals, Metals Park, OH, 1983.

Kulkarni, K. M. (ed.), *Powder Metallurgy for Full Density Products*, Metal Powder Industries Federation, Princeton, NJ, 1987.

Lall, C., *Soft Magnetism: Fundamentals for Powder Metallurgy and Metal Injection Molding*, Metal Powder Industries Federation, Princeton, NJ, 1992.

APPENDIX I

Definitions of Common Terms Used in P/M Processing of Iron and Steel

absolute pore size The maximum opening in a porous material, such as a filter, through which no larger particle will pass.

acicular powder Needle-shaped particles.

activation energy The energy required to break atomic bonds to move an atom to a new lattice site or to evaporate an atom.

activated sintering An enhanced sintering process based on a treatment that reduces the activation energy for atomic motion. The increase in sintering rate allows faster sintering, lower sintering temperatures, or improved properties.

adaptive control Control strategy for watching production and making a correction after a deviation occurs, with some loss of good quality until the process is back into specification.

admixed powder A small, discrete powder mixed with another powder for lubrication, bonding, or alloying.

agglomeration A tendency for fine particles to stick together and appear as larger particles.

aggregate A hard, difficult-to-disperse cluster of particles held together by strong adhesion forces, possibly even sinter bonds.

aging or **ageing** Heating a material to a temperature where precipitation reactions come to completion, usually accompanied by hardening.

air classification Separation of moving particles into size fractions by means of differential settling in a high-velocity air stream.

alloy powder A powder in which each particle is composed of the same alloy of two or more metals.

amorphous A material lacking crystallinity. For alloys, amorphous structures are obtained by rapidly quenching molten droplets or mechanical alloying.

angle of repose The angle from the horizontal a pile of loose powder will assume when freely poured through an orifice.

anneal A heat treatment designed to soften and remove residual strains by heating to a high temperature followed by slow cooling in the furnace.

annealed powder After powder fabrication, the powder is heated to an intermediate temperature to relieve any residual strain, giving a soft powder for compaction.

apparent density The mass of a unit volume of powder in the loose condition, usually expressed in grams per cubic centimeter (g/cm^3).

apparent hardness The value obtained by testing a sintered material with standard indentation hardness equipment. Since the reading is a composite of pores and solid material, it is usually lower than that of solid material of the same composition and condition (heat treatment).

Arrhenius relation A mathematical description of the probability for atomic motion based on an assumed energy distribution. It leads to the exponential temperature dependence for many of the sintering and coarsening events apparent in sintering, creep, hot pressing, and heat treatment.

aspect ratio The ratio of the maximum particle dimension to the minimum dimension.

atomization The dispersion of molten metal into droplets by a rapidly moving stream of gas or liquid, or by centrifugal force.

atomized metal powder Metal powder produced by the disintegration and subsequent solidification of a molten metal stream.

attrition A mechanical milling or grinding process that typically employs a stirred, vibrated, or tumbled container filled with small balls that collide against particles mixed with the balls.

austenite The face-centered cubic crystal structure of iron, also known as *gamma iron*. It is stabilized by certain alloying additions and is characterized by good ductility and a high solubility for carbon.

austenitic stainless steel A stainless steel, usually containing at least 18% chromium, where the face-centered cubic crystal structure is stable to room temperature. Usually this stabilization is via alloying by nickel, but might involve nitrogen, manganese, or other austenite stabilizers.

automatic press A compaction press that runs without an attendant, giving repeated die compaction or sizing.

axial loading Force applied along the pressing direction through the axis of symmetry.

bainite A form of iron and carbide produced by a special cooling cycle that is initially fast but subsequently slow, allowing transformation into ferrite and cementite without the formation of the platelet microstructure known as *pearlite* or the formation of martensite.

ball mill A simple means of deagglomeration or attritioning a powder by rolling a jar partially filled with balls on its side, allowing the balls to fall and impact on the powder.

batch furnace Closed-box-type furnace that accepts green P/M compacts and performs a desired heating and cooling cycle, such that only a limited number of components can be processed at a given time.

BET surface area The specific surface area as measured by gas adsorption according to the Brunauer–Emmett–Teller theory.

bimodal A particle size distribution exhibiting two mode sizes. Such a powder can result from mixing two different particle size distributions. Also a mixture of pores of two distinct sizes.

binder A cementing and lubricating polymer agent; either a material added to the powder to increase the fluidity during forming or to improve the green strength of the compact (which is expelled during sintering), or a polymer added to a powder for the specific purpose of holding particles together.

blank A pressed, presintered, or fully sintered compact, usually in the unfinished condition, requiring cutting, machining, or some other operation to give a final shape.

blending The thorough intermingling of powders of the same nominal composition.

blind pore A dead-ended pore connected to the surface.

bonding The joining of compacted or loose particles or components into a single component by the application of heat, usually termed *sintering*.

bridging The formation of arched cavities in a powder mass due to poor packing.

brittle A steel that fails without bending or stretching under load.

bubble point The pressure to force gas through an open-pore structure saturated with a wetting liquid, giving a rough measure of the pore size.

bulk density The density a powder assumed in a shipping container.

burnoff or **burnout** The removal of an additive by preheating prior to sintering.

cake A coalesced mass of unpressed metal powder.

capillary rheometer A device for measuring viscosity by applying a pressure on molten feedstock and determining the flow-rate dependence on applied stress as flow occurs through a small capillary tube.

carbide A compound phase consisting of a metal and carbon, with common examples involving iron, vanadium, chromium, tungsten, or other metals.

carbon control A measure of the ability to remove a processing polymer or oxygen without loss of desired carbon level. Carbon control is quantified by the final carbon level and the uniformity of that level between parts, days, and feedstock batches. For some materials the desire is no final carbon, while others require a precise final carbon level.

carbon potential Atmosphere measure of the equivalent carbon in steel at the sintering temperature.

carbonitriding Introduction of carbon and nitrogen into a ferrous alloy by holding above the temperature at which austenite forms.

carbonyl powder A metal powder prepared by the thermal decomposition of a metal carbonyl molecule. The resulting particle size is typically in the size range from 1 to 10 μm.

carburize Introduction of carbon into a solid ferrous alloy by holding above the temperature at which austenite forms.

case-harden Use of a packed powder or atmosphere gas to diffuse carbon into the exterior surface of the component to increase surface hardness.

casting An object or finished shape obtained by solidification of a substance in a mold.

catalytic debinding The partial removal of the filler polymer via heating in an atmosphere containing an agent that induces depolymerization. The classic process involves extraction of polyacetal using a nitrogen atmosphere doped with nitric acid.

cemented carbide A solid composite consisting of a metal carbide and a matrix phase, usually cobalt or nickel aluminide. The composite is formed by liquid-phase-sintering a mixture of the carbide and matrix metal powders.

cementite The carbide of iron where there are three iron atoms for every carbon atom. It is a hard, brittle phase that provides good wear resistance to a steel.

centrifugal atomization The formation of spherical particles by combining a melt with a centrifugal force such that the melt is thrown off into droplets that spheroidize prior to solidification.

cermet A body consisting of ceramic particles bonded with a metal, used for cutting and tooling.

Charpy test The standard impact energy test for determining the energy needed to fracture a steel. Most conventional tests place a V-notch in the test bar to localize stress, but because many powder metallurgy steels are poor in impact testing, the convention is to use unnotched bars to obtain higher values.

chemically precipitated powder Powder produced as a precipitate by chemical displacement, typically with a small particle size.

clamping force The force used to hold two halves of a tool cavity together, used as a measure of molding machine capacity.

classification Separation of a powder into fractions according to particle size.

closed-loop control Use of in situ sensors in a fabrication process and constantly monitoring the state of affairs to make decisions on component quality, essentially capable of making every component the same.

closed pore An isolated pore not linked to the external surface.

coalescence The merging of two objects into a larger object, as can be observed in particles during fabrication, and pores and grains during sintering.

coarse fraction That portion of the powder that will not pass through a given screen. Usually the largest particles sizes in a lot of powder.

coarsening The progressive enlargement of the grain size or pore size during sintering due to diffusion, coalescence, or solution–reprecipitation processes.

coated powder Powder treated with a polymer coating or conformal coating by vapor phase, solvent, fluid-bed reaction, or chemical routes, largely to passivate or change the surface properties.

coercive force The magnetizing force that must be applied in the direction opposite that of the previous magnetizing force in order to reduce the net magnetization to zero. It measures the resistance to demagnetization.

coining The final pressing of a sintered compact to obtain a definite surface configuration.

cold isostatic pressing The compaction of a powder under isostatic pressure conditions at room temperature using a flexible mold and a high-pressure hydrostatic pressure in a hydraulic pressure chamber.

cold pressing Forming a compact at a temperature low enough to avoid sintering—usually at room temperature

cold welding Cohesion between two surfaces of metals, generally under the influence of externally applied shear stress at room temperature.

comminution Pulverization, milling, or breakup of a material via mechanical energy.

compact An object produced by the compression of metal powders.

compacted angle of repose Angle-of-repose test performed by tilting a powder vibrated to the tap density, which is a measure of the ability to hold shape after a binder is extracted.

compaction The shaping, deformation, and densification of a powder by the application of pressure through a tool material.

component Constituent or discrete object produced as a portion of a system or assembly. Most of powder metallurgy is involved in the production of components that make up larger systems, such as automobiles.

composite A mixture of two or more powders to form a multiple-phase structure, typically designed to deliver properties that are a hybrid of the constituents' properties.

compressibility The reciprocal of the compression ratio. It is a measure of the ease of compaction. The compact is made following a procedure in which the die, pressure, and pressing speed are specified.

compression ratio The ratio of the volume of the loose powder to the volume of the compact made from it.

contact angle The wetting angle formed at the intersection of liquid, solid, and vapor phases.

contiguity A microstructural measure of the relative interface area of solid–solid bonds versus the total interface area during liquid-phase sintering.

continuous furnace A pusher, belt, or walking-beam furnace where components are always entering and exiting, so that the temperature remains constant at each location while the work passes through a sequence of temperature zones on a conveyor system.

coordination number The number of touching nearest neighbors for a given particle or grain in a powder compact.

core rod The separate member of the compacting tool set that forms a hole in the compact.

creep The time dependent strain occurring under stress at high temperatures.

critical loading The maximum volume fraction of solid particles that can be incorporated in a polymer binder without forming pores.

cut The fraction of powder above or below a given screen size; a selective particle size range.

debinding A step between molding and sintering where the majority of the binder used in molding is extracted by heat, solvent, or other techniques.

deformation The permanent change in shape due to stress, usually with a stress that exceeds the material strength.

delamination The cracking of a green compact during ejection from a tool set due to a high elastic springback and low tensile strength.

dendritic powder Particles having a structure composed of primary stems with several side branches, usually formed by electrolytic techniques.

dense random packing The packing density for a powder at its tap density, where the particles are in a random but tight structure.

densification The change in porosity divided by the initial porosity due to pressing or sintering.

density The mass divided by the volume, and sometimes given as a ratio to theoretical density.

density ratio The ratio of the determined density of a compact to the absolute density of a metal of the same composition.

dew point A temperature that provides a relative measure of the moisture content in an atmosphere.

die The part or parts making up the confining form in which a powder is pressed.

die body The stationary or fixed part of a die supporting the die insert or liner.

die compaction The consolidation of powder through the application of uniaxial stress while the powder is constrained in rigid tooling.

die insert A removable liner of a die body or punch; also known as a *die liner*.

die lubricant A lubricant mixed with the powder or applied to the die walls to facilitate the pressing and ejection of compacted powder while minimizing tool wear.

die set The parts of a press that hold and locate the die in proper relation to the punches.

diffusion The progressive motion of atoms via single steps but that, if added up over a length of time, leads to measurable component shape and strength changes. Used mostly to describe sintering.

dihedral angle The angle formed by a grain boundary with a pore or liquid during sintering. It is a measure of the relative interfacial energies.

dilatometry Measurement of dimensional change during thermal processing to determine the sources of sintering densification, phase transformation, or other causes of dimensional control problems. Usually dilatometry is conducted with a pushrod that makes contact with the specimen in a furnace.

dimensional control High-pressure compaction and controlled sintering to ensure uniform compact dimensions.

dislocation A linear crystal defect that moves when the stress exceeds the yield strength and is the cause of plastic flow.

dispersion strengthen The incorporation of a distribution of small, inert second-phase particles throughout the microstructure to provide high-temperature strength and creep resistance.

dissociated ammonia A reducing atmosphere produced by the thermal decomposition of anhydrous ammonia over a catalyst, resulting in a gas of 75% hydrogen and 25% nitrogen; also termed *cracked ammonia*.

distortion Nonuniform dimensional change measured versus the average dimensions, measured by the standard deviation in dimensions point to point over the component exterior as referenced to the mean dimensions.

double-action pressing A method by which a powder is pressed between opposing punches where the combined motion of all punches is toward the die center.

double press–double sinter A variant on traditional die-compaction powder metallurgy where the sintered compact is passed through compaction and sintering a second time to increase density and precision.

dry-bag cold isostatic pressing Similar to traditional cold isostatic pressing, but the rubber container that holds the powder remains in the die cavity and only the compact is ejected.

ductile A steel with considerable ability to stretch or bend prior to breaking.

ductile–brittle transition temperature Steels often change from ductile failure to brittle failure on cooling, and the critical temperature is denoted as the DBTT.

ductility A measure of the permanent stretching or deformation a steel can take prior to failure. Brittle materials have no ductility, while materials such as stainless steel exhibit large ductilities, often measured at 60% stretch or elongation prior to failure.

dwell The time period during which maximum pressure is applied to a compact.

dynamic compaction Explosive or gas-gun compaction of powder at shock-wave velocities.

ejection The final stage of die compaction where the pressed powder compact is forced out of the die.

ejector pin In injection molding, an ejector pin pushes the compact out of the die cavity after cooling and often leaves a slight impression on the compact surface.

elastic modulus A material constant that measures the proportionality between stress and strain during the initial, elastic portion of the tensile test.

elastic strain Nondimensional stretch, or length change, divided by the original length, fully recoverable strain when the stress is removed.

electrolytic powder Powder produced by electrolytic deposition and subsequent pulverization. The resulting powder has a dendritic or sponge shape.

elemental powder Powder of a single chemical species like iron; a powder with no alloying ingredients.

elongation A measure of the ductility as determined by the total stretch divided by the initial length of the sample test section prior to tensile testing.

elutriation Classification of powder particles by size using a rising stream of gas or liquid acting against the settling due to gravity, similar to air classification.

endothermic atmosphere A reducing-gas atmosphere used in sintering. It is produced by the reaction of a hydrocarbon fuel gas and air over a catalyst with the aid of an external heat source. The resulting atmosphere is low in carbon dioxide and water vapor with relatively large percentages of hydrogen and carbon monoxide.

equiaxed powder Particles with approximately the same size in all three (perpendicular) dimensions.

erosion The dissolution of the metal compact at the surface where a liquid infiltrant flows into the part.

eutectic An alloy having the composition indicated by the eutectic point on an equilibrium phase diagram, meaning that two solid phases crystallize from one homogeneous liquid at a single temperature.

eutectoid steel Plain carbon steels that contain approximately 0.8% carbon, or similar values that depend on the alloying level, such that on slow cooling austenite transforms directly into ferrite and cementite.

exothermic atmosphere A reducing-gas atmosphere used in sintering, produced by partial or complete combustion of a hydrocarbon fuel gas and air. The maximum combustible content is 25%.

expansion The increase in dimensions of a compact due to unbalanced chemical reactions, pore formation, or pore growth during sintering.

extrusion The conversion of a billet into lengths of uniform cross section by forcing metal powder, or a metal powder–plastic mixture, through a die orifice of desired cross section.

exude The expulsion of a low-melting-temperature alloying ingredient from the pore structure due to poor wetting or excess liquid.

fatigue strength The resistance to repeated loading without failure. With loading to the fatigue strength, 50% of the samples will fail in 10 million cycles.

feed shoe The powder conveyance (usually a flat hopper) that places loose powder at the apparent density into the die cavity between compaction strokes.

feedback control Control systems that employ embedded sensors to make continuous corrections to the operation to ensure proper quality.

feedstock The mixture of powder and binder used in injection molding or other binder-assisted shaping processes.

ferrite The body-centered cubic phase of iron stable at room temperature, also known as alpha iron. Also used to describe a form of iron oxide applied to ceramic magnetics.

ferrous Indicating a material based on iron, including all steels, tool steels, stainless steels, iron, and similar high-iron-content alloys.

fill position The position of the lower punch in the die cavity when powder is filled. The fill position and apparent density determine the mass of the final component.

fill ratio The height of powder placed in a die cavity divided by the final compact height.

fines The portion of a powder composed of particles that are smaller than 45 μm; also known as the subsieve size fraction.

finishing operations The steps applied to a P/M material after sintering to tailor the dimensions, properties, or attributes to the application; examples include machining, heat treatment, and electroplating steps.

flake powder A flat or scale-like powder that is relatively thin with a large aspect ratio.

flame spray A coating process that uses a gas torch or plasma gun with injected metal powder to create a spray of molten droplets. The deposition of a coating from these droplets is the basis for corrosion, oxidation, or wear coatings.

flash The thin extra leaf on the surface that forms because of feedstock penetration along a parting line in a tool set, leaving a surface blemish on the powder compact.

floating die A tool set designed to allow transmission of force from the upper punch through the powder into the die wall. As compaction occurs, the die moves in proportion to the upper-punch motion to effectively place the lower punch near the compact centerline.

flow time The time required for a powder sample to flow through an orifice in a standardized test. The flow time gives a measure of the interparticle friction.

fluid bed A technique for powder levitation in an upward-flowing gas stream. Usually the gas is heated and contains polymers or other agents to surface treat the powder.

forging The plastic deformation of a metal into a desired shape. The deformation is usually performed hot at high strain rates and may be done in constraining dies.

fraction That portion of a powder sample that lies between two stated particle sizes or mesh sizes.

fracture strength The final failure stress of a tensile sample, giving a measure of the strength and resistance to deformation.

fracture toughness For a given combination of stress and defect size, this is the energy needed to cause a crack to grow with catastrophic failure of the component.

gas atomized powder A rounded or spherical powder formed by the disintegration of a melt stream by a gas expansion nozzle. The particles solidify during free flight after atomization.

gate The constricted opening into the die cavity in an injection-molding tool set.

gelation A binder-setting process where a macromolecule grows in a binder solution to form a highly interlinked structure with most of the binder trapped in cells formed by the long-range molecule.

glove box A controlled- or inert-atmosphere box used for working with pyrophoric, high-purity, or toxic powder. The operator stands outside the box and works with the powder through rubber gloves.

grain A single crystal of material. Particles might be composed of many grains, but once consolidated and bonded, the particle becomes a grain in a polycrystalline structure.

grain boundary The interface between two crystals where atomic bonding is disrupted along the surface layer. Grain boundaries are common sites for impurities and are active in sintering.

grain shape accommodation The progressive shape change of neighboring grains during liquid-phase sintering such that the grains attain a high packing density, nonspherical shape to release liquid to fill remaining voids.

granular powder Particles having approximately equidimensional, nonspherical shapes and a coarse particle size.

granulation A general term for the production of coarse powder agglomerates using binders, heat, moisture, or liquid metals, often performed by tumbling and drying.

green density The powder density after compaction.

green strength The strength of the as-pressed powder compact.

growth The increase in dimensions of a powder compact that may occur during sintering; the dimensional change opposite to shrinkage.

hard metal Another name for the cemented carbides that reflects their high hardness after sintering.

hardenability The measure of heat treatment ease in forming martensite via the Jominy test. It largely reflects the ease in forming martensite at large depths.

hardness The resistance to penetration as measured by impressing a small point or ball into the surface of a steel and measuring the size of the resulting imprint based on a controlled loading.

hot forging The application of a high-strain-rate impact to a preform to densify the heated powder preform, usually performed in a closed die.

hot isostatic pressing A process combining temperature and high-pressure gas to densify a material into a net-shape component using a pressuretight outer envelope.

hot pressing The high-pressure, low-strain-rate forming of a compact at a temperature high enough to induce sintering and creep processes.

hot repressing Similar to hot forging, where a heated sintered compact is contained in a closed die and pressurized to increase density and possibly to improve surface finish or dimensional precision.

hot working The plastic deformation of a metal at a temperature and strain rate that minimize work hardening.

hydraulic press A compaction press that derives the tool motions from a hydraulic pressure system, where a sequence of valves determines the events and motions in compaction.

hydrogen loss The loss in weight of metal powder caused by heating for a specified time and temperature in a hydrogen atmosphere.

hydrogen-reduced powder Powder produced by the hydrogen reduction of a compound, usually an oxide. The resulting powder often has a high surface area or large amount of internal porosity.

hypereutectoid steel A steel with more than approximately 0.8% carbon, over the eutectoid composition.

hypoeutectoid steel A steel with less than approximately 0.8% carbon, under the eutectoid composition.

hysteresis The gap between the forward and reverse paths in any process. Magnetic materials working in an oscillating field will experience a difference in magnetic state depending on the direction of applied field. This gap is a measure of energy loss on each magnetization cycle.

impact strength An energy to cause failure as measured by the energy extracted from a swinging beam that breaks a standard test bar, better known as *toughness*.

impact toughness Usually measured by the energy needed to break a small test sample. Sometimes called the *Charpy test*, although most powder metallurgy samples are not notched, unlike typically ferrous metallurgy.

impregnation A process of filling the pores of a sintered compact with a polymeric material such as oil, wax, or resin.

infiltration The process of filling the pores of a compact with a lower-melting-temperature metal or alloy, typically copper or bronze.

injection molding A hydrostatic technique for shaping powders using plastic binders and relatively low temperatures and pressures.

interconnected porosity The network of contiguous pores connected to the surface of a sintered compact. The term is usually applied to materials where the interconnected porosity is determined by impregnating the specimen with oil.

interparticle friction The friction between powders that limits sliding, packing, and densification.

invar Low-thermal-expansion alloys of iron–nickel–cobalt where a martensitic phase transformation is balanced against the thermal expansion coefficient to give a near-zero thermal expansion coefficient over a range of temperatures.

irregular powder A powder that lacks shape symmetry in the individual particles.

isostatic pressing The compaction of a powder by subjecting it to equal pressure from all directions; hydrostatic compaction.

jet mill A self-impact attritioning technique where powder is shot at high velocities into circulating powder. The milling process requires no moving parts and is suitable for reducing the particle size of smaller powders.

jetting A condition that arises with the rapid filling of an injection mold where the feedstock shoots across the mold and fills back toward the gate.

Jominy test Quenching test performed on an austenitized steel that is chilled at one end; the hardness profile is measured versus distance from the quench end to determine ease of forming hard martensite.

knit line A linear defect occurring where feedstock streams merge from two gates or from flow around a core or other solid portion of the die.

kovar A glass-to-metal sealing alloy used for microelectronic packaging and other situations where a matched or graded structure is required.

lamination A layered structure or cracking in the pressed compact resulting from ejection stresses exceeding the green strength.

liquid-phase sintering Sintering at a temperature where a liquid and solid coexist as a result of chemical reactions, partial melting, or eutectic liquid formation.

loading The filling of a die cavity with loose powder. Also related to the volume percentage of powder in a feedstock, termed the solids loading.

log-normal A statistical distribution useful for dealing with particle size data, where the standard deviations of the cumulative distribution are plotted against the logarithm of the size.

loose-powder sintering Sintering of uncompacted powder with no external pressure. Used for weak structures where porosity is desirable, such as in filters and air distribution plates.

loose random packing The lowest packing density attainable through the random filling of a container, corresponding to approximately the apparent density.

low-alloy steel A mixture of iron and carbon with a few additions to improve strength and hardenability in heat treatment.

low-carbon steel Alloy consisting of largely iron and carbon, with the carbon content below 0.8%.

lower punch The member of the compaction tool set or die that determines the volume of powder fill and forms the bottom of the part being produced. Multiple-segment lower punches may be necessary to aid in uniform filling, densification in pressing, or defect-free ejection of multilevel parts.

lubricant An organic additive that is mixed with a powder to minimize die wear and aid in ejection after compaction.

martensite A metastable form of iron and carbon that forms at lower temperatures when austenite is alloyed with carbon and rapidly cooled. It is a hard phase that often proves too brittle for use in powder metallurgy.

master alloy powder When alloys are created in sintering, some of the ingredients might be partially alloyed with iron to assist in diffusion. For example, instead of directly adding chromium, the additive might be an iron–chromium alloy.

matrix metal The continuous phase in a polyphase alloy or mechanical mixture; the physically continuous metallic constituent in which separate grains of another constituent are embedded.

mean size The average value from the particle size distribution.

mechanical alloying The formation of an alloy powder by milling elemental powders for a prolonged time; frequently used to create amorphous or dispersion-strengthened alloy powders.

mechanical press A compaction press where motions of the tool set in compaction are derived from a flywheel or rotating crank, such that, when engaged, a powder compact is formed on each rotation.

median size The centroid of the particle size distribution, where half of the particles are larger and half are smaller; not necessarily the mean, but easily identified as the 50% value.

melt index A measure of flow at low shear strain rates, where a capillary tube is used to extrude molten feedstock under a dead load. The melt index depends on the capillary tube diameter and applied load, but is always reported as the grams of feedstock collected from the tube in 10 min.

mercury porosimetry A technique for measuring the pore size distribution of open pores using high-pressure intrusion of mercury.

mesh The screen number representing the number of wires per inch in a square grid. The higher the mesh number, the smaller the opening size.

metering Controlled forward extrusion of molten feedstock past the screw tip and check ring to ensure that the proper shot volume is ready for the next filling event.

milling Mechanical agitation used to break particles or agglomerates into smaller particles.

minus sieve The portion of a powder sample that passes through a sieve of a specified mesh.

mixing The thorough intermingling of powders of two or more different compositions.

mode size The most common particle size. The mode corresponds to the size at the peak in the interval particle size distribution and may differ from the mean size for skewed distributions.

moldability A relative measure of the ease of filling out a tool cavity during injection molding. It can be determined by the length of filling for a long, narrow passage.

molding press A press used to form compacts from powder.

monodisperse A powder system where the dispersion is quite small; all the particles have essentially the same size.

monosize A powder sample where all the particles have the same size.

multimodal A powder that exhibits several modes, possibly generated by blending several monosized powders.

nanoscale Powders or microstructures with sizes that can be measured in nanometers. Typically the powders are less than 100 nm or 0.1 μm in size.

near-net shape A compact that has the general shape of the final product, but is oversized and requires final machining.

neck The bond between contacting particles developed during sintering.

needles Elongated, rodlike particles.

net-shape A compact manufactured to the final density and dimensions without the need for machining.

neutral zone The region at the approximate center of a compact where the balanced particle motions occur; motion from upper punch and lower punch (or die float) give near-zero relative motion.

nodular powder Irregular particles with knotted, rounded shapes.

nonferrous Alloys that are not based on steel or iron, such as copper, aluminum, and zinc.

oil impregnation Filling of open pores with oil to provide lubricity during function.

open-loop control Processing without any monitoring of the product quality, simply using approximate machine settings to produce components that might be sorted for quality after production.

open pore A pore extending completely through a compact from one surface to another.

ordered packing A collection of powder particles in a regular pattern, similar to atomic structures, where each particle is in precise positioning with the same number of neighbors.

Ostwald ripening The progressive coarsening of a microstructure (both grains and pores) at high temperatures, leading to a larger-size scale with fewer objects per unit volume.

overfill An excess of powder in the die cavity in filling, some of which should be ejected prior to compaction.

oversinter A condition where the compact has been heated too long or heated at too high a temperature, usually resulting in distortion, blisters, or loss of mechanical properties.

oversize powder Particles larger than the maximum permitted by a given specification of particle size.

packing A semicohesive collection of particles in point contact which exhibits resistance to flow, typically formed by filling a container under gravitational forces.

particle A discrete solid object with a size below 1 mm.

particle hardness Hardness of the particle as measured by scratch, indentation, microhardness, on nanohardness tests. This hardness is the true material hardness as used in wear testing as compared to the aggregate hardness, which is often lower because of porosity.

particle size The controlling linear dimension of an individual particle, as determined by analysis with screens or other instruments.

particle size analyzer An automated device for determination of the particle size distribution.

partially prealloyed powder Mixed powders are heated to a temperature where incipient sinter bonds form, but the alloying ingredients have not interdiffused. This procedure is used to minimize segregation or separation of alloying ingredients while keeping the high compressibility associated with elemental powders.

parting line The linear mark on a compact indicating where two separate tool or die pieces mated during shaping. In injection molding, it is where the two halves of the die joined together.

pearlite A mixture of ferrite and cementite in a lamellar structure, consisting of alternating layers of alpha iron and iron carbide. It forms on slow cooling austenite that contains carbon.

pelleting or **pelletizing** The formation of discrete chunks of feedstock with repeated sizes and shapes that allow easy flow and filling of the molding machine.

permeability A measure of the resistance to gas flow through an open-pore structure; or for a magnetic material, the initial magnetization response to an applied magnetic field.

plain carbon steel Simple steels consisting of iron and carbon, with a few possible minor alloying additions.

planetary mixer Usually these have two mixing blades that rotate around individual shafts and the two blades further rotate around a center axis. The net effect is intermixing and stirring and shear to produce PIM feedstock, usually in a heated vessel.

plasma spray Injection of powder into a plasma torch and the deposition of the molten droplets as a coating.

plastic deformation Flow of a steel that is permanent, meaning that once the stress is removed, the old shape is not recovered since the stress exceeded the yield strength.

plasticizer An additive to the binder system to lower viscosity, largely by lubricating flow of molecules past one another.

plates Powder particles with a flat shape and considerable thickness, in contrast to flakes that are thin with respect to the width.

plunger molding A hydraulic plunger is used to push molten feedstock into a die cavity, where motion and pressure are controlled by the applied hydraulic pressure.

plus sieve The portion of a powder sample retained on a sieve of specified mesh.

pneumatic molding Use of an air pressure head over molten feedstock to push it into a die cavity. This is the lowest-cost and least-precise form, yet is widely employed in the fabrication of large ceramic structures.

Poisson ratio The negative of the radial contraction divided by the axial elongation in a tensile test, giving a measure of the shape change during stressing.

polydisperse Implies a broad powder size distribution, covering a wide range of particle sizes with no clear mode size.

polymorph Multiple crystal structures, as with iron, which changes from alpha to gamma to delta iron during heating from room temperature.

pore A void between particles. In the green body the pore size should be much smaller than the particle size.

pore former A substance included in a powder mixture that volatilizes during sintering and thereby produces a desired pore size and shape in the finished compact.

pore size The size of the holes or voids between powder particles.

porosimeter A device for measuring the pore size of the open pores using high pressures and mercury intrusion techniques.

porosity The amount of void space in a powder compact.

powder Particles of solid matter characterized by a small (less than 1 mm) size. A collection of particles.

powder flowmeter An instrument for measuring the rate of flow of a powder according to specified procedure.

powder metallurgy The art and science of producing metal powders and utilizing metal powders for the production of massive materials and shaped objects.

powder rolling The progressive compaction of metal powders by the use of a rolling mill; also known as *roll compaction.*

prealloyed powder Each particle contains an intimate mixture of two or more elements in a prescribed ratio to form an alloy; examples include brass, bronze, steel, and stainless steel.

precipitated powder Powder formed by a chemical reaction that nucleates solid particles from solution or out of a vapor.

precipitation harden Strengthening by growing precipitates after sintering using an isothermal hold at a temperature where a second phase forms in the microstructure from a supersaturated solid solution.

preform An initial compact of metal powder subjected to subsequent pressing or forging operations.

preheat First heating of a powder compact as it enters the furnace, used to burn off contaminants prior to entry into the high-temperature portion of a furnace.

premix A powder system composed of two or more ingredients that are mixed and bonded together to avoid segregation in handling.

presintering The heating of a compact to a temperature lower than the normal sintering temperature to gain strength for subsequent handling, including machining.

press A machine for compaction via controlled tool motions and pressures.

pressure-assisted sintering Sintering with the application of an external pressure. It is often performed by initially sintering in vacuum and subsequently pressurizing the furnace to densify any remaining closed pores.

proeutectoid A phase that forms prior to reaching the eutectoid temperature when cooling from austenite. For a hypereutectoid steel, this would be cementite; for a hypoeutectoid steel, ferrite.

punch The part of a tool set that delivers pressure to powder in the die cavity.

pycnometer A device for measuring the theoretical density of a loose powder or preform.

pyrophoric powder A powder unstable in air that spontaneously burns when ignited because of a high surface area and exothermic oxidation reaction.

quenching Rapid cooling to preserve a high-temperature microstructure.

radial crush strength The relative capacity of a ring-shaped specimen to resist fracture induced by a force applied between flat parallel plates in a direction perpendicular to the axis of the specimen.

random packing A collection of contacting particles with no symmetry or order to the particle positions, showing little predictable structure over distances of a few particle diameters.

rapid solidification The extraction of heat from molten material at cooling rates of 10^4 °C/s or higher, resulting in new microstructures, compositions, or phases.

rate-controlled sintering A nonisothermal sintering process where the compact heating rate is optimized to give excellent sintering while minimizing microstructural coarsening.

reactive sintering A novel sintering process where an exothermic reaction is initiated in a mixture of dissimilar (elemental) powders. The reaction produces a compound (carbide, boride, aluminide, or other compound) and the heat from the reaction is used to simultaneously sinter the product phase.

rearrangement The instantaneous motion of particles into new packing positions that occurs in liquid phase during sintering and the early portion of compaction.

reciprocating screw Injection molding with a screw located in a heated barrel. During compression, metering, melting, and forward advance of the feedstock, the screw is turning, while during mold filling it becomes a plunger to quickly fill the cavity.

recovery The relaxation of stresses and strains from deformation as a first event in annealing or sintering.

recrystallization Formation of new, smaller grains in a deformed steel as an early event in heating.

reduced powder Metal powder produced by the chemical reduction of a compound, most typically a metal oxide.

reduction in area A measure of ductility, where the change in cross-sectional area of the tensile sample is divided by the initial area.

refractory A metal or ceramic having a high-melting-point temperature, usually over 1700°C. Example refractory metals are tungsten, molybdenum, rhenium, and zirconium.

remanent magnetization The amount of magnetization remaining in a ferrous component after removal from an applied magnetic field.

repressing The application of pressure to a previously pressed and sintered compact for the purpose of improving dimensions or properties.

restrike Additional compaction or coining after sintering.

rheology The study of elastic, plastic, and viscous flow of polymers and feedstocks used in injection molding.

roll compaction The formation of continuous sheet material by compacting powder in a rolling mill.

runner A portion of the feed path for filling an injection-molding die. The runner is between the sprue and gate; the latter is the inlet to the actual die cavity.

saturation A relative measure of the fraction of void space or porosity filled by a liquid phase.

saturation magnetization The peak level of magnetization in an applied field.

screen analysis Particle size distribution information generated by segmenting or classifying a powder sample with a stack of vibrating screens of decreasing opening sizes.

secondary operations Those activities performed to adjust dimensions or properties of a compact after sintering.

sedimentation Classification or sizing of a powder by settling in a gas or liquid. Separation of sizes occurs by differences in settling rates as described by Stokes law.

segregation Nonuniform distribution of ingredients, such as powder separation by size, shape, or density, or chemical separation in the microstructure of a solidified material.

self-propagating high-temperature synthesis An exothermic reaction of mixed dissimilar powders that forms a compound and liberates heat. The heat further induces a reaction between unreacted particles to give a continual reaction wave once ignited.

setter The tray or shaped substrate for support of P/M compacts during sintering.

shape accommodation Grain growth with deviations from spherical shape to allow efficient fitting together of the growing grains to fill space and eliminate porosity.

shaping Usually involves a slurry of powder and processing aids where pressures are low, so the die cavity is used to hold the slurry until the shape is attained.

shot peening Progressive impacts from small balls on a compact surface to build up compressive stresses for prolonged fatigue life.

shot size The maximum amount of powder–binder feedstock that can be injected in a single stroke, which is a measure of molding machine capacity.

shrinkage A decrease in dimensions of a compact that occurs during sintering.

sieve analysis Particle size distribution data as expressed by the weight percentage of powder retained on each of a series of sieves of decreasing sizes (increasing mesh).

sieve fraction That portion of a powder sample that passes through a sieve and is retained by a finer sieve.

single-action pressing Compaction with only one moving upper punch and fixed die and lower punch.

sinter bonding The joining of two green P/M components as part of the sintering cycle.

sinter-harden The generation of bainite or tempered martensite via controlled cooling from the sintering temperature, where the final microstructure is usually composed of a mixture of martensite and bainite.

sintering A thermal process that increases the strength of a powder mass by bonding adjacent particles via diffusion or related atomic level events. Most of the properties of a powder compact are improved and frequently density changes with sintering.

sintering diagram A process map showing the interaction of the key variables with respect to the densification of a powder. The variables include particle size, grain size, temperature, time, and applied pressure.

sizing A final pressing of a sintered compact to secure a desired size or dimension, or to minimized scatter in the final compact size.

slip casting A forming process applicable to small powders where the powder is dispersed in a low-viscosity fluid to allow shaping by casting the slurry into a porous mold that extracts the fluid.

solids loading The relative powder volume in feedstock designed for binder-assisted shaping. A common injection-molding feedstock will have a solids loading of 0.6 or 60 vol% powder.

solutionize A heat to a single-phase temperature that allows alloying ingredients to enter into solid solution, possibly followed by a quench to freeze in the high-temperature structure; also termed *homogenization* and *austenitization* in steels.

solution–reprecipitation The dissolution of small particles with subsequent deposition on large particles via diffusion through a liquid phase, leading to overall coarsening of the microstructure.

solvent debinding The extraction of the binder or some portion of the binder by leaching into a solvent. This is performed by immersion or exposure to solvent vapors. The solvent might be flammable, toxic, or simply water or ethanol.

specific surface area The surface area of a powder divided by the weight of the powder, usually expressed in units of area per mass (m^2/g).

spherical powder Powder with a uniform spherical shape and a size that can be characterized by a diameter. Gas and centrifugally atomized powders are often spherical.

sponge powder A porous powder; it is usually observed in chemically reduced powders where there is a volume change in the solid-state reduction reaction.

spray drying Procedure used to form agglomerates by mixing powder, binder, and solvent into a slurry that is sprayed into a heated chamber where the solvent evaporates from the droplets during flight.

springback The elastic relaxation of a pressed powder on removal of the compaction pressure or ejection from the die set.

sprue The initial inlet into the die set for injection molding feedstock. The sprue is usually filled by the nozzle and is tapered for easy extraction after mold filling.

stainless steel Steels that contain at least 12% chromium to improve corrosion resistance.

statistical process control Continuous monitoring of an operation to ensure that the output is in a tolerable range of variables.

steam treatment A heat treatment typically above 500°C (932°F) in an atmosphere containing steam, used to grow a blue-black Fe_3O_4 tenacious surface oxide on steels.

steel Alloys based on iron, often containing carbon, providing most of the metallic materials and constituting the largest segment of ferrous metallurgy.

stoking Presintering or sintering where the compacts are advanced through a furnace at a fixed rate as applied to continuous sintering.

strain Nondimensional stretching of a steel during loading. Initial deformation is elastic or recoverable with removal of the stress, but at high strains the deformation is permanent or plastic.

strain hardening The ability of a steel to become stronger and harder to resist further deformation as it is deformed.

strength A measure of the resistance to fracture or deformation.

stress relief An annealing treatment designed to remove residual stresses from machining or working without altering the basic microstructure, followed by controlled cooling to avoid new stresses.

submicron Powder with a size below one micrometer, smaller than 1 μm.

subsieve fraction The fraction of a powder that passes through a 325-mesh screen; thus, the powders are less than 45 μm.

superalloy The highest-performance alloys used for the most demanding applications. Typically these are formulated with a large number of components for tailored properties that allow use in moving components in jet turbines.

superfines The portion of a powder composed of particles smaller than 10 μm.

supersolidus sintering A liquid-phase sintering process applied to prealloyed powders where sintering occurs over the solidus temperature, thereby nucleating liquid within the particles.

surface area The external powder area as measured by gas adsorption or gas permeability.

swelling The growth in compact dimensions due to pore formation, alloying, or liquid penetration of boundaries. A low-melting-point additive or trapped gas can form pores during sintering and thereby give swelling.

tap density The density of a powder obtained when it is vibrated for a prolonged period. The tap density represents the highest packing density possible for a powder without the application of pressure.

tape casting A means to fabricate sheets from a powder and binder, where the slurry is formed on a moving sheet as a thin layer. After a solvent evaporates the sheet is separated from the support.

temper A postquenching heat treatment designed to partially transform or modify the quenched martensitic microstructure via precipitation of cementite to improve toughness at the expense of hardness.

tensile strength The maximum strength attainable prior to failure during a uniaxial tension test.

tensile test A fracture test for measuring mechanical properties, where a standard geometry sample is pulled to failure and the stress and strain are recorded during testing.

tetrakaidecahedron A 14-sided polyhedron used to represent the grain shape in powder compacts.

theoretical density The true crystal density for a material corresponding to the limit attainable through full-density products without pores.

thermal debinding Extraction of the polymer phase in debinding by the application of heat. The classic processes relied on slow heating in air to evaporate the binder over a period of days.

thermogravimetric analysis A fancy term that means measuring the sample weight change versus temperature while heating. It is used for determination of polymer evaporation temperatures by performing constant heating rate experiments and recording the sample weight versus temperature.

thermoplastic Binders or polymers that soften on heating, but stiffen on cooling, and can repeat the heating and cooling process without undergoing any long-range chemical bonding changes.

thermosetting Binder or polymer hardening due to heating to a temperature where cross-links form between the polymer chains. Unlike a thermoplastic that can be softened by reheating, a thermosetting binder becomes hard on the first heating cycle and cannot be subsequently softened.

through-harden A heat treatment that involves solutionizing and quenching an alloy to form a hard phase throughout the component.

tool set Combination of the die, punches, core rods, and other components needed to form a part. It is usually assembled and inserted as a package into the press.

tool steel A high-alloy steel used for metalworking, cutting, and compaction purposes. Tool steels contain refractory metals and high carbon contents to form stable carbide dispersions.

toughness Energy required to fracture a steel in tensile testing or impact.

transient liquid-phase sintering A sintering cycle characterized by the formation and disappearance of a liquid phase during heating. The initial compact has at least two differing chemistries, and the first liquid must be soluble in the remaining solid.

transition temperature A temperature where phase changes or other reactions occur.

transverse rupture strength A three-point fracture test applied to brittle materials or green compacts to assess relative strength.

twin screw processor A mixer or extruder consisting of two intermeshing screws that convey, shear, and mix a powder or powder–binder mixture during transport through the mixer.

ultrafines Particles with a size generally below 1 μm.

ultrasonic inspection The use of high-frequency sound waves for detection of cracks or other defects. The most common procedure is to use through- or back-transmission waves out of the die cavity, but new technology allows guided-wave inspection in the die cavity.

ultimate tensile strength The peak stress encountered in a tensile test.

undersize powder Particles finer than the minimum permitted by a given specification of particle size.

uniaxial compaction Pressurization of a powder compact with punches and pressure on only one axis, also known as *die compaction*.

upper punch The member of the compacting tool set that closes the die and forms the top of the compact.

upset forging Forging of a powder compact that is smaller in size than the die cavity, such that the compact undergoes radial flow prior to coming into contact with the die wall.

vacancy A missing atom from the normally ideal periodic structure.

vacuum sintering Sintering at high temperatures in a furnace that constantly removes the gases to effectively create no atmosphere.

vent A thin relief in tooling to allow air to escape during filling.

viscosity The resistance to flow of a feedstock or polymer. Formally it is the proportionality between the stress and applied shear strain rate. A high viscosity makes molding difficult, while a low viscosity is easier to mold; however, there are often difficulties with flashing.

viscous flow The time-dependent deformation or densification of a powder structure under stress; similar to high-temperature creep.

void A pore or hole in the powder structure or sintered part.

warpage Distortion in a compact that occurs during sintering or heat treatment typically due to nonuniform heating or density gradients in the initial compact.

water-atomized powder Irregular or ligamental particles formed by impacting a molten metal stream with high pressure water.

wax–polymer A general term for polymers used as binders that consist of low-molecular-weight polymers and wax fillers. The polymers are usually simple molecules such as polyethylene, polypropylene, or copolymers such as ethylene vinyl acetate. These are the classic binders used in molding.

weld line A linear defect or mark on a compact surface where two portions were bonded, either in molding or subsequently, such as in sinter bonding.

wetting angle The equilibrium angle formed by a liquid–solid–vapor combination; it provides a relative measure of spreading and capillary forces.

wicking debinding A popular means of binder extraction for wax-based binders, where the compact is packed in a small wicking powder and heated to a temperature where the wax melts and flows by capillary action into the surrounding packing powder.

withdrawal A process in compaction where the die descends over a fixed lower punch to facilitate ejection of the pressed powder.

work hardening The ability of a steel to become stronger and resist deformation as it is deformed.

wrought A material fabricated by conventional fusion metallurgy techniques, with a final stage of plastic deformation to improve mechanical properties.

yield strength A measure of first permanent deformation in a tensile test. It is the stress needed to initiate permanent shape changes.

Young's modulus Another term for the elastic modulus, which is the proportionality between elastic stress and strain.

zeta potential A electrochemical measure of the surface charge or surface chemistry of a fine powder in a suspension.

APPENDIX II

Standards Applicable to Powder Metallurgy

Abbreviations

MPIF = Metal Powder Industries Federation

ASTM = American Society for Testing Materials

ISO = International Standards Organization

Testing Metal Powders

Acid Insoluble Matter in Iron and Copper Powders—MPIF 06, ASTM E194, ISO 4496

Analysis by Microscopical Methods for Particle Size Distribution of Particulate Substances of Subsieve Size—ASTM E20

Apparent Density of Non-Free Flowing Metal Powders Using the Hall Apparatus—MPIF 04, ASTM B212, ISO 3923/1

Apparent Density of Non-Free Flowing Metal Powders Using the Carney Apparatus—MPIF 28, ASTM B417, ISO 3923/1

Apparent Density of Metal Powders Using the Arnold Meter—MPIF 48, ASTM B703

Average Particle Size of Metal Powders Using the Fisher Subsieve Sizer—MPIF 32, ASTM B330, ASTM C72 (see also ISO 10070)

Compactibility (Compressibility) of Metal Powders—MPIF 45, ASTM B331, ISO 3927

Determination of Envelope Specific Surface from Measurement of Air Permeability of a Powder Bed under Steady Flow—ISO 10070

Determination of Particle Size Distribution by Gravitational Sedimentation in a Liquid and Attenuation Measurement—ISO 10076

Flow Rate of Free-Flowing Metal Powders Using the Hall Apparatus—MPIF 03, ASTM B213, ISO 4490

Iron Content and Iron Oxide Content of Iron Powder—MPIF 07

Loss of Weight in Hydrogen for Metal Powders (Hydrogen Loss)—MPIF 02, ASTM E159, ISO 4491/2

Oil Absorption of Pigments by Spatula Rub-out—ASTM D281

Particle Size Distribution of Metal Powders and Related Compounds by Light Scattering—ASTM B822

Sampling Finished Lots of Metal Powders—MPIF 01, ASTM B215, ISO 3954

Sieve Analysis of Metal Powders—MPIF 05, ASTM B214, ISO 4497, ISO 2591

Subsieve Analysis of Granular Metal Powder by Air Classification—ASTM B293

Tap Density of Metal Powders—MPIF 46, ASTM B527, ISO 3953

Testing of P/M Products

Carburized Case Hardness and Case Depth of Sintered Parts—MPIF 37, ISO 4507

Compression Testing of Metallic Materials at Room Temperature—ASTM E9

Copper-Base Sintered Metal Powder Bearings (Oil Impregnated)—MPIF 35, ASTM B438, ISO 5755

Copper Infiltrated Sintered Carbon Steel Structural Parts—ASTM B303

Density of Compacted or Sintered Metal Powder Products—MPIF 42, ASTM B328, ASTM B376, ASTM B311, ISO 2737, ISO 2738, ISO 3369

Dimensional Change from Die Size of Sintered Metal Powder Specimens—MPIF 44, ASTM B610, ISO 4492

Green Strength of Compacted Metal Powder Specimens—MPIF 15, ASTM B312, ISO 3995

Hardness of Sintered Metal Powder Products—MPIF 43, ISO 4498

Impact Strength of Sintered Metal Powder Specimens—MPIF 40, ISO 5754

Iron-Base Sintered Metal Powder Bearings (Oil Impregnated)—ASTM B439

Iron–Copper Sintered Metal Powder Structural Parts—MPIF 35, ASTM B222, ISO 5755

Iron–Copper–Carbon Sintered Metal Powder Structural Parts—ASTM B426

Nickel-Iron Powder Metallurgy Soft Magnetic Alloys—ASTM A904

Phosphorous Iron Fabricated by Powder Metallurgy Techniques—ASTM A839

Safety Requirements for the Construction, Safeguarding, Care and Use of P/M Presses—MPIF 47

Sintered Austenitic Stainless Steel Structural Parts—ASTM B525

Sintered Carbon Steel Structural Parts—ASTM B310

Sintered Metal Brushes, Determination of Radial Crushing Strength—ISO 2729

Sintered Metal Materials and Hardmetals Determination of Young's Modulus—ASTM E111, ISO 3312

Sintered Metal Materials, Excluding Hardmetals, Determination of Transverse Rupture Strength—ISO 3928

Sintered Nickel Steel Structural Parts—ASTM B484

Soft Magnetic Iron Fabricated by Powder Metallurgy—ASTM A811

Tension Test Specimens for Pressed and Sintered Metal Powders—MPIF 10, ASTM E8, ISO 2740

Transverse Rupture Strength of Sintered Metal Powder Test Specimens—MPIF 41, ASTM B528, ISO 3325

Terminology

Terms for Metal Powder Compacting Presses and Tooling—MPIF 31

Terms Used in Powder Metallurgy—MPIF 09, ASTM B243, ISO 3252

INDEX